Lecture Notes in Computer Science 4806

Commenced Publication in 1973
Founding and Former Series Editors:
Gerhard Goos, Juris Hartmanis, and Jan van Leeuwen

T0145097

Robert Meersman   Zahir Tari
Pilar Herrero (Eds.)

On the Move to Meaningful Internet Systems 2007: OTM 2007 Workshops

OTM Confederated International Workshops and Posters
AWeSOMe, CAMS, OTM Academy Doctoral Consortium,
MONET, OnToContent, ORM, PerSys, PPN,
RDDS, SSWS, and SWWS 2007
Vilamoura, Portugal, November 25-30, 2007
Proceedings, Part II

 Springer

Volume Editors

Robert Meersman
Vrije Universiteit Brussel (VUB), STARLab
Bldg G/10, Pleinlaan 2, 1050 Brussels, Belgium
E-mail: meersman@vub.ac.be

Zahir Tari
RMIT University, School of Computer Science and Information Technology
Bld 10.10, 376-392 Swanston Street, VIC 3001, Melbourne, Australia
E-mail: zahir.tari@rmit.edu.au

Pilar Herrero
Universidad Politécnica de Madrid, Facultad de Informática
Campus de Montegancedo S/N, 28660 Boadilla del Monte, Madrid, Spain
E-mail: pherrero@fi.upm.es

Library of Congress Control Number: 2007939492

CR Subject Classification (1998): H.2, H.3, H.4, C.2, H.5, I.2, D.2, K.4

LNCS Sublibrary: SL 3 – Information Systems and Application, incl. Internet/Web
and HCI

ISSN   0302-9743
ISBN-10   3-540-76889-0 Springer Berlin Heidelberg New York
ISBN-13   978-3-540-76889-0 Springer Berlin Heidelberg New York

Springer is a part of Springer Science+Business Media

springer.com

© Springer-Verlag Berlin Heidelberg 2007
Printed in Germany

Typesetting: Camera-ready by author, data conversion by Scientific Publishing Services, Chennai, India
Printed on acid-free paper   SPIN: 12193530   06/3180   5 4 3 2 1 0

Volume Editors

Robert Meersman
Zahir Tari
Pilar Herrero

AWeSOMe

Pilar Herrero
Gonzalo Méndez
Rainer Unland

CAMS

Annika Hinze
George Buchanan

OTM Academy Doctoral Consortium

Antonia Albani
Torben Hansen
Johannes Maria Zaha

MONET

Fernando Ferri
Maurizio Rafanelli
Arianna D'Ulizia

OnToContent

Mustafa Jarrar
Andreas Schmidt
Claude Ostyn
Werner Ceusters

RDDS

Eiko Yoneki
Pascal Felber

ORM

Terry Halpin
Sjir Nijssen
Robert Meersman

PerSys

Skevos Evripidou
Roy Campbell
Anja Schanzenberger

PPN

Farid Naït-Abdesselam
Jiankun Hu
Azzedine Boukerche

SSWS

Achille Fokoue
Yuanbo Guo
Thorsten Liebig
Bijan Parsia

SWWS

John Mylopoulos
Elizabeth Chang
Ernesto Damiani
Yoke Sure
Tharam S. Dillon

OTM 2007 General Co-chairs' Message

OnTheMove 2007 held in Vilamoura, Portugal, November 25–30 further consolidated the growth of the conference series that was started in Irvine, California in 2002, and then held in Catania, Sicily in 2003, in Cyprus in 2004 and 2005, and in Montpellier last year. It continues to attract a diversifying and representative selection of today's worldwide research on the scientific concepts underlying new computing paradigms that of necessity must be distributed, heterogeneous and autonomous yet meaningfully collaborative.

Indeed, as such large, complex and networked intelligent information systems become the focus and norm for computing, it is clear that there is an acute and increasing need to address and discuss in an integrated forum the implied software and system issues as well as methodological, semantical, theoretical and application issues. As we all know, e-mail, the Internet, and even video conferences are not sufficient for effective and efficient scientific exchange. This is why the OnTheMove (OTM) Federated Conferences series has been created to cover the increasingly wide yet closely connected range of fundamental technologies such as data and Web Semantics, distributed objects, Web services, databases, information systems, workflow, cooperation, ubiquity, interoperability, mobility, grid and high-performance systems. OnTheMove aspires to be a primary scientific meeting place where all aspects of the development of Internet- and Intranet-based systems in organizations and for e-business are discussed in a scientifically motivated way. This sixth 2007 edition of the OTM Federated Conferences event, therefore, again provided an opportunity for researchers and practitioners to understand and publish these developments within their individual as well as within their broader contexts.

Originally the federative structure of OTM was formed by the co-location of three related, complementary and successful main conference series: DOA (Distributed Objects and Applications, since 1999), covering the relevant infrastructure-enabling technologies, ODBASE (Ontologies, DataBases and Applications of SEmantics, since 2002) covering Web semantics, XML databases and ontologies, CoopIS (Cooperative Information Systems, since 1993) covering the application of these technologies in an enterprise context through, e.g., workflow systems and knowledge management. In 2006 a fourth conference, GADA (Grid computing, high-performAnce and Distributed Applications), was added as a main symposium, and this year the same happened with IS (Information Security). Both started off as successful workshops at OTM, the first covering the large-scale integration of heterogeneous computing systems and data resources with the aim of providing a global computing space, the second covering the issues of security in complex Internet-based information systems. Each of these five conferences encourages researchers to treat their respective topics within a framework that incorporates jointly (a) theory , (b) conceptual design and development, and (c) applications, in particular case studies and industrial solutions.

Following and expanding the model created in 2003, we again solicited and selected quality workshop proposals to complement the more "archival" nature of the main conferences with research results in a number of selected and more "avant garde" areas related to the general topic of distributed computing. For instance, the so-called Semantic Web has given rise to several novel research areas combining linguistics, information systems technology, and artificial intelligence, such as the modeling of (legal) regulatory systems and the ubiquitous nature of their usage. We were glad to see that no less than eight of our earlier successful workshops (notably AweSOMe, CAMS, SWWS, ORM, OnToContent, MONET, PerSys, RDDS) re-appeared in 2007 with a second or third edition, and that four brand-new workshops emerged to be selected and hosted, and were successfully organized by their respective proposers: NDKM, PIPE, PPN, and SSWS. We know that as before, workshop audiences will productively mingle with another and with those of the main conferences, as is already visible from the overlap in authors! The OTM organizers are especially grateful for the leadership and competence of Pilar Herrero in managing this complex process into a success for the fourth year in a row.

A special mention for 2007 is to be made of third and enlarged edition of the OnTheMove Academy (formerly called Doctoral Consortium Workshop), our "vision for the future" in research in the areas covered by OTM. Its 2007 organizers, Antonia Albani, Torben Hansen and Johannes Maria Zaha, three young and active researchers, guaranteed once more the unique interactive formula to bring PhD students together: research proposals are submitted for evaluation; selected submissions and their approaches are presented by the students in front of a wider audience at the conference, and are independently and extensively analyzed and discussed in public by a panel of senior professors. This year these were once more Johann Eder and Maria Orlowska, under the guidance of Jan Dietz, the incumbent Dean of the OnTheMove Academy. The successful students only pay a minimal fee for the Doctoral Symposium itself and also are awarded free access to all other parts of the OTM program (in fact their attendance is largely sponsored by the other participants!).

All five main conferences and the associated workshops share the distributed aspects of modern computing systems, and the resulting application-pull created by the Internet and the so-called Semantic Web. For DOA 2007, the primary emphasis stayed on the distributed object infrastructure; for ODBASE 2007, it became the knowledge bases and methods required for enabling the use of formal semantics; for CoopIS 2007, the topic as usual was the interaction of such technologies and methods with management issues, such as occur in networked organizations; for GADA 2007, the topic was the scalable integration of heterogeneous computing systems and data resources with the aim of providing a global computing space; and last but not least in the relative newcomer IS 2007 the emphasis was on information security in the networked society. These subject areas overlap naturally and many submissions in fact also treated an envisaged mutual impact among them. As for the earlier editions, the organizers wanted to stimulate this cross-pollination by a shared program of famous keynote speakers: this year we were proud to announce Mark Little of Red Hat, York Sure of SAP Research, Donald Ferguson of Microsoft, and Dennis Gannon of Indiana University. As always, we also encouraged multiple event attendance by providing all authors, also those of workshop papers, with free access or discounts to one other conference or workshop of their choice.

We received a total of 362 submissions for the five main conferences and 241 for the workshops. Not only may we indeed again claim success in attracting an increasingly representative volume of scientific papers, but such a harvest of course allows the Program Committees to compose a higher-quality cross-section of current research in the areas covered by OTM. In fact, in spite of the larger number of submissions, the Program Chairs of each of the three main conferences decided to accept only approximately the same number of papers for presentation and publication as in 2004 and 2005 (i.e., average one paper out of every three to four submitted, not counting posters). For the workshops, the acceptance rate varied but was much stricter than before, consistently about one accepted paper for every two to three submitted. Also for this reason, we separate the proceedings into four books with their own titles, two for main conferences and two for workshops, and we are grateful to Springer for their suggestions and collaboration in producing these books and CD-Roms. The reviewing process by the respective Program Committees was again performed very professionally and each paper in the main conferences was reviewed by at least three referees, with arbitrated e-mail discussions in the case of strongly diverging evaluations. It may be worthwhile to emphasize that it is an explicit OnTheMove policy that all conference Program Committees and Chairs make their selections completely autonomously from the OTM organization itself. Continuing a costly but nice tradition, the OnTheMove Federated Event organizers decided again to make all proceedings available to all participants of conferences and workshops, independently of one's registration to a specific conference or workshop. Each participant also received a CD-Rom with the full combined proceedings (conferences + workshops).

The General Chairs are once more especially grateful to all the many people directly or indirectly involved in the set-up of these federated conferences, who contributed to making them a success. Few people realize what a large number of individuals have to be involved, and what a huge amount of work, and sometimes risk, the organization of an event like OTM entails. Apart from the persons mentioned above, we therefore in particular wish to thank our 12 main conference PC Co-chairs (GADA 2007: Pilar Herrero, Daniel Katz, María S. Pérez, Domenico Talia; DOA 2007: Pascal Felber, Aad van Moorsel, Calton Pu; ODBASE 2007: Tharam Dillon, Michele Missikoff, Steffen Staab; CoopIS 2007: Francisco Curbera, Frank Leymann, Mathias Weske; IS 2007: Mário Freire, Simão Melo de Sousa, Vitor Santos, Jong Hyuk Park) and our 36 workshop PC Co-chairs (Antonia Albani, Susana Alcalde, Adezzine Boukerche, George Buchanan, Roy Campbell, Werner Ceusters, Elizabeth Chang, Antonio Coronato, Simon Courtenage, Ernesto Damiani, Skevos Evripidou, Pascal Felber, Fernando Ferri, Achille Fokoue, Mario Freire, Daniel Grosu, Michael Gurstein, Pilar Herrero, Terry Halpin, Annika Hinze, Jong Hyuk Park, Mustafa Jarrar, Jiankun Hu, Cornel Klein, David Lewis, Arek Kasprzyk, Thorsten Liebig, Gonzalo Méndez, Jelena Mitic, John Mylopoulos, Farid Nad-Abdessalam, Sjir Nijssen, the late Claude Ostyn, Bijan Parsia, Maurizio Rafanelli, Marta Sabou, Andreas Schmidt, Simão Melo de Sousa, York Sure, Katia Sycara, Thanassis Tiropanis, Arianna D'Ulizia, Rainer Unland, Eiko Yoneki, Yuanbo Guo).

All, together with their many PC members, did a superb and professional job in selecting the best papers from the large harvest of submissions.

We also must heartily thank Jos Valente de Oliveira for the efforts in arranging facilities at the venue and coordinating the substantial and varied local activities needed for a multi-conference event such as ours. And we must all be grateful also to Ana Cecilia Martinez-Barbosa for researching and securing the sponsoring arrangements, to our extremely competent and experienced Conference Secretariat and technical support staff in Antwerp, Daniel Meersman, Ana-Cecilia, and Jan Demey, and last but not least to our energetic Publications Chair and loyal collaborator of many years in Melbourne, Kwong Yuen Lai, this year vigorously assisted by Vidura Gamini Abhaya and Peter Dimopoulos.

The General Chairs gratefully acknowledge the academic freedom, logistic support and facilities they enjoy from their respective institutions, Vrije Universiteit Brussel (VUB) and RMIT University, Melbourne, without which such an enterprise would not be feasible.

We do hope that the results of this federated scientific enterprise contribute to your research and your place in the scientific network.

August 2007   Robert Meersman
                                                          Zahir Tari

Organization Committee

The OTM (On The Move) Federated Workshops aim at complementing the more "archival" nature of the OTM Federated Conferences with research results in a number of selected and more "avant garde" areas related to the general topic of distributed computing. In 2007, only 11 workshops were chosen after a rigourous selection process by Pilar Herrero. The 2007 selected international workshops were: AWeSOMe (International Workshop on Agents, Web Services and Ontologies Merging), CAMS (International Workshop on Context-Aware Mobile Systems), OTM Academy Doctoral Consortium, MONET (International Workshop on MObile and NEtworking Technologies for social applications), OnToContent (International Workshop on Ontology content and evaluation in Enterprise), ORM (International Workshop on Object-Role Modeling), PerSys (International Workshop on Pervasive Systems), PPN (International Workshop on Peer-to-Peer Networks), RDDS (International Workshop on Reliability in Decentralized Distributed Systems), SSWS (International Workshop on Scalable Semantic Web Knowledge Base Systems), and SWWS (IFIP WG 2.12 and WG 12.4 International Workshop on Semantic Web and Web Semantics). OTM 2007 Federated Workshops were proudly **supported** by RMIT University (School of Computer Science and Information Technology) and Vrije Universiteit Brussel (Department of Computer Science).

Executive Committee

OTM 2007 General Co-chairs	Robert Meersman (Vrije Universiteit Brussel, Belgium), Zahir Tari (RMIT University, Australia), and Pilar Herrero (Universidad Politécnica de Madrid, Spain)
AWeSOMe 2007 PC Co-chairs	Pilar Herrero (Universidad Politécnica de Madrid, Spain), Gonzalo Médez (Universidad Complutense de Madrid, Spain), and Rainer Unland (University of Duisburg-Essen, Germany)
CAMS 2007 PC Co-chairs	George Buchanan (University of Wales Swansea, UK) and Annika Hinze (University of Waikato, New Zealand)
OTM 2007 Academy Doctoral Consortium PC Co-chairs	Antonia Albani (Delft University of Technology, The Netherlands), Torben Hansen (German Research Center for Artificial Intelligence, Germany) and Johannes Maria Zaha (University of Duisburg-Essen, Germany)

MONET 2007 PC Co-chairs	Fernando Ferri (National Research Council, Italy), Maurizio Rafanelli (National Research Council, Italy) and Arianna D'Ulizia (National Research Council, Italy)
OnToContent 2007 PC Co-chairs	Mustafa Jarrar (Vrije Universiteit Brussel, Belgium), Andreas Schmidt (FZI, Germany), Claude Ostyn (IEEE-LTSC, USA), and Werner Ceusters (University of Buffalo, USA)
ORM 2007 PC Co-chairs	Terry Halpin (Neumont University, USA), Sjir Nijssen (PNA, The Netherlands), and Robert Meersman (Vrije Universiteit Brussel, Belgium)
PerSys 2007 PC Co-chairs	Skevos Evripidou (University of Cyprus, Cyprus), Roy Campbell (University of Illinois at Urbana-Champaign, USA), Anja Schanzenberger (Middlesex University, UK)
PPN 2007 PC Co-chairs	Farid Naït-Abdesselam (University of Science and Technology of Lille, France), Jiankun Hu (RMIT University, Australia), and Azzedine Boukerche (University of Ottawa, Canada)
RDDS 2007 PC Co-chairs	Eiko Yoneki (University of Cambridge, UK) and Pascal Felber (Université de Neuchâtel, Switzerland)
SSWS 2007 PC Co-chairs	Achille Fokoue (IBM T.J. Watson Research Center, USA), Yuanbo Guo (Microsoft Corp, USA), Thorsten Liebig (Ulm University, Germany), and Bijan Parsia (University of Manchester, UK)
SWWS 2007 PC Co-chairs	John Mylopoulos (University of Toronto, Canada), Elizabeth Chang (Curtin University of Technology, Australia), Ernesto Damiani (Milan University, Italy), Yoke Sure (University of Karlsruhe, Germany), and Tharam Dillon (University of Technology Sydney, Australia)
Publication Co-chairs	Kwong Yuen Lai (RMIT University, Australia) and Vidura Gamini Abhaya (RMIT University, Australia)
Local Organizing Chair	José Valente de Oliveira (University of Algarve, Portugal)
Conferences Publicity Chair	Jean-Marc Petit (INSA, Lyon, France)
Workshops Publicity Chair	Gonzalo Mendez (Universidad Complutense de Madrid, Spain)
Secretariat	Ana-Cecilia Martinez Barbosa, Jan Demey, and Daniel Meersman

AWeSOMe (Agents, Web Services and Ontologies Merging) 2007 Program Committee

M. Brian Blake
José Luis Bosque
Juan A. Botía Blaya
Paul Buhler
Jose Cardoso
Isaac Chao
Adam Cheyer
Ian Dickinson
Jorge Gómez
Dominic Greenwood
Jingshan Huang
Margaret Lyell
Dan Marinescu
Gregorio Martínez
Viviana Mascardi
Michael Maximilien
Barry Norton
Julian Padget
Mauricio Paletta

Juan Pavón
José Peña
María Pérez
Ronald Poell
Omer Rana
Paul Roe
Marta Sabou
Manuel Salvadores
Alberto Sánchez
Weisong Shi
Marius-Calin Silaghi
Ben K.M. Sim
Hiroki Suguri
Henry Tirri
Santtu Toivonen
Sander van Splunter
Julita Vassileva
Yao Wang

CAMS (Context-Aware Mobile Systems) 2007 Program Committee

Pilar Herrero
George Buchanan
Trevor Collins
Keith Cheverst
Dan Chalmers
Gill Dobbie
Tiong Goh
Annika Hinze

Reto Krummenacher
Johan Koolwaaij
Diane Lingrand
Gero Muehl
Michel Scholl
Goce Trajcevski
Katarzyna Wac

OTM Academy (International Doctoral Consortium) 2007 Program Committee

Antonia Albani
Domenico Beneventano
Jaime Delgado
Jan Dietz
Schahram Dustdar

Johann Eder
Torben Hansen
Jörg Müller
Maria Orlowska
Johannes Maria Zaha

MONET (MObile and NEtworking Technologies for Social Applications) 2007 Program Committee

Russell Beale
Yiwei Cao
Tiziana Catarci
Richard Chbeir
Karin Coninx
Simon Courtenage
Juan De Lara
Anna Formica
Patrizia Grifoni
Otthein Herzog
C.-C. Jay Kuo
Irina Kondratova
David Lewis
Stephen Marsh

Rebecca Montanari
Michele Missikoff
Nuria Oliver
Marco Padula
Andrew Phippen
Tommo Reti
Tim Strayer
Henri Ter Hofte
Thanassis Tiropanis
Yoshito Tobe
Riccardo Torlone
Mikael Wiberg

OnToContent (Ontology Content and Evaluation in Enterprise) 2007 Program Committee

Bill Andersen
Keith Baker
Ernst Biesalski
Paolo Bouquet
Simone Braun
Christopher Brewster
Michael Brown
Yannis Charalabidis
Ernesto Damiani
Aldo Gangemi
Fausto Giunchiglia
Giancarlo Guizzardi
Mohand-Said Hacid
Martin Hepp
Stijn Heymans
Christine Kunzmann
Jens Lemcke

Stefanie Lindstaedt
Alessandro Oltramari
Jeff Pan
Paul Piwek
Christophe Roch
Pavel Shvaiko
Miguel-Angel Sicilia
Barry Smith
Silvie Spreeuwenberg
Armando Stellato
Andrew Stranieri
Karl Stroetmann
Sergio Tessaris
Robert Tolksdorf
Francky Trichet
Luk Vervenne

ORM (Object-Role Modeling) 2007 Program Committee

Guido Bakema
Herman Balsters
Linda Bird

Anthony Bloesch
Scott Becker
Peter Bollen

Lex Bruil
Andy Carver
Dave Cuyler
Necito dela Cruz
Aldo de Moor
Olga De Troyer
Jan Dietz
David Embley
Ken Evans
Gordon Everest
Mario Gutknecht
Henri Habrias
Pat Hallock
Terry Halpin
Hank Hermans
Stijn Hoppenbrouwers
Mike Jackson

Mustafa Jarrar
Inge Lemmens
Robert Meersman
Tony Morgan
Maurice Nijssen
Sjir Nijssen
Baba Piprani
Erik Proper
Jos Rozendaal
Gerhard Skagestein
Peter Spyns
Deny Smeets
Sten Sundblad
Jos Vos
Theo van der Weide
Jan Pieter Zwart

PerSys (Pervasive Systems) 2007 Program Committee

Xavier Alamá
Jalal Al-Muhtadi
Susana Alcalde Bagüés
Christian Becker
Michael Beigl
Alastair Beresford
Roy Campbell
Antonio Coronato
Thanos Demiris
Hakan Duman
Bob Hulsebosch
Hesham El-Rewini
Skevos Evripidou
Alois Ferscha
Nikolaos Georgantas
Alex Healing

Markus Huebscher
Cornel Klein
Jelena Mitic
Andrew Rice
Philip Robison
Das Sajal
George Samaras
Anja Schanzenberger
Gregor Schiele
Behrooz Shirazi
Sotirios Terzis
Issarny Valerie
Gregory Yovanof
Apostolos Zarras
Arkady Zaslavsky

PPN (Peer-to-Peer Networks) 2007 Program Committee

Marinho Pilla Barcellos
Jalel Ben-Othman
Brahim Bensaou
Tarek Bijaoui
Jean Carle

Song Ci
Pilar Herrero
Ashfaq Khokhar
Nouredine Melab
Alberto Montresor

Florent Nolot
Aris Ouksel
Douglas Reeves
Ahmed Serhrouchni
Orazio Tomarchio

Kurt Tutschku
Carlos Becker Westphall
Sherali Zeadally

RDDS (Reliability in Decentralized Distributed Systems) 2007 Program Committee

Licia Capra
Paolo Costa
Simon Courtenage
Patrick Eugster
Pascal Felber
Ludger Fiege
Christos Gkantsidis
Michael Kounavis
Marco Mamei
Gero Muehl

Jonathan Munson
Maziar Nekovee
Andrea Passarella
Peter Pietzuch
Matthieu Roy
Francois Taiani
Einar Vollset
Eiko Yoneki

SSWS (Scalable Semantic Web Knowledge Base Systems) 2007 Program Committee

Pascal Hitzler
York Sure
Kavitha Srinivas
Takahira Yamaguch
Raúl García Castro
Aditya Kalyanpur
Oscar Corcho
Jeff Heflin
Ralf Möller
Ian Horrocks

Boris Motik
Pierre-Antoine Champin
Ying Ding
Marko Luther
Timo Weithöner
Andy Seaborne
Ulrike Sattler
Jan Wielemaker
Volker Haarslev

SWWS (Semantic Web and Web Semantics) 2007 Program Committee

Aldo Gangemi
Amandeep Sidhu
Amit Sheth

Angela Schwering
Avigdor Gal
Birgit Hofreiter

Carlos Sierra
Carole Goble
Chris Bussler
Claudia d'Amato
David Bell
Elena Camossi
Elisa Bertino
Elizabeth Chang
Ernesto Damiani
Farookh Hussain
Feng Ling
Grigoris Antoniou
Hai Zhuge
Jaiwei Han
John Debenham
John Mylopoulos
Katia Sycara
Krzysztof Janowicz
Kokou Yetongnon
Kyu-Young Whang
Ling Liu
Lizhu Zhou
Lotfi Zadeh
Manfred Hauswirth
Maria Andrea Rodriguez-Tastets

Masood Nikvesh
Mihaela Ulieru
Mohand-Said Hacid
Monica De Martino
Mukesh Mohania
Mustafa Jarrar
Nicola Guarino
Paolo Ceravolo
Peter Spyns
Pieree Yves Schobbens
Pilar Herrero
Qing Li
Rajugan Rajagopalapillai
Ramasamy Uthurusamy
Riccardo Albertoni
Robert Meersman
Robert Tolksdorf
Stefan Decker
Susan Urban
Tharam Dillon
Usuama Fayed
Wil van der Aalst
York Sure
Zahir Tari

Table of Contents – Part II

Location Sensing and Management

Mobility in Pervasive Systems

Workshop on Peer to Peer Networks (PPN)

P2P Security

P2P Routing

Content Distribution

General

Workshop on Reliability in Decentralized Distributed Systems (RDDS)

Distributed Algorithms

Self Adaptive Systems

Fault Tolerant Systems

Workshop on Scalable Semantic Web Knowledge Base Systems (SSWS)

Large Scale Knowledge Bases and Data Integration

Scalable Reasoning

Query Languages and Semantic Mapping

IFIP WG 2.12 and WG 12.4 International Workshop on Semantic Web and Web Semantics (SWWS)

Ontology Development, Management and Evolution

Process Semantics and Mining

Semantic Interoperability

Ontology and Knowledge Matching

Biomedical Ontologies

Workshop on Pervasive Systems (PerSys)

PerSys 2007 PC Co-chairs' Message

We welcome our readers to a very interesting collection of papers. We received many good papers and had to limit the acceptance rate to 50% of those submitted. The submissions cover a range of topics from infrastructure, service discovery, personalization, environments, security, privacy, to wireless and sensor networks, showing the scope of interests and activities represented by the PerSys participants. We would like to thank everyone who submitted a paper. The reviewers of the papers did a terrific job in helping authors prepare better final versions and deserve special thanks for completing the job promptly. Last, we would like to thank the organizers of the conference for the hard work they have put in to providing us with the opportunity to gather, present, listen, and discuss our research.

August 2007

Roy Campbell
Skevos Evripidou
Anja Schanzenberger

COMITY - Conflict Avoidance in Pervasive Computing Environments

Verena Tuttlies, Gregor Schiele, and Christian Becker

Universität Mannheim
Schloss, 68131 Mannheim, Germany
{verena.tuttlies,gregor.schiele,christian.becker}@uni-mannheim.de

Abstract. Pervasive Computing leads to novel challenges of application coordination. Applications may interfere with each other since they affect and share the same physical environment. A user that executes a navigation application utilizing a loudspeaker as output may annoy a user listening to music. To ensure a conflict-free execution of multiple applications, approaches are needed to avoid conflicts at runtime. In this paper we identify typical conflicts arising in Pervasive Computing. We propose how to detect potential conflicts by requiring applications to communicate their effects on the physical environment. After the detection, a suitable strategy is applied resolving the potential conflict. As a practical validation, we discuss how to model and avoid conflicts using our component-based software system PCOM.

1 Introduction

Pervasive Computing aims at the ubiquitous provision of context-aware services. Distributed applications are executed on devices present in a specific environment, offering functionality to users. In case the context changes, applications adapt to the modified environment, e.g., by adjusting the data presented to the user or by reselecting used devices.

When multiple applications are executed in the same environment, they can interfere with each other as they affect and share the same physical environment. A prominent example are mobile phones that interfere with nearly every other application and user, e.g., in meetings, concerts, dinners, etc. Vice versa, a person on a mobile phone may feel disturbed by a user passing by, utilizing loudspeakers for his music application. To handle such conflicts, applications must be able to detect conflicts before they actually occur. Thus, they can coordinate themselves to avoid the conflicting situation. Our goal is to provide a middleware that enables applications to avoid conflicts by detecting and resolving potential conflicts. In earlier work [1] we discussed this problem space and gave initial directions to address it.

In this paper, we propose an extension to component/service-based Pervasive Computing infrastructures that requires applications to specify their influence on the physical environment and allows them to define context situations they consider to be conflicts. Based on this information, potential conflicts can be

R. Meersman, Z. Tari, P. Herrero et al. (Eds.): OTM 2007 Ws, Part II, LNCS 4806, pp. 763–772, 2007.

detected by the system and a negotiation between conflicting applications can be induced. Finally, we propose an integration into PCOM, our component model for Pervasive Computing.

The paper is structured as follows. In Section 2 we introduce our system model. Conflicts and their avoidance are discussed in Section 3. Our conflict avoidance middleware is presented in Section 4. After a discussion on related work in Section 5, we give a short conclusion and an outlook on future work in Section 6.

2   System Model

We assume a Pervasive Computing environment to consist of a number of instrumented spaces. Such spaces combine stationary and mobile devices as well as embedded devices such as sensors and actuators. Applications can make use of the functionality provided by a space.

Instrumented spaces can be realized by smart peer groups or smart environments possibly employing different operating systems. For the management of an office building, for instance, we assume that all – potentially competing – companies will provide their own infrastructure. Even in common areas, such as foyers or shared meeting rooms, a diversity of systems will exist. To add to this diversity, visitors will be roaming through such environments with their devices, e.g., mobile phones, navigation systems, PDAs, etc. As a consequence, conflict avoidance has to be conducted among a number of instrumented spaces.

With respect to conflict detection and resolution, we assume a cooperative behavior of applications. For this purpose, applications are not bound to a specific application model, e.g., component- or task-based, or a common service specification. The only requirement we have for applications is that they specify their side effects on the physical environment by using a contract model. This can be done at different levels of granularity. A fine grained model would specify the effects for each individual part of an application, e.g., each component or object, while coarse grain models would specify the side effects for each participating node or only for the complete application. Later, we will see that an integration of our approach into a fine grained model can provide means for adaptation to conflicts based on the available mechanisms of the platform. However, in general, we only assume that an application is aware of its side effects and that some means for conflict resolution will be provided.

3   Conflict Avoidance in Pervasive Computing Environments

In this chapter we define conflicts and discuss their characteristics. Subsequently, we analyze the system components required for the detection of potential conflicts and sketch our approach.

3.1 Conflicts

The execution of an application in Pervasive Computing may affect its physical environment, namely its *context*. For example, a music application playing music influences its context by changing the audio level in the surrounding area. When multiple applications are executed simultaneously, their effects on the context may interfere. The influence of two music applications, each playing a different CD in the same room, will obviously cause a conflict for both applications.

In this paper we denote a conflict as follows: A *conflict* occurs for an application or a user, if a context change – caused by an application – leads to a state of the environment which is considered inadmissible by the application or user[1].

Consider the following scenario of a phone call and a simple "follow-me" application. User Bob is in the living room, talking on his phone. As Bob likes to talk hands-free, the phone call application uses the microphone and a speaker installed in the living room. Meanwhile, user Anne listens to music provided by her "follow-me" application. As she moves from the bedroom to the living room, her application automatically detects the changes in its context. It discovers an unoccupied speaker in the living room, adapts to the output device and plays the music in the living room.

As the resource requirements of both – the "follow-me" and the phone call application – are satisfied, the two applications can be executed. However, Bob feels disturbed by Anne's application, because the music interferes with his phone call. As he considers the situation to be inadmissible, a conflict for Bob has occurred. In contrast, Anne may not consider Bob's phone call to be disturbing while listening to music. This example shows that conflicts are not necessarily symmetric.

In general, the existence of a conflict depends on the preferences of the respective user. Thus, for conflict avoidance, conflicts need to be defined by applications and users. In the subsequent section we discuss how conflicts can be specified and which components are needed to avoid conflicts at runtime.

3.2 Conflict Avoidance

In an open dynamic system environment – as addressed in this paper – the number of applications which will be executed in parallel is unknown. In order to ensure a conflict-free execution, conflicts need to be detected before they actually happen. However, the consideration of arbitrary context changes in order to predict conflicts, can be very complex and resource consuming. Instead, we require every application to state how it will affect the context in order to be executed. Based on this information, the system can reliably and efficiently "predict" potential conflicts.

In this section we describe the basic idea of conflict avoidance in Pervasive Computing in detail. We discuss necessary system components and show how they interact at runtime.

---

[1] Note, that a change of the context, e.g., by a user entering a room requires an application, for instance a personal tracking service, in order to affect the context.

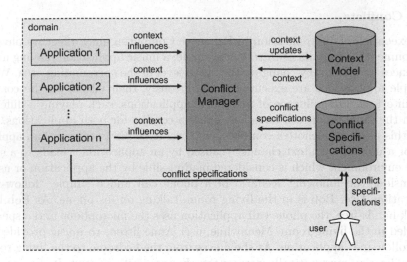

Fig. 1. General Approach

Figure 1 depicts our approach. The system consists of applications, domains, a conflict manager, a context model and a conflict specification database, which we describe in the following:

Applications may be executed by arbitrary – possibly different – operating systems in a given domain. A *domain* refers to a spatial area of the physical environment. Consequently, the domain of an applications is defined by the physical environment in which its user is located and whose devices are utilized by the application, e.g., speakers, displays, etc. Each domain is associated with a conflict management component, the so-called *conflict manager*. This component is responsible for the detection of potential conflicts and their resolution.

In order to detect potential conflicts, the conflict management component relies on the information stored in the *context model* and the *conflict specification database*. The context model holds information on how the physical environment is currently affected by active applications. The conflict specification database contains descriptions of situations that are considered a conflict by some application or user in the domain.

Both, context modifications as well as conflict specifications, are provided by applications, using suitable specification languages. In addition, conflicts may be specified by users independent of an application. As an example, a user may configure a living room to respect siesta. To realize the restriction, a conflict specification would be added to the database, stating that no audio output above a given level is allowed between 12 a.m. and 2 p.m.

Before an application is executed, it communicates its effects on the context to the conflict manager. The conflict manager then determines, if the defined effect will lead to a conflict. For this purpose, it alters the current context model by the specified context change and compares the result with the set of conflict specifications. In case a specification matches, a potential conflict has been detected. The application is informed that its execution will cause a conflict and a conflict resolution is induced. If the effect does not lead to a conflict, the application can be executed and the context model is modified by the application effects.

In case a conflict has been detected, the system should try to resolve the conflict automatically. A simple approach to resolve the conflict could be to prohibit the execution of the respective application. A more sophisticated resolution strategy is the adaption of the applications encountering and causing the conflict. An alternative application configuration could be inferred enabling all applications to be executed conflict-free. In Section 4.2 we will show that the integration into our component system PCOM will automatically provide means for such an adaptation.

Note that due to contradicting user preferences automatic conflict resolution may be impossible. Users requiring different levels of brightness in a room will always be in conflict with each other. In such cases, the users must be notified to resolve the conflict manually. Clearly, other resolution strategies are conceivable and are subject to future work.

In the next section we present the realization of the proposed approach as an extension of our previous work in Pervasive Computing system software, the PCOM system [2]. First, we give an introduction to PCOM and then present the additions to enable conflict avoidance in our system.

4 COMITY

This section describes the COMITY system, our realization of conflict avoidance in Pervasive Computing environments. Although our discussion focuses on COMITY as an extension of PCOM, applications running on other platforms can also participate in the conflict avoidance by specifying their influences and conflict specifications.

First, we briefly sketch the application model and automatic adaptation mechanisms of PCOM. After that, the extension of PCOM's contract model for conflict management is presented.

4.1 PCOM

PCOM is a minimal component system that is designed specifically for Pervasive Computing systems with resource restricted devices. It provides a component-based application model that allows to specify semantically enriched contracts between components. Using this model, PCOM is able to compose applications dynamically from currently available components and to adapt them automatically to changing environments, e.g., fluctuating resource and device availability. In this section we give a brief overview of the features of PCOM that are needed to understand the integration of conflict avoidance into the system. More details about PCOM can be found, e.g., in [2].

The basic building block of an application in PCOM is a component. Components are atomic with respect to their distribution and provide a contract model

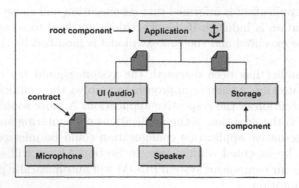

Fig. 2. PCOM Application Tree

that specifies i) the functionality offered to other components, ii) the functionality required from other components, and iii) the resources required from the local execution platform, such as needed memory or a special sensor.

An application is modeled as a tree of components, as depicted in Figure 2 where a designated component – the root – identifies the application's life cycle. In order to execute the application, the system searches candidates for every dependency, i.e., for each required component, starting top down. Once a complete configuration is found, the application can be executed.

If a component becomes unavailable or the contract changes due to resource constraints, the application must be adapted. To do so, the component container is notified. The container provides a runtime environment for components and manages their life cycles. Upon notification, the container pauses the application and starts searching for an appropriate replacement. If it is successful, it resumes the application. Otherwise the container also marks the component needing the replacement as unavailable and repeats the process for its parent component. If the root component is reached and no configuration could be found, the container may abort the application.

To manage the available resources in the system, the container relies on so-called resource managers. Each manager is responsible for monitoring a specific resource and for coordinating components accessing it. As an example, if a resource manager for energy detects that the battery on a given device is running low, it adapts the contracts of all components that requested a certain level of energy to work. Consequently, contracts may become invalid and applications need to adapt to the changes.

4.2   PCOM Extensions

In Section 3.1 we discussed how an application can state its influence on and its requirements towards the physical environment using so-called influence and conflict specifications. Based on this information, a *Conflict Manager* can determine if the execution of an application will lead to a conflict in the environment.

In order to realize this approach with PCOM, two steps must be taken. At first, the influence and conflict specifications of an application have to be integrated into PCOM and an interface to the Conflict Manager has to be defined. For this purpose, we model both kinds of specifications as extensions of PCOM's contract model, creating so-called *context contracts*.

In the second step, the Conflict Manager itself needs to be realized. The Conflict Manager is responsible for negotiating the context contracts. In consideration of the state of the physical environment and a set of relevant conflict specifications, the Conflict Manager must determine, if the execution of an application will lead to a conflict in the environment.

Context Contract. The basic idea of the integration of the context contract in PCOM is to model the physical environment as a set of resources. Each resource abstracts a certain aspect of the physical environment that can be affected by applications, e.g. light or audio level. A classification of such aspects can be found in [3].

In order to be executed, an application has to acquire the context resource which reflects the impact it will have on the physical environment. The application shown in Figure 2, for example, outputs audio to the environment using a speaker. Thus, it must specify that it requires the context resource "AudioOutput" in its contract, indicating how it will affect the environment if executed.

As applications are built from components in PCOM, the effect on the physical environment depends on the current component composition. For example, if the speaker cannot be bound at runtime, the application could adapt to a set of headphones. Using the headphones will not affect the physical environment. Thus, context resources are not allocated to the application but to the component which is responsible for the impact on the physical environment.

We specify conflicts using first order predicate logic. Every expression may address the state of one or more context resources. Conflict specifications can be defined in any component contract in the application tree.

The interface between the context contracts and the Conflict Manager is realized by so-called *Context Resource Managers*. As discussed before, PCOM automatically forwards a resource requirement to the local resource manager that is responsible for the respective resource type at runtime. If the manager grants the resource, the component can be executed. Otherwise PCOM searches for alternative application configurations. To reuse this mechanism, we include special resource managers for context resources on every device, the Context Resource Managers. If a Context Resource Manager receives a context contract, it forwards the request to the Conflict Manager responsible for the current domain. The Conflict Manager then checks if the allocation of the required context resource will lead to a conflict. In case a conflict is detected, the Context Resource Manager is notified that it cannot allocate the resource. It denies the context resource to the component and the conflict is automatically avoided by searching for an alternative configuration.

Conflict Manager. The avoidance of conflicts in PCOM is the task of the *Conflict Manager*. It is responsible for managing conflicts for a given domain

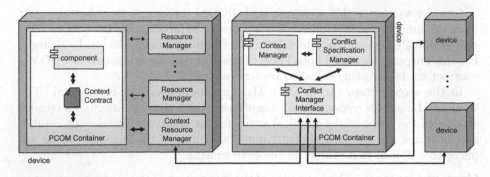

Fig. 3. Conflict Management with COMITY

and is automatically discovered by the Context Resource Managers using the PCOM discovery service.

As depicted in Figure 3, the Conflict Manager consists of three PCOM components, the Conflict Manager Interface, the Conflict Specification Manager, and the Context Manager.

The *Conflict Manager Interface* acts as the interface between the Conflict Manager and the remote Context Resource Managers. It receives context contracts, extracts the context influences and conflict specifications, and forwards them to the Conflict Specification Manager and the Context Manager respectively. In addition, it handles concurrent requests from several Context Resource Managers to ensure a consistent system state, e.g., by serializing incoming requests.

The *Conflict Specification Manager* manages the set of conflict specifications for an administrative domain. It keeps the set of conflict specifications of currently active applications and components updated. If a contract becomes valid, it adds the contained conflict specifications to the set and removes them if the contract is terminated. The Conflict Specification Manager uses a soft state approach to be notified if a device becomes unavailable. This is necessary as we cannot assume that devices will always be able to unsubscribe from the system.

The *Context Manager* is responsible for administrating the context resources and grants Context Resource Managers permits to allocate them to components at runtime. It monitors, how context resources are distributed among the applications and which resources are not in use. The state of the entire set of context resources is referred to as the *state* of the physical environment. Based on this information and the current set of conflict specifications, the Context Resource Manager determines if a context resource can be allocated to a requesting component.

For this purpose, two different checks have to be performed. In case a component requests a context resource, the Context Resource Manager checks if the application's specified context modifications will cause a conflict for another application in the environment. In case the test is negative, it checks if the execution of the application will lead to a conflict for the application itself. If the second test proves to be negative as well, the Context Resource Manager allocates the required context resource to the component.

However, in case the execution of an application will cause a conflict in the environment, the respective resource is denied. As a consequence, the application cannot be executed, as the requirement towards the context resource cannot be satisfied. Thus, the unavailability of the resource induces the application to adapt. This process resolves the conflict before an actual interference of the two applications happens. However, the resolution through adaptation as presented, implies a first come – first served strategy. This is not always desirable, as priorities of applications or a compromise between applications are conceivable, for example. The development of suitable resolution strategies is subject to future work.

5 Related Work

Conflicts and their detection and resolution have been examined in a number of research projects.

In Gaia, conflicts can be detected based on a set of space-level policies [4]. This set has to be defined by an administrator, and enables the detection of conflicts based on the state and the context of the environment. Park *et al.* [5] consider conflicts as actions of services which have contradictory effects on context attributes and service state variables. Based on action semantics, they determine if two actions have encountered a conflict and resolve the conflict by using a resolution policy. These approaches concentrate on the detection of occurred conflicts. In contrast to this, we aim at the detection of conflicts before they actually occur and try to avoid them.

Further approaches addressing conflicts exclusively focus on the resolution of conflicts. Shin *et al.* [6] assume a prior detection of conflicts and discuss different resolution strategies for conflicts. Similarly, CARISMA [7] solves existing conflicts by using a micro-economic approach based on a sealed-bid auction.

Most similar to our approach, Syukur *et al.* discuss conflict detection and avoidance strategies assuming a policy-based application model in [8]. Such policies are used to explicitly control the behavior of an application. In contrast to this, our approach is independent of a specific application model.

6 Conclusion and Future Work

In this paper we presented an approach to avoiding conflicts in Pervasive Computing environments. The general idea of our approach is that applications specify their intended influence on the physical context before they can be executed. Based on this information, a set of conflict specifications and the current state of the environment the system is able to detect potential conflicts at runtime. In order to avoid the actual conflict, the potential conflict is resolved using a suitable strategy. Subsequent, we showed how to integrate the proposed approach into our component system PCOM.

Currently we are developing our specification languages for conflicts and are implementing the necessary extensions for PCOM. In addition, we are working on an extended definition of conflicts in order to factor in context changes that are not caused by applications. In future work we will examine different resolution strategies in more detail. Finally, we plan to extend our system to handle non-exclusive context resources.

References

1. Schiele, G., Handte, M., Becker, C.: Good Manners for Pervasive Computing–An Approach Based on the Ambient Calculus. In: PERCOMW 2007: In: Proceedings of the Fifth IEEE International Conference on Pervasive Computing and Communications Workshops, White Plains, USA (2007)
2. Becker, C., Handte, M., Schiele, G., Rothermel, K.: PCOM - A Component System for Pervasive Computing. In: Proceedings of the Second IEEE International Conference on Pervasive Computing and Communications (PerCom 2004), Orlando, USA (2004)
3. Korpipää, P., Mäntyjärvi, J., Kela, J., Keränen, H., Malm, E.J.: Managing Context Information in Mobile Devices. IEEE Pervasive Computing 02(3), 42–51 (2003)
4. Ranganathan, A., Chetan, S., Al-Muhtadi, J., Campbell, R.H., Mickunas, M.D.: Olympus: A High-Level Programming Model for Pervasive Computing Environments. In: International Conference on Pervasive Computing and Communications (PerCom 2005), Kauai Island, Hawaii (2005)
5. Park, I., Lee, D., Hyun, S.J.: A Dynamic Context-Conflict Management Scheme for Group-Aware Ubiquitous Computing Environments. In: Proceedings of the 29th Annual International Computer Software and Applications Conference (COMPSAC 2005), Edinburgh, Scotland (2005)
6. Shin, C., Woo, W.: Conflict Resolution Method utilizing Context History for Context-Aware Applications. In: Proceedings of ECHISE 2005 - 1st International Workshop on Exploiting Context Histories in Smart Environments, Munich, Germany (2005)
7. Capra, L., Emmerich, W., Mascolo, C.: CARISMA: Context-Aware Reflective mIddleware System for Mobile Applications. IEEE Transactions on Software Engineering 29(10), 929–945 (2003)
8. Syukur, E., Loke, S.W., Stanski, P.: Methods for Policy Conflict Detection and Resolution in Pervasive Computing Environments. In: Policy Management for the Web Workshop in conjunction with the 14th International World Wide Web Conference, Chiba, Japan (2005)

An OSGi-Based Semantic Service-Oriented Device Architecture

Panagiotis Gouvas, Thanasis Bouras, and Gregoris Mentzas

Information Management Unit, National Technical University of Athens, Iroon
Polytexneiou 9, 15780 Athens Greece
{pgouvas,bouras,gmentzas}@mail.ntua.gr

Abstract. The implementation of service-oriented device architectures (SODA) suffers from restrictions that are imposed by the use of existing syntactic technologies. Related problems include data and message-level heterogeneities among interoperating services, insufficient search and discovery of exposed services and inadequate web process composition. In this paper we propose an approach for introducing semantics in a SODA environment. Specifically we examine the introduction of data, functional and behavioural semantics and the role they play in a semantically-enabled SODA (SeSODA) pervasive environment. We put special emphasis in data semantics. We present the architecture of an OSGi-based SeSODA implementation, analyse the most critical components of this architecture and discuss implementation issues.

Keywords: Service Oriented Device Architecture, OSGi, Semantic SODA.

1 Introduction

The Service-Oriented Device Architecture (SODA) is an adaptation of the Service-Oriented Architecture (SOA) to the world of devices, in the sense that programmers deal with devices (sensors or actuators) just as business services are used in today's enterprise SOAs [2]. A SODA implementation provides an *abstract service model* of a device by providing interface to proprietary and standard device interfaces and presenting device services as SOA services over a network through a bus adapter. Consequently, a SODA implementation is a *'conceptual'* trend that may be interpreted ambiguously by service providers and consumers.

The underlying technical standards of web services such as WSDL[1], SOAP[2] etc can be considered as the basic cornerstones of enabling a service-oriented architecture. In the present paper we refer to these standards as *syntactic standards* in order to distinguish them from semantic ones. The enablement of SOA, based on these standards, has been a major step forward regarding the management and evolution of IT systems. In parallel, the wide adoption of SOA has led to an increased need for *application integration* and *service composition*. This necessity has been

---

[1] http://www.w3.org/TR/wsdl20/
[2] http://www.w3.org/TR/soap/

R. Meersman, Z. Tari, P. Herrero et al. (Eds.): OTM 2007 Ws, Part II, LNCS 4806, pp. 773–782, 2007.

partially satisfied by additional syntactic standards such as WSCDL[3] and BPEL[4]; but the *overhead of manual intervention* tends to be huge as IT systems evolve and the number of services increases.

As a solution to this problem, the use of semantic technologies has been proposed. *Semantic technologies* include software standards and methodologies that are aimed at providing more explicit meaning for the information that is at our disposal. This takes different forms depending on where in the information cycle the semantic technology is applied and which area of the problem it is addressing. Examples include data and message level heterogeneities, insufficient service search and discovery and inadequate web process composition.

In parallel, in the area of device management and service orientation, OSGi[5] technology provides a service-oriented architecture that enables device specific components to discover and bind each other. The OSGi Alliance has developed many standard component interfaces for common functions like HTTP handling, configuration, logging, security, user administration, XML handling and many more. Plug-compatible implementations of these components can be obtained from different vendors with different optimizations and costs. Additionally, OSGi low level interfaces are being developed on a proprietary basis. We claim that OSGi is the *most mature* technology that supports traditional SODA systems. Consequently the choice of using OSGi as a starting point for developing a Semantically-enabled SODA (SeSODA) implementation is justifiable.

In this paper we identify the types of semantics that are required in the framework of a SODA, inspired by the restrictions that are imposed by syntactic technologies Furthermore we present an OSGi-based architecture that overcomes these restrictions. The rest of the paper is organised as follows. Section 2 refers to our motivations and related work. Section 3 overviews the proposed SeSODA principles while Section 4 outlines an OSGi based architecture and its implementation. Finally, section 5 provides an example of overcoming data level heterogeneities and section 6 concludes the paper.

2 Motivations and Related Work

2.1 Motivations

Current Application Integration trends and syntactic technologies are up to now quite mature. However, if we try to increase the level of automation we confront several problems and challenges. We examine these problems below.

Data and message level heterogeneities between interoperating services: Between two exposed services a lot of semantic heterogeneities may occur [8]. Maybe the most crucial one refers to message level heterogeneities, according to which an output of one service is *semantically equivalent* with the input of another service (e.g. Temperature) but these services cannot be executed sequentially because of syntactic

---

[3] http://www.w3.org/TR/ws-cdl-10/
[4] http://www.ibm.com/developerworks/library/ws-bpel/
[5] http://www.osgi.org/osgi_technology/

incompatibility (e.g. Temperature expressed in Celsius Degrees versus Temperature expressed in Fahrenheit Degrees). It is obvious that we need a vehicle through which the exposed device services may resolve their message level heterogeneities. Such a need can be satisfied with the introduction of *Device Specific Data Semantics*.

Insufficient search and discovery of exposed services in a common registry: Assuming the existence of several exposed services it is very crucial to classify each service under a semantic categorization model. Such a semantic categorization of services combined with ontology-driven match-making algorithms supports efficient and effective search and discovery of exposed services. Hence there is a need for introducing *Device Specific Functional Semantics*.

Inadequate Web Process composition: Some services have deterministic observable States (e.g. Switch_Is_Open) based on given inputs (i.e. events). Furthermore the behavior of some services can be expressed as a transition map between observable states (also called state space). The a-priori knowledge of these transition maps is sufficient enough in order to compose two or more services in order to achieve a desirable observable state (a.k.a. desired goal). But these states must be meaningful across the separate transition maps (i.e. Switch_Is_Open must have the same meaning for two different transition maps that belong to two different services). Therefore there is a need for introducing shared *Device Specific Behavioral Semantics*.

2.2 Our Contribution

Having already identified what kinds of semantics are necessary for implementing a Semantically-enabled SODA (SeSODA) approach, in this paper we propose an overall approach for introducing semantics in a SODA environment and we present the architecture of an actual SeSODA implementation, analyse the most critical components of this architecture and discuss implementation issues.

Specifically, automatic, dynamic data mediation can be enabled by providing a priori the mappings and transformations for all service message elements (inputs and outputs of correspondent device services) to a common-reference conceptual, ontological model [8]. A sophisticated SeSODA implementation should facilitate automatic, dynamic data mediation during execution time by making use of these device specific *data semantics*. Furthermore, the power of web services derived from device services can be achieved only when appropriate services are discovered based on the functional requirements. As a step towards representing the functionality of the service for better discovery and selection, Services can be annotated with *functional semantics* [9]. Furthermore lot of research has been performed in the area of specification, design and analysis of service-oriented distributed systems regarding service composition involving stateful partner services, such as specification of behavioral interfaces (also called conversation protocols) and compositions [12]. The a priori knowledge and annotation of the state spaces of given services is sufficient in order to identify if these services are composable. Such (state specific) annotations can be considered as *behavioral semantics*.

2.3 Related Work – State of the Art

Since the need of moving from syntactic SODA approaches to semantic SODA approaches has been established, a lot of work has been accomplished in two tightly bounded domains; *a)* the evolution of semantic languages and their expressiveness such as OWL[6], WSML[7] etc and *b)* the evolution of Semantic Web Services (SWS) Frameworks such as OWL-S[8], WSMO[9], SAWSDL[10] etc.

The research of Ontologies in the domain of ubiquitous and pervasive systems has generated significant outcomes. Standard Ontology for Ubiquitous and Pervasive Applications (SOUPA) [1] is perhaps the most representative Ontology in the domain. Most of these ontologies target the shared conceptualisation of the functional aspects of services that exist in a pervasive environment. However, we claim that the use of ontologies should not target only the functional aspects of services since in such a case the utilisation of semantics is restricted [5]. Furthermore the research for providing flexible composition of smart device services [11] has also evolved. As a result many research projects (e.g. Amigo[11]) are targeting the dynamic reconfiguration of a pervasive environment. This target also imposes the need for utilising different kind of semantics.

We rely our SeSODA implementation on the SAWSDL specification and the main reasons for that are that *a)* SAWSDL respects and builds upon syntactic (web services) standards *b)* The mechanism for annotating web services with semantics supports user's choice of the semantic representation language[10], *c)* The mechanism for annotating Web services with semantics allows the association of multiple annotations written in different semantic representation languages, *d)* Supports semantic annotation of web services whose data types are described in XML schema and *e)* Provides support for rich mapping mechanisms between web service schema types and ontologies.

We mentioned in our motivation that we want to resolve *"Data and Message level heterogeneities"* between interoperating services during runtime. We will do that in the frame of SeSODA by uplifting syntactically exposed services to semantically enriched ones that are hosted in a centralized web service repository. We will address these uplifted services as mediator services. These mediator services communicate with each other throughout the exchange of interoperable Ontological concepts; see also [4] for a similar use of "semantically uplifted" services in Enterprise Application Integration.

Concerning *service composition* issues with regard to the desired functionality and operational requirements in the scope of SeSODA, research is wide. There are many different approaches for applying composition [7]. However these approaches are based on a common technique according to which initially the service has to be expressed in a mathematical model (which implies the definition of a graph, etc) then

---

[6] http://www.w3.org/2004/OWL/
[7] http://www.wsmo.org/wsml/
[8] http://www.w3.org/Submission/OWL-S/
[9] http://www.wsmo.org/
[10] http://www.w3.org/2002/ws/sawsdl/
[11] http://www.hitech-projects.com/euprojects/amigo/

a desired goal has to be defined in the terms of the mathematical model that is formulated and lastly the composition algorithm has to take place.

Finally, search and discovery in the OSGi framework is limited to a request for an implementation of a given service interface, with an optional LDAP query over the service properties of each instance, for discriminating between multiple implementations. In a pervasive environment, it is likely that many services will be controlling devices within the environment, where context (state) information about the services and devices would be more important for service discovery rather than properties of the service implementation. Current service descriptions also lack the ability to describe contextual information that is not specifically attributed to the service implementation. Such problems can be solved by using Functional Semantics that were introduced earlier. Some initial work for defining a Semantic Service Description Bundle [3] has been accomplished by the research community.

3   SeSODA Architecture

The proposed Semantically-enabled SODA (SeSODA) architecture is layered in three distinct layers (device layer, OSGi layer and semantic layer) which are presented in the following paragraphs (see Figure 1).

The Device Layer is agnostic to semantics. It contains various devices that act as sensors or actuators. In an ideal SODA world these devices would be able to provide a rich description of their capabilities but the restrictions that are imposed by the

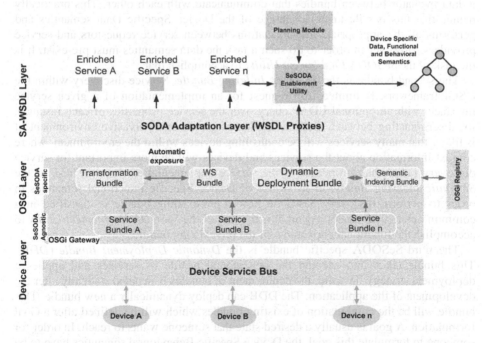

Fig. 1. SeSODA Architecture

various deployment architectures and the resource-constrained nature of the devices does not makes it feasible for the majority of devices.

The OSGi layer 'captures' the first level of interaction with the physical world. It provides the initial modeling which is essential for the semantic enrichment which will take place afterwards. The OSGi layer contains various bundles. These bundles can be grouped in two categories a) Bundles that do not make use of semantic descriptions (SeSODA agnostic bundles) and b) bundles that utilize semantic description in design time or in run time (SeSODA specific bundles).

The SeSODA agnostic bundles provide the essential "uplifting" of device components to services. Of course this uplifting is not automated since the OSGi developer has to undertake 1) the communication with the device which most of the times is proprietary and 2) the exposure of the device's internal business logic as services' operations. Both of these tasks are essential and cannot be automated.

The SeSODA specific bundles are responsible for two major tasks:

1) Semantic uplifting of exposed device services to a semantic level. This is essential since the semantic enablement will take place upon the description of these web services. This procedure can be fully automated with the use of *WS-Bundle* (see Figure 1).

2) make use of semantically uplifted services during the runtime. We examine these two tasks separately. Once the deployed bundles are (automatically) uplifted to semantic web services there are three different kinds of functionalities than can be exploited. Each of these functionalities is implemented in SeSODA specific bundles.

The first bundle is the *Transformation Bundle* whose main functionality is to act as a data mediator between bundles that communicate with each other. This practically means that this bundle takes advantage of the Device Specific Data semantics and performs all the appropriate transformations between service requestors and service providers. However in order to do such a task the data semantics must pre-exist. It is the task of the *SeSODA Enablement Utility* to accomplish that.

The second bundle is the *Semantic Indexing Bundle*. Service discovery within the OSGi framework is limited to a request for an implementation of a given service interface, with an optional LDAP query over the service properties of each instance, for discriminating between multiple implementations. In a pervasive environment, it is likely that many services will be controlling devices within the environment, where context information about the services and devices will be more important for service discovery rather than properties of the service implementation. It is the task of the *Semantic Indexing Bundle* to make use of Device Specific Functional semantics in order to perform discovery based on how a pre-existing bundle is classified in a common ontological model. Again, it is the task of the *SeSODA Enablement Utility* to accomplish the functional enrichment of the pre-existing bundle.

The third SeSODA specific bundle is the *Dynamic Deployment Bundle (DDB)*. This bundle takes into account the behavioural profile of services and applies a deployment strategy. Usually, the deployment of bundles is decided statically after the development of the application. The DDB can deploy dynamically a new bundle. This bundle will be the combination of existing bundles which will be defined after a Goal formulation. A goal is usually a desired state that someone wants to reach. In order for someone to formulate this goal, the Device Specific Behavioural semantics have to be utilised (to create awareness of the state space of a given functionality of a device e.g. Temperature of Room Regulated at 25°C).

Regarding the Semantic Layer, we already saw that the WS-Bundle is responsible for automatically uplifting the exposed device services to web services. Consequently the semantic enrichment of device bundles is *substituted* by the semantic enrichment of web services that represent these bundles. Not only this approach is efficient, but it is also imperative since web service exposure out of existing bundles in the first step for the SeSODA enablement. The second step is to semantically enrich the description (WSDL) that has been automatically generated.

The final step is to substitute the actual web service that has been created automatically (which actually is a WSDL wrapper of the low level device operation) with a *Proxy Web Service*. The difference between the web service and its proxy web service is that the latter can be triggered only by ontological concepts and all generated responses also correspond to ontological concepts. This step is also fully automated. We also address this layer as SA-WSDL Layer because we use the SA-WSDL recommendations as guidance for semantic placeholders.

The *SeSODA Enablement Utility* is the utility that provides the Device Specific Data, Functional and Behavioral annotations to the service descriptions that exist in the SA-WSDL Layer. This utility manages dynamically the above mentioned descriptions.

The Planning Module is the module that uses the behavioral semantics of exposed services in order to formulate a goal. This goal will be the primary input for the Dynamic Deployment Bundle. For example, let's assume that we have two exposed services that have a certain behavioral profile. The behavioral profile is actually the capturing of all possible states that these services may end up along with a transition map that describes the transitions from one possible state to another possible state. A formulated goal may be one of all of these possible states (or a combination of them). In order to achieve this goal many services may need to be incorporated. The system is aware of these services if it knows the superset of the transition maps. It is the task of the Planning Module to identify automatically how many device services must be used and how (i.e. the order with which they must be used) based on goal definition.

4   Implementation Issues

The implementation of the architecture described above is summarized in the implementation of the four SeSODA specific bundles along with the implementation of the SeSODA Enablement Utility and the Planning Module. Concerning the OSGi engine we are using the knopflerish[12] OSGi R4 reference implementation while for the experimental testing of some bundles we used Prosyst's[13] OSGi R4 engine.

Regarding the *WS-Bundle* we use Knopflerfish's embedded Axis servlet. When the Axis bundle is running, objects registered with the OSGi lookup with the property "SOAP.service.name" set are automatically made available as Web services. We use Axis 2.0, however we provide support only to the WSDL 1.1 standard.

---

[12] Knopflerish OSGi Framework http://www.knopflerfish.org
[13] Prosyst OSGi Framework http://www.prosyst.com/products/osgi_framework

Concerning the *Transformation Bundle* we are using an OWL2XSD library and a SAXON XSLT transformation engine to undertake data transformations. The *Semantic Indexing Bundle* is currently implemented over the Semantic Service Discovery Bundle [3] (SSDB). Finally the *Dynamic Deployment Bundle* and the *Planning Module* are still under development. Our main effort is focused in using a Petri-net Engine as CPNets in order to extract the state space of a given device service. Then the annotated State Spaces will be composed by the planning tool.

5 Example

In order to illustrate the use of data semantics in the frame of SODA we use an example scenario. Let's imagine that we have a temperature regulation scenario, according to which a passive device i.e. a temperature sensor and an active device i.e. a heat actuator must cooperate in a sequential execution manner. Let us also imagine that these devices are OSGi enabled. This practically means that there are already implemented OSGI bundles hosted in an OSGi repository that expose the functionality of these sensors and actuators. Our SeSODA realization relies on the fact that the functionality is also *automatically* exposed as web services through an OSGi plugin. The example scenario can be depicted in Figure 2.

First of all at the OSGi layer there are two exposed methods; the *readTemperature* and the *regulateTemperature* method. For the shake of simplicity we assume that the output of *readTemperature* and the input of *regulateTemperature* consist of basic types (*float* and *int* respectively). This is rather indicative since a complex data structure (with a given XSD) could also be used. The first step is to provide an association of this interchangeable structure with an ontological concept. We use an existing weather ontology[14] where the temperature concept exists.

```
<owl:Class rdf:ID="Temperature"/>
<owl:DatatypeProperty rdf:ID="hasTemperatureReading">
  <rdfs:range rdf:resource="http://www.w3.org/2001/XMLSchema#float"/>
  <rdfs:domain rdf:resource="#Temperature"/>
</owl:DatatypeProperty>
```

According to the proposed SeSODA implementation, firstly the ontological class related to the interchangeable structure has to be converted automatically to a syntactic structure i.e. an XSD schema. Of course this automation is not trivial at all. Semantic languages and XML standards have a lot in common concerning their expressivity [6], however there are substantial differences between them. In scientific literature there are a lot of RDFS2OWL and vice versa converters however in the OWL-DL level (that we work) conversion is not straightforward since existing object properties may result in schema explosion.

In our case a conversion guideline would be a) Generate complex types that derive by the name of ontology classes, b) Enrich the complex types with all the datatype properties that the instances of the class may have c) Let the user choose if an object property is crucial or not and then include it or exclude it form the complex type.

---

[14] http://www.csd.abdn.ac.uk/research/AgentCities/WeatherAgent/weather-ont.daml

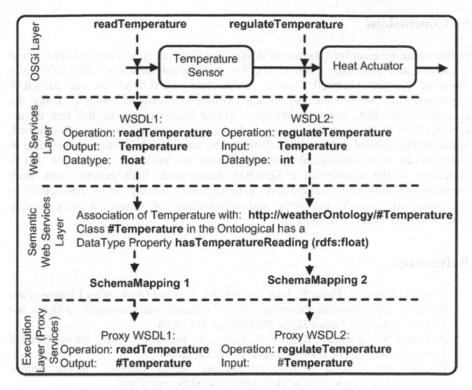

Fig. 2. Temperature regulation example scenario

The ontological Concept would result in the creation of the following XSD:

```
<xs:schema xmlns:xs="http://www.w3.org/2001/XMLSchema"
elementFormDefault="qualified"
attributeFormDefault="unqualified">
<xs:element name="Temperature">
   <xs:complexType>
     <xs:sequence>
       <xs:element name="hasTemperatureReading"
type="xs:float"/>
     </xs:sequence>
   </xs:complexType>
 </xs:element>
</xs:schema>
```

The next step is the provision of up-cast and down-cast information which is actually an XSLT transformation between the XSD of the correspondent ontological concept and the XSD of the input and output structures. These XSLT transformations are embedded in the WSDL description of a new Proxy Services. Proxy Services now can communicate only by interchanging the Ontological Concept *Temperature* (transformations will be handled automatically).

6 Conclusions

In this paper we examined the use of device-specific data, functional and behavioural semantics and proposed a framework for the implementation of an OSGi-based Semantic Service Oriented Device Architecture (SESODA). In our SESODA implementation we semantically enrich existing syntactic technologies using the upcoming SAWSDL standard. We give special emphasis on the fact that device services have to be automatically exposed to web services and semantically uplifted. Semantically uplifted services are substituted by service proxies that are then able to communicate with ontological concepts. Finally we provided an example that is indicative of the benefits of a SESODA deployment. Such benefits vary from automatic resolution of message level heterogeneities (during device interaction) to semi-automatic search, discovery and composition of device services (during programming of pervasive spaces).

References

1. Chen, H., Perich, F., Finin, T., Joshi, A.: SOUPA: Standard Ontology for Ubiquitous and Pervasive Applications. In: Proceedings of the 1st International Conference on Mobile and Ubiquitous Systems August 22-26, 2004 Boston, MA (2004)
2. de Deugd, S., Carroll, R., Kelly, K.E., Millett, B., Ricker, J.: SODA: Service-Oriented Device Architecture. In: IEEE Pervasive Computing (2006)
3. Docherty, L.: A guide to using the Semantic Service Discovery Bundle, Stirling University, http://www.cs.stir.ac.uk/ lsd/tutorials/SSDB_tutorial.pdf
4. Friesen, A., Alazeib, A., et al.: Towards semantically-assisted design of collaborative business processes in EAI scenarios. In: 5th IEEE International Conference on Industrial Informatics, INDIN 2007, Vienna, Austria, July 23-27 (2007)
5. Helal, S.: Programming Pervasive Spaces. IEEE Pervasive Computing 04(1), 84–87 (2005)
6. Klein, M., Fensel, D., van Harmelen, F., Horrocks, I.: The relation between ontologies and XML schemas (2001)
7. Milanovic, N., Malek, M.: Current Solutions for Web Service Composition. IEEE Internet Computing 8(6), 51–59 (2004)
8. Nagarajan, M., Verma, K., Sheth, A., Miller, J., Lathem, J.: Semantic Interoperability of Web Services - Challenges and Experiences. In: 2006 IEEE International Conference on Web Services (ICWS 2006) (2006)
9. Sivashanmugam, K., Sheth, A., Miller, J., Verma, K., Aggarwal, R., Rajasekaran, P.: Metadata and Semantics for Web Services and Processes. Datenbanken und Informationssysteme. Publication Hagen (October 2003)
10. Sivashanmugam, K., Verma, K., Sheth, A., Miller, J.: Adding Semantics to Web Services Standards. In: Proceedings of the 1st International Conference on Web Services (ICWS 2003), Las Vegas, Nevada, pp. 395–401 (June 2003)
11. Vallée, M., Ramparany, F., Vercouter, L.: Flexible composition of smart device services. In: The 2005 International Conference on Pervasive Systems and Computing (PSC 2005), June 27-30, 2005 Las Vegas, Nevada, USA (2005)
12. Yi, X., Kochut, K.J.: Specification and Analysis of Service-Oriented Distributed Systems using Colored Petri Nets: Algorithms and Tools. UGA computer science department technical report (November 2004)

A Semantic Framework for Priority-Based Service Matching in Pervasive Environments

Ayomi Bandara[1], Terry Payne[1], David De Roure[1], and Tim Lewis[2]

[1] University of Southampton, Southampton, UK
{hmab02r,trp,dder}@ecs.soton.ac.uk
[2] Telecommunications Research Laboratory, Toshiba Research Europe Ltd, Bristol, UK
Tim.Lewis@toshiba-trel.com

Abstract. The increasing popularity of personal wireless devices has raised new demands for the efficient discovery of heterogeneous devices and services in pervasive environments. The existing approaches such as Jini [1], UPnP [8], etc., describe services at a syntactic level and the matching mechanisms in these approaches are limited to syntactic comparisons based on attributes or interfaces. In order to overcome the limitations in these approaches, there has been an increased interest in the use of semantic description and matching techniques to support effective service discovery. This paper proposes a semantic matching approach which facilitates the discovery of device-based services in a pervasive environment; the approach provides a ranking facility that orders services according to their suitability and also considers priorities placed on individual requirements in a request during the matching process. The evaluation studies have shown that the matcher results correlate reasonably well with human judgement.

1 Introduction

With the current trends in the electronic world, devices of increasing heterogeneity are being introduced into pervasive environments. These vary from headsets to mobile phones, PDAs and laptops; each of which offer a plethora of services. This has raised new demands for the discovery of devices and their services in a dynamically changing environment. The existing device description and discovery solutions (such as UPnP [8], Jini [1], etc.), characterise the services by using predefined service categories and fixed attribute value pairs; also the matching mechanisms in these approaches are limited to string comparisons or key-word based searches. Since discovery is not supported by any form of inferencing, such approaches will be unable to identify a match between logically equivalent services that have syntactically different descriptions.

With recent trends in the Semantic Web, there has been an increased interest in the use of ontologies to describe services and the use of logical reasoning mechanisms to support service matching. The advantage of such frameworks include the ability to extend and adapt the vocabulary used to describe services and to harness the inferential benefits of logical reasoning over such descriptions. Recently, a number of semantic matching approaches have been developed (targeted at different domains), which try to address various limitations in traditional discovery techniques in order to come up with a pragmatic solution to meet the challenges in the service discovery arena.

The work described in [4] and [2] propose Semantic Matching approaches for pervasive environments. Both these approaches use ontologies to describe the services and

R. Meersman, Z. Tari, P. Herrero et al. (Eds.): OTM 2007 Ws, Part II, LNCS 4806, pp. 783–793, 2007.
© Springer-Verlag Berlin Heidelberg 2007

a Prolog-based reasoning engine to facilitate the semantic matching. They provide 'approximate' matches if no exact match exists for the given request. However, the criteria used for judging the 'closeness' between the service advertisements and the request is not clear from the literature. In both these approaches, the matching process does not perform any form of match ranking. There have also been a number of efforts that use description logic (DL) based approaches for semantically matching web services. For example the matchmaking framework presented in [5] uses a DAML-S based ontology for describing the services. A DL reasoner has been used to compute the matches for a given request. The matches are classified into one of its five "degrees of match" (namely Exact, Plug-In, Subsume, Intersection and Disjoint) by computing the subsumption relationship of the request description w.r.t. all the advertisement descriptions. No ranking is performed in the matching process, although the match class suggesting the 'degree of match' gives an indication of how 'good' a match is.

The above mentioned semantic matching approaches provide important directions in overcoming the limitations present in traditional syntactic approaches to service discovery. However, these solutions still have limitations and overlooked issues that need to be addressed; particularly these approaches do not have an effective ranking criterion to facilitate the ordering of potential matches, according to their suitability to satisfy the request under concern. Also these approaches do not facilitate priorities/ weights on the individual requirements and thus the matching process will consider that all requirements have equal priority.

In this paper we propose a pragmatic approach, that facilitates the effective matching of resource requests and advertisements in pervasive environments. This semantically compares the request against the available services and provides a ranked list of most suitable services. The rank will indicate the appropriateness of a service to satisfy a given request and thus provides a valuable decision support for the service seeker, in selecting the most suitable service. The matching process also considers the priorities/ weights on the individual requirements of a request; this helps capture any context dependencies involved and subjective preferences of the resource seeker. This is an important facility in any matching system since in many practical scenarios certain service requirements will be more important than others and therefore priority-based matching can produce results that can better meet the users expectations. An OWL ontology is used to describe the services and a Description Logic reasoner is used to support the background reasoning tasks in the matching process. The remainder of this paper is organised as follows: Section 2 discusses the motivation behind the proposed matching framework and identifies the requirements of a pragmatic approach for matching pervasive resources. Section 3 briefly describes the methodology behind the matchmaking framework. Section 4 discusses the prototype implementation of the service matching approach in a pervasive scenario and presents the initial evaluation results. The concluding remarks and future directions of this work are discussed in section 5.

2   Motivating Factors

There are several desirable properties that must be present in an effective service matching framework. In this section we discuss these along with the motivating reasons behind them.

Semantic Description and Matching: An ontological approach for the description of services coupled with reasoning mechanisms to support service discovery and matching enables logical inferencing over these descriptions and therefore offers several benefits over the traditional syntactic approaches. It is often the case, that the service providers usually describe devices in terms of lower-level properties, and the service seekers or clients usually prefer to describe service requests using more abstract or higher level concepts. Semantic matching approaches supported by logical reasoning mechanisms will be able to identify a match between logically equivalent services that have syntactically different descriptions and therefore can offer flexibility in how the service advertisements and requests are described.

Match Ranking: Ranking refers to the ordering of the available advertisements in the order of their suitability to satisfy the given request. In the absence of an exact match, a requester might be willing to consider other advertisements that are closer to the request and thus the ranking will be useful in gaining an understanding of the appropriateness of the advertisement. Most existing matchmaking solutions do not have an effective criterion to rank the available services according to their suitability. Providing a ranking mechanism that will rank the advertisements on the basis of how well it satisfies the properties specified in the request, is one of the main objectives behind the proposed matching framework.

Approximate Matching: Offering approximate matching, is one of the core objectives of semantic matching. i.e. services that deviate from the request in certain aspects should not be discarded but must be ranked or classified appropriately to indicate the suitability. In the current semantic matching approaches [5], [6], the suitability of the advertisement have been determined using subsumption reasoning based on the taxonomic relation between the concepts. However we argue that subsumption reasoning alone is not sufficient in determining similarity for the purpose of resource matching. Depending on the concept involved, reasoning based on the taxonomy alone, will not accurately reflect the similarity between concepts. For example, consider the concept *Processor*; assume there is a request for a computer that has processor *Pentium4* and advertisements of computers with processors *Pentium3* and *Pentium1*. Both *Pentium3* and *Pentium1* will be disjoint from the originally requested concept of *Pentium4*, but a requester will consider *Pentium3* to be a better match than *Pentium1* and will be ranked higher. Thus the type of attribute involved in the individual requirement of a request will have to be considered in approximating and ranking of advertisements. Section 3 describes the types of attributes and the approach taken in judging the similarity between them for the purpose of ranking.

Priority-Based Matching: The current matchmaking researches do not consider any priorities or preferences that a user/agent may be having with respect to various attributes or properties of a service (except in [6]). In many practical scenarios certain requirements/ attributes in a request will be more important than others, either due to the context involved or the subjective preferences of the user. In such cases, facilitating priority-handling in the matching process will produce match results that are more relevant and suitable for the context involved. For example, consider two users looking for a printer; considering the time to service and quality properties of the printer, both may want to take the printouts as quick as possible and with the highest quality possible. But a user who wants to rush off to a meeting in the next five minutes will definitely be more concerned about the time factor and be willing to compromise on quality. But a user who is working at leisure, will not mind waiting in order to obtain a more quality print.

Thus in cases like this it is vital to consider the importance placed on the properties of the service by a user, by taking into account the priorities of the attributes.

Mandatory requirements or strict matching requirements have to be considered when, the resource seekers requires a certain individual property requirement in a request, to be strictly met by any potential resource advertisement; i.e. they will not want to consider any advertisements that will have even a minor deviation, with respect to that property. For example consider the case where a resource seeker needs to utilise a computer to run an application which will only run on the operating system WindowsXP; he will specify the operating system requirement in the request along with the other desirable characteristics. In the context involved the operating system property is a mandatory requirement and hence the resource seeker will not need to consider any available computers which deviates with respect to the operating system requirement (no matter how good it is with respect to other attributes). Hence this needs to be taken into account in the matching process and the available resources that deviate from this strict requirement must not be included in the result set (or ranked as the worst matches). Priority matching is applicable when a resource seeker has varying importance placed on the individual property requirements of the request. Strict matching can in fact be considered as a specific case of priority matching.

This factor will be taken into account in the proposed work by giving a service requester the option of placing priorities/ weights on the specified attributes of the service request. These weights will be considered in the matching process during the ranking of advertisements.

3   The Semantic Matching Methodology

3.1   Service Description

In order to facilitate the use of logical reasoning over the service descriptions, we describe the requests and advertisements in the Web Ontology Language (OWL).

A request will typically consist of several sub requirements to be satisfied. Each individual requirement will specify: the description of the requirement (which is the resource characteristic the resource seekers expect in a resource, for the their needs to be satisfied) and the priority or weight of that individual requirement, which will be a decimal value that indicates the relative importance of the particular requirement. The priority value can also be used to indicate if the requirement considered is a mandatory requirement; i.e. if the requirement should be strictly satisfied in an advertisement for the requester to consider it as a potential match. The description of an individual requirement will include the property or attribute the requesters are interested in and the ideal value desired.

The request will take the form of:

$Request \equiv (Req_1) \sqcap (Req_2) \sqcap \ldots \sqcap (Req_n)$

where Req_i is an individual requirement. The requirement in turn can take the form of:

$Req \sqsubseteq (= 1hasDescription.RD) \sqcap (= 1hasPriority.PriorityValue)$

where RD is the requirement description, which can be either a named concept or an existential restriction of the form, $\exists p.C$ where p is a role and C is a named concept or a complex concept. For describing each RD, an ontology that describes the services in the domain concerned can be used. The $PriorityValue$ indicates the relative importance of the individual requirement in the request. This is a decimal value defined between 0 and 1. In addition, to indicate that the requirement is a mandatory requirement that must be strictly met in any potential match, the $PriorityValue$ is defined as 2. The resource seeker must pick the appropriate $PriorityValue$ (according to these pre-defined values) for each individual requirement, to indicate its relative importance.

The resource provider will specify all the relevant characteristics of the available resource in the resource advertisement. The advertisement can take the form of: $Advertisement \equiv (r_1) \sqcap (r_2) \sqcap \ldots \sqcap (r_n)$; where r_i is either a named concept or an existential restriction describing a characteristic of the resource.

3.2   Ranking Process

As mentioned previously a request will consist of a number of individual requirements along with their priority values. The presence of any mandatory requirements that must be fully satisfied by any potential match will also be indicated by using the appropriate priority value as mentioned in 3.1. In the matching process, the available resource will be checked to see if each mandatory individual requirement (RD) is satisfied in the advertisement description. If the mandatory requirement(s) are met, then the advertisement will be evaluated through approximate matching.

In approximate matching, the available resources should be evaluated according to how well it satisfies each individual requirement specified in a request; i.e. the matching engine should quantify the extent to which each individual requirement description (RD) is satisfied by the resource advertisement. For this, the matching engine will check how similar the advertisement is with respect to each non-mandatory requirement (RD) specified in the request; the similarity will be determined depending on the semantic deviation of the expected value in request and the available value in advertisement for the same requirement, and a score will be assigned accordingly ($Score_i$).

Each characteristic specified in the request (RD) can be a named concept(C_R) or an existential restriction ($\exists p.C_R$). If it is a named concept, similarity will be compared between the corresponding concepts in request and advertisement ($Similarity(C_R, C_A)$). If it is an existential restriction, the corresponding existential restriction(s) will be found in the advertisement ($\exists p.C_A$) and the similarity will be compared between the corresponding concepts in request and advertisement. If it is a composite concept the similarity will be judged recursively. The score ($Score_i$) for each individual characteristic in the request will be assigned depending on this similarity.

The degree of similarity between concepts will be determined depending on the type of concept or attribute involved; determination of similarity between concepts will be discussed later in this section. A score ($Score_i$) is assigned for each sub-requirement (RD) specified in the request. The score for the advertisement (match score) will be determined by using the weighted average of these individual scores (the weight will be the corresponding priority value of each individual requirement).
$MatchScore = \sum_{i=1}^{n} w_i.Score_i \div \sum_{i=1}^{n} w_i$
where w_i and $Score_i$ is the priority value and the score of the sub-requirement RD_i. The overall score for the advertisement provides an indication of how good the advertisement is in satisfying the given request. The score for an advertisement will in turn be used as the basis for ranking; the highest score will receive the highest rank and so on.

The attributes in a resource description are categorized into three types, for the purpose of approximating and judging similarity within individual requirements. These are:

Type 1 Attributes: Attributes involving symbolic concepts for which judging similarity using the taxonomic relation is sufficient. In this case the matching engine will make use of a reasoner to judge the similarity by subsumption relation. Say the advertisement specifies that it has concept C_A as its value for a particular property or attribute and request specifies it has concept C_B. When a description logic reasoner is used to find the subsumption relation between these two concepts, it could fall under one of four types. These types, and the scores assigned are represented in Table 1. In the case where C_A is a super concept of C_B, the score assigned must be a value between 0 and 1; the ideal value of t can be determined through a human user study. However, for the purpose of this implementation of the matching system we use a value of 0.6.

Type 2 Attributes: Attributes involving symbolic concepts for which judging similarity using the taxonomic relation is not sufficient. For example *Processor* and *Display Technology* concepts fall into this category. Hence in our work, if we wanted to find similarity between different Processor Types for example, the features/ properties of the Processors such as Clock Speed, Manufactured-By etc. will have to be used in measuring the similarity. However measuring similarity between concepts is out of the scope of the current work and we assume that the knowledge of similarities between such concepts is available to us (measured by using an available similarity measurement approach such as [7], and available as domain knowledge in the ontology) and can be accessed by the matchmaking process.

Type 3 Attributes: When the attributes are numeric (integer or decimal) the degree of similarity between the requested and advertised attribute values must be determined depending on the level of deviation. This deviation can be determined by using either a fuzzy membership function or by computing the percentage deviation from the original requested value.

Table 1. Assignment of similarity scores when Subsumption Relation is considered

Subsumption Relation	Similarity Score
C_A and C_B refer to the same concept	1.0
C_A is a super concept of C_B	t (where $0 < t < 1$)
C_B is a super concept of C_A	1.0
C_A, C_B are not equivalent and do not have a subsumption relation	0.0

4   Application of the Matching Framework in a Pervasive Context

The matching framework presented in the previous sections has been implemented in a pervasive context for matching of device based services. The service requesters seek to utilise specific devices and their services depending on their functionality. The advertisements and the individual requirements in a request are described using the Device Ontology presented in [3] (available at   http://www.ecs.soton.ac.uk/ ~hmab02r/DeviceOnt/DevOntology.owl. This facilitates the description of features and functionalities of the devices and their services. The necessary ontologies were developed with the Protégé ontology editor. The matching engine was implemented in Java and the Pellet DL reasoner in combination with the Pellet-API is used to facilitate the necessary reasoning tasks during the matching process.

Figure 1 illustrates the architecture of the matching system. Once the matching system receives the OWL descriptions of the advertisements and request, it checks for the consistency of the descriptions. If they are consistent the matching process begins. Each advertisement is compared with the request using the matching mechanism presented before and depending on the suitability of the advertisement to satisfy the request a score is assigned to the advertisement. Once all the advertisements are compared and scored, the advertisements are ranked on the basis of the score they have received. Then the system returns the advertisements along with their rankings.

As emphasised in Section 2, semantic approaches to service discovery can clearly provide many benefits over syntactic approaches. However, we have to bear in mind the fact that certain resources in pervasive environments (small mobile devices such as mobile phones and PDA's), are heavily constrained in terms of computing power and therefore the standard semantic web tools and technologies can be too heavy-weight for such resources. Hence a feasible architecture has to be chosen for the discovery process, while facilitating the use of semantic descriptions and reasoning mechanisms to provide effective description and matching of services. For example, the matching process could always run centrally on the network and the devices could communicate through the network as appropriate.

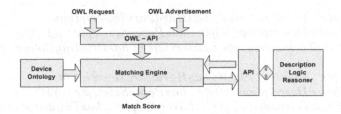

Fig. 1. The Matching System

4.1 Matching Example

We illustrate the application of the proposed matching approach with the use of the following example. We assume a scenario where a user in a pervasive environment seeks a printer with certain characteristics. The request concerned is a *Colour, Laser* Printer, that can print on the paper size *A2*. We also assume that the Paper Size *A2* is the most important (highest priority) attribute under the context involved and that the other two attributes (the Print Technology being *Laser* and the *Colour* printing capability) are of lesser priority. The priority values or weights for the three attributes are assigned as 0.6 for the paper size attribute and 0.2 for both the other attributes (the printing colour and printer technology). Since the request is for a *Printer*, this is stated as a mandatory requirement with priority value 2.0. The request will be described in description logic notation as:

$Request \sqsubseteq$
$\exists\,hasRequirement\,(Requirement \sqcap \exists hasPriority. = 2.0 \sqcap$
$\exists hasRequirementDescription.RD1) \sqcap$
$\exists\,hasRequirement\,(Requirement \sqcap \exists hasPriority. = 0.6 \sqcap$
$\exists hasRequirementDescription.RD2) \sqcap$

$\exists\, hasRequirement\, (Requirement \sqcap \exists hasPriority. = 0.2\ \sqcap$
$\exists hasRequirementDescription.RD3)\ \sqcap$
$\exists\, hasRequirement\, (Requirement \sqcap \exists hasPriority. = 0.2\ \sqcap$
$\exists hasRequirementDescription.RD4)$

$RD1 \equiv ptr : Printer$

$RD2 \equiv \exists dev : hasHardwareDescription$
$(dev : HardwareDescription \sqcap \exists ptr : hasPaperSize\,.\,ptr : A2)$

$RD3 \equiv \exists dev : hasHardwareDescription$
$(dev : HardwareDescription \sqcap \exists ptr : hasPrintTechnology\,.\,ptr : Laser)$

$RD4 \equiv \exists dev : hasHardwareDescription$
$(dev : HardwareDescription \sqcap \exists ptr : hasPrintingColour\,.\,ptr : Colour)$

For the purpose of this evaluation experiment we assume the availability of twelve advertisements (with varying characteristics) of which we include the descriptions of four. The available advertisements are:

$Advert1 \sqsubseteq ptr : Printer \sqcap \exists dev : hasHardwareDescription$
$(dev : HardwareDescription \sqcap \exists ptr : hasPaperSize\,.\,ptr : A2\ \sqcap$
$\exists ptr : hasPrintTechnology\,.\,ptr : Laser \sqcap \exists ptr : hasPrintingColour\,.\,ptr : BW)$

$Advert2 \sqsubseteq ptr : Printer \sqcap \exists dev : hasHardwareDescription$
$(dev : HardwareDescription \sqcap \exists ptr : hasPaperSize\,.\,ptr : A2\ \sqcap$
$\exists ptr : hasPrintTechnology\,.\,ptr : Laser \sqcap \exists ptr : hasPrintingColour\,.\,ptr :$
$Colour)$

$Advert3 \sqsubseteq ptr : Printer \sqcap \exists dev : hasHardwareDescription$
$(dev : HardwareDescription \sqcap \exists ptr : hasPaperSize\,.\,ptr : A2\ \sqcap$
$\exists ptr : hasPrintTechnology\,.\,ptr : Inkjet \sqcap \exists ptr : hasPrintingColour\,.\,ptr :$
$BW)$
$\vdots$

$Advert12 \sqsubseteq ptr : Printer \sqcap \exists dev : hasHardwareDescription$
$(dev : HardwareDescription \sqcap \exists ptr : hasPaperSize\,.\,ptr : A4\ \sqcap$
$\exists ptr : hasPrintTechnology\,.\,ptr : Inkjet \sqcap \exists ptr : hasPrintingColour\,.\,ptr :$
$BW)$

Considering the attributes involved in this example: The Paper Size attribute is a Type-2 attribute and we assume that the similarity values between A2, A3 and A2, A4 are given as 0.6 and 0.25. Printer Technology attribute is also a Type-2 attribute and the similarity values between Laser and Inkjet is 0.7. Printing Colour is a Type-1 attribute where the concept Colour is defined as a subclass of the concept Black_White (since all colour printers can print black & white as well). Therefore considering Advert12, this satisfies the mandatory requirement of being a *Printer* and therefore will proceed through to the approximate matching process. This will get subscores of 0.25, 0.7 and 0.6 for the attributes of Paper Size, Printer Technology and Printing Colour. By considering the weighted average of these subscore values (using the corresponding priority values of the attributes as weights), the Advert will get a match score of .41. Similarly the other advertisements can be evaluated in the same way and by considering the match score, the advertisements could be ranked accordingly.

4.2 Evaluation

The effectiveness of the proposed matching approach is evaluated by comparing the results of the matching system with human perception. This is done by comparing the results obtained through the matching system with the rankings provided by domain users that rank the available resources in the same scenario.

We conducted a study involving human subjects to obtain the human rankings for this evaluation exercise. For each experiment, a scenario or use case (in a pervasive context) is devised that will involve a resource seeking situation where the seeker raises a query for a resource with certain property requirements. For each use case we construct a questionnaire which specifies: the device and the property/ functionality requirements that the resource seeker is interested in, the context that has given rise to the need of the device and the available devices and their properties. We hand out the questionnaire to the subjects involved and request them to assume that they are the resource seeker in the given context and rank the available devices specified, in the order they would consider them for utilising for the specified need. For each experiment, at least 10 subjects were involved and the rankings provided were averaged (to minimise the effects of subjective judgements) for the purpose of comparison with the matchmaker results. Each experiment was designed to test a different aspect of the matching framework.

To judge the degree of conformance of the match results to human perception, the matcher ranking is compared with the average human ranking (obtained for the same experiment) using graphical illustration of the plots and the measurement of standard deviation. It was generally observed in all the experiments that the matcher results were reasonably close to the average human ranking.

Due to space limitations, the detailed results of all the experiments in this evaluation exercise will not be presented in this paper. However, to give an indication of how well the matcher ranking conforms to human judgement, and to show that considering priorities on individual requirements during the matching process improves the effectiveness of the matching system, we present the results obtained for the example scenario discussed in Section 4.1. The average human ranking has been obtained for this scenario through a human study and compared with the results obtained from the proposed priority-based matcher. To illustrate how close the matcher ranking is to the average human ranking, we have computed the difference between the matcher rank and the average human rank and this is graphically illustrated in Figure 2; the y axis of the figure depicts the difference in the rankings. For the sake of comparison, the results were also obtained from the matcher when no priorities are considered (i.e. when all the requirements are assumed to have equal priority/ weight). The difference between this and the human rank is also computed and is illustrated in the Figure 2. Through the observation of these plots we can see that the matcher rank with priority consideration is closer to the human ranking than in the case where priority is not considered. The standard deviation between the matcher ranking (without priority consideration) and the average human ranking is 7.19%. The same value for the matcher rank with priority consideration is 3.35%. This shows that including the priorities in the individual

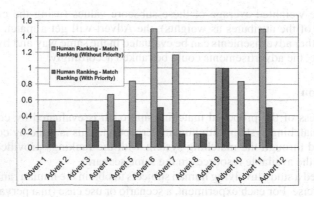

Fig. 2. The Difference Between the Averaged Human Ranking and the Matcher Rankings

requirements and considering them in the matching process will allow the matcher to produce results that better suit the context involved and that conform to human judgement.

5   Conclusions and Future Work

In this paper we have proposed a semantic approach that provides effective matching of resources in a pervasive environment. The approach facilitates the ranking of potential matches in the order of their suitability to satisfy the request, which aids the users of the matching system to identify the order in which they should consider the returned matches. The ranking mechanism overcomes the limitations present in matchers which uses subsumption reasoning alone. The matching framework also facilitates the specification of priorities in the service request and incorporates priority handling in the matching system; this helps to identify the relative importance of the individual requirements in a request and also to indicate whether certain requirements are mandatory and thus should be strictly met in any potential match. Hence the matching system can produce results that better suit the context involved and the subjective preferences of service seekers. The involvement of match ranking and the priority handling are both important and useful additions to the existing work on service matching.

We have implemented the proposed solution in a pervasive context and results have been obtained. The effectiveness of the solution has been evaluated through the use of a human user study and initial results indicate that the matcher results correlate well with human perception. As part of the future work of this research, we plan to formally evaluate the performance of the proposed approach to justify that any compromise in performance resulting from the involvement of an ontological description and reasoning mechanism, is outweighed by the benefits gained from semantic matching. Specifically we plan to investigate the scalability and the performance of the matching approach, in terms of the number of advertisements involved in the matching process and the size of the resource request (in terms of the number of individual requirements).

Acknowledgements. This research is funded and supported by the Telecommunications Research Laboratory of Toshiba Research Europe Ltd.

References

1. Arnold, K., OSullivan, B., Scheifler, R.W., Wollrath, A., Waldo, A.W.J.: The Jini Specification. Addison-Wesley, Reading (1999)
2. Avancha, S., Joshi, A., Finin, T.: Enhancing the bluetooth service discovery protocol. Technical report, University of Maryland Baltimore County (2001)
3. Bandara, A., Payne, T.R., de Roure, D., Clemo, G.: An ontological framework for semantic description of devices. In: Poster Abstracts of the Third Int. Semantic Web Conference (2004)
4. Chakraborty, D., Perich, F., Avancha, S., Joshi, A.: Dreggie: Semantic service discovery for m-commerce applications. In: Workshop on Reliable and Secure Applications in Mobile Environment, Symposium on Reliable Distributed Systems (2001)
5. Li, L., Horrocks, I.: A software framework for matchmaking based on semantic web technology. In: Proc. of the Twelfth Int. World Wide Web Conference, pp. 331–339. ACM Press, New York (2003)
6. Paolucci, M., Kawamura, T., Payne, T.R., Sycara, K.P.: Semantic matching of web services capabilities. In: Int. Semantic Web Conference, pp. 333–347 (2002)
7. Tversky, A.: Features of similarity. Psychological Review 84, 327 (1977)
8. Upnp (2000), http://www.upnp.org/download/UPnPDA10_20000613.htm

Enhancing Privacy by Applying Information Flow Modelling in Pervasive Systems

Steffen Ortmann, Peter Langendörfer, and Michael Maaser

IHP microelectronics, Im Technologiepark 25, 15236 Frankfurt (Oder)
{ortmann,langendoerfer,maaser}@ihp-microelectronics.com

Abstract. In today's working and shopping environment a lot of sources are present that collect data of people located in those environments. The data gathered by devices such as video cameras, RFID tags, use of credit cards etc. can be combined in order to deduce information which cannot be "measured" directly. In this paper we introduce deduction rules that help to describe which information can be inferred from which sources. Using these rules all information that can be gathered by a pervasive system can be identified and linked to the sources of the raw input data. By that the pervasive system is represented as an information flow graph. In order to enhance privacy we use this graph to determine the data sources, e.g. video cameras or RFID tags, that need to be switched off to adapt a given system to privacy requirements of a certain person. Due to the fact that we do not consider an individual device a data source but cluster those devices into a single source of a certain type, our approach scales well even for large sensor networks. Our algorithms used to build and analyze the information flow graph offer low calculation complexities. Thus, they are well suited to be executed on mobile devices giving the end user back some control over her/his data. Even if she/he cannot influence the system, she/he at least knows which information is exposed to others.

Keywords: Privacy, Pervasive Computing, Information Flow Modelling, Sensor Networks.

1 Introduction

The advances in the areas of semi conductor industry, computer science and sensing devices indicate that the realization of Mark Weiser's vision of ubiquitous computing is becoming reality in the near future [1]. On the one hand this vision looks like heaven, users do no longer need to bother with unfamiliar devices, get immediate support from their environment etc. On the other hand such systems look like hell from a privacy perspective. If accidentally or intentionally misused pervasive system can become the perfect surveillance tool [2] compared to which the scenario of George Orwell's novel 1984 looks like good old times.

We consider privacy an essential for the general acceptance of pervasive systems. We also believe in the way pervasive systems comfort their user, i.e. they somehow adapt to the users needs and requirements. Therefore we are investigating means that

R. Meersman, Z. Tari, P. Herrero et al. (Eds.): OTM 2007 Ws, Part II, LNCS 4806, pp. 794–803, 2007.

allow pervasive systems to adapt to user's privacy requirements. In this paper we are discussing means that allow processing a configuration of a given pervasive system that obeys the privacy of a certain user. As a basic means we introduce deduction rules that allow modelling the information flow of a pervasive system. Here we take into account the sensing devices and their provided raw data as well as information that can be inferred from these sensor readings. After a pervasive system is modelled that way, a setting that fits to the user's privacy requirements can be determined. This is done by analysing the information flows in the direction from sink to source. After detecting the offspring of a data item that shall not be revealed, the user can advise the pervasive system to switch off at least one of the sources that are essential for deducing this special information.

The rest of this paper is structured as follows. We start with a short overview on privacy related research. Section 3 introduces our approach, including our applied information flow model. In Section 4 we present how to use our model to determine privacy related environment configurations. Finally our paper concludes with a summary and outlook.

2 Related Work

Privacy and pervasive systems have been hot research topics for several years and there has been a lot of work done in these areas. This section presents some recent advancement in sensor technologies and privacy regarding approaches for pervasive systems.

Today's sensor technologies provide a wide variety of skills and applications and can often be used from a distance without any contact to the measured objective. For example an installed system of cameras is not only capable of recording video data. They can analyse user's behaviour [3], monitor positions and movement profiles of users and detect single events like remaining at an advertising display [4]. Even "hidden" features like heart rate detection by thermal imaging is possible [5, 6]. A research project in office monitoring showed that special events like intrusion detection can also be discovered by certain combinations of general environment data measured by light-, sound-, movement- and other sensors [7]. Also the usage of RFID tags provides an enormous amount of potential applications. These tags are capable to store and process data and answer data requests within a distance of 7 to 100 meter [8, 9]. Pervasive environments equipped with RFID readers are able to gather the data of these tags even without any notice of their owner. In addition to the stored data all tags contain a unique serial number that allows recognizing a certain tag easily and most likely its respective owner. Since RFID tags and also other items like credit- and bonus-cards can import additional data into a system this fact must also be considered when modelling a pervasive environment.

Also some privacy related research projects for pervasive environments are in progress. Almost every platform for location based services comes with its own privacy protection approach [10, 11, 12, 13]. The EU is funding the Privacy and Identity Management for Europe (PRIME) project [14] which aims at giving the user control over his/her data e.g. by use of pseudonyms. Several architectural concepts for privacy protection in mobile or pervasive environments have been published [11, 15, 16, 17]. All these approaches are valuable steps into the right direction, but they focus on data that is directly revealed, i.e. information such as current position, age, gender etc. Our approach attempts to analyse privacy settings also for information that can be inferred from raw data provided by the user herself and/or sensors deployed as part of the pervasive system.

3 Modelling a Pervasive Environment

To model a pervasive environment we apply the idea of information flow modelling. Information flow modelling in its original form [18] classifies all available data into sets and defines directed relations between those sets. The relations define all directions where a flow of information is allowed. Mostly such approach is used for hierarchical ranking, for example in military applications where different security levels like "confidential", "secret" and "top-secret" are defined. Thus a flow of information is exclusively allowed towards higher ranked security classes. We also use information flow modelling but allow building the directed relations on single data and sources instead of relations between certain sets. Therefore we use special deduction rules that describe which information can be inferred from which sources and their given raw input data. Additionally we regard secondary deducible information that can be gathered by combinations of certain information or raw input data. As mentioned before, we also consider additional data that can be imported into an environment by external sources like RFID tags or credit cards for example. Based on the deduction rules the pervasive system can be represented as an information flow graph. This section presents how to use our approach to model a pervasive environment and exemplifies the procedure by an application scenario.

3.1 Application Scenario

In order to illustrate our approach, we first give an application scenario we will build a model for later in this section. Let us assume a supermarket with installed cameras. By using systems like BehaviourIQ [4], such a supermarket becomes capable to determine positions of any person in the market. By collecting multiple positions of a certain person over a certain time interval they can be assembled to a movement profile. Equipped with face detection methods and thermal imaging, the cameras could also provide identity and recognition functionality and vital monitoring. All that data can be gathered about any customers without their knowledge and without their assistance. Also customer may identify themselves to the system when using their credit card to pay. Thus a credit card must be considered as data source providing additional data, i.e. the name and details of the account of a client. If a customer is uniquely identified, all gathered information about that client can be stored into a customer profile that may already exist and contain information about former shopping's.

3.2 Deduction Rules

Due to space limitation we only model some parts of the introduced application scenario to give an impression of our approach. To describe a flow of information in a system, we use special deduction rules. They describe which information can be inferred from which sources as well as from already deduced information. Using these rules all information directly and indirectly available in a pervasive system can be identified and linked to the sources of the raw input data. Here we differ between internal and external data sources. An internal data source is a permanent part of the pervasive system and its gathered raw data is always available for further processing. In contrast to that external data sources are imported to the system, remember RFID tags and credit cards, and provide additional data that is not given generally. Note that it is also possible to gather the same data from different sources. The name of a customer could be gathered from a credit or bonus card or from a passport for example. To model both kind of sources and their raw input data we use the following deduction rules:

1. Internal(Cameras) → Image

2. External(Credit Card) → Name

As given in the application scenario, the first rule describes the raw input data "Image" that is retrieved by the cameras in the supermarket. Following to the second rule, a credit card delivers the "Name" of a customer as additional input to the system. Please note if a customer would pay in cash the system would not be capable to gather her name. After linking the available raw data to their sources more deduction rules have to be defined that describe further processing on available data. We need three additional kinds of rules to describe a pervasive system and its further processing on available data. The first one is a deduction rule that is defined from data to data directly. We call it a 1-to-1 deduction which announces an existing function that allows deducing a certain data from another one without any additional input data. According to the application scenario, deriving a customer's "Position" out of an "Image" from a camera is such a 1-to-1 deduction. Given as a deduction rule it appears as follow:

3. Image → Position

The second kind of data processing we consider is to look at a certain kind of data in an interval. I.e. to gather new information by concatenating repeated readings of a certain type of data. Consequently we call it a N-to-1 deduction, symbolized by 'o'. In our scenario gathering many "Positions" of a customer enables the system to generate a "Movement Profile" for that customer that comes out by the following deduction rule:

4. o(Position) → Movement Profile

The last kind of possible flow of information we consider is to gain new information by the combination of certain data. Since all data items are needed to correctly infer the additional information we call this the AND rule which is symbolized by '&'. In our scenario we mentioned that the "Movement Profile" and the "Name" of a customer are needed to render a "Customer Profile". This conclusion applies exactly in the following rule:

5. Movement Profile & Name → Customer Profile

Having identified all possible flows of information the rule based model of the pervasive system is constructed. Based on all deduction rules we are able to model the pervasive system as an information flow graph that contains exactly the same information as the real world system. The next paragraph explains how to use the mentioned rules to build such a graph.

3.3 Information Flow Graph

To visualize our model we decided to represent the deduction rules graphically. In order to show the potential information flow we decided to use a directed graph as the final symbolization of the pervasive system. The nodes represent the sources and the data items whereas all flows of information are represented as directed edges. According to the deduction rules three types of nodes are needed, named internal source, external source and information. The same applies to the edges where each of the 1-to-1, N-to-1 and AND deduction rules get their own graphical counterpart. Equivalent to their rules, the edges for N-to-1 and AND deductions are also symbolized by 'o' and '&'. Due to the meaning of the AND edge it must have at least two starting points and exactly one end point to represent the combination of given data. Additionally the gathering of raw input data from sources is visualised by directed edges as they are also used for 1-to-1 deduction rules. Figure 1 displays an information flow graph that shows the example system mentioned earlier. It contains all deduction rules introduced in paragraph 3.2. The next section exemplifies how to use this graph to determine privacy-aware environment configurations.

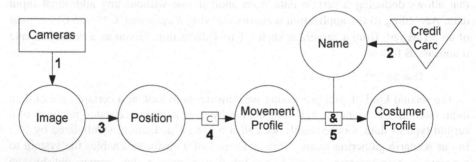

Fig. 1. Information flow graph of the application scenario representing the deduction rules introduced in paragraph 3.2

4 Determine Privacy-Aware Environment Configurations

An information flow graph represents the relevant information of a pervasive system, when it comes to privacy issues. It contains all data – measured and inferred – and therefore it is the ideal basis to brief the use about the information gathered. It can also be used to determine which information or data sources need to be disabled to adapt the modelled system to given privacy requirements. Details on both evaluation schemes are given in the next paragraphs.

4.1 Forward Analysis

The forward analysis is executed in direction of information flows and determines all directly or indirectly gatherable information that may result from given information or sources as well as sets of those. It may be used by the system provider to determine all information that is collectable by a certain source. The other way around, users may use it to calculate all further information that could be gained if they allow the system to collect a certain set of data. Thus this method calculates all reachable nodes in the graph from a given set of starting nodes. Since this is the most intuitive way to use the model we go not further into detail. According to the application scenario we exemplify the forward analysis at the source node "Cameras". Referring to the graph in figure 1 the result of that analysis is:

Forward Analysis (Cameras) = {Image, Position, Movement Profile}

This means that the cameras themselves are capable to gather the information "Image", "Position" and "Movement Profile". If the system wants to collect more information, additional data sources would be needed.

4.2 Generating Derivation Trees

To determine which information or data sources need to be switched off to assert given privacy requirements, any information must be linked to all its potential sources. To determine all possible sources of certain information, we build a derivation tree where this information item is set as the root node of the tree. Therefore the information flow graph is backward traversed from that information along all incoming edges down to the sources. Each branch of the tree represents a certain derivation path that consequently ends up in an internal or external data source. During the traversing any edge in the graph is used at most once even if the graph is cyclic. Already followed edges don't need to be processed again because the respective parent node is already a part of the tree. Hence the generation of derivation trees offers a complexity of O(n), whereas n is the amount of edges or the available deduction rules respectively.

Figure 2 displays the derivation tree for the "Customer Profile" of the application scenario. Attention should be paid to the branch that represents the combined deduction of the "Movement Profile" and the "Name". The special node that is symbolized with '&' only represents the AND combination of the child nodes. It does not represent an own data item.

4.3 Calculate Environment Configurations

The first step to enforce a person's privacy requires calculating environment configurations that avoid forbidden information to be gathered at all. For all forbidden information we first construct the derivation trees and convert them into computation formulas. These formulas connect all possible derivation paths by logical operations and can be exploited to calculate environment configurations that cannot gain the considered information. In the following we exemplify this feature by calculating configurations that are not capable to gather a "Customer Profile". Listing 1 specifies the computation formulas for the "Customer Profile" of the application scenario.

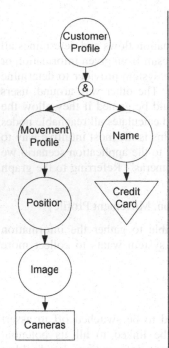

1. $\overline{\text{Customer Profile}} = \overline{\text{Movement Profile} \wedge \text{Name}}$
   $\overline{\text{Customer Pr ofile}} = \overline{\text{Movement Pr ofile} \wedge \text{Name}}$
   $\overline{\text{Customer Pr ofile}} = \overline{\text{Movement Pr ofile} \vee \text{Name}}$

2. $\overline{\text{Name}} = \overline{\text{CreditCard}}$

3. $\overline{\text{Movement Pr ofile}} = \overline{\text{Position}} = \overline{\text{Im age}} = \overline{\text{Cameras}}$

Proposals to avoid the Customer Profile:
- a. prohibit Movement Profile
- b. prohibit Position
- c. prohibit Image
- d. switch off Cameras
- e. prohibit Name
- f. don't use Credit Card

Fig. 2. Derivation tree of the information "Customer Profile"

Listing 1. Computation formulas for environment configurations that avoid deriving a "Customer Profile"

According to the derivation tree in figure 2, the first formulas represent the combination of the "Movement Profile" and the "Name" that is needed to retrieve the information "Customer Profile". Hence, eliminating the "Customer Profile" can be achieved by prohibiting either the "Movement Profile" (Proposal a.) or the "Name" (e.) or certainly both of them. Formula (2) and (3) represent the derivation trees for "Name" and "Movement Profile". Both end at their physical sources i.e. the "Credit Card" and the video "Cameras" respectively. Thus, these derivation trees provide two additional solutions to avoid a detailed and personalized "Costumer Profile". These solutions are switching off the cameras and/or avoiding the use of the credit card. Even if a user is not allowed to influence the camera system, a mobile device may at least advise his owner not to pay with a credit card. The example shows, that in general several configurations exist that fit the user's privacy preferences.

The most intuitive way to configure a pervasive environment surely is to switch certain sources on or off. However also technical solutions exist that allow to eliminate certain information in such a system. Using policies or changes in access rights may allow locking certain parts of a data base where the information is regularly stored.

4.4 Scalability of Our Approach

We do not consider individual devices as data sources but cluster those into a single source of a certain type. For example we do not deal with the existence of 100 separate temperature sensors - but we cluster them into one device that is capable of measuring temperature. The same applies to the cameras in the mentioned scenario. All of them are capable of gathering images, but we do not consider cameras single devices since they all provide the same functionality. According to that, we only need to consider the capabilities of the clustered devices when modelling a pervasive environment.

Even extending an existing pervasive system with new devices of already existing types does not increase the computation time needed to calculate configurations. Integrating devices with new functionality extends the information flow graph with new edges. These edges represent additional deduction rules embedding newly available information into the model of the system. That is the only case that produces additional calculation effort which grows only linearly with the number of edges in the graph. It is much more likely that systems are extended with devices of existing types which does not result in growing calculation effort than to add new functionalities. So our approach scales well even for large pervasive environments or for large sensor networks.

4.5 Simulation Tool

The information flow model introduced in this paper is used to calculate the preferred environment configuration of single users based on their privacy demands. These configurations can be visualized by our simulation tool. It allows flexible configurations of simulated pervasive environments according to the privacy requirements of several users at the same time. All positions of users as well as their corresponding privacy demands can be changed during runtime. The graphical user interface of the simulation tool visualizes the simulated environment and a number of users as well as their respective configurations. Please refer to [19] for more details on the simulation tool and its basic architecture. We measured that the calculation of user configurations and needed adaptations in the environment take less than a half second per modification. These measurements were executed on a Pentium 4 1.6 GHz PC with 512 MB RAM.

5 Summary and Outlook

An approach that enhances privacy in pervasive environments has been introduced. Therefore the idea of information flow modelling was applied to pervasive systems. I.e. to identify all directly and indirectly available data in a pervasive system and link them to the sources of the raw input data. All flow of information is modelled by deduction rules and represented as information flow graph graphically. That graph can be used to determine which information or data sources need to be disabled to adapt the modelled system to given privacy requirements. Therefore forbidden information is linked to all its potential sources by analyzing the graph from the information to the edges by generating a derivation tree. That tree can be converted into computation formulas, which can be exploited to calculate environment configurations unable to provide the forbidden information. Due to the low complexity of the used algorithms, our approach is also well suited for large pervasive systems. Furthermore it is lightweight enough for execution on mobile devices giving the end user some control over her/his data.

We intend to test our approach in real applications with a real sensor network. Therefore we plan to build a pervasive system as demonstrator for tests in public places like universities where the everyday user is offered to determine a preferred environment configuration. The demonstrator is not only for purely testing our approach. Additionally the demonstrator should allow collecting statistical data about how much importance the everyday user attributes to her/his privacy.

References

1. Weiser, M.: The Computer for the 21st Century. In: Hot Topic: Ubiquitous computing, pp. 71–72. IEEE Computer, Los Alamitos (1993)
2. Bohn, J., Coroama, V., Langheinrich, M., Mattern, F., Rohs, M.: Living in a World of Smart Everyday Objects – Social, Economic, and Ethical Implications. Journal of Human and Ecological Risk Assessment 10(5), 763–786 (2004)
3. Wren, C.R., Azarbayejani, A., Darrell, T., Pentland, A.P.: Pfinder: Real-Time Tracking of the Human Body. IEEE Transactions on Pattern Analysis and Machine Intelligence 19(7) (1997)
4. BehaviorIQ, Brickstream, http://www.brickstream.com
5. Chekmenev, S.Y., Rara, H., Farag, A.A.: Non-contact, Wavelet-based Measurement of Vital Signs using Thermal Imaging. In: The first international conference on graphics, vision, and image processing (GVIP), Cairo, Egypt, pp. 107–112 (December 2005)
6. Garbey, M., Sun, N., Merla, A., Pavlidis, I.: Contact-Free Measurement of cardiac pulse based on the analysis of thermal imagery. Technical report number UH-CS-04-08, Department of computer science, University Houston, Texas (December 2004)
7. Bissig, M.: Office Monitoring with Sensor Networks. Project report, Department of computer science, University Bern (April 2006)
8. TAUCIS – Technikfolgenabschätzung: Ubiquitäres Computing und Informationelle Selbstbestimmung. German Federal Ministry of Research and Technology
9. Intelligent Long Range Tags, IDENTEC SOLUTIONS GmbH, Hertzstrasse 10, 69469 Weinheim, Germany
10. Gruteser, M., Grunwald, D.: Anonymous Usage of Location-Based Services Through Spatial and Temporal Cloaking. In: ACM/USENIX International Conference on Mobile Systems, Applications, and Services (MobiSys) (2003)
11. Langendörfer, P., Kraemer, R.: Towards User Defined Privacy in location-aware Platforms. In: Proceeding of the 3rd international Conference on Internet computing, CSREA Press, USA (2002)
12. Synnes, K., Nord, J., Parnes, P.: Location Privacy in the Alipes platform. In: Proceedings of the Hawai'i International Conference on System Sciences (HICSS-36), Big Island, Hawai'i, USA (January 2003)
13. Wagealla, W., Terzis, S., English, C.: Trust-based Model for Privacy Control in Context-aware Systems. In: Dey, A.K., Schmidt, A., McCarthy, J.F. (eds.) UbiComp 2003. LNCS, vol. 2864, Springer, Heidelberg (2003)
14. PRIME: Privacy and Identity Management for Europe, http://www.prime-project.eu.org/
15. Brar, A., Kay, J.: Privacy and Security in Ubiquitous Personalized Applications. In: UM 2005 Workshop on Privacy-Enhanced Personalization (2005)

16. Langheinrich, M.: A Privacy Awareness System for Ubiquitous Computing Environments. In: Borriello, G., Holmquist, L.E. (eds.) UbiComp 2002. LNCS, vol. 2498, pp. 237–245. Springer, Heidelberg (2002)
17. Robinson, P., Beigl, M.: Trust Context Spaces: An Infrastructure for Pervasive Security. In: First International Conference on Security in Pervasive Computing (2003)
18. Denning, D.: Cryptography and Data Security. Addison-Wesley, Reading (1982)
19. Ortmann, S., Langendörfer, P., Maaser, M.: A Self-Configuring Privacy Management Architecture for Pervasive Systems. In: 5-th ACM International Workshop on Mobility Management and Wireless Access (MobiWAC), October 22, 2007 Chania, Crete Island, Greece (2007)

Managing Pervasive Environment Privacy Using the "fair trade" Metaphor

Abraham Esquivel, Pablo A. Haya, Manuel García-Herranz,
and Xavier Alamán

Escuela Politécnica Superior, Universidad Autónoma de Madrid,
C. Francisco Tomás y Valiente. 11, 28049 Madrid, Spain
{Abraham.esquivel@estudiante.uam.es},{Pablo.Haya,
Manuel.garciaherranz,Xavier.alaman}@uam.es

Abstract. This article presents a proposal for managing privacy in pervasive environments. These environments are capable of sensing personal information anywhere and at anytime. This implies a risk to privacy that might not be assumed if a clear and trustable privacy management model is not provided. However, since this kind of environments posses a set of highly heterogeneous sensing techniques, even basic privacy policies require a great management effort. Therefore, there is a tradeoff between providing automatic privacy configuration mechanisms and granting trustable privacy management models.

Following the "fair-trade" metaphor, this paper presents a privacy solution dealing with user's privacy as a tradable good for obtaining environment's services. Thus, users gain access to more valuable services as they share more personal information. This strategy, combined with optimistic access control and logging mechanisms, enhances users' confidence in the system while encouraging them to share their information.

Keywords: privacy, pervasive environments, "fair-trade".

1   Motivation

Pervasive computing is an emerging field of research with everyday growing possibilities over active environments. Initially, most problems found in pervasive environments appeared in technical areas. Several research projects have been oriented on that direction. In particular, great efforts have been made to accomplish a seamless integration between devices operating in such environments [1]. Although technological achievements in this area have been of outstanding importance, social factors will also be key roles in the success or failure of pervasive environments. In this sense, a major step in leveraging such environments comes from overcoming their ethical and psychological issues [2]. In fact, ethical and psychological problems are not strictly from pervasive environments and can be found in many software applications involving social interaction e.g. instant messengers or shared calendars. Clearly, the use of such applications entails new social problems, in particular those concerning user privacy.

R. Meersman, Z. Tari, P. Herrero et al. (Eds.): OTM 2007 Ws, Part II, LNCS 4806, pp. 804–813, 2007.

Privacy is a dynamic phenomenon; its configuration has as many variations as variations has context i.e. a single change on context can trigger a change on privacy preferences. In consequence, privacy management should be a continuous negotiation in which the definition of the public and private boundaries will depend directly on user's context. Therefore, privacy management is a constant process of limits regulation. Those limits to accessibility of personal information determine the "sincerity" or "openness", or "distrust" or "closeness" characterizing the user and his/her current context.

But, how can the user establish those limits? A first approximation can be to manually configure privacy, assigning the desired level to each source of information. The key problem of this solution relays on the nature of privacy and pervasive environments: the degree of privacy desired for a source of information depends not only on the user and the source but also on the context. In other words, for every source and every person there can be as many different privacy configurations as different contexts they can be involved in. To configure a priori each possible arising situation for every source of information can be an overwhelming task that justifies the use of automatic management solutions.

On the other hand, the success of such automatic management solutions depends directly on the trust the user place on them. This trust depends directly on the following requirements: a) The model must be simple enough to be understood by a non-technical user, and b) The user must be able to modify the automatic configuration at any moment. In developing our framework we focused particularly on the importance of usability. Especially in this case, confidence and trust are synonyms of usability. Following this premises, we present a privacy solution dealing with user's privacy as a tradable good for obtaining environment's services. Thus, users gain access to more valuable services as they share more personal information. This strategy enhances users' confidence in the system while encouraging them to share their information.

This article is structured as follows: we will present first the privacy framework of our pervasive environments. Next, we will introduce a semi-automatic privacy management approach based on the "fair trade" metaphor. In this section, we will motivate this proposal and its suitable for privacy-aware pervasive environments. Finally, we will compare our work with others' and will extract the conclusions and future work.

2 Privacy Framework

Privacy can be seen as a matter of information, entities and their relations. Every privacy solution has to deal with these factors and their characteristic problems. As other environments, pervasive environments has its own fauna. In this section we will analyze the different elements and factors involved in our privacy world.

2.1 Taxonomy of Personal Information

Information is probably the most important element of privacy. In fact, privacy can be summarized as protecting information loosely enough to permit interaction but sufficiently tight to preserve confidentiality. When talking about pervasive environments, in which all the information is digital -and thus storable- it is important to think about the use that can be made with some information not only in the present but also in the future. In this sense, we can classify information in two categories: **static information** and **dynamic information**. The former comprises information with a low changing rate -e.g. telephone number, social security number, name, surname or postal direction- The latter, on the other hand, contains information of changing nature -e.g. location, activity or cohabitants. These two types of information have their own strengths and weakness. Thus, static information is object of being used in a future for what one only unwanted access can compromise at all the information. In this way, we can categorize this information as especially sensible in a time line. Conversely, dynamic information has no value in the future (or less value, at least) for what it could seem less sensible. On the other hand, it is precisely this kind of information what really defines a person's way of living -e.g. where you are, with who or doing what- so.This variety -and idiosyncrasy- of types of information present in the environment will determine the privacy solution designed to deal with them.

2.2   Pervasive Environment Entities Involved in Privacy Management

From the privacy management point of view, several entities can be distinguished:

- **User.** An entity with administrative rights. It can be a single person or an organization.
- **Agent.** A software module that acts on behalf of others persons or organizations.
- **Space.** A physical or virtual area.
- **Data Source.** A resource providing information about environment's entities. Data can vary from a single value such as temperature to a multimedia streaming.
- **Data Source Owner.** The user with administrative rights on the data source.
- **Inquirer.** Agent or user inquiring a data source.
- **Receiver.** Agent or user receiving the answer of the inquirer. In some instances, it is possible that the receiver and the inquirer are not the same entity, as for example when the inquirer's answer is broadcasted to a group of receivers.

In Figure 1, we provide a high-level illustration of our design framework. As shown, we consider multiple spaces organized hierarchically (used for granularity of location, for instance). Each of them represents a physical or conceptual space serving as container of data sources [3].

Additionally, every data source is represented in the virtual world, where it can also store information (see Figure 1). This is the case of the RFID card reader

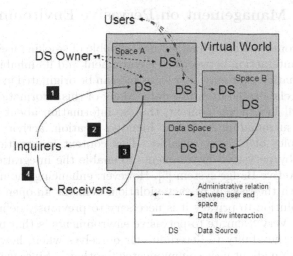

Fig. 1. Order in the information flows for interaction with the Data Sources in the virtual world

installed in our laboratory[1]. This card reader generates location information and stores it in the virtual world representation.

2.3 Interaction Between Information and Users

Besides spatial and administrative relationships, Figure 1 also shows an example of interaction between data sources and agents playing different roles. The procedure is as follows:

1. The inquirer requests some information to the data source.
2. The data source, according to privacy rights, returns the answer to the inquirer. The answer may contain the whole data, or it can be filtered. In the latter, the inquirer will only receive a partial view of the data.
3. When the answer is delivered, other agents or users may receive it. This may happen with or without an explicit indication of the inquirer. For instance, an answer may be broadcasted to unsolicited receivers if a speaker is used as output device.
4. Once a user obtains data from a source, he may freely distribute it to other users.

The information produced in active environments can be protected in two ways: by a binary model or by a precision scale based model. Depending on the identity and access rights of the consultant, the binary model classifies the information like secret or public. According to the second model, the information is shown with different degrees of precision corresponding to the access rights.

---

[1] http://amilab.ii.uam.es

3   Privacy Management on Pervasive Environments

Pervasive environments are infrastructures capable of tracking user activity. Additionally, the interaction between the environment and its inhabitants provides personal information that may be private and can be originated in multiple physical sensors, or elaborated from raw data. Many of this information corresponds to what Dey [4] identify as context, that is, information about the particular circumstances surrounding an entity. Identity, location, activity and time are classical examples of context variables. This context information is acquired autonomously by pervasive environments to enable the integration of adaptive services as in Active Badge system [5]. However, enhancing the user experience and dealing with the risks of privacy violations remains an open issue. In order to develop a solution to privacy it is necessary to previously define what is understood by it. Very accurate to pervasive environments is that of Alan Westin [6]: privacy is "the ability to determine for ourselves when, how, and to what extent information about us is communicated to others". Since environments are populated by heterogeneous technologies, multiple and varied sensors in multiple and varied situations, applying this definition to pervasive environments is a challenging problem: To configure manually every possible solution to every possible menace can overflow the user. Thus, the importance of having means to automatically configure them.

For these reasons, we consider that the privacy management model must fulfill the following three requirements: comprehensibleness, usefulness and flexibleness.

- **Comprehesive.** Privacy management must be simple enough to be done by a non-technical user. Additionally a simple mean to manage privacy enforces the understanding and trust the user places on the system.
- **Usefull.** Since the system extracts benefit from users' information, users must extract benefit from sharing their information too. In other words, to configure privacy in the less possible restrictive level must be useful to users i.e. we must give sense to the openness-services relation.
- **Flexible.** The model must be flexible in the sense that their variables should be powerful enough to adapt and define the wide range of privacy preferences arising in the environment. There are two aspects users consider when setting up their privacy: the users with whom they will share their information and the services for which that information will be used. Any change on those aspects may trigger a change on users'privacy neccessities.

3.1   Privacy Management Using the "fair trade" Metaphor

Dealing with the requirements stated above, we present a model based on the "fair trade" metaphor. Our proposal relies on the following assumption:

Users will accept to harm their privacy if, on return, they receive valuable services and they are able to track the flows of their disclosed information.

Our information management protocol is among those refered to as "fair trade" policy, in other words, a user can see another user's context variable (identity, location, activity, time) -with the degree of privacy established by the owner- if he harms his privacy equally. Thus, our privacy model is based on the following premise: users will obtain services from the active environment inversely proportional to the degree of protection they give to their own information. To alleviate the risk of sharing information, the system provides a logging service, where the user can check (or be notified) when and who consulted which information.

Privacy can be configure in three levels; from restrictive to without privacy based on the users' will to harm their privacy in exchange of more and better services.

- **Restrictive.** Completely restricted dynamic information (identity = anonymous, location = unavailable, activity = unavailable).
- **Customizable.** Partially restricted dynamic information: context information is shown with some degree of granularity. In other words, the real identity is replaced by an alias, the location is given at "floor" level, and the activity takes a descriptive value ("Possible working/resting").
- **Without privacy.** The dynamic information is shown with its highest precision level. It provides the identity registered in the virtual world data structure, the location at room level and the activity directly acquired by the sensors (for instance, notify if the user's computer is switched on and busy).

The information retrieved by the inquirer is filtered depending on the data source privacy level. This level is automatically updated according to the inquirer's and user's privacy levels following the equation 1.

$$PrivacyLevel_{(data)} = Min(PrivacyLevel_{(inquirer)}, PrivacyLevel_{(user)}). \quad (1)$$

The model make stress in "fair trade". Thus, a user will set his privacy according to the precision level in which he wants to see others' information, assuming that others will be able to see his/her own equally. For instance, a user setting his privacy level to *Customizable*, will only have access to others' information with the same degree of granularity as much, even if they have declared it as "without privacy". In table 1, we provide several examples how different variables are filtered depending on the resulting privacy level.

3.2 Supporting Mechanisms

Four different mechanisms are used to support our privacy control system. These mechanisms, when together, are able to deal with all the factors identified in previous sections and are the following:

Optimistic Access Control. Access control in active environments must stimulate the cooperation between users and groups. In addition it must guarantee

Table 1. Context variables are filtered using different granularity depending on its privacy level

Privacy Levels	Context variables		
	Identity	Location	Activity
Restrictive	Anonimous	Unavailable	Unavailabe
Customizable	Alias	Floor level	Possibly working-resting
Without_Privacy	Real	Office room	Raw data sensor

that individual's access rights are not lost when joining a group in a collaboration process. Now, most systems use pessimistic access controls, based on the principle that the user restricts his information from the beginning, granting access later to those considered reliable. This mechanism becomes inflexible to support less rigid policies like those taking place in active environments as it will be seen in section 4. Conversely, an optimistic access control allows free access to information, with strict registration of whom, what and when acceded to it. This registration acts as a dissuasive mean to prevent abuses within the system i.e. The only fact that user B is aware that A can know if he consults A's information, and how many times, prevents B from abusively consult A's information.

Templates to Model Degrees of Privacy. Templates, combined with default configurations, help users in controlling their privacy and modifying it when necessary. The goal is to promote user's interest in protecting his/her static information in a practical and simple way. Thus, if a user wants to show his/her telephone number he/she will explicitly specify, through a template, to which user or group of users must be shown. Contrary to dynamic information, templates for static information use a restrictive configuration.

Punitive Measures. As mentioned in the subsection 3.2, optimistic access control provides a transactions loggin service to maintain informed users about who accesses their information and in what degree. This mechanism provides a way to detect inappropriate, suspicious or abusive conducts (for instance, to detect a user consulting activity and location of others excessively in the course of the day). The punitive measures can consist in applying a restrictive privacy to that specific user, expelling him/her from users' groups of confidence and in consequence, limiting the quality of services he/she may receive. Additionally, a voting system could be applied in which those conforming the community maintain a list of unpleasant users to which deny services in the environment.

Manual Override. The value of personal information depends on each person. Thus, independently of the information nature, a manual procedure must be provided to the user for modifying his/her privacy.

4   Related Work

Many pervasive computing proyects are trying to solve the problem of data privacy. Hengartner and Steenkiste suggest in [7] a location system, for the AURA Project. Their services protect privacy using distributed access control policies based on digital certificates. Policies are applied to users as well as to the environment in order to avoid unauthorized access. To control privacy, they show location with diverse granularity levels. In the same way, CoBrA [2] works over a server of policies developed with an extension of the Rei language [8]. Rei allows to especify rules authorizing or rejecting access to contextual information as well as define structures encoding granularity parameters of access information.

Those projects provide a strict control through policies. While these policies permit to control specific situations and devices in an active environment, every situation must have its own policy. Thus, if not carefully managed, conflicts may arise. On the other hand, having to define policies for every situation, echoes directly in a poor scalability of the environment. Additionally these systems need technical management, i.e. a central and technical authority managing the different policies of environment and users.

On the other hand, Gruteser and Grunwald [9] propose to protect user's location information through reducing the resolution of the information provided by sensors. For example, providing an approximate count of the users located on an specific area. The degree of approximation will depend on the user's access rights. Mist Project [10] employs a protocol of anonymity and confidentiality for securing communications, instead of using access control policies. Thus communication flows are driven through "Mist Routers" in charge of hiding the source and destiny devices' identities.

An important work of privacy is of Langheinrich[12]. He works the privacy from the technical, social and legal point view, through configurables policies. Even though the system cannot contemplate the usability, create a sense of accountability over the request.

It is important to remember that pervasive environments require privacy solutions especially flexible. Too much detail or too many restrictions in every situation configuration can end up oppressing the end-user. In this sense, we developed PrivacyPAQ, an administrative tool for privacy, based on the fair trade metaphor of dynamic context information (identity, location, activity and time), granting user's control over his/her static and highly sensible information. Our goal is to motivate users to configure their privacy in order to obtain more and better services from the environment, propitiating the creation of affinity groups from where to obtain feedback about what, when and to whom information is delivered. In our approach, context information is filtered in a privacy server in charge of adjusting the granularity with which information is delivered. Due to the "fair trade" policy, filtering is a matter of two factors: the access granted by the owner and the privacy settings of the inquirer. The use of anonymity

---

[2] Context Broker Architecture.

and pseudonyms have been influenced by the work of Lederer [11] about the development of a user interface to control privacy in an active environment.

5   Conclusions and Future Work

At the time of this writing, a first prototype for privacy managment has been developed. This prototype deals with the optimistic access control as well as with the fair trade metaphor, remaining as future work the use of punitive measures to abusers. This prototype can also be combined with popular privacy solutions currently developed for electronic communications, such as anonymizing tools and encription schemes.

Informal tests of the prototype with laboratory members anticipate the following conclusions about the its potentials and qualities:

- **Promote privacy configuration.** This protocol encourages users to set up their privacy in order to obtain more and better environment's services.
- **Detect users abusing the system.** Due to its optimistic access control and transactions loggin, abusive users can be identified.
- **Discover sensible variables.** Identifying the most disclosed information allows to discover sensible points in privacy and improve the model.
- **Punitive measures.** Combined with the identification of abusive users, punitive measures can prevent system's abuses from malicious users.
- **Groups of confidence.** Since the protocol was thought to distinguish users with similar interests it is possible to automatically form groups of confidence and used them, between other things, to dynamically establish punitive measures through voting policies.

These conclusions must be contrasted with formal tests of non-technical users and diverse scenarios. Formal tests will also allow a deeper study of the possible punitive measures to apply on abuses -from voting alternatives to access denial- as well as to identify the most sensible variables, from the user perspective -those he/she feels as more private- as well as from the abuser one -those being most attacked.

The "fair trade" metaphor deals only with what we called dynamic information (see section 2.1). Meanwhile opening or closing the static one to some, none or all others is up to the information's owner through manually configuring so. Furthermore, while an optimistic access control (see section 3.2) supports the "fair trade" model of dynamic information, using templates are a good option to deal with static information.

Additionally, we assume that every data source providing information of a user is controlled by him/her. Conversely, this is not completely certain in all active environments, since some of them are public. This is the case of, for example, a camera in a public space. In these cases, our approach does not work properly.

Finally, we should mention our current work on integrating two far apart pervasive environments, this brings the opportunity of studying privacy in environments in which all communication are digital and, thus, sensible to our privacy control.

References

1. Hightower, J., Borriello, G.: Location Systems for Ubiquitous Computing. IEEE Computer 34(8), 57–66 (2001)
2. Stone, A.: The Dark Side of Pervasive Computing. IEEE Pervasive Computing 2(1), 4–8 (2003)
3. Dix, A., Rodden, T., Davies, N., Trevor, J., Friday, A., Palfreyman, K.: Exploiting space and location as a design framework for interactive mobile systems. ACM Trans. Comput.-Hum. Interact. 7(3), 285–321 (2000)
4. Dey, K.A.: Understanding and Using Context. Personal Ubiquitous Computing 5(1) (2001)
5. Want, R., Hoper, A., Falcao, V., Gibbons, J.: The Active Badge location system. ACM Trans. on Information Systems 10(1), 91–102 (1992)
6. Westin, A.F.: Privacy and Freedom. Atheneum, NY (1967)
7. Hengartner, U., Steenkiste, P.: Exploiting Hierarchical Identity-Based Encryption for Access Control to Pervasive Computing Information. In: 1st IEEE/CreateNet Intl. Conf. on Security and Privacy for Emerging Areas in Communication Nerworks (IEEE/CreateNet SecureComm 2005) (2005)
8. Kagal, L., Finin, T., Joshi, A.: A Policy Language for a Pervasive Computing Environment. In: IEEE 4th Intl. Workshop on Policies for Distributed Systems and Networks (2003)
9. Gruteser, M., Grunwald, D.: Anonymous Usage of Location-Based Services Through Spatial and Temporal Cloaking. In: Proc. of the 1st Intl. Conf. on Mobile Systems, Applications, and Services MobiSys (2003)
10. Al-Muhtadi, J., Campbell, H.R., Kapadia, A., Dennis, M.M., Yi, S.: Routing Through the Mist: Privacy Preserving Communication in Ubiquitous Computing Environments. In: 22th Intl. Conf. on Distributed Computing Systems (22th ICDCS 2002), Vienna, Austria (2002)
11. Lederer, S.: Designing Disclosure: Interactive Personal Privacy at the Dawn of Ubiquitous Computing. Research Project Report, Master of Science, Computer Science Division, University of California, Berkeley (2003)
12. Langheinrich, M.: A Privacy Awareness System for Ubiquitous Computing Environments. In: Borriello, G., Holmquist, L.E. (eds.) UbiComp 2002. LNCS, vol. 2498, pp. 237–245. Springer, Heidelberg (2002)

Spontaneous Privacy Policy Negotiations in Pervasive Environments

Sören Preibusch

German Institute for Economic Research
Mohrenstraße 58, 10117 Berlin
spreibusch@diw.de

Abstract. Privacy issues are a major burden for the acceptance of pervasive applications. They may ultimately result in the rejection of new services despite their functional benefits. Especially excessive data collection scares the potential user away. Privacy Negotiations restore respect for the user's privacy preferences because the kind and amount of personal data to be disclosed is settled individually. This paper outlines the advantages of negotiable privacy policies for both the user and the service provider. Special requirements for the implementation of privacy policy negotiations in pervasive environments are discussed, notably ease-of-use. Having sketched how individualized privacy policies can be realized within the privacy policy language P3P, two case-studies illustrate how the negotiation can be integrated seamlessly in existing usage patterns to enhance privacy: first, for mobile interactions in urban spaces, and second, for social network sites.

1 Introduction and Related Work

Ubiquitous computing has interwoven our personal life with technology usage. On the one hand, private interactions are no longer restricted to physical face-to-face contacts. Instead, they have spread throughout the Web, making it "my space" to interact with friends. On the other hand, we use technology to interact with our relatives, our friends, and our colleagues. The mobile phone and other ubiquitous devices are embedded in personal interaction patterns. In parallel, the increasing personal involvement in technology-mediated interactions has augmented the users' concern to keep some aspects of their lives private. The more connected one gets to her mobile phone, the more she perceives an intrusion in or via this mobile phone as an intrusion into the nucleus of her private lifestyle [4].

Privacy has been largely studied in Web-based environments. Personalization efforts have been implemented in the vast majority of popular sites and they are much-valued by the users. Yet, these systems' inherent need for personal information and the long-time storage of these data drew attention to the induced privacy issues. Still, this Privacy-Personalization trade-off is only one example for increased awareness of privacy issues. Personal identifiable information is collected during virtually every Web interaction.

R. Meersman, Z. Tari, P. Herrero et al. (Eds.): OTM 2007 Ws, Part II, LNCS 4806, pp. 814–823, 2007.

As of today, privacy related initiatives are often limited to the communication of data-handling practices by means of "privacy policies" posted on the Web site. Though, this approach is too inflexible and impractical: the need for personalization demonstrates that the users are not all the same [2] – also with regard to privacy preferences. The need for individualized privacy policies is therefore obvious.

The problem how to balance between the users' quest for privacy and the service providers' personalization efforts has escalated towards pervasive environments. This question is crucial because privacy concerns may ultimately result in the rejection of new services despite their functional benefits. Little, however, is known about how a good equilibrium can be achieved in the context of ubiquitous interactions. Solution approaches from the Web need to be revised and adapted if need be [1].

The contribution of this paper is twofold. First, on a theoretical level, we will examine how the promising approach of negotiable privacy policies can be adapted to resolve the Privacy-Personalization trade-off in pervasive computing environments and generate benefits for either contracting party. Second, on a practical level, two case-studies illustrate the beneficial implementation of privacy negotiations.

The remainder of this paper is organized as follows: Section 2 outlines the advantages of negotiable privacy policies for both the user and the service provider. Special requirements for the implementation of privacy policy negotiations in pervasive environments are discussed in Section 3, notably ease-of-use. Section 4 describes how P3P is suited to implement privacy negotiations and fulfill the requirements compiled beforehand. Finally, two case-studies in Section 5 demonstrate the rewarding and intuitive set-up of individually concluded privacy policies: the examples are interactions with mobile devices and interactions inside social network sites, both corresponding to typical pervasive applications.

2 Advantages of Privacy Policy Negotiations

Thompson defines negotiations as an "interpersonal decision-making process necessary whenever we cannot achieve our objectives single-handedly" [10]. Distributive negotiations are carried out in case the amount to be distributed is fixed and the stakeholders only discuss how to allocate it. Due to the scarcity of goods, distributive negotiations lead to win-lose situations. Multi-attribute negotiations, however, offer the opportunity to find an integrative solution which is beneficial for all parties involved. Such mutually beneficial bargains increase welfare and unleash additional economic potential [9]. When applied to privacy levels, integrative negotiations between a service provider and a user can overcome two major shortcomings of existing online privacy handling mechanisms:

– The first shortcoming is the "take-it-or-leave-it" principle. The user can only accept or refuse the provider's privacy policy proposal as a whole. Even a minor disagreement results in a cancellation of the potential transaction.

– The second shortcoming is the "one-size-fits-all" principle. Once the service provider has designed his privacy policy, it will be proposed to all potential customers – regardless what their individual privacy preferences are. There may be users who would have accepted offers with less privacy protection and would have agreed to the provider's proposal even if more personal data would have been asked. This is a lack in "meta-personalization" as the methods and parameters determining how the personalization is carried out are not flexible.

Due to the limned shortcomings, the provider fails to tap the users' full potential. Using static privacy policies is not efficient.

In a privacy policy negotiation, the amount and the kind of data to be collected is individually settled between the service providers. The data disclosure level is "gauged" along relevant and negotiable privacy dimensions. We define a *privacy dimension* as one facet of the multi-dimensional concept 'user privacy'. Privacy dimensions can be identified at different degrees of granularity [5]. The four top-level privacy dimensions are the recipient of the data, the purpose for which the data are collected, the time they will be stored, and the kind of data. These four dimensions (recipient, purpose, retention time, and data) are in accordance with European privacy legislation and the OECD Fair Information Practices. The 'data'-dimension will be the most flexible privacy dimension in most applications [7].

For each privacy dimension, different discrete revelation levels exist, monotonously associated with the user's willingness to reveal the data. The revelation levels correspond to increasing detailedness and data quality; they usually can be deduced intuitively from the nature of the dimension. For instance, the dimension 'birth date' may have the sensible revelation levels 'none', 'year', 'year and month', 'year, month, and day' [6]. For a location-based service, the dimensions could 'time zone', 'country', 'region' / 'state', 'city', 'ZIP code', and 'street'. All of these levels can be expressed in P3P, from `timezone` to `street`, and extensions to P3P to include other context-aware data elements have been proposed [14]. Depending on the nature of the service, the service provider sets a revelation threshold corresponding to the least of data the service needs: a restaurant information service is likely to require at least a ZIP code for appropriate accuracy whereas a currency converter will only need the country information. The user's consent to provide the data will also differ along with the intended usage.

The service provider's preference is for higher detailedness, as it allows for better targeting the customer. The user's preference is for lower detailedness as to keep her privacy secured. By enlarging the scope of the negotiation, latter becomes an integrative negotiation: the user will get a (monetary or non-monetary) compensation for disclosing personal information. The rationale is that private information is of value to both the information holder and the information seeker. The integrative potential comes from the parties' antagonistic preference structures for 'privacy' and 'compensation'. Examples for the reward include rebates, vouchers, and better service quality.

3 Special Requirements in Pervasive Environments

Pervasive Environments impose specific requirements on privacy policy negotiations, mostly concerning their ease-of-use. This section highlights three manifestations of usable privacy negotiations: real-time requirements, low perturbation of well-practised interaction patterns, and tool support.

Real-time requirements. The usage of mobile devices is characterized by its spontaneity. The need to use a functionality provided by the device often comes ad-hoc, following an external, unforeseen stimulus. In addition, the user expects that she can complete the operation in a timely manner. Both usage characteristics can be deduced from the fact, that a mobile device (with a smaller keyboard, a smaller screen, lower performance) is preferred over a stationary device: if the task had been foreseeable, the user would have scheduled it to a stationary device. If the task had been delayable, the user would have postponed it to a stationary device. Regarding privacy negotiations, this means that they have to be carried out quickly, without delaying the task completion. A time-consuming overhead has to be avoided.

Low perturbation. In a similar way, the task completion path must not be deviated by the negotiation. It must seamlessly integrate in the flow of actions that would have been taken in case of static privacy policies. Considering the case of a social network site, the introduction of negotiable privacy policies must not hamper the intuitive usage of the site or diminish the "convenience factor". The spontaneity of the interactions has to be preserved because quick interactions and the exchange of messages are key feature of social network sites.

Tool support. The user has to be provided with adequate tool support. Ideally – following the requirement of low perturbation – the support for privacy negotiations is included in the tools ordinarily used, e.g. Internet browsers. Adequacy of tool support includes robust and usable interfaces, low computing requirements, responsiveness, and assisted decision-making, i.e. helping the user to make a good choice with regard to her privacy preferences. For mobile devices, small screens, limited input mechanisms, and low computing power are the most restricting context factors. It may be sensible to design the negotiation interfaces towards specific delivery channels.

The design of privacy negotiations in pervasive computing environments should be aimed at meeting (at least) these requirement. The coding of negotiable privacy policies should facilitate the implementation of adhering software. The next section sketches the coding of negotiable privacy policies with P3P in a complying manner.

4 Implementing Privacy Negotiations with P3P

The Platform for Privacy Preferences (P3P), is a protocol designed by the W3C and industry partners to inform Web users about the data-collection practices of Web sites. It provides a way for a Web site to encode its data-collection and data-use practices in a machine-readable XML format known as a P3P policy [13]. Moreover, P3P enables Web users to understand what data will be collected by sites they visit, how that data will be used, and what data/uses they may "opt-out" of or "opt-in" to [13].

The communication of the data processing practices follows the four top-level privacy dimensions (recipient, purpose, retention time, and data). P3P policies express the service provider's data processing practices using STATEMENTs; each of those statements having child elements indicating the RECIPIENT of the data, the PURPOSE for which the data will be used, the RETENTION time and what kind of DATA will be collected.

P3P includes a built-in extension mechanism by which privacy policies can be enriched by elements newly defined in referenced namespaces. To code a negotiable privacy policy, two extensions are added: a NEGOTIATION-GROUP-DEF element at the global policy level, and several NEGOTIATION-GROUP elements at the level of the STATEMENTs. The mechanism is comparable to the tandem of STATEMENT-GROUP-DEF and STATEMENT-GROUP in P3P 1.1 and follows the design principles of P3P [13]. Negotiating at the level of statements that bundle several privacy dimensions instead of negotiating along a single privacy dimension again extends the negotiation space and increases the integrative potential of the negotiation. It also accounts for the fact that privacy dimensions can interact, as depicted for a location-based service in the previous section: the intended use of personal information influences the detailedness of this information that a user will accept to provide.

A NEGOTIATION-GROUP-DEF element defines an abstract pool of alternative usage scenarios. One or several statements (identified by the attribute id) code a possible usage scenario; the pool membership is expressed by the NEGOTIATION-GROUP extension in the statement (attribute groupid), which describes relevant parameters of the given scenario, including the benefits for the user. The fallback contract can be indicated via the fallback-attribute of the NEGOTIATION-GROUP-DEF element. The contract alternative described by the fallback-attribute will be set in case the user cancels the negotiation prematurely. The standard-attribute indicates which scenario is offered as default. By means of the selected-attribute, the currently active alternative can be marked, as for persistent storage of a negotiation state or for the final agreement [7].

The coding of negotiable privacy policies in P3P as sketched in the preceeding paragraphs fulfills the requirements that were compiled in Section 3. The data organisation allows quick parsing of negotiable policies. All information is available at once, making timely request-response exchanges unnecessary. Instead, as for standard P3P policies, the user confirms the acceptance of the policy by fetching the URI resource to which the policy is associated. The serviceuri-attribute specifies the respective URI. Standard P3P policies may be bound to these URIs for providing backward-compatibility. Moreover, a heavy-weight request-response-protocol would contravene the design principles of P3P and ultimately result in an exchange of numerous temporary policies, in turn needing an overhead for securing the exchange. The user agent can guide the user sequentially through the available options or display all alternatives at once. Hence, different presentation styles can be developed, targeting small or standard screens. As the negotiation alternatives are coded in P3P, privacy preferences expressed in languages as APPEL or – more likely – XPref can be evaluated by tools to support the user's decision.

Negotiable privacy policies link alternative negotiation outcomes to URIs. Relying on this base technology allows for simple development of tool support within existing technologies. In addition, any user action resulting in a URI retrieval can be incorporated in a user interface. As such, various designs can be used, including links, forms with drop-down-lists and choice sets with radio-buttons, or even visuals. XML schema definitions are available for negotiable privacy policies [7], so that the processes can be tested prior to the deployment.

5   Usage Scenarios

5.1   Usage Scenario 1: Mobile Interactions in Urban Spaces

Ubiquitous computing has reshaped the way we use and interact with and within urban spaces [11]. Information technology provides townspeople with increasing flexibility. Occasions where previously a physical movement would have been necessary, are now substituted by information exchanges. The newly acquired spare time can be used for leisure activities, also involving communication in many cases. Using mobile devices, information can not only be retrieved from Web sources, but several urban installations are providing information services and communicate with the user. One of the early examples for communicating billboards was the 2005 Christmas special by the European mail-order firm Quelle (see Figure 1). 1500 interactive billboards were installed throughout Germany, that called attention by a blinking red light [3]. Interested passers-by were informed that they could participate in a lottery and win vouchers amounting up to 100 euro. They needed to activate the infra-red (IrDA) interface on their mobile phone, that would then be triggered by the billboard to send an SMS to participate in the lottery. Alternatively, an SMS could be send manually.

The two participation channels in the lottery were associated with different levels of privacy. On the one hand, the interaction via IrDA targeted technology-savvy customers only marginally concerned about their privacy. The disclosed data included the phone number, the mobile phone type, and the time and the date associated with the location, because every billboard had a unique identifier. The compensation was monetary (the expected value of the voucher) and non-monetary (fun, convenience, participation in a cutting-edge technology). On the other hand, the manual sending of an SMS allowed profile-averse customers to participate. The only data collected was the sender's phone number. It was not possible to record time and location reliably, as the sender could have deferred the sending of the SMS. In return, these participants did not enjoy the non-monetary benefits of the participation.

Fig. 1. Communicating Billboard, Quelle AG, December 2005 (The headline says: "This heavenly poster conjures Christmas vouchers on your mobile phone")

This example illustrates how the conceptual framework of privacy negotiations can be applied to mobile interactions. Here, the negotiable privacy dimension is the amount of data. The compensation mainly differs in the non-monetary components.

5.2   Usage Scenario 2: Friendship in Social Networks

Several new privacy challenges emerge with the raise of social network sites [8]. One of the most remarkable developments is that privacy protection is not a private decision anymore. Due to the highly interactive nature of social networks, other members inside the network may – accidentally or deliberately – reveal data about oneself that was intended to be kept private. This section considers the case of friendship relations inside the network. Usually, the network site operator enforces the symmetry of such relationships, for instance at MySpace or Xing. Consequently, if user A is a friend of user B, user B is also a friend of user A. Both propositions are equivalent and thus one proposition may be inferred from the other one. If A decides to keep her contacts hidden, but B reveals her contacts, this contravenes A's decision. Data disclosure has side effects [8].

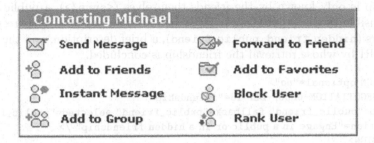

Fig. 2. Screenshot from MySpace (http://profile.myspace.com/)

Using the methods provided by privacy policy negotiations, one may formally express the difference between a public and a hidden friendship. Users may decide in which kind of friendship they want to engage. And by means of the concluded policies, the social network operator can proceed a semantics-driven enforcement in its back-end infrastructure [8]. This may include the control of data-base access or the deployment of intra-organisational privacy policies [12].

Consider the following example from the social network site "MySpace". In Figure 2 is shown a screenshot from one member's profile site. Visitors of this profile may opt to become a friend of Michael by clicking on "Add to Friends". This friendship will then be public.

The following altered screenshot, however, provides evidence, that privacy negotiations can be seamlessly integrated in the interaction patterns of a site (Figure 3). A variety of interface designs is possible. Especially the requirement of low perturbation articulated in Section 3 is thereby fulfilled. A new link "Add to Secret Friends" has been added. If a visitor uses this link, the newly established friendship will not be visible to the public.

These alternative usage scenarios can be coded in a single negotiable privacy policy as shown in the following listing. Using a NEGOTIATION-GROUP-DEF element, a new pool of alternatives is defined under the id "friendship". Two STATEMENTs follow that differ in the scope of the RECIPIENT. Whereas a hidden

Contacting Michael

✉ Send Message	↪ Forward to Friend
Add to Friends	Add to Favorites
Instant Message	Block User
Add to Group	Rank User
Add to Secret Friends	

Fig. 3. Altered Screenshot from MySpace (http://profile.myspace.com/)

friendship is only known by the friends themselves (`<ours/>`), a public friendship is visible to the (`<public/>`). These two alternatives have their respective identifiers (`hidden_friend`, `public_friend`), a brief description, and they differ in the URI by whose retrieval the friendship is concluded.

```
<EXTENSION optional="no">
<PRINT:NEGOTIATION-GROUP-DEF id="friendship"
standard="public_friend" fallback="public_friend" selected="public_friend"
description="Engage in a public or in a hidden friendship" />
</EXTENSION>

<STATEMENT> <EXTENSION optional="no">
  <PRINT:NEGOTIATION-GROUP id="public_friend" groupid="friendship"
  serviceuri="/index.cfm?fuseaction=invite.addfriend"
  description="Make this user a public friend of yours" /> </EXTENSION>
 <CONSEQUENCE>Other visitors will see that you are friends </CONSEQUENCE>
 <RECIPIENT> <ours/> <public/> </RECIPIENT>
 <PURPOSE> <contact/> <other-purpose> friendship </other-purpose> </PURPOSE>
 <RETENTION> <indefinitely/> </RETENTION>
 <DATA-GROUP> <DATA ref="#user.login.id"/> </DATA-GROUP>  </STATEMENT>

<STATEMENT> <EXTENSION optional="no">
  <PRINT:NEGOTIATION-GROUP id="hidden_friend" groupid="friendship"
  serviceuri="/index.cfm?fuseaction=invite.addfriend_hidden"
  description="Make this user a hidden friend of yours" /> </EXTENSION>
 <CONSEQUENCE>Other visitors will not see that you are friends</CONSEQUENCE>
 <RECIPIENT> <ours/> </RECIPIENT>
 <PURPOSE> <contact/> <other-purpose> friendship </other-purpose> </PURPOSE>
 <RETENTION> <indefinitely/> </RETENTION>
 <DATA-GROUP> <DATA ref="#user.login.id"/> </DATA-GROUP>  </STATEMENT>
```

6   Conclusion

Privacy concerns of users are currently mainly addressed by posting static privacy policies on the respective Web site. Yet, these privacy policies are too inflexible as they overlook the diverging privacy preferences of the customers. Potential transactions may be missed due to privacy-induced cancellations, other users may continue the transaction but feel uncomfortable. Still other users would have agreed in providing even more personal information for a more personalized service in return.

Privacy policy negotiations resolve this privacy-personalization trade-off. Additional requirements regarding the negotiation need, however, to be taken into account in the context of pervasive computing environments. These are mainly aspects of ease-of-use, including real-time requirements, low perturbation of well-practised interaction patterns, and adequate tool support. The coding of negotiable privacy policies in P3P allows for implementations of negotiations that fulfill these requirements.

As an illustration, this paper provided evidence how privacy negotiations can be implemented rewardingly in two typical applications of ubiquitous computing. Privacy policy negotiations can be executed on-the-fly and can integrate seamlessly into the overall usage patterns of a service. Future work will follow the pragmatic approach to privacy policy negotiations and aim at developing usable and effective interfaces for the negotiation.

References

1. Barkuus, L., Dey, A.: Location-Based Services for Mobile Telephony: a Study of Users' Privacy Concerns. In: Proceedings of the INTERACT 2003, 9th IFIP TC13 International Conference on Human-Computer Interaction (2003)
2. Forsyth, J., McGuire, T., Lavoie, J.: All visitors are not created equal. McKinsey marketing practice. McKinsey & Company. Whitepaper (2000)
3. innovativ in - der Business-Club. Neue Werbeform: Vor Plakaten gewinnen! (2005), http://www.innovativ-in.de/c.2880.htm
4. Kiesler, et al.: 1994 in Katz, J.E.: Connections Social and Cultural Studies of the Telephone in American Life. Transaction Publishers, New Brunswick (1999)
5. Preibusch, S.: Implementing Privacy Negotiations in E-Commerce. In: Zhou, X., Li, J., Shen, H.T., Kitsuregawa, M., Zhang, Y. (eds.) APWeb 2006. LNCS, vol. 3841, pp. 604–615. Springer, Heidelberg (2006)
6. Preibusch, S.: Privacy Negotiations enhance Data Collection for CRM. In: Proceedings of CollECTeR Europe 2006, pp. 11–20 (2006)
7. Preibusch, S.: Privacy Negotiations with P3P. In: W3C Workshop on Languages for Privacy Policy Negotiation and Semantics-Driven Enforcement (2006), http://www.w3.org/2006/07/privacy-ws/papers/24-preibusch-negotiation-p3p/
8. Preibusch, S., Hoser, B., Gürses, S., Berendt, B.: Ubiquitous social networks – opportunities and challenges for privacy-aware user modelling. In: Proceedings of the Workshop on Data Mining for User Modelling at UM 2007 (2007), http://vasarely.wiwi.hu-berlin.de/DM.UM07/Proceedings/05-Preibusch.pdf
9. Ströbel, M.: On Auctions as the Negotiation Paradigm of Electronic Markets Success Factors, Limitations and Research Directions. Journal of Electronic Markets 10(1), 39–44 (2000)
10. Thompson, L.L.: The Mind and Heart of the Negotiator, 3rd edn. Pearson Prentice Hall, Upper Saddle River, New Jersey (2005)
11. Townsend, A.M.: Life in the Real-Time City: Mobile Telephones and Urban Metabolism. Journal of Urban Technology 7(2), 85–104 (2000)
12. World Wide Web Consortium. W3C Workshop on the long term Future of P3P and Enterprise Privacy Languages (2003), http://www.w3.org/2003/p3p-ws/
13. Wenning, R., Schunter, M.: The Platform for Privacy Preferences 1.1 (P3P1.1) Specification, 2006. W3C Working Group Note November 13 2006 (2006), http://www.w3.org/TR/P3P11/
14. Martijn, Z., Filho, G., van Sinderen, J.M.: Marten. Using P3P in a web services-based context-aware application platform. In: Proceedings of EUNICE 2003 9th Open European Summer School and IFIP WG6.3 Workshop on Next Generation Networks, pp. 238–243 (2003), http://eprints.eemcs.utwente.nl/7625/

Combining Pragmatics and Intelligence in Semantic Web Service Discovery

Electra Tamani and Paraskevas Evripidou

Department of Computer Science, University of Cyprus
5 Kallipoleos St., T.K. 537, 1678 Nicosia, Cyprus
{electrat,skevos}@cs.ucy.ac.cy

Abstract. In this paper we present an architecture that augments existing semantic-based web service discovery solutions. The idea is to combine principles from the semantic and pragmatic web in order for the web service selection process to become more efficient and effective. We utilize offer and request xml files to describe the functional properties (inputs/outputs) and context parameters of web services. Our approach distinguishes from others in that it 1) applies to any semantic web service language and 2) provides the means for web service providers and requestors to make ontological commitments and establish common understanding in an intelligent way.

1 Introduction

The goal of the Semantic Web is to develop the basis for intelligent applications that enable more efficient information use by not just providing a set of linked documents but a collection of knowledge repositories with meaningful content and additional logic structure [SdMD06]. The main components for implementing the Semantic Web are ontologies. Ontologies however, as pragmatic web researchers argue, should not be exploited as fixed conceptualizations of some domains, as they are, but rather as dynamic structures which co-evolve with their communities of use. Members of a community have to communicate and continuously negotiate on their shared background/context. To do so, they need to understand each other especially when they 1) come from different professional, social, or cultural backgrounds and 2) use different terminology from other organizations/entities.

Web services is one widely used standard which has been established as the de facto for interoperability. The last years, the Semantic Web researchers proposed semantic enhancements to web services descriptions in order to facilitate more efficient discovery and composition in intelligent systems. Services should not be described independently of how they are used, because communities of practise use services in novel, unexpected ways [Mo05, Si02]. *Semantic web services should not assume context-independent annotations*. For these reasons service-oriented systems require social mechanisms that consider the contexts and interactions.

In this paper, we recognize these limitations and propose an architecture which augments existing semantic-based web service discovery techniques by combining pragmatic and semantic web principles. Our approach aims to make the web service

R. Meersman, Z. Tari, P. Herrero et al. (Eds.): OTM 2007 Ws, Part II, LNCS 4806, pp. 824–833, 2007.

selection process even more efficient and effective. We capture personal, functional and contextual information of the providers/requestors [TE06] in offer/request profile repositories, perform the matching between the various offer/request parts and produce ontologies with the ontological commitments made by the agents of both parties (provider and requestor).

The rest of the paper is organized as follows: Section 2 describes how the web service discovery can be improved, thus making the process more effective and efficient, and presents an example. Section 3 introduces the patterns of pragmatic web and operationalizes these patterns for the web service discovery scenario. Section 4 presents the proposed architecture that matches requests against offers, generates a matching degree based on which the selection is made and produces an ontology describing the common (agreed-upon) concepts between providers and requestors. Section 5 introduces related work. Finally, we conclude in Section 6.

2 Improving Web Service Discovery

Web Service discovery is of paramount importance for Service Oriented Systems. Consequently, if the services identified are not the most appropriate then neither the requestor will be satisfied nor the intelligent agents could be able to construct meaningful real life processes (workflows of web services). For these reasons it is essential that the discovery phase should be as effective as possible and meet the expectations of humans.

Semantic Web researchers proposed enhancements, through ontologies usage[1], to web services' definitions to enable some form of automation of the services in the semantic web. To this extend, they base the "matching" on various meaning resolution techniques. Particularly, the OWL-S (Ontology Web Language for Services) Matchmaker utilizes an algorithm for finding the best match between two descriptions (offer and request) and identifying the different relations (matching degree) (exact-match, subsumes-match, plug-in-match and fail-match) [PKP02, PSS04]. The WSDL-S (Web Service Semantics) Matcher is a combination of two types of matching algorithms (element level and schema) which complement one another in order to find all possible mappings between WSDL and ontology concepts [SVM05]. The WSMO (Web Service Modeling Ontology) Matcher offers almost the same type of matching (exact-match, subsumes-match, plug-in-match, intersection and non-match) with OWL-S Matchmaker but distinguishes itself in the fact that it separates a web service from a service [D.04]. Finally, other matchmaking techniques utilize algorithms that compare stateless web services (those that do not alter the state of world and thus do not require pre- and post- conditions processing) with different intentions [HZB06].

Although the semantic solutions allowed some form of automation, still the matching can greatly be improved. The most cumbersome assumption lies in the fact that web services' annotations (ontological references) are context-independent. However context should not be neglected especially in such highly intelligent environments as those involved with web services and agents (s/w applications) of humans coming from different professional/social/cultural backgrounds or utilizing different terminology. The

---

[1] Knowledge repositories of meaningful and structured content.

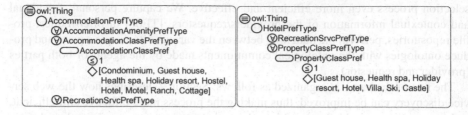

Fig. 1. Domain-specific Ontologies

matchmaking technique should therefore additionally consider the context. Augmenting the matcher functionality to check for context dependencies is a step towards a more effective and efficient web service discovery. The following example demonstrates the need for context-dependent annotations.

For the sake of this example, let us assume that the agents represent humans (the provider and the requestor in this case) coming from different cultural backgrounds and the web service (being offered and requested) returns a hotel list given some hotel search criteria. Figure 1 shows the ontologies utilized by the provider (ontology on the left) and the requestor (ontology on the right) to annotate their web services. The classes HotelPrefType and AccomodationPrefType take all values from classes Recreation-SrvcPrefType PropertyClassType and AccomodationClassPrefType AccomodationA-menityPrefType RecreationSrvcPrefType respectively on some properties. In the case of PropertyClassPref it can take any one value from the list [Guest house, Health spa, Holiday resort, Hotel, Villa, Ski, Castle] and in the case of AccomodationClassPref any one from [Condominium, Guest house, Health spa, Holiday resort, Hostel,Hotel, Motel, Ranch, Cottage]. Since each party used his own domain specific context-dependent ontology to annotate the web service, the traditional plug-in, subsume and exact matches between the request and the offer are not enough. A context-depended ontology is the one that is defined and/or altered according always to the context of web service usage. If for example the provider decides to add another subclass to the Accomo-dationPrefType class in order for his ontology to become more accurate and his web service input annotation more meaningful, still the traditional matchers would have just checked the schema of the two ontologies to decide the type of match. The addition of the new subclass may correctly make sense for the provider but the requestor may have his own perception of the concept which does not agree with such modification. Thus both the provider and the requestor must share common understanding and mutually agree in any modifications made to the context-specific ontologies.

3 The Patterns of Pragmatic Web

The Pragmatic Web can greatly assist in augmenting the web discovery because it is more or less a set of pragmatic contexts about the semantic resources. To model these contexts and operationalize the vision of Pragmatic Web, scientists have introduced some core pragmatic patterns [Mo05]. These pragmatic patterns model the context of interest to the community as a whole, the context of the individual members of the community and the common context established during a negotiation process. In the following section we revisit the example introduced previously and adopt the pragmatic patterns in order to: 1) show how these can help adequately capture the contextual information of the parties and their actions in a web service discovery scenario and 2) resolve possible meaning conflicts during the matching process (assume an intelligent agent environment).

3.1 Operationalize the Pragmatic Patterns for the Web Service Discovery Scenario

In order to operationalize our scenario we choose to use xml and xpath queries due to their wide acceptability and versatility. The pragmatic patterns, as being adjusted to our scenario, are the following:

- *Pragmatic context*: a pattern that captures the individual members, namely the providers and the requestors[2], of the web service discovery community. The pattern also contains identifiers to the individual and common contexts of the community.
- *Individual context*: a pattern that defines an individual community member, individual context parameters and identifiers to the individual context ontology. The individual context ontology in our case is captured in offer and request xml files (extensive analysis can be found in [TE06, TE07]) as follows:

```
<request>
    <Who Identity="Y" Role="requestor"/>
    <Why Purpose="..." Goal="..."/>
    <What>
        <Input name="HotelSearchInformation"
               annotation="AccommodationPrefType"/>
        <Output name="HotelCatalogue" annotation="..."/>
    </What>
</request>
<offer>
    <Who Identity="X" Role="provider"/>
    <Why Purpose="..." Goal="..."/>
    <What>
        <Input name ="HotelSearchCriteria"
               annotation="HotelPrefType"/>
        <Output name="HotelList" annotation="..."/>
    </What>
</offer>
```

- *Common context*: a pattern that defines the common context parameters and an identifier to the common context ontology of a community (to be generated dynamically).
- *Individual pragmatic pattern*: a meaning pattern relevant to individual community member. These patterns in our case are expressed as xpath queries.
- *Common pragmatic pattern*: a meaning pattern relevant to the community as a whole. The common context ontology contains all common pragmatic patterns (to be generated dynamically).

---

[2] The terms provider and requestor refer to intelligent applications acting on behalf of provider and the requestor respectively.

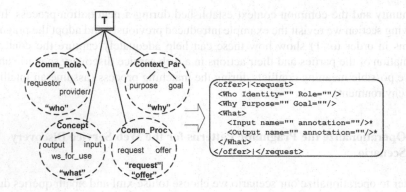

Fig. 2. The Web Service Discovery Context Ontology

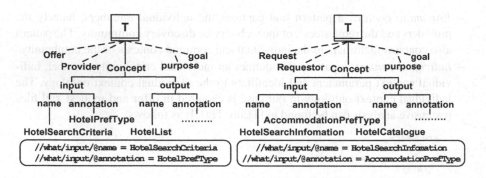

Fig. 3. The Offer and Request Context Ontologies

As you can see in Figure 2 the context ontology of the web service discovery community, consists of the concept *web service (WS)*. Two important properties of this concept are the *input* and *output* parameters. These parameters specify what the service does i.e what it outputs given a certain input. The two *communication roles (Comm_Role)* encountered in a web service discovery are the provider and the requestor. The community distinguishes two types of *communication processes (Comm_Proc)*: the *offer* of a web service, initiated by the provider and the *request* for a web service, initiated by the requestor. The *context parameters (Context_Par)* considered in the community, although other parameters can be included in the future (i.e temporal), are the *purpose* and the *goal*. The purpose indicates the reason this web service is being offered/requested and the goal indicates the targeted goal for its provision/offering.

The provider X and the requestor Y defined their individual context (ontologies) to contain among the others the input/output parameters of the web services being offered/requested respectively. These ontologies, which are shown in Figure 3, present in a greater detail the "what" part of the offer/request xml files. The individual pragmatic patterns that allow us to retrieve the input parameters of the "what" part from an offer/request file correspond to the xpath queries shown below each ontology (the same applies for the output parameters).

Let us now present the architecture we propose to cope with the matching of offers and requests during the web service discovery.

4 Matching: The Architecture

The idea behind the architecture we propose is to allow providers/requestors define their offers/requests asynchronously and intelligently select/return the offer which is the most appropriate for each request. At first, the provider and the requestor specify asynchronously their context information (who, why, what parameters) in a user-friendly web-enabled application (not focus of this paper). The system converts this information, in a completely transparent way, into the xml structure presented in Section 3 and stores it in the server of the requestor or the provider depending on the type of the xml (request/offer). Subsequently the matchers retrieve the requests and the offers and compare them against each other until the most appropriate offer is found for each request. A key feature of the matchers is that they can traverse any xml/rdf-based ontology (owl, daml etc).

The components of proposed architecture, as depicted in Figure 4, are in short the following: 1) The **Central Manager (CenMan)** is a centralized component responsible for the dispatching of requests to the distributed components and the coordination of the distributed components' activities. The CenMan utilizes a request profile, a profile manager and common profile repositories and 2) The **Profile Manager (ProfMan)** is a distributed component responsible for carrying out the matching between the offers and the requests for web services. It can be considered as the agent of the provider. The ProfMan consists of the **What Matcher** and the **Why Matcher** and utilizes an offer profile repository and trusted ontologies (i.e WordNet, OTA) to carry out its tasks. In the following sections the various components are analyzed in detail.

4.1 Central Manager

The Central Manager is the core component of the architecture. It is responsible for the transparent coordination of the various components. In general the Central Manager's

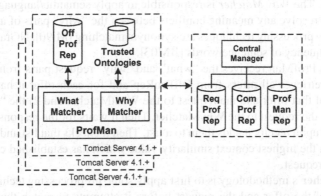

Fig. 4. The proposed architecture

methodology is to retrieve the request from the Request Profile Repository (ReqProfRep) as well as the registered $ProfMan_s$ (providers) from the Profile Manager Repository (ProfManRep), and in a threaded fashion forward to each ProfMan the "what" and "why" parts of the request. Subsequently it will wait for the result from each ProfMan and upon receipt of all results (from all $ProfMan_s$) it will select the offer which successfully established common context and has the highest context similarity value.

Request Profile Repository: The Request Profile Repository (ReqProfRep) is a repository where all request xml files are stored. (contains the *identifiers* to the Individual Context Ontologies of the requestors).

Common Profile Repository: The Common Profile Repository (ComProfRep) is a repository where all the request and offer xml files that shared common context are stored(Common Context Ontology of the community as a whole).

Profile Manager Repository: The Profile Manager Repository (ProfManRep) is the repository where the information for each registered ProfMan component are kept (URL address, registration date etc).

4.2 Profile Manager

The Profile Manager (ProfMan) is a distributed agent component that executes on the machine of each provider. It the most important part of the architecture because it carries out all the **matching** between the offers and requests and subsequently hands out the results to the Central Manager. The reasons we have considered a distributed solution for the matching are 1) it is less costly than centralized approaches since the machine resources utilized are minimal and 2) the Central Manager is freed from carrying out messy time-consuming matching tasks and therefore focused in its coordination activities.

In order to carry out its activities the ProfMan invokes the **Why Matcher** and the **What Matcher**. The *What Matcher* is responsible to apply a combination of language processing techniques and pragmatic principles between the "what" parts of an offer and a request. The *Why Matcher* is responsible to apply semantic/language processing techniques to resolve any meaning conflicts between the "why" parts of an offer and a request. These processing techniques are synonym matching [Mi90], NGram [AFW83], and order/frequency of sentence words [BM03].

After each ProfMan receives the "what" and "why" request parts from the Central Manager, it retrieves all offers from OffProRep and for each offer it handles: 1) the "what" part of that offer and the request to the WhatMatcher and 2) the "why" part of that offer and the request to the WhyMatcher. Then it waits for the responses from both matchers and upon receipt it adds them to a list. Then it checks that list and finds/selects the offer with the highest context similarity value which has established common context with the request.

WhyMatcher's methodology is to first apply some preprocessing techniques on the "why" part of the offer and the request before it attempts to match them. This preprocessing involves tokenizing the "why" parts based on punctuation and capitalization,

$$
\text{matching\_degree} \begin{cases} \emptyset & \text{if result(WhatMatcher)} = \emptyset \\ \text{result(WhyMatcher)} & \text{otherwise} \end{cases}
$$

$$
\text{result(WhyMatcher)} \begin{cases} 1 & \text{if (alg1 } \lor \text{ alg2 } \lor \text{ alg3)} = 1 \\ \emptyset & \text{if (alg1} = \text{alg2} = \text{alg3)} = \emptyset \\ \text{avg(alg1,alg2,alg3)} & \text{if 0} < \text{(alg1 } \lor \text{ alg2 } \lor \text{ alg3)} < 1 \end{cases}
$$

where alg1 = Ngram, alg2 = Synonym Matching, alg3 = Position Checking

Fig. 5. Formula for the calculating matching degree

converting everything to lowercase and removing unnecessary words (i.e the, of etc). If after the preprocessing the matching was unsuccessful, it stems the tokens using porter stemmer algorithm and tries to match them using NGram technique [AFW83]. Also it checks for synonymity of tokens by referencing WordNet [Mi90]. Moreover it applies some order/frequency algorithms that check the position of the tokens in the overall "why" parts of the request/offer [BM03]. The result of the WhyMatcher algorithm, which ranges from 0 to 1, will be the average of the three techniques.

WhatMatcher's role on the input/output parameters of the "what" parts between an offer and a request is dual. At first it performs similarity checks on the name of an input/output and at second it applies the pragmatic principles on the annotation part of an input/output. The similarity checks applied on the name part are the same as the synonymity checking applied by the WhyMatcher. The synonymity checks reference trusted ontologies as OTA and WordNet (explained below) to resolve meaning ambiguities and establish common context. The checks performed on the annotation part are a mixture of the synonymity checking mentioned above and the schema comparison of the ontology part/class involved with the annotation (including their subclasses, restrictions etc). If the schema comparison does not produce an exact match (even after meaning resolution occurred) i.e the request parameter (input or output) is a subClassOf of the offer parameter (subsume match) or the opposite (plug-in match), then extra measures must be taken. In either case the provider and the requestor, especially when social/professional/cultural backgrounds apply, must agree on common context: the ontologies they reference must be understandable by both. Therefore the agents of both parties must apply the pragmatics patterns, discussed earlier, to establish common context. If a concept exists in one ontology but not in the other one, the WhatMatcher has to reference OTA (trustful for both parties) to examine whether the concept exists there. If it does then the concept/common pragmatic pattern is saved in the common context ontology. If it does not, then the parties are said not to share common context so nothing is added in the common context ontology. If a common context have been established for all ontology concepts identified with subsume or plug-in match, then both parties are said to commit on the common context ontology derived.

If common context on the "what" part of a request and an offer is established, the matching degree equals that of the "why" part or $\emptyset$ otherwise. The Figure 5 summarizes the formulas used for calculating the matching degree.

Offer Profile Repository: The Offer Profile Repository (OffProfRep) is a repository where all offer xml files are stored (contains the *identifiers* to the Individual Context Ontologies of the providers).

Trusted Ontologies: The Open Travel Alliance (OTA) (www.opentravel.org) and the WordNet (http://wordnet.princeton.edu/) are two widely used ontologies in our framework. The OTA ontology is well accepted ontology in the travel industry which helps business sectors to build common understanding (exchange, share, use information) in the area of travel. WordNet, is also a widely accepted lexical database (dictionary) which organizes words into sets of cognitive synonyms (synsets) and thus makes it easy for applications to process natural language.

5   Related Work

The idea of this paper is inspired from Pragmatic Web principles as these have been defined in [Mo05, SdMD06]. The Pragmatic Web is a new research direction which aims to extend the Semantic Web by capturing the context of ontologies' usage, from the participants' point of view, in a given communication process. We have attempted to provide a rigorous application of pragmatic principles in the web service discovery scenario.

In order to effectively discover web services we have proposed the maintenance of offer and request files. These xml files contain information regarding the functionality of the web service (i.e what the service does: input and output parameters), who is requesting/offering the service and finally why this service is being offered/requested in the first place. The structure of the xml files posses a lot of similarities with the profile schema file proposed by OWL-S scientists [DMJ04]. Specifically, the actor details maintained in the OWL-S profile correspond to "who" part and and the service parameters (functional properties) correspond to "what" part of our methodology. The differences are 1) in our solution we additionally maintain context parameters, the why part, to capture valuable information about web services usage and 2) the xml specification we utilize is independent of the semantic web service language used underneath thus it is more expressive.

The matching techniques facilitated by the "WhyMat" component in our architecture are very similar to the Element Level Schema matching techniques applied in METEOR-S [POS04]. We, as in METEOR-S, consider a combination of techniques ranging from preprocessing of texts, tokenization, and stemming algorithms to synonymity checking. However in our solution we additionally consider text position checking and utilize all the senses (synonyms usage rankings) of the texts for more accurate matching results.

6   Conclusion

In this paper we have proposed an architecture which augments existing semantic-based web service discovery techniques with pragmatic web principles in order to make the selection process more efficient. As we have presented in the paper, the context of web services usage (context of ontological annotations) must be taken into consideration when we semantically match web services. To operationalize the principles of pragmatic web and tackle the problem, we have considered both the functionality/IO parameters (what part) and the context (why part) of web services offers and requests and proposed an architecture capable of performing the matching of the various request/ offer parts.

The technique utilized by the WhyMatcher is based on natural language/semantic processing. However it is also possible, at this level, to restrict the matching by 1) classifying the relevance of concepts and 2) applying the pragmatics as we did with WhatMatcher to resolve any meaning conflicts. It is possible to modify/enrich the ontologies utilized for the annotation of web services using the common context ontology produced. We plan to offer this kind of extensions in our architecture in the future.

The disadvantages of the approach lie on the need for 1) a centralized architecture and 2) the acceptance of providers to install software on their machines, an action that might raises security, trust and other problems. We plan to examine these issues at a later stage.

References

[AFW83]   Angell, R.C., Freund, G.E., Willett, P.: Automatic Spelling Correction Using a Tri-gram Similarity Measure. Inf. Process. Manage. 19(4), 255–261 (1983)
[BM03]   De Boni, M., Manandhar, S.: The Use of Sentence Similarity as a Semantic Rele-vance Metric for Question Answering. In: New Directions in Question Answering, pp. 138–144 (2003)
[D.04]   Roman, D.: WSMO: Current status and open points (November 2004)
[DMJ04]   Martin, D., Burstein, M., Hobbs, J., Lassila, O., McDermott, D., McIlraith, S., Narayanan, S., Paolucci, M., Parsia, B., Payne, T., Sirin, E., Srinivasan, N., Sycara, K.: OWL-S: Semantic Markup for Web Services (November 2004)
[HZB06]   Hull, D., Zolin, E., Bovykin, A., Horrocks, I., Sattler, U., Stevens, R.: Deciding Semantic Matching of Stateless Services. In: AAAI, AAAI Press (2006)
[Mi90]   Miller, G.: WordNet: An on-line lexical database. International Journal of Lexicog-raphy Special Issue (1990)
[Mo05]   De Moor, A.: Patterns for the Pragmatic Web (invited paper). In: Sunderam, V.S., van Albada, G.D., Sloot, P.M.A., Dongarra, J.J. (eds.) ICCS 2005. LNCS, vol. 3514, pp. 1–18. Springer, Heidelberg (2005)
[PKP02]   Paolucci, M., Kawmura, T., Payne, T., Sycara, K.: Semantic Matching of Web Ser-vices Capabilities (2002)
[POS04]   Patil, A., Oundhakar, S., Sheth, A., Verma, K.: METEOR-S Web service Annotation Framework. In: Proceedings of the 13th international conference on World Wide Web, ACM Press, New York (2004)
[PSS04]   Paolucci, M., Soundry, J., Srinivasan, N., Sycara, K.: A Broker for OWL-S Web Services. In: AAAI Symbosium (2004)
[SdMD06]   Schoop, M., de Moor, A., Dietz, J.L.G.: The pragmatic web: a manifesto. Commun. ACM 49(5), 75–76 (2006)
[Si02]   Singh, M.P.: The Pragmatic Web: Pleliminary thoughts. In: NSF-OntoWeb Work-shop on Database and Information Systems Research for Semantic Web and Enter-prises, April 2002, pp. 82–90 (2002)
[SVM05]   Sheth, A., Verma, K., Miller, J., Rajasekaran, P.: Enhancing Web Service Descrip-tions using WSDL-S (2005)
[TE06]   Tamani, E., Evripidou, P.: A Pragmatic and Pervasive Methodology to Web Service Discovery. In: OTM Workshops, pp. 1285–1294 (2006)
[TE07]   Tamani, E., Evripidou, P.: A Pragmatic Methodology to Web Service Discovery. In: ICWS, pp. 1168–1171 (2007)

Modelling Context-Aware Multimedia Community Content on Mobile Devices

Diana Weiß

Mobile & Distributed Systems Group, Department of Computer Science,
Ludwig-Maximilians-Universität München, Germany
diana.weiss@ifi.lmu.de

Abstract. Community services play a steadily growing role in the current World Wide Web. Especially services that provide video or other multimedia content - in particular user-generated content - enjoy great popularity. With the integration of multimedia technology like video cameras or -players into mobile devices these services could also be used while on the move. Some mobile devices are already equipped with GPS receivers, compasses, thermometers and so on and therefore can be used for so called context-aware services. This allows for a variety of new, content-based community services.

This paper introduces a metadata model that takes important features of context-aware multimedia community content on mobile devices into account. The model is used for automatic preselection and filtering of content based on context. Besides the annotation of multimedia content with context, it supports the linkage between content and associated comments and the definition of access rights for different communities and its members. The proposed model is mapped on the MPEG-7 standard, a generic description tool for multimedia metadata.

1 Introduction

A new trend can be found within the World Wide Web in the last few years: Web 2.0 applications offer much bigger involvement of the user. One important field here are so called community services like YouTube [1] or Wikipedia [2]. These services allow users to interact with each other or to create own content and provide it to other users (user-generated content). With the development of mobile devices that are more advanced than personal computers were a few years ago, community services cannot only be used on classical desktop computers but also while on the move on devices like smart phones, pocket PCs or PDAs.

When using these services in a mobile environment, the context of content and user can be considered more thoroughly. Context of content refers to the context in which the content is to be used. In some cases this is identical to the context at creation time (e.g. the location where the content was created), in other cases it is different (e.g. the content was created in the morning but should only be accessed in the afternoon). Today, an increasing number of mobile devices is able to determine context information: all have knowledge about

R. Meersman, Z. Tari, P. Herrero et al. (Eds.): OTM 2007 Ws, Part II, LNCS 4806, pp. 834–844, 2007.
© Springer-Verlag Berlin Heidelberg 2007

the current time, a built-in GPS-receiver provides the location, an electronic compass directs the bearing, the surrounding temperature is measured by built-in thermometers and so on. Since most devices have internet access, web-based services can be used as well to derive context information based on the current location. Unfortunately, this increasing amount of information is not yet sufficiently used. Some projects already consider multimedia in a context-aware environment, but focus on context-aware delivery [3], context-aware access control [4] or the context *location* only [5]. But context can also be used to preselect and filter content based on its context and the context of the user of a community service respectively. For this, the content needs to provide metainformation to allow automatic processing. In this paper, a metadata model is proposed to support context-aware multimedia community content on mobile devices. The model is mapped on MPEG-7, a standard for multimedia descriptions.

This paper is structured as follows: an example scenario is given in Section 2. The scenario shows requirements that need to be fulfilled by the proposed metadata model. Section 3 gives a short overview of the MPEG-7 standard and several extensions. The developed model is introduced, mapped on the MPEG-7 standard and discussed in Section 4. In the last section, the paper provides a short conclusion and overview about future work.

2   Scenario

In the following, an example is given that shows a possible application for context-aware multimedia community content on mobile devices. To simplify matters, the multimedia content contemplated is mainly video content.

MobiGuide, a mobile tourist guide, provides tourists with context-dependent videos on different topics, e.g. *important sights, food and drink,* or *shopping*. The user selects one or more topics and MobiGuide displays a map containing all locations nearby that are described by videos associated with these topics. The service offers guided tours or allows users to walk at random alerting them whenever they approach a location that has videos assigned to it. Both is supported: official and professionally created videos showing interesting places or videos created by users and provided to all other users or just to a subset like their own community. Content is not only filtered by location or topic but also by further context information. A beer garden for example is only displayed when the weather is dry and sunny.

MobiGuide allows users to comment all available videos after watching them. They can create video, audio, images or text as comments. After a user watched a video, all available comments to this videos are filtered according to topics, context and access restrictions similar to user-generated videos and displayed as list. The user can choose a comment and can annotate it with a further comment after display.

Requirement Analysis. On closer examination of the given scenario, one can find several requirements regarding metadata that need to be fulfilled. In the following, I assume a good market penetration of mobile devices with built-in location receivers (e.g. GPS or Galileo) and compasses, internet access for further context retrieval, video camera and microphone, as well as high processing power, which will be very likely in the near future [6].

First of all, an association of the content with its context has to be included to the metadata model (**Req. 1**). In the scenario, a video recommending a beer garden should only be provided when the weather is dry and sunny, so this information has to be added to the content metadata.

Since context information can be expressed in different formats, its scale has to be included in the model as well (**Req. 2**). For example, the temperature can be measured in degree Celsius or degree Fahrenheit, or positioning systems have different formats as UTM or WGS-84 etc.

To restrict access to videos to specific user groups, the allowed groups and their members need to be described in the model (**Req. 3**). Several categories of user groups have to be predefined (e.g. family, friends or colleagues) and user profiles within the mobile phone have to assign other users to one of these categories.

Further on, a classification of all videos within the system is necessary, to assign them to the different topics, e.g. *important sights*, *food and drink*, or *shopping* (**Req. 4**).

At last, all user-generated comments need to be linked to the associated video content (**Req. 5**).

To realize Req. 1-5, an approach is proposed that is based on multimedia metadata for efficient use in a multimedia information management system. Because of its universality, the MPEG-7 Multimedia Content Description Interface was used as a basis for realization.

3   Multimedia Content Description Interface (MPEG-7)

The Multimedia Content Description Interface (MPEG-7) [7,8,9] provides a set of tools to describe all kinds of multimedia content. It has been specified by the Moving Picture Expert Group (MPEG). In order to stay as generic as possible, all application-related issues are left outside the scope of the standard and it is solely focused on describing multimedia content.

3.1   Overview

Description Tools in MPEG-7 are defined by so called Descriptors (Ds) and Description Schemes (DSs). Ds (e.g. the color within an image) represent the features of audio-visual content and define its syntax and semantics. They can be classified in high- and low-level Ds. Structure and semantic of relations between MPEG-7 entities are specified by DSs (e.g. relating different colors to a single image). The so related MPEG-7 entities can either be Ds or other DSs.

Ds, DSs and their structural relations are specified by the Description Definition Language (DDL). It allows the creation of new Ds and DSs and the extension of existing ones. The MPEG-7 DDL is an extension of W3C's XML Schema [10]. It describes features specific to video and audio as well as multimedia descriptions and generic entities.

MPEG-7 was not designed for a specific purpose or application. Mostly it is used for information retrieval and queries in multimedia databases. But because of its flexibility, it is suitable for all kinds of multimedia application domains. Thus, MPEG-7 can be extended to easily combine context and communities with multimedia content. This is why this standard was chosen to be extended.

3.2 MPEG-7 Extensions

Since MPEG-7 is a generic standard, it has already been used for a large number of applications. Extensions can be found e.g. in the area of context, structuring of elements, or semantics. Because of the number of approaches, the following provides just a small excerpt of the whole range of related work.

Hwang et al. [11] designed an MPEG-7 metadata scheme to be used within a video-based Geographic Information System. Information about 3D geo-coordinates are embedded for each video frame using newly created elements within MPEG-7's `FreeTextAnnotation` element. This way, the position of spatial objects can be calculated and corresponding videos searched.

A metadata and user profile model was proposed by Tsekeridou [12]. It is used within an enhanced digital TV environment for semantic annotation and personalized access to multiple sports. New types describing e.g. viewer profiles or sports events are created that support the given scenario.

Vakali et al. [13] describe a multimedia modeling and description scheme based on multi-level modeling and semantic classification for videos. DSs and Ds are designed that represent large collections of video data, instead of single multimedia documents as provided by MPEG-7. They define e.g. `VideoCollection` DS or `ClusterSet` DS that aggregate different documents.

Gomez et al. [14] discuss ways of describing music constructs in MPEG-7. They identify limitations in the standard regarding melodic, rhythmic and instrumental information of music and enhance the standard with the appropriate types (e.g. `NoteSegmentType`) and elements (e.g. `NoteDS`).

An approach that extends the MPEG-7 standard to - partly or totally - support context-aware multimedia community content on mobile devices cannot be found. Therefore, the proposed extension is based on the original standard.

4 Supporting Context-Aware Multimedia Community Content on Mobile Devices

To meet the requirements set in Section 2, a model is created that realizes the scenario. Afterwards, this model is mapped onto the MPEG-7 standard to existing types and elements or newly created extensions.

4.1   A Model for Context-Aware Multimedia Community Content

Since multimedia content is a collection of multiple forms of content, the *content* is modeled as an aggregation of different *contentItem*s. These items can be *video*, *audio*, *image*s or *text*s (see Figure 1). For filtering, *contentItem*s can be associated with specific *genre*s like e.g. "important sights", "food and drink" or both. Relating the *genre* to the *contentItem* instead of the *content* itself has the advantage that e.g. video content can have different audio files belonging to different *genre*s associated to it and is combined on the mobile device with the specific file that matches the *user*'s preferences. For the same reason, the *context* is associated to the *contentItem* instead of the *content*. An entity's context is composed of different kind of context information, e.g. location, bearing, time, temperature etc. Thus, the *context* is modeled as an aggregation of *contextInformation*. It can be measured by the mobile device or retrieved using the mobile's internet access. Every *contextInformation* has a *scale*.

Content can be annotated by *comments*. As seen in the scenario, *comments* have genre, context and access restrictions assigned to it and can be commented themselves. Thus, they are modeled as *content* and are therefore a collection of *contentItem*s as well.

Communities consist of one or more persons, so *communityGroup*s are modeled as a collection of *user*s. Since *user*s can also create *contentItem*s, the corresponding association was added to the model. The *right* to use *contentItem*s is restricted to specific *communityGroup*s. This is modeled as a ternary association between those entities. Similar to the modeling of *context* and *genre*, the association was added to the *contentItem* instead of the *content*. This way, e.g. a video can be forbidden to watch, but its audio is allowed to be played. *User*'s can have the right to just display the *contentItem* or can additionally be admitted to annotate it with *comments*. Thus, the entities *readAccess* and *addComments* were modeled, whereas *addComments* includes all rights of *readAccess*.

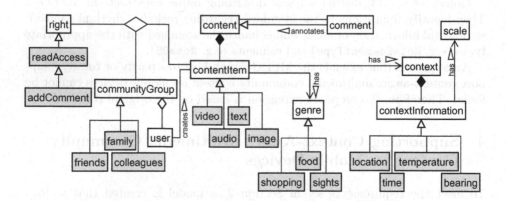

Fig. 1. Context-aware Multimedia Community Content Modeled in UML. Grey rectangles refer to concrete values.

4.2 Mapping the Model to the MPEG-7 Standard

In the following, the model is mapped onto the description tools of the MPEG-7 standard. Since the standard does not fully support the given scenario and model, some extensions are made. In the shown code examples underlined elements and attributes are new, others are already supported by the MPEG-7 standard or have to be slightly adjusted.

Content and Content Items. A simple MPEG-7 file may look like seen in Listing 1.1. The Mpeg7 element is the root tag of any MPEG-7 file. It contains the metadata of one or more descriptions. The Description tag may include one or more MultimediaContent elements that contains exactly one specific multimedia content as video, audio etc. Thus *content* and *contentItem* entities of the model can easily be represented by the MPEG-7 standard: a Description represents a single *content* and the MultimediaContent the *contentItem*. For all subtypes of the *contentItem* entity in the model, an appropriate type is available. For example, *audio* is mapped to the Audio tag and *video* to the Video tag.

```
1  <Mpeg7>
2      <Description>
3          <MultimediaContent>
4              <Video>   ...   </Video>
5          </MultimediaContent>
6      </Description>
7  </Mpeg7>
```

Listing 1.1. Mapping *Content* and *ContentItems*

```
1  <Video>
2      <CreationInformation>
3          ...
4          <Classification>
5              <Genre href="
                  urn:mpeg:mpeg7:cs:ContentCS:2007:1">
6                  <Name xml:lang="en">Interesting Sights</
                      Name></Genre>
7              <Genre href="
                  urn:mpeg:mpeg7:cs:ContentCS:2007:2">
8                  <Name xml:lang="en">Food And Drink</Name><
                      /Genre>
9          </Classification>
10     </CreationInformation>
11 </Video>
```

Listing 1.2. Mapping *Genre*

Genre. All *contentItem*s of the model are associated to one ore more specific *genre*s. This information can be mapped to the `Genre` element found within the `CreationInformation` tag of a specific `MultimediaContent` tag, e.g. `Video` (see Listing 1.2). Genres are specified by a uniform ressource name (urn) and an included `Name` tag stating name and language. This way, the standard allows to be easily extended with the urns and name values of all *genre*s needed.

Context and Scale. MPEG-7 supports the description of context information at creation time as e.g. the creation location. The context of the here proposed model refers to the context in which the content is to be used, but this is not always identical to the context at creation time. Thus, a new `Context` tag was added within the `UsageInformation` tag (see Listing 1.3) where information about the media usage like access rights or the availability is specified. The `Context` tag contains one or more generic `ContextInformation` elements. According to the Aspect-Scale-Context (ASC) model introduced by Strang et al. [15], it contains the attributes `type` to specify the context type (e.g. temperature), `scale` to set the used *scale* (e.g. `degreeCelsius` for temperature measured in degree Celsius) and `value` to provide the value. This allows easy extension with new types of context information. The way of describing the creation location in MPEG-7 mentioned before is very inflexible and no information about the actual location format is given. Thus, the generic `ContextInformation` tag was used for describing the location as well.

```
1  <Video>
2     <UsageInformation>
3        <Context>
4           <ContextInformation type="longitude" scale="
              WGS-84" value="48,85854"/>
5           <ContextInformation type="latitude" scale="WGS
              -84" value="11,353832"/>
6           <ContextInformation type="temperature" scale="
              degreeCelsius" value="25"/> ...
7        </Context>
8     </UsageInformation>
9  </Video>
```

Listing 1.3. Mapping *Context* and *Scale*

Comments. In MPEG-7, content can be annotated with e.g. free text annotations or with key words within the `TextAnnotation` tag. To realize multimedia *comment*s as specified by the model, the new `ContentAnnotation` tag was defined as seen in Listing 1.4. Within this tag one or more `Comment` tags can be embedded. A comment is given by the `MediaUri` tag that specifies the media type of the comment (e.g. audio) and points to the location of the multimedia comment. According to the model, a *comment* has also *context*, *genre* etc. so it also has an MPEG-7 file describing this information. This description file is localized by the `DescriptionUri` element. Because of the sparse air interface, it is more efficient to base the filtering of the comments according to context, genre, etc. on the small description files and retrieve the multimedia comments themselves only when they are relevant and explicitly queried by the user.

```
1  <Video>
2      <TextAnnotation>
3          <ContentAnnotation>
4              <Comment>
5                  <MediaUri type="audio" uri="http://www.
                       mobiGuide.de/C1.aiff"/>
6                  <DescriptionUri uri="http://www.mobiGuide.
                       de/C1.xml"/>
7              </Comment> ...
8          </ContentAnnotation>
9      </TextAnnotation>
10 </Video>
```

Listing 1.4. Mapping *Comments*

Access Rights & Community Groups. MPEG-7 supports mechanisms for different kinds of rights management, but it just describes where to access current rights owners information, without dealing with the information and negotiation. Further, it does not support community access. Communities have to be globally defined within the description because they are equal for all *contentItems* building a *content*. For this, the new tag `CommunityGroups` was added within the `DescriptionMetadata` element where all global information is set. The `CommunityGroups` tag contains one or more `CommunityGroup` tags that are identified by a unique `groupID` and contain one or more `Member` tags. The `Member` tag references to a specific user using `mref`. The `Users` tag is also inserted within the `DescriptionMetadata` element. MPEG-7 already supports user descriptions with unique `id` by specifying the `Role` of the user, the `Agent` type as `PersonType` (other types refer e.g. to organizations) and naming information. For mapping the model, just the new role `MEMBER` has to be added (Listing 1.5).

As mentioned before, rights information of single *contentItems* is defined within the `UsageInformation` tag. The `Rights` tag includes one or more `Rights-ID` elements. To link the rights to a specific `CommunityGroup`, the `groupRef` attribute was added (see Listing 1.6). According to the model, the two new types `readAccess` and `addComment` were created.

```
1   <Mpeg7>
2       <Description>
3           <DescriptionMetadata>
4               <CommunityGroups>
5                   <CommunityGroup groupID="gr1">
6                       <Member mref="us1"> ...
7                   </CommunityGroup> ...
8               </CommunityGroups>
9               <Users>
10                  <User id="us1">
11                      <Role href="
                            urn:mpeg:mpeg7:cs:RoleCS:MEMBER"/>
12                      <Agent type="PersonType">
13                          <Name> <GivenName> John </
                                GivenName>
14                              <FamilyName> Doe </FamilyName>
                                </Name>
15                      </Agent>
16                  </User> ...
17              </Users>
18          </DescriptionMetadata>
19      </Description>
20  </Mpeg7>
```

Listing 1.5. Mapping *CommunityGroup*

```
1   <Video>
2       <UsageInformation>
3           <Rights><RightsID type="readAccess" organization="
                MobiGuide" groupRef="gr1">
4                   Read Access </RightsID> ... </Rights>
5       </UsageInformation>
6   </Video>
```

Listing 1.6. Mapping *Rights*

4.3   Discussion

In Section 2, several requirements are defined that the model must support. The association of the content with its context is to be modeled (Req. 1). Since multimedia content consists of multiple forms of *contentItems* as explained before, the *context* entity is directly linked the *contentItem* entity in the model. Therefore, the *context* of *content* consists of the summary of the *context* of its *contextInformation*. The *scale* of the context information is integrated to the model as well, to avoid mistakes (Req. 2). The association between the content and its *genre* is modeled similar to the context, so the *genre* is linked to the *contentItems*

(Req. 4). The *content* may be annotated with *comments* (Req. 5). Thus, a link between the *content* entity and all of its associated *comments* is added to the model. Finally, to represent access restrictions (Req. 3), the *right* entity and its subtypes, the *communityGroup* with its *users* and the ternary relationship to the *contentItem* where added to the model.

To show the applicability of the given model and the mappings described in Section 4.2, the MPEG-7 DDL was extended according to the details explained before. Additionally, a small prototype application was designed that allows the creation of MPEG-7 files including all introduced extensions. This prototype automatically combines information retrieved from the user profile (e.g. user's name) and derived positioning information to an appropriate MPEG-7 file with the described extensions. A user interface is provided that allows the user to choose what context should be included to the description and to enter further information or restrictions. The extension of this prototype to realize the given scenario is part of the future work. So far, it helped to improve and adjust the model and its mapping.

5 Conclusion and Future Work

This paper combines the concept of context-aware services with the new idea of community services and the new multimedia capabilities of mobile devices. A metadata model has been proposed that contains relevant entities to model context-aware community content on mobile devices and fulfills the given requirements. Subsequently, the model is mapped on the MPEG-7 standard.

Future work includes a closer validation of the proposed metadata model. For this, the model will be integrated into a fully functional prototype implementation according to the given scenario. This prototype will deal with different aspects, e.g. multimedia storing and retrieval. Additionally, further considerations will be made to extend the model to support even more personalized or interactive context-aware multimedia content.

References

1. YouTube - Broadcast Yourself: (last access: 09.01.2007)www.youtube.com
2. Wikipedia - The Free Encyclopedia: (last access: 09.01.2007)www.wikipedia.org
3. Dynamic and distributed Adaptation of scalable multimedia coNtent in a context Aware Environment (DANAE): (last access: 09.01.2007) danae.rd.francetelecom.com
4. Bouna, B.a., Chbeir, R., Miteran, J.: Mca2cm: Multimedia context-aware access control model. In: ISI (2007)
5. Pan, P., Kastner, C., Crow, D., Davenport, G.: M-studio: an authoring application for context-aware multimedia. In: MULTIMEDIA 2002: Proceedings of the 10th ACM international conference on Multimedia, ACM Press, New York (2002)
6. canalys.com ltd.: Smart mobile device and navigation trends 2006/2007 (2006)
7. ISO/IEC MPEG Group: www.chiariglione.org/mpeg last access: 09.01 (2007)

8. International Organisation for Standardisation: Martínez, J.M. (ed.) ISO/IEC JTC1/SC29/WG11N6828: MPEG-7 Overview (version 10) (2004)
9. Manjunath, B.S.: Introduction to MPEG-7, Multimedia Content Description Interface. John Wiley and Sons, Ltd. Chichester (2002)
10. Walmsley, P., Fallside, D.C.: XML schema part 0: Primer second edition. W3C recommendation, W3C (October 2004), http://www.w3.org/TR/2004/REC-xmlschema-0-20041028
11. Hwang, T.H., Choi, K.H., Joo, I.H., Lee, J.H.: Mpeg-7 metadata for video-based gis applications. In: Proceedings of the Geoscience and Remote Sensing Symposium, vol. 6 (2003)
12. Tsekeridou, S.: MPEG-7 MDS-Based Application Specific Metadata Model for Personalized Multi-Service Access in a DTV Broadcast Environment. In: Multimedia and Expo, 2005. IEEE International Conference on Volume (2005)
13. Vakali, A., Hacid, M.S., Elmagarmid, A.: MPEG-7 based description schemes for multi-level video content classification. Image and Vision Computing (6) (2004)
14. Gomez, E., Gouyon, F., Herrera, P., Amatriain, X.: Using and enhancing the current mpeg-7 standard for a music content processing tool. In: Proceedings of the 114th Audio Engineering Society Convention (2003)
15. Strang, T., Linnhoff-Popien, C., Frank, K.: Cool: A context ontology language to enable contextual interoperability. In: 4th IFIP WG 6.1 International Conference on Distributed Applications and Interoperable Systems (DAIS 2003), Springer, Heidelberg (2003)

Towards a Versatile Problem Diagnosis Infrastructure for Large Wireless Sensor Networks*

Konrad Iwanicki[1,2] and Maarten van Steen[2]

[1] Vrije Universiteit, Amsterdam, The Netherlands
[2] Development Laboratories (DevLab), Eindhoven, The Netherlands
{iwanicki,steen}@few.vu.nl
http://www.few.vu.nl/~iwanicki/lupa-web/

Abstract. In this position paper, we address the issue of durable maintenance of a wireless sensor network, which will be crucial if the vision of large, long-lived sensornets is to become reality. Durable maintenance requires tools for diagnosing and fixing occurring problems, which can range from internode connectivity losses, to time synchronization problems, to software bugs. While there are solutions for fixing problems, an appropriate diagnostic infrastructure is essentially still lacking. We argue that diagnosing a sensornet application requires the ability to dynamically and temporarily extend the application on a selected group of nodes with virtually any functionality. We motivate this claim based on deployment experiences to date and propose a highly nonintrusive solution to dynamically extending a running application on a resource-constrained sensor node.

"During the spring of 2004, 80 mica2dot sensor network nodes were placed into two 60 meter tall redwood trees in Sonoma, California. [. . .] One month later, initial examination of the gathered data showed that the nodes in one tree had been entirely unable to contact the base station. Of the 33 remaining nodes, 15% returned no data. Of the 80 deployed nodes, 65% returned no data at all, from the very beginning. [. . .] One week into the Sonoma deployment, another 15% of the nodes died [. . .] and no records exist of the events that may have caused this failure. [. . .]"

G. Tolle and D. Culler [2].

"[. . .] In 2004, [Dutch] researchers [. . .] teamed up in an ambitious project to use 150 wireless sensor nodes in a three-month pilot deployment on precision agriculture. [. . .] Out of 97 nodes running for [only] three weeks generating 1 message per 10 minutes, we received only 5874 messages, which amounts to 2%. [. . .]"

K. Langendoen et al. [3].

1 Introduction

Wireless embedded sensing provides real-time, on-the-spot surveillance of the physical environment. The progress in miniaturization, energy conservation, radio technology, and algorithm design, enables building larger and longer-lived wireless sensor networks

---

* This paper is based on an earlier technical report [1], available online.

R. Meersman, Z. Tari, P. Herrero et al. (Eds.): OTM 2007 Ws, Part II, LNCS 4806, pp. 845–855, 2007.
© Springer-Verlag Berlin Heidelberg 2007

(WSNs) which run increasingly complex applications. Their potential has been widely recognized and we are now seeing evidence that embedded sensing in the large can provide new scientific tools, maximize productivity, limit expenses, improve security, and enable disaster containment [4].

However, if such WSNs of hundreds or even thousands of nodes operating over months, if not years, are to become reality, it is time we started paying much more attention to their deployment and subsequent durable maintenance. Both these aspects require the ability to first diagnose, and then fix occurring problems. There are stable solutions for fixing problems, once detected, but an infrastructure for inspecting and diagnosing large, operational WSNs is essentially still lacking. In other words, what we need is a *network debugging system*.

One challenge that needs to be faced is the difficulty of predicting during application development, which features may need to be examined for diagnosing yet unknown, post-deployment problems. First, applications for sensor nodes are no longer simple "sense-and-send." Currently, they often exceed thousands of lines of code and can provide SQL support [5], decentralized topographic mapping [6], or primitive visual object recognition [7]. Second, the exposure of the system to the physical environment and complex internode interactions, together often result in unanticipated situations, which are the main cause of post-deployment problems [2,3,8,9,10]. Third, because there are still so many unexplored issues of WSNs, or issues unique to a particular environment, many occurring problems have not yet been fully studied [8,11].

Another challenge is the infeasibility of performing exhaustive and/or on-the-spot diagnosis. Limited accessibility to nodes combined with the scale of a system often rule out direct inspection of individual nodes. Hardware and resource constraints of sensor nodes preclude intensive debugging over large regions. Moreover, for critical applications, diagnosis of a problem in one part of the network should not disrupt the correctly functioning parts.

In this paper, we take the position that *diagnosing a WSN application requires the ability to dynamically and temporarily extend the application on a selected group of nodes with virtually any functionality*. Being able to dynamically extend the functionality of a running application provides the means to react to unanticipated situations, which is crucial considering all the deployment experiences to date. Combining this with temporal and spatial locality ensures scalability and resource conservation. In particular, there is no need for a node to perform extra functions when the application is behaving apparently correctly. A major issue that needs to be addressed is that extending a running application for diagnostic purposes should be as unintrusive as possible. As we discuss in this paper, this additional requirement calls for special solutions.

We further motivate our position based on existing work in Sect. 2. Section 3 explains our proposal, while Sect. 4 discusses the architectural support it requires. In Sect. 5, we conclude and present our future research agenda.

2   Motivation

Maintaining any large network is difficult. It is even more demanding in the case of a network composed of tiny, cheap, resource-constrained devices embedded in and interacting with the surrounding environment. Deployments to date were literally plagued by a number of rarely foreseen technical problems including, amongst others, losses of connectivity with a number of nodes, duty cycle desynchronization, sensor decalibration, too rapid power consumption, and software bugs in seemingly mature and stable code [2,3,8,9,10]. Drawing from that experience, we predict that, like corporate or campus networks, large WSNs will ultimately be run by specialized personnel rather than proverbial "herbologists."[1] Equipping such personnel with the ability to deal with unanticipated situations without exhaustive and/or on-the-spot inspections imposes a number of requirements on a diagnostic infrastructure.

To begin with, the infrastructure should enable remote observation of crucial application metrics, that is, a network administrator operating from a remote base station should be able to inspect selected metrics of the application run by particular nodes.

Because of the complexity of current WSN applications and resource limitations of sensor nodes, it is infeasible to define all application variables as observable metrics. Due to the lack of prior knowledge, inherent to novel deployments, and the unpredictability of the majority of problems, a programmer simply cannot foresee which of the variables may need monitoring. Consequently, the infrastructure should enable on-demand access to arbitrary application variables. Additionally, some situations require the ability to trace the changes of a variable's state in a particular part of code. For instance, unstable connectivity can present an unexpected combination of inputs that triggers bugs in untested control paths of the application-level routing code [8].

Diagnosing certain problems (e.g., related to time synchronization or sensor calibration) often requires local coordination and collaboration of nodes in reading metric values. For example, although a clock synchronization algorithm was embedded in the application, a few nodes in the Sonoma deployment died due to time synchronization issues and resulting energy depletion [2]. To diagnose such a problem, one needs to compare clocks of neighboring nodes. Because of huge multi-hop routing delays, the comparison should be performed locally by the neighboring nodes (e.g., using Cristian's timestamp exchange [12]). Otherwise, it is intractable to determine whether the time difference is authentic or due to the multi-hop delays incurred by the network.

The resource constraints of nodes combined with possible network sizes prevent each node from constantly reporting the values of selected metrics. Taking a single snapshot of a thousand-node multi-hop network may involve hundreds of thousands of messages. Such periodic traffic could drain the energy of the nodes and seriously disturb regular application communication. Instead, by exploiting spatial and temporal characteristics of problems, the network debugging infrastructure should minimize intrusiveness. It should be possible to run diagnostic code only as long as it takes to pinpoint a problem and only on the nodes directly affected by this problem.

Finally, communication in WSNs consumes several orders of magnitude more energy than computation. Therefore, if possible, the processing involved in problem diagnosis should be performed by the nodes rather than at a base station. This way, the expensive reporting of raw metric values to a base station is replaced with much cheaper computation and local communication.

Existing problem diagnosis infrastructures for WSNs [2,8,13] proved invaluable for deploying and maintaining relatively small, real-world networks running simple, well-studied applications. However, they were not designed for rapidly emerging

---

[1] At least, until we fully master (if ever) the technology of self-diagnosing and self-healing systems. The real-world experience to date, however, evidences that in the case of large WSNs we are still quite far away from this goal.

systems, characterized by the increasing complexity and scale, and often operating in unfamiliar settings. Although all solutions enable remote monitoring of application metrics, they require the programmer to foresee the important ones prior to the deployment [2,8], or put constraints on how the metric values are obtained [2,13]. Thus, many unanticipated problems cannot be diagnosed. Moreover, metric values are collected per node, and then processed at a base station [2,8,13]. This lack of support for in-network processing and dynamic collaboration of nodes limits scalability and restricts applicability to certain problems. Consequently, as large WSNs pose many challenges unencountered before, an appropriate problem diagnosis infrastructure is necessary.

In contrast to the aforementioned software-based solutions, a recent alternative approach to deploying and maintaining WSNs [11] involves specialized hardware. It requires a second support network to be deployed together with the network running the actual application. The advantage of this approach is that the support network provides a separate reliable wireless backbone for the transport of debug and control information from and to the nodes running the application. The major disadvantage is the cost. The solution we propose in this paper is orthogonal to using a support network. In particular, the support network can be used for transporting application extensions to and retrieving results from the nodes.

3   Principal Operation

Diagnosing a problem involves a human administrator. We assume that he operates from a (possibly mobile or remote) base station capable of communicating with the sensor network. The base station runs the client-side software for interaction with the node-side software of our infrastructure. In other words, the administrator uses the base station to post diagnostic code to the nodes, and to retrieve and analyze information gathered by this code.

Problem detection is often application specific. For example, in a typical data-gathering application, the fact that nodes in some part of the network do not return data implies a problem in that part. In contrast, detecting a problem in a system for reactive tasking (i.e., based on their readings, sensor nodes trigger nearby actuators) looks completely different. For this reason and due to space limitations, here we focus only on those aspects of our solution that are application independent. However, we recommend the application to be designed in a way that facilitates problem detection. For instance, a simple, lightweight generic problem detection mechanism can be implemented by defining a few basic "health" metrics of a node and using multi-resolution aggregation on those metrics [14,15]. Abnormal aggregate metric values indicate a problem, which is then diagnosed using the proposed infrastructure. Alternatively, the infrastructure itself can be used to periodically install primitive diagnostic extensions, which can detect if there are any problems. In either case, remote diagnosis of a problem requires some nodes not to be harmed by the problem, so that they can diagnose other nodes.

We illustrate the basic idea behind our solution with the example use-case from Fig. 1. Consider an application, like TinyDB [5], that gathers sensor data at the base station. The nodes are organized in a tree that is used for routing. On their way up the tree (to the base station), the data can be aggregated based on accompanying timestamps.

Fig. 1. An example illustrating the operations involved in a problem diagnosis using our approach

At some point the administrator notices that data from the nodes in the black area are missing [Fig. 1(a)]. This may indicate, for instance, that the nodes in this area ran out of batteries or that they lost connectivity with the rest of the network. To diagnose the problem, the administrator first installs an application extension on the nodes at the perimeter of the black region — marked with gray [Fig. 1(b)]. The task of the extension is to check whether the black region is connected to the rest of the network [Fig. 1(d)]. This can be accomplished by examining the neighbor table maintained by the application on the nodes at the perimeter or by snooping incoming packets at the perimeter nodes and analyzing their sender. All nodes running the extension aggregate a single connectivity result (YES or NO) and route it to the base station [Fig. 1(c)]. When the administrator has gathered requested information, the extension is removed from the application [Fig. 1(e)].

Problem diagnosis is performed by gradually narrowing the set of possible causes. Extending the application on a group of nodes can be viewed as "zooming into this group," and the extension itself as "a magnifying glass." Multiple magnifying glasses can be used to progressively "zoom into the problem." The increasing level of detail of subsequently used glasses (in terms of the amount of data analyzed and reported) and the decreasing range (in terms of the number of nodes affected by a glass) correspond to pinpointing the cause of the problem. For instance, if the black region was not connected, a further on-the-site inspection would be the only option. However, if it turned out that the black region was connected, the administrator could install another extension, but this time in the black region itself, since the nodes there were obviously running. The new extension would get the routing subtree used by the application. If the subtree was correct, then the problem was likely caused by its root node, and thus, an even more specific extension could be installed on that node and its neighbors. This extension could check if the root node did not have time synchronization problems, such that it rejected the data from its subtree as stale. The extension could even force resynchronization or turn off the faulty node if the problem persisted. Using a sequence of application extensions allows for precisely diagnosing occurring problems, and in some cases (e.g., time desynchronization, sensor decalibration, software bugs), eliminates the need for on-the-site inspections.

The proposed approach explicitly deals with the unpredictability of problems. By extending the application in an arbitrary place in the code with virtually arbitrary diagnostic functionality, we can pinpoint and react to the problems that were unanticipated by the application developers, even if the application is extremely complex. We are also able to temporarily coordinate a group of nodes to perform collaborative debugging. The examples include the aforementioned clock synchronization problem, or local isolation of nodes reporting erroneous sensor readings.

Furthermore, our solution addresses the scale and resource limitations of large WSNs, and minimizes intrusiveness. First, the diagnosis is performed in reaction to a problem, only for the time necessary to determine the cause, and only on the nodes directly affected by the problem. Second, diagnostic extensions can perform local inspection and aggregation of results. In the example from Fig. 1, the nodes in the perimeter do not individually report their neighbor tables to the base station. Instead, they determine the connectivity on their own, collaboratively aggregate the value locally, and report a single YES/NO answer. This way resource-critical processing and communication are limited to the vicinity of the problem rather than stretched over the whole network. In extreme cases, the extension can alter the application (e.g., reconfigure it) to fix the problem, without even resorting to the base station.

None of the aforementioned diagnostic infrastructures for WSNs [2,8,11,13] addresses all these issues. The issues are, however, crucial for large-scale systems.

The idea of monitoring an application's health using dynamic extensions was explored previously in the context of wired networks [16]. That work proposed on-demand delegation of diagnostic code to a monitored server, to minimize the volume of data transferred between the server and a monitoring host. However, because of a completely different setting, our approach is distinctly novel. Not only does it introduce internode collaboration as a technique for problem diagnosis in distributed systems, but also encourages employing the characteristic properties of WSNs for the benefit of debugging. In the example from Fig. 1, the spatial dependencies between nodes enable the nodes in the perimeter of the black region to collaboratively determine whether that region is connected to the network. Likewise, the broadcast nature of the wireless medium allows a node to snoop all packets sent by its neighbors. Furthermore, because of the hardware constraints of sensor nodes, the implementation of our infrastructure must face challenges not encountered in traditional systems, as explained in the next section.

4   Infrastructure Support

An infrastructure for problem diagnosis according to the presented approach requires resolving a number of issues: delivering a diagnostic extension to selected nodes; ensuring authenticity and integrity of the delivered code; extending the application running on a node in a nonintrusive way; or securely patching a bug in the application. Most of these issues already have stable solutions. In particular, for delivering an extension to the selected nodes, one of the existing routing and multicasting protocols [17] or mobile-agent middleware [18] can be employed. Code authenticity and integrity can be ensured at a minimal energy cost incurred by the nodes using a finite-stream signing technique, adopted for WSNs [19]. To patch a bug in the application, a secure high-throughput network reprogramming protocol [20,19] can be employed.

Here, we concentrate on a core aspect — the architectural support for extending an application running on a node. This is extremely challenging because, despite the hardware constraints of a sensor node and application complexity, the extension installation should be as nonintrusive as possible while giving the extension full access to the application state in the face of concurrency. In particular, it is infeasible for the hardware and the OS to provide traditional mechanisms, like `ptrace`. We also do not want to modify the compilers, which is not always possible, or redesign the operating system, which would preclude using an already immense existing contributed code-base.

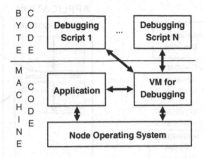

Fig. 2. Architecture overview

Due to space limitations, in the rest of the paper, we assume a network of homogeneous nodes, each running the same application. Although we use TinyOS as our base platform, most of the solutions presented in this section are generally applicable.

4.1 Extending the Application

To extend the application at runtime, the node OS should support dynamic code linking, which is usually not the case due to hardware constraints. A node's MCU executes a single static binary combining the application and OS. The binary is executed directly from EEPROM to minimize the amount of expensive RAM and to ensure preservation across node reboots. Reprogramming EEPROM, however, drains a lot of power and is difficult, time consuming, and error-prone, especially if the application is constantly running.[2] Moreover, the number of times EEPROM can be written is always limited. Finally, many architectures do not support position-independent code, so dynamically installing an extension could require patching addresses in the whole binary.

For these reasons, we propose extensions based on interpreted bytecode scripts. Apart from the application and OS, a node's software contains a virtual machine (VM) for interpreting diagnostic scripts (see Fig. 2). Since the scripts are executed from RAM, installation and uninstallation is easy and inexpensive. By using an application-specific VM with a custom instruction set the execution overhead can be minimized [21,22]. Moreover, a bytecode script is usually much smaller than its machine-code counterpart (tens *versus* hundreds of bytes), which allows for delivery through the network and minimizes the RAM footprint to a few hundred bytes [21,22,23].

A script is installed as a breakpoint handler in the application. When the execution of the application reaches a breakpoint the control is transfered to the VM which interprets the script associated with this breakpoint (see Fig. 4). Interrupting the application on a breakpoint can be implemented either in the hardware or in the software. The hardware implementation involves no changes to the application and no runtime overhead, but not every architecture supports it. The software solution requires modifying the application binary (at runtime or compile time) and imposes some runtime overhead. Preliminary experiments for our implementation, however, indicate that the overhead per compile-time breakpoint is very small (6-7 MCU cycles). Thus, we believe this is a viable solution for the architectures lacking the necessary hardware support.

---

[2] Since the OS and the application are merged into one address space, during erroneous reprogramming we may damage not only the application, but also the internal OS code.

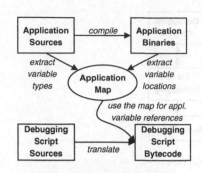

Fig. 3. Application mapping

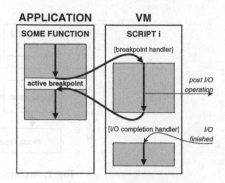

Fig. 4. Invoking breakpoints

4.2 Accessing the Application State

The diagnostic scripts must be able to access the state of the application. For instance, a script for checking the connectivity of a network region may need to examine neighbor tables maintained by the application. In general, internally the application uses a number of variables declared by the programmer. To read (and write) those variables, the script needs their precise mapping to the node's RAM. Problems arise because the application is a machine-code binary, generated by an arbitrary (third-party) compiler with arbitrary optimizations.

To cope with this, we concentrate on mapping globally accessible variables (i.e., heap and static variables). Embedded applications store the critical state (e.g., message buffers, caches, neighbor tables, timer counters) statically, and many OSes, including TinyOS, enforce this approach. Heap-allocated variables can also be mapped as the heap is declared statically. Only stack variables are not mapped, because this would require the knowledge of the compiler's internals and would preclude applying many crucial code optimizations.

The application map is generated at compile time (see Fig. 3). From the application sources we extract variable type information (e.g., the types of a structure's fields). From the symbol table in the application binary we extract variable locations in RAM. The combined information constitutes the application map. The map is used to replace symbolic references to application variables in a diagnostic script with memory access instructions. This is performed at the base station, during the translation of the diagnostic script from the scripting language to the bytecode to be interpreted by the VM. Thus, an application map, once generated, is utilized for translating many scripts, as long as the application is not modified. Moreover, such an approach also enables accessing internal OS structures, offering superb debugging capabilities.

Our implementation of generating application maps is nonintrusive with respect to the original compiler of the application. It bears some similarities to Marionette [13], which uses maps for remote procedure calls.

To cope with possible future memory protection of a sensor node and a separation between the application and OS address spaces, the VM can be a library linked to the application. This way it can freely access the application's address space.

4.3 Handling Concurrency (TinyOS-specific)

An application for TinyOS is composed of tasks triggered by events.[3] Once started, a task runs to completion, that is, it cannot be preempted by any other task. This concurrency model simplifies synchronization and allows for using the same stack for all tasks. It also enables a breakpoint handler to be run in the context of the task that triggered it (see Fig. 4). Upon completion, the handler returns the control to the application, which resumes from the instruction following the breakpoint — again, in the context of the same task. Such a synchronous, nonpreemptable handler execution inherently ensures mutual exclusion of operations on application variables.

A breakpoint handler, however, can perform long I/O operations (e.g., sending a message or logging a value to the flash memory). To avoid blocking the whole application, I/O operations are asynchronous. A breakpoint handler posts an I/O operation (see Fig. 4), which is executed later, in the context of a different task. When the operation has finished, the I/O completion handler of the diagnostic script is invoked. This way the diagnostic script can perform more processing outside the breakpoint handler. Such an asynchronous I/O model with completion handlers is the same as the one used by TinyOS, which reduces the learning curve of our solution and simplifies implementation.

The concept of handlers is further generalized to other event handlers (e.g., for receiving a message sent by a diagnostic script on a neighbor node). Therefore, in fact, a diagnostic script consists not only of the breakpoint handlers, but also of a number of other handlers, each corresponding to a well-specified event.

5 Work Status

Considering the research results on mobile VM scripts for WSNs and our preliminary experience with an ongoing implementation, we suspect that the overhead and memory footprint incurred due to the interpreted scripts will be outweighed by the benefits provided by the dynamic diagnostic extensions when deploying our solution on a large scale. It is also possible to further reduce the delivery overhead by caching the scripts in the flash memory of the nodes, and then only sending activation/deactivation commands. In the future, such an approach would allow us to investigate to what extent the network can autonomously help with diagnosing occurring problems. The latter goal constitutes a particularly exciting research agenda, but requires substantially more insight into the nature of such problems.

Our project received much attention from a Dutch consortium planning to build and deploy a very large WSN (see http://www.devlab.nl/myrianed/). The established collaboration will enable us to validate whether nonintrusive interpreted dynamic extensions and collaborative problem diagnosis are the tools for industry to maintain future ubiquitous embedded systems.

Acknowledgments

Discussions with S. Sivasubramanian, M. Szymaniak, E. Ogston, and H. Bos helped to refine this paper. The authors would also like to thank anonymous reviewers for their perceptive comments.

---

[3] Concurrency handling may need to be implemented differently if some other OS is used.

References

1. Iwanicki, K., van Steen, M.: Sensor network bugs under the magnifying glass. Technical Report IR-CS-033, Vrije Universiteit, Amsterdam, the Netherlands (2006), Available at, http://www.few.vu.nl/~iwanicki/

2. Tolle, G., Culler, D.: Design of an application-cooperative management system for wireless sensor networks. In: Proc. 2nd EWSN, Istanbul, Turkey (2005)

3. Langendoen, K., Baggio, A., Visser, O.: Murphy loves potatoes: Experiences from a pilot sensor network deployment in precision agriculture. In: Proc. 20th IPDPS, Rhodes Island, Greece (2006)

4. Culler, D., Hong, W., (eds.): Wireless Sensor Networks. Communications of the ACM 47 (2004)

5. Madden, S.R., Franklin, M.J., Hellerstein, J.M., Hong, W.: TinyDB: An acquisitional query processing system for sensor networks. ACM Transactions on Database Systems (TODS) 30(1), 122–173 (2005)

6. Hellerstein, J.M., Hong, W., Madden, S., Stanek, K.: Beyond average: Toward sophisticated sensing with queries. In: Proc. 2nd IPSN, Palo Alto, CA, USA (2003)

7. Rahimi, M.H., Baer, R., Iroezi, O.I., García, J.C., Warrior, J., Estrin, D., Srivastava, M.B.: Cyclop: In situ image sensing and interpretation in wireless sensor networks. In: Proc. 3rd SenSys, San Diego, CA, USA, pp. 192–204 (2005)

8. Ramanathan, N., Kohler, E., Estrin, D.: Towards a debugging system for sensor networks. International Journal of Network Management 15(4), 223–234 (2005)

9. Krishnamurthy, L., Adler, R., Buonadonna, P., Chhabra, J., Flanigan, M., Kushalnagar, N., Nachman, L., Yarvis, M.: Design and deployment of industrial sensor networks: Experiences from a semiconductor plant and the north sea. In: Proc. 3rd SenSys, San Diego, CA, USA, pp. 64–75 (2005)

10. Glaser, S.D.: Some real-world applications of wireless sensor networks. In: Proc. 11th SPIE Symposium on Smart Structures and Materials, San Diego, CA, USA (2004)

11. Dyer, M., Beutel, J., Kalt, T., Oehen, P., Thiele, L., Martin, K., Blum, P.: Deployment support network. In: Proc. the 4th EWSN, Delft, The Netherlands, pp. 195–211 (2007)

12. Cristian, F.: Probabilistic clock synchronization. Distributed Computing 3, 146–158 (1989)

13. Whitehouse, K., Tolle, G., Taneja, J., Sharp, C., Kim, S., Jeong, J., Hui, J., Dutta, P., Culler, D.: Marionette: Using RPC for interactive development and debugging of wireless embedded networks. In: Proc. 5th IPSN, Nashville, TN, USA, pp. 416–423 (2006)

14. Ganesan, D., Estrin, D., Heidemann, J.: DIMENSIONS: Why do we need a new data handling architecture for sensor networks? ACM SIGCOMM Computer Communication Review 33(1), 143–148 (2003)

15. Iwanicki, K., van Steen, M.: The PL-Gossip algorithm. Technical Report IR-CS-034, Vrije Universiteit, Amsterdam, the Netherlands (2007), Available at: http://www.few.vu.nl/~iwanicki/

16. Goldszmidt, G., Yemini, Y.: Distributed management by delegation. In: Proc. 15th ICDCS, Vancouver, Canada, pp. 333–340 (1995)

17. Akkaya, K., Younis, M.: A survey on routing protocols for wireless sensor networks. Ad Hoc Networks 3(3), 325–349 (2005)

18. Fok, C.L., Roman, G.C., Lu, C.: Rapid development and flexible deployment of adaptive wireless sensor network applications. In: Proc. 25th ICDCS, Columbus, OH, USA, pp. 653–662 (2005)

19. Dutta, P., Hui, J., Chu, D., Culler, D.: Securing the Deluge network programming system. In: Proc. 5th IPSN, Nashville, TN, USA (2006)

20. Hui, J.W., Culler, D.: The dynamic behavior of a data dissemination protocol for network programming at scale. In: Proc. 2nd SenSys, Baltimore, MD, USA, pp. 81–94 (2004)
21. Levis, P., Gay, D., Culler, D.: Active sensor networks. In: Proc. 2nd NSDI, Boston, MA, USA (2005)
22. Koshy, J., Pandey, R.: VM⋆: Synthesizing scalable runtime environments for sensor networks. In: Proc. 3rd SenSys, San Diego, CA, USA, pp. 243–254 (2005)
23. Levis, P., Culler, D.: Maté: A tiny virtual machine for sensor networks. In: Proc. 10th ASPLOS, San Jose, CA, USA, pp. 85–95 (2002)

Processing Location-Dependent Queries with Location Granules*

Sergio Ilarri, Eduardo Mena, and Carlos Bobed

IIS Department, University of Zaragoza
Maria de Luna 1, 50018, Zaragoza, Spain
{silarri,emena,cbobed}@unizar.es

Abstract. Existing approaches for the processing of location-dependent queries implicitly assume location data expressed at maximum precision (e.g., GPS). However, there exist applications where managing location data with this precision is not required and may even be inconvenient. Thus, for example, a user could just be interested in the city that a train is currently traversing. In this situation, retrieving the precise geographic coordinates would lead to an unnecessary overhead in terms of the data communications needed to track the continuously changing current location. Moreover, the user would need some mechanism to translate the coordinates into the corresponding city.

In this paper, we stress the importance of a query processing approach that adapts itself to the needs of the user and the level of resolution required: the user should be able to express queries and retrieve results according to his/her own terminology for locations (GPS, cities, states, provinces, or any other geographical area). We have implemented a prototype to test the functionality and the interest of location granules and show the independence between the query processing approach and different ways of presenting the answers.

1 Introduction

Nowadays, there is a great interest in mobile computing, motivated by an ever-increasing use of mobile devices, that aims at providing data access anywhere and at anytime. Recently, there has been an intensive research effort in *location-based services* (LBS) [1]. These services provide value-added by considering the locations of the mobile users in order to offer more customized information.

How to efficiently process a continuous location-dependent query (a query whose answer depends on the locations of objects and is refreshed automatically) is one of the greatest challenges in location-based services. Thus, these queries require a continuous monitoring of the locations of relevant moving objects in order to keep the answer up-to-date efficiently. For example, a user with a PDA (*Personal Digital Assistant*) may want to locate available taxi cabs that are near him/her while he/she is walking home on a rainy day. The answer to this query must be continuously refreshed because it can change immediately, as the

---

* This work is supported by the CICYT project TIN2004-07999-C02-02.

R. Meersman, Z. Tari, P. Herrero et al. (Eds.): OTM 2007 Ws, Part II, LNCS 4806, pp. 856–865, 2007.

user and the taxi cabs move independently. Moreover, even if the set of taxis satisfying the query condition does not change, their locations and distances to the user do change continuously, and therefore the answer to the query must be updated with the new location data (e.g., to update the locations of the taxis on a map that is shown to the user).

Existing works on location-dependent query processing implicitly assume GPS locations for the objects in a scenario [2,3]. However, some applications do not require location data at GPS resolution, and a coarser representation may be more appropriate for them. For example, a train tracking application could just need to consider in which city the train currently is, and its exact coordinates may be irrelevant. For such applications, we consider useful to define the concept of *location granule* as a set of physical locations. In our previous example, every city would be a location granule. Other examples of location granules could be: freeways, buildings, offices in a building, etc. The idea is that the user should be able to express queries and retrieve results according to the idea of "location" that he/she requires, whether he/she needs to talk in terms of GPS locations or locations at a different *resolution*. The use of location granules can have an impact on:

- *The presentation of results.* The user may want to retrieve the geographic coordinates of the objects in the answer to the query. Alternatively, he/she may prefer a different location granularity (e.g., the cities where the objects are) because it is more appropriate for his/her context. Notice that, in some cases, he/she may even be unable to interpret an answer in terms of GPS locations; for example, if he/she only knows about "cities", then returning GPS locations is useless because he/she does not know how to obtain which city corresponds with a GPS location. Independently of the required location granularity, the answer can be presented to the user using different mechanisms (e.g., different types of graphical, sound-based, textual, etc., representations).
- *The semantics of the queries.* Location-dependent constraints can be interpreted in terms of location granules. As an example, the user may be interested in the cars that are near the city where another car is currently present. Notice that the user may have no idea about geographic coordinates or the boundaries of the cities.
- *The performance of the query processing.* In a general case, the continuous query processing will demand less resources when coarse location granules are used. For example, if a mobile user is interested in the GPS locations of certain cars, the location data of the relevant cars should be communicated very frequently to the mobile device (the GPS locations of the cars are continuously changing). However, less wireless communications are required if he/she is just interested in the part of the city (e.g., north, south, west, east, center) where the cars are.

In this paper, we prove the utility of dealing with locations expressed at different levels of granularity when processing location-dependent queries. In Sect. 2

we define the architecture and concepts that will be used along the paper. In Sect. 3, we justify the importance of location granules and how they can be used to express queries with the semantics that is useful for the user. In Sect. 4, we present some related works. Finally, conclusions and future work appear in Sect. 5.

2 Location Granules and Architecture

We focus on the processing of location-dependent queries with location granules, where we consider the following three datatypes: *Object*, *Location Granule*, and *Location Granule Mapping*.

Definition 1. *Datatype Object*
We call object to any moving entity (e.g., a person or a car) of interest. A value of type *Object* (in the following, denoted by O) is defined by a tuple with the structure <**id, loc, class**>, where *id* is the identifier of a moving object, *loc* is the GPS location of the object, and *class* indicates the type of object (e.g., *vehicle*); besides, there may be other attributes specific to every object class (e.g., an object of class *vehicle* may have an attribute *maxSpeed*). A **showObject** operation is also defined for an object, which represents the object (by using graphics, sounds, text, etc.).

Definition 2. *Datatype Location Granule*
A location granule is a geographical area (not necessarily contiguous). A value of type *Location Granule* (in the following, denoted by G) is defined by a tuple <**id, Fs**>, where *id* is the granule identifier and *Fs* is a set of bidimensional figures $Fs = \{F1, F2, ..., Fn\}$ (e.g., polygons, circles, etc.). The following operations are defined for location granules:

- **inGranule: G x GPS → boolean**, with GPS = set of possible GPS locations
  $inGranule(g, loc) = true \Leftrightarrow \exists\, Fi \in g.Fs \mid contains(Fi, loc)$
  $contains(Fi, loc) = true \Leftrightarrow loc$ is within Fi
- **distanceCentroidGranule: G x GPS → real**
  $distanceCentroidGranule(g, loc) = min\{distance(loc, centroid(Fi)), \forall\, Fi \in g.Fs\}$
- **distanceLimitsGranule: G x GPS → real**
  $distanceLimitsGranule(g, loc) = min\{distance(loc, Fi), \forall\, Fi \in g.Fs\}$

Besides, a **showGranule** operation is defined for a granule, which shows the location granule to the user by using graphics, sounds, text, etc.

Definition 3. *Datatype Location Granule Mapping*
A location granule mapping is a data structure that, given a GPS location, allows to obtain one or more location granules containing that GPS location. A value of type *Location Granule Mapping* (in the following, denoted by M) is defined by a tuple with the structure <**id, granules**> where *id* is the granule mapping identifier and *granules* is the set of location granules a certain GPS location can be translated to using that mapping. The following operations are defined for location granule mappings:

- **getGranules: M x GPS → {G}**
  $getGranules(m, loc) = \{g \mid g \in m.granules \land inGranule(g, loc)\}$
- **getNearestGranule: M x GPS x D → G**
  with $D = \{distanceCentroidGranule, distanceLimitsGranule\}$ (see Def. 2)
  $getNearestGranule(m, loc, d) = g \mid g \in m.granules \land inGranule(g, loc) \land$
  $\neg \exists\ g' \in m.granules \mid inGranule(g', loc) \land d(g', loc) < d(g, loc)$
- **getGranulesOfObject: M x O → {G}**
  $getGranulesOfObject(m, o) = getGranules(m, o.loc)$
- **getNearestGranuleOfObject: M x O x D → G**
  $getNearestGranuleOfObject(m, o, d) = getNearestGranule(m, o.loc, d)$

The basic architecture that we propose for the processing of continuous location-dependent queries with location granules is shown in Fig. 1. A *Mobile User* launches queries using his/her PDA, which are processed by a *Query Processor* executing on a *Server*. The user can reference his/her own location granule mappings (*User Mappings*) in the queries (User Mappings in use by current queries are stored on a *Mapping Cache* by the Server) or use predefined mappings (*Server Mappings*). The Query Processor interacts with a *Location Server*, which handles location data about moving objects and is able to answer standard SQL-like queries about them (e.g., it could be a moving objects database [4]). Finally, a *Query Table* is used by the Query Processor to store the active queries: changes in the answers to such queries must be communicated to the corresponding mobile devices.

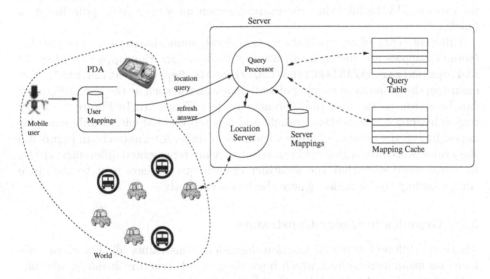

Fig. 1. General architecture for location query processing using location granules

3 Use of Location Granules

In this section, we will use an SQL-like syntax to express the queries; although we could adopt an existing query language (e.g., *SQL/MM Spatial* [5] or FTL [4]), this syntax allows us to emphasize the use of location granules and state our queries concisely.

Specifications of location granules can appear in the SELECT and/or in the WHERE clause of a query, depending on whether the location granules must be used for visualization of results and/or processing of constraints, respectively. A GPS location will be considered for an object if a location granule is not specified instead. In the following, we first explain how location granules allow the user to specify the way the results must be presented. Then, we show that location granules can be also used to specify queries with the required semantics. Of course, both usages can appear simultaneously in the same query. For simplicity, we will use *getNearestGranuleOfObject* (instead of the more general *getGranulesOfObject*, see Sect. 2), which for the sake of brevity will be denoted by *gr* in the queries, and omit the parameter *d* that indicates the distance function that must be used.

3.1   Granules in Query Projections

The user can decide to retrieve location granules as part of the query projections. For example, "SELECT gr(bus25, region) FROM Bus" retrieves the location granule of type *region* corresponding to the current location of *bus25*. When a location granule is retrieved as part of a query answer, it is automatically shown to the user by executing the implemented *showGranule* operation (see Sect. 2). Thus, in the previous sample query, the retrieved granule may be represented, for example, by highlighting the current region on a map (e.g., painting it in black).

Different types of representations (graphical, sound-based, etc.) are possible (some examples are shown in an interactive demonstration applet at http:// sid.cps.unizar.es/ANTARCTICA/LDQP/granulesRepresentation.html). For example, the granules of moving objects in a scenario such as the one in Figure 2 can be shown to the user in different ways (see Figure 3). In Figure 3.a every region in France is a location granule and it is represented with a different color depending on the number of buses within the granule. Alternatively, in Figure 3.b the granules are also regions of France but they are represented differently: circles in a bus route such that the identifier of each bus appears next to the circle corresponding to the region where the bus is currently.

3.2   Granules in Query Constraints

There are different types of location-dependent constraints [6]. For clarity, we focus on *inside* constraints, which have the general structure *inside(r, obj, target)*, where *r* is called the *relevant radius*, *obj* is the *reference object* of the constraint, and *target* is called the *target class*. Such a constraint would retrieve the objects of the class *target* which are within distance *r* of the object *obj*. Moreover, we can specify location granules in an *inside* constraint instead of *obj*, *target*, or both. We describe in the following these three cases (an interactive demonstration of how these cases are processed is available as a Java applet at http://sid.cps.unizar.es/ANTARCTICA/LDQP/granules.html).

Fig. 2. Map of France with some buses

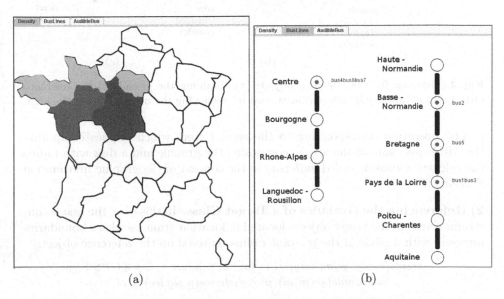

(a)   (b)

Fig. 3. Different presentation mechanisms: (a) painting the regions with a different color depending on the number of buses within, and (b) highlighting the current region on the bus routes

1) Referencing the Granule of a Reference Object. In this case, the *inside* constraint retrieves the *target objects* (objects of the target class) whose distance from the granule of the reference object is not greater than the relevant radius:

$$inside(r,\ gr(m,\ obj),\ target) = \{oi \mid oi \in target \land \exists\, p \in GPS \mid$$
$$inGranule(gr(m, obj), p) \land distance(p, oi.loc) \leq r\}$$

For example, the query "SELECT Car.id FROM Car WHERE inside(130 miles, gr(province, car38), Car)" retrieves the identifiers of the cars within 130 miles of the province where *car38* (the reference object of the constraint) is located.

To obtain the objects that satisfy this type of *inside* constraint, the following operations are performed (see Fig. 4 for an example, where we consider the map of Spain divided in provinces): 1) the granule of the reference object (in the example, the province of *car38*) is obtained (see Fig. 4.a); 2) the area/s corresponding to such a granule is/are enlarged in order to obtain the *relevant area/s* according to the radius relevant to the query (in the example, 130 miles, see Fig. 4.b); and 3) the target objects within that area are retrieved (see Fig. 4.c).

(a)   (b)   (c)

Fig. 4. Granules for the reference object: (a) obtaining the granule of the reference object, (b) obtaining the relevant area, and (c) retrieving the objects within the area

The operation corresponding to the second step, which implies computing the Minkowski sum of the area/s composing the granule and a disk with radius the relevant radius, is called *buffering* in the context of Geographic Information Systems [7].

2) Referencing the Granules of a Target Class. In this case, the *inside* constraint retrieves the target objects located in location granules whose boundaries intersect with a circle of the relevant radius centered on the reference object:

$$inside(r,\ obj,\ gr(m,\ target)) = \{oi \mid oi \in target \land \exists\, p \in GPS \mid$$
$$inGranule(gr(m, oi), p) \land distance(p, obj.loc) \leq r\}$$

For example, the query "SELECT Car.id FROM Car WHERE inside(130 miles, car38, gr(province, Car))" retrieves the cars located in provinces whose boundaries are (totally or partially) within 130 miles of *car38*.

To obtain the objects that satisfy this type of *inside* constraint, the following operations are performed: 1) a circular area with the relevant radius (in our example, 130 miles) centered on the current GPS location of the reference object (in the example, *car38*) is computed (see Fig. 5.a); 2) the granules intersected by such an area are determined (see Fig. 5.b); and 3) the target objects within any of those granules are obtained (see Fig. 5.c).

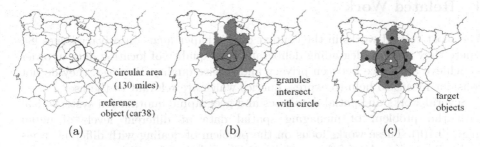

Fig. 5. Granules for the target class: (a) obtaining the circular area, (b) obtaining the granules that intersect, and (c) retrieving the objects within the granules

3) Referencing the Granules of the Reference Object and the Target Class. In this case, the *inside* constraint retrieves the target objects located in location granules whose boundaries are within the relevant radius from the granule of the reference object:

$$inside(r,\ gr(m1,\ obj),\ gr(m2,\ target)) = \{oi \mid oi \in target \wedge \ \exists\, p1 \in GPS \ \wedge$$
$$\exists\, p2 \in GPS \mid \ inGranule(gr(m2, oi), p1) \wedge inGranule(gr(m1, obj), p2) \wedge$$
$$distance(p1, p2) \leq r\}$$

For example, the query "SELECT Car.id FROM Car WHERE inside(130 miles, gr(province, car38), gr(province, Car))" retrieves the cars located in provinces whose borders are (total or partially) within 130 miles of the province where *car38* is located.

In this case, to obtain an answer, the granule of the reference object is obtained and the area/s corresponding to such a granule is/are enlarged in order to obtain the *relevant area/s* according to the radius relevant to the query (as in case 1, see Fig. 6.a). Then, the set of granules intersected by such an area are determined (as in case 2, see Fig. 6.b). Finally, the target objects within those granules are retrieved (see Fig. 6.c).

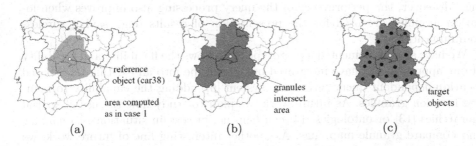

Fig. 6. Granules for the reference object and the target class: (a) obtaining the relevant area, (b) obtaining the granules that intersect it, and (c) retrieving the objects inside

4   Related Work

The most related works in the context of this paper focus on privacy issues [8], where they propose degrading deliberately the quality of location data in order to achieve a balance between privacy and service quality. However, they do not consider means to help to express queries which are relevant to the user.

Several works on spatial databases and geographic information systems deal with the problem of managing spatial data at different levels of detail (e.g., [9,10]). These works focus on the problem of dealing with different levels of detail/specification of the spatial entities, and therefore they do not consider the use of locations at different granularities to enhance the expressiveness of queries.

It is also worth mentioning the so-called *location granularity mismatch*, which occurs when the granularity of the locations stored on a database and the granularity of the locations specified in a location-dependent query are different [11]. This problem is orthogonal to our work, as we focus on query processing and rely on granule mappings to perform the required translations.

Finally, we would like to mention the work in [12], where the *Nimbus* framework is proposed to provide locations with the appropriate granularity. As opposed to ours, it deals with location granularity at the database level; therefore, for example, the user is not able to define his/her own granule mappings.

5   Conclusions and Future Work

In this paper, we have proposed an approach to consider the appropriate location granularity for the processing of location-dependent queries. The use of location granules greatly increases the expressiveness of location-dependent queries and the range of applications that can benefit from the query processing. On the one hand, it allows the user to specify the queries with the needed semantics (i.e., he/she can talk about locations "on his/her own terms"). On the other hand, the results can be represented in a way that is useful for the user; different representation mechanisms can be applied independently of the query processing. Moreover, the performance of the query processing also improves when location granules are used for the presentation of results (e.g., saving wireless communications).

We have also implemented a prototype that shows the flexibility and feasibility of our approach. We have integrated our prototype into an existing GPS-based location-dependent query processing system [6], adding the capability to manage location granules. As future work, we plan to study how the use of spatial hierarchies [13] or ontologies [14] can help us, in certain situations, to manage and compare granule mappings. As another interesting line of future work, we think that our system could also be easily extended to deal with imprecise locations [15] (e.g., cells in a cellular network), as these could be considered just as a special kind of location granules.

References

1. Schiller, J., Voisard, A. (eds.): Location-Based Services. Morgan Kaufmann, San Francisco (2004)
2. Cai, Y., Hua, K.A., Cao, G., Xu, T.: Real-time processing of range-monitoring queries in heterogeneous mobile databases. IEEE TMC 5(7), 931–942 (2006)
3. Gedik, B., Liu, L.: MobiEyes: A distributed location monitoring service using moving location queries. IEEE TMC 5(10), 1384–1402 (2006)
4. Sistla, A.P., Wolfson, O., Chamberlain, S., Dao, S.: Modeling and querying moving objects. In: 13th Intl. Conference on Data Engineering (ICDE 1997), Birmingham, United Kingdom, pp. 422–432. IEEE Computer Society Press, Los Alamitos (1997)
5. Stolze, K.: SQL/MM spatial - the standard to manage spatial data in a relational database system. In: Datenbanksysteme für Business, Technologie und Web (BTW 2003), February 2003. Lecture Notes in Informatics (LNI), vol. 26, pp. 247–264 (2003)
6. Ilarri, S., Mena, E., Illarramendi, A.: Location-dependent queries in mobile contexts: Distributed processing using mobile agents. IEEE TMC 5(8), 1029–1043 (2006)
7. Kreveld, M.: Computational geometry: Its objectives and relation to GIS. Nederlandse Commissie voor Geodesie (NCG), 1–8 (2006)
8. Duckham, M., Kulik, L.: A formal model of obfuscation and negotiation for location privacy. In: Gellersen, H.-W., Want, R., Schmidt, A. (eds.) PERVASIVE 2005. LNCS, vol. 3468, pp. 152–170. Springer, Heidelberg (2005)
9. Fonseca, F., Egenhofer, M., Davis, C., Câmara, G.: Semantic granularity in ontology-driven geographic information systems. Annals of Mathematics and Artificial Intelligence 36(1-2), 121–151 (2002)
10. Camossi, E., Bertolotto, M., Bertino, E., Guerrini, G.: Issues on modeling spatial granularities. In: Kuhn, W., Worboys, M.F., Timpf, S. (eds.) COSIT 2003. LNCS, vol. 2825, Springer, Heidelberg (2003)
11. Seydim, A.Y., Dunham, M.H., Kumar, V.: Location dependent query processing. In: 2nd ACM Intl. Workshop on Data engineering for Wireless and Mobile Access (MobiDe 2001), pp. 47–53. ACM Press, New York (2001)
12. Roth, J.: The role of semantic locations for mobile information access. In: 35th Annual GI Conference, Bonn, Germany, September 2005. Lecture Notes in Informatics (LNI), vol. 2, pp. 538–542 (2005)
13. Malinowski, E., Zimányi, E.: Spatial hierarchies and topological relationships in the spatial MultiDimER model. In: Jackson, M., Nelson, D., Stirk, S. (eds.) Database: Enterprise, Skills and Innovation. LNCS, vol. 3567, pp. 17–28. Springer, Heidelberg (2005)
14. Staab, S., Studer, R. (eds.): Handbook on Ontologies. Intl. Handbooks on Information Systems. Springer, Heidelberg (2004)
15. Trajcevski, G., Wolfson, O., Hinrichs, K., Chamberlain, S.: Managing uncertainty in moving objects databases. ACM TODS 29(3), 463–507 (2004)

A Model Checking-Based Approach for Location Query Processing in Pervasive Computing Environments

Christian Hoareau[1,2] and Ichiro Satoh[1]

[1] National Institute of Informatics
2-1-2 Hitotsubashi, Chiyoda-ku, Tokyo 101-8430, Japan
[2] The Graduate University for Advanced Studies
Shonan Village, Hayama, Kanagawa 240-0193, Japan
{hoareau,ichiro}@nii.ac.jp

Abstract. We present a new approach to handle location query processing in pervasive computing environments. We extend the commonly used hierarchical model of space to a semantic model for hybrid logics, i.e. a Kripke structure, and thus map location query processing into a model checking framework. Our approach is built on theoretical foundations that show the soundness of our query framework, explores the connection between location modelling and location query processing, and provides a hybrid logic-based query language that enables efficient search over a decentralised space repository. A prototype implementation is presented and will be discussed.

1 Introduction

Pervasive computing devices, such as sensors and embedded processors, enable emerging applications to handle locations of miscellaneous entities, and to adapt accordingly. Current research has been focused on the design and implementation of application-specific location-aware services, like smart rooms and navigation systems. Besides, these are often dedicated to particular tracking systems that capture location information about moving users and objets. As a result, the task of data management for location information has attracted little attention so far. The underlying connection between location modelling and location query processing in pervasive systems has yet to be explored.

1.1 Symbolic Location Models

Many location-aware applications [4,10] reason about *places*. Places are labels that identify either geometric volumes, such as buildings and rooms, or entities contained in geometric volumes, such as people and portable devices. The set of all interrelations between places thus defines the overall location information. This information can be represented by a hierarchical space tree (Fig. 1), whose nodes represent places, and edges represent containment relations between

R. Meersman, Z. Tari, P. Herrero et al. (Eds.): OTM 2007 Ws, Part II, LNCS 4806, pp. 866–875, 2007.

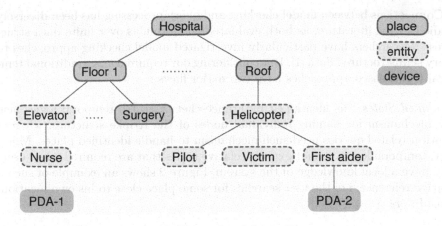

Fig. 1. Hierarchical Space Tree

places. Containing (resp. contained) places are called *parents* (resp. *children*). Besides representing static places, the hierarchical space tree can also represent mobile objects (e.g. "Elevator", "Helicopter"), people (e.g. "Nurse", "First Aider") and computing devices (e.g. "PDA-1", "PDA-2").

1.2 Location Query Processing

As presented by Becker et al. [3], location-related queries encompass simple position queries (e.g. "where is the surgery ?"), nearest neighbour queries (e.g. "where is the closest computer to the elevator ?"), navigation queries (e.g. "how to reach the roof ?"), and range queries (e.g. "who are the victims located at the first floor ?").

In this paper, we aim to provide a query language for location-based services. We believe that the theoretical foundations of such a language should be investigated. Indeed, modern database systems and standard query languages such as SQL rely on the first-order logic, that is efficiently implemented using relational algebra [9]. Our approach is based on the hierarchical model of space, that we extend to a semantic model for modal logics. Location query processing is then regarded as dynamic graph search over decentralised data structures, and tackled using model checking techniques.

2 Query Processing Via Model Checking

Model checking [5] is a well founded approach that has been successfully applied for both hardware and software verification. It aims to algorithmically check if a given system S satisfies a specification P. The inputs to a model checking algorithm are the specification P written in a suitable temporal logic formula, and a finite transition graph, called *Kripke structure* [8], that captures all the possible states of S.

Connections between model checking and query processing has been diversely studied in the literature, as both evaluate logic formulas over finite data structures. Researchers have particularly investigated model checking approaches to query semistructured data [1]. But considering our requirements, traditional temporal logics-based approaches have two major flaws :

Unnamed States : as mentioned by Franceschet et al. [7], temporal logics lack any mechanism for naming individual nodes of the Kripke structure, whereas location-related queries need such mechanism to handle identified places. Moreover, temporal logics cannot express relative nodes that are required when users only have a local knowledge of the system. Figure 2 shows an example of such a relative reference, i.e. the user searches for some place close to his own location, named *here*.

Top-to-bottom Query Routing : Because temporal logics define operators for current and future states, location queries could be routed downward the containment hierarchy only. Due to the innate decentralised location information and the many query provenances and targets, our query language should provide both downward and upward operators to parse the space tree.

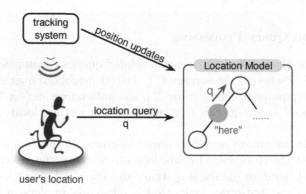

Fig. 2. Upward query routing and relative location

As a consequence, we assume from now on an extension of temporal logics called *Hybrid Logics* [2]. Hybrid logics introduce the concept of nominal. Nominals are propositional variables that are true at exactly one node in the Kripke structure, i.e. if p is a nominal, the formula p holds if and only if the current node is called p. Thus, it is easy to capture the notion of place : a place corresponds to a nominal. Besides, hybrid logics define the following operators :

- access operator $@_p$: it gives random access to the place named p. The formula $@_p\phi$ holds if and only if ϕ holds at the place p.
- downarrow binder $\downarrow x$: it creates a brand new name x and assigns it to the current place. The formula $\downarrow x.\phi$ holds if and only if ϕ holds whenever the current place has been named x.

3   Extending Hierarchical Location Model

In this section, we extend the hierarchical space tree to a Kripke-like structure, that serves as both a semantic and data model.

Definition 1 (Hierarchical Space Graph). *A Hierarchical space graph is a 4-tuple* $G = \{N, R_\downarrow, R_\uparrow, L : N \rightarrow \mathcal{P}(N)\}$ *in which :*
- *Places are represented by the set N of nodes,*
- *Transitions $R_\downarrow \subseteq N \times N$ and $R_\uparrow \subseteq N \times N$ define the containment relation between parents and children,*
- *Places are labelled with their children by the function $L : N \rightarrow \mathcal{P}(N)$ ($\mathcal{P}(N)$ is the power set of N).*

The hierarchical space graph of Fig. 3 extends the space tree depicted in Fig. 1. Transitions $R_\downarrow$ and $R_\uparrow$ represent the reachability between parents and children. Labels are Boolean variables that refer to contained places. For example, the helicopter coming back from a rescue operation may be either on the roof of the hospital or still on its way back. In the former case, the place *roof* would be labelled by its child *helicopter*, whereas in the latter case, it would have no label *helicopter*, as the place *helicopter* would not be its child.

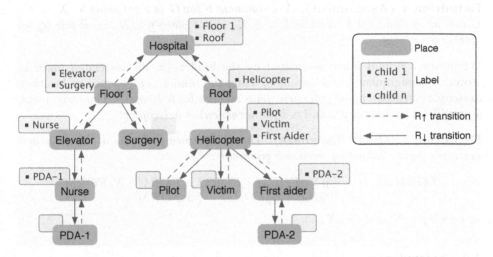

Fig. 3. Hierarchical space graph extending space tree of Fig. 1

Definition 2 (Transition relations). *We define $R_\downarrow$ and $R_\uparrow$ as follows :*
1. *$R_\downarrow$ and $R_\uparrow$ are irreflexive, intransitive and asymmetric*
2. *$\forall p_1, p_2 \in N : (p_1, p_2) \in R_\downarrow \Leftrightarrow (p_2, p_1) \in R_\uparrow$*
3. *$\forall p_1, p_2, p_3 \in S : (p_1, p_2) \in R_\uparrow \wedge (p_3, p_2) \in R_\downarrow \Rightarrow p_1 = p_3$*

Property (1) ensures that a place is neither its own parent nor own child (irreflexivity). Intransitivity and asymmetry are trivial. Property (2) defines the relation between a parent p_1 and its child p_2. Finally, a place cannot have multiple parents, i.e. there is no overlapping area (3).

4 Hybrid Logics-Based Query Language

In the previous section, we defined a semantic model to express hybrid logic-based queries on the location data. "Where is the nurse ?" will consist of exploring the hierarchical space graph and finding the place labelled by *nurse*. The answer would be the *elevator*.

4.1 Syntax

Definition 3 (Language components). *The query language $\mathcal{Q}$ contains :*
- *a countable set $N = \{p_1, p_2, \ldots, p_n\}$ of places*
- *a countable set X of variables $x_1, x_2, \ldots, x_n$ used for bound names*
- *standard logical symbols $\vee$, $\wedge$*
- *spatial modalities $E_\downarrow$ and $E_\uparrow$*
- *hybrid logics operators $\downarrow x$ (binder) and $@_p$ (random access)*

The operator $E_\downarrow$ is the spatial counterpart of the CTL temporal operator EF [6], and $E_\uparrow$ is its backward analogue. They are routing triggers that move queries along the space graph's hierarchy, into both directions.

Definition 4 (Assignment). *A assignment b for G is a mapping $b : X \to N$. Given an assignment b, a variable $x \in X$, and a place $p \in N$, we define b_p^x by setting $b_p^x(x) = p$.*

Whenever a bound name is created by the binder $\downarrow x$, its associated place is stored by assignment b. For example, using the binder $\downarrow emergency$ at location *helicopter* creates a name *emergency*, i.e. an alias for *helicopter*. The assignment b provides a lookup function, i.e. $b_p^x(emergency) = helicopter$.

Definition 5 (Query Language). *The well-formed formulas of our language are given by the following recursive grammar :*

$$FORM ::= true \mid p \mid \neg FORM \mid FORM \wedge FORM \mid E_\uparrow FORM$$
$$\mid E_\downarrow FORM \mid @_p FORM \mid \downarrow x.FORM$$

in which $p \in N$ and $x \in X$.

4.2 Semantics

Hybrid-logics formulas are formally interpreted in terms of hierarchical space graphs. The satisfaction relation $G, b, l \models \varphi$ gives the meaning of formula φ. Its left part, named the *context*, captures the parameters required to evaluate the query. A context encompasses the hierarchical space graph G, the bound names (stored by assignment b) and the current location l where the query is evaluated. We give below its semantics and some examples (referring to Fig.3).

1. $G, b, l \models true$
   This is a standard assumption. True is valid everywhere.

2. $G, b, l \models p$ iff p is true at l
Place p is contained in location l. For example, the relation $G, b, elevator \models nurse$ is satisfied, i.e. the nurse is in the elevator.

3. $G, b, l \models \neg\varphi$ iff $G, b, l \not\models \varphi$
This is the standard definition of $\neg$ operator. For example, the relation $G, b, surgery \models \neg victim$ is satisfied, i.e. the victim is not in the surgery.

4. $G, b, l \models \varphi_1 \wedge \varphi_2$ iff $(G, b, l \models \varphi_1) \wedge (G, b, l \models \varphi_2)$
This is the standard definition of $\wedge$ operator. For example, the relation $G, b, helicopter \models victim \wedge first\_aider$ is satisfied, i.e. both victim and first aider are inside the helicopter.

5. $G, b, l \models \varphi_1 \vee \varphi_2$ iff $(G, b, l \models \varphi_1) \vee (G, b, l \models \varphi_2)$
This is the standard definition of $\vee$ operator. For example, the relation $G, b, helicopter \models nurse \vee first\_aider$ is satisfied, i.e. either the nurse or the first aider is inside the helicopter.

6. $G, b, l \models E_\uparrow \varphi$ iff $\exists l' \in N$, such that l' is reachable from l following $R_\uparrow$ transitions and $(G, b, l' \models \varphi)$

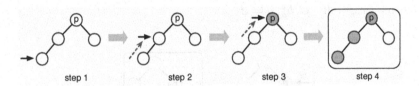

step 1   step 2   step 3   step 4

Fig. 4. Upward operator $E_\uparrow$

Operator $E_\uparrow$ routes the query upward, searching for a place that satisfies formula φ (steps 1 and 2). Place p is found (step 3), so all the nodes between p and the source of the query also satisfy φ (step 4). For example, the relation $G, b, nurse \models E_\uparrow floor\_1$ is satisfied, i.e. the nuse can enter the first floor by leaving the elevator.

7. $G, b, l \models E_\downarrow \varphi$ iff $\exists l' \in N$, such that l' is reachable from l following $R_\downarrow$ transitions and $(G, b, l' \models \varphi)$
Operator $E_\downarrow$ routes the query downward, searching for a place that satisfies formula φ (steps 1 and 2). Place p is found (step 3), so all the nodes between

step 1   step 2   step 3   step 4

Fig. 5. Downward operator $E_\downarrow$

p and the source of the query also satisfy φ (step 4). For example, the relation $G, b, hospital \models E_\downarrow surgery$ is satisfied, i.e. the surgery room is located at the first floor of the hospital.

8. $G, b, l \models @_p\varphi$ iff $\begin{cases} G, b, p \models \varphi \text{ for } p \in N \ (1) \\ G, b, g(p) \models \varphi \text{ for } p \in X \ (2) \end{cases}$

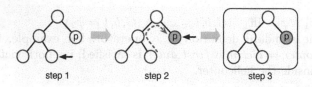

step 1   step 2   step 3

Fig. 6. At operator $@_p$

Operator $@_p$ straightly forwards the query to the specific place p (step 1), where it is evaluated (step 2). Name p is either a place identifier (1) or a bound name (2). In the latter case, a name resolution is first performed. For example, the relation $G, b, hospital \models @_{roof} helicopter$ is satisfied, i.e. the helicopter is on the roof.

9. $G, b, p \models \downarrow x.\varphi$ iff $G, b_p^x, p \models \varphi$

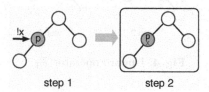

step 1   step 2

Fig. 7. Binder operator $\downarrow x$

Operator $\downarrow x$ binds the variable x to the place p, where the query is evaluated (step 1). Within the same query, x can then be used as an alias for p (step 2). For example, the formula $(G, b, hospital \models (E_\downarrow \downarrow emergency.victim) \land (@_{emergency} first\_aider))$ is satisfied, i.e. there is an emergency place in the hospital where a first aider is assisting a victim. *Emergency* place is an alias for *helicopter*.

5   Implementation

To evaluate our approach, we built an implementation of the query processing framework, named Chequery. The language itself is independent of any programming languages but the current implementation uses OCAML (version 3.09 or later).

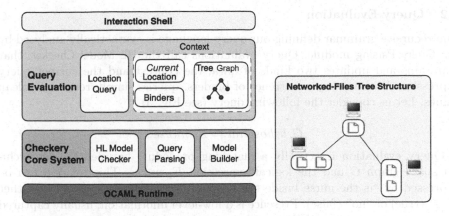

Fig. 8. Checkery Architecture

Chequery works in interactive mode : users can change the provenance of the query by moving along hierarchical space graphs[1] and execute location queries based on the syntax specified in Sect. 4.

5.1 Architecture

Tree-Based Database. The first version of our system supports networked-files tree structures, that are mapped to hierarchical space trees. Files correspond to places, and hierarchical links correspond to containment relations between places.

Hierarchical Space Graph. Model Builder processes the input files and extracts their spatial organisation to produce a corresponding modal structure, i.e. a hierarchical space graph, which :

1. associates files to places,
2. links the places together according to the containment structure by using the transition relations,
3. labels each place with its contained places. Thus, nodes of the hierarchical tree graph are augmented places, and we call them *worlds*[2].

```
type world =
    {mutable container  : place ;
     mutable valid_prop  : place list ;
     mutable nested_loc  : place list ;}
```

The data (or Kripke) structure has the form ((place world (worlds))...). It is implemented in a Hashtable, whose keys are worlds' labels and values are worlds. Note that in the current implementation, after Model Builder is executed, functions implemented in HL Model Checker use the same modal structure for all queries.

---

[1] This navigation-based system may be related to the navigation inside tree-based file systems (i.e. enter and leave a folder).

[2] "World" is the appropriate term in model checking terminology.

5.2 Query Evaluation

The recursive grammar defining our query language is syntactically analysed by the Query Parsing module. Query evaluations call the HL Model Checker, that maintains and updates two kinds of data, the context and the current query expression. Without any occurrence of binders, queries simply returns Boolean values. Let us consider the following query (see Fig. 3) :

$$G, b, hospital \models E_\downarrow \ nurse$$

Query evaluation is actually a matching procedure between the hierarchical space graph G and the logical expression $E_\downarrow \ nurse$. This relation can be expressed by "is the nurse inside the hospital ?". The answer would be either "yes" (true) of "no" (false). Presence is a low-level information, usually captured by a sensing system like RFID-tags. Location-based services are rather searching for objects or people location. Binder operator $\downarrow x$ enriches the relations, so that results of the model checking algorithm are either failures or bound places where formulas are satisfied. For example :

$$G, b, hospital \models E_\downarrow (\downarrow place.nurse)$$

The result is the nurse's current location bound to *place*. The meaning of the relation changes from "is there a place inside the hospital where the nurse is ?" to "where is the nurse ?". Module Binders maintains the bound variables in a list of pairs {bound variable, real place}. Real places' names are resolved after a computation step, carried out by a query evaluator's sub-routine.

6 Conclusion and Future Works

We presented a novel approach to handle location query processing in pervasive computing environments. Starting from hierarchical space trees, we defined an extended Kripke-like structure, as both a semantic and data model for a hybrid logic-based query language. The logic is defined by a small set of operators and provides good expressiveness for common location queries. We designed and implemented a prototype system based on the framework. Moreover, our approach is general and may not be limited to location query processing. Actually, it has a broad application domain and can be applied to any hierarchical data structure.

We would like to point out further issues. First, this paper assumes static location models and focuses on the spatial model of space. However, mobility of users, objects and services is one of the standard requirement of pervasive systems, that makes the spatial model truly dynamic. Thus, we need to extend our language to cope with this temporal aspect. Second, privacy is a critical issue for any information-based application, and particularly for pervasive computing applications, where information about users is increasingly available. As our query language deals with global references and direct accesses that may lead to privacy violations, security mechanisms such as access controls should be integrated in the core of the query processing framework.

References

1. Afanasiev, L., Franceschet, M., Marx, M., de Rijke, M.: CTL Model Checking for Processing Simple XPath Queries. In: Proc. 11th International Symposium on Temporal Representation and Reasoning (TIME 2004), pp. 117–124. IEEE Computer Society, Los Alamitos (2004)
2. Areces, C., ten Cate, B.: Hybrid Logics. In: Blackburn, P., Wolter, F., van Benthem, J. (eds.) Handbook of Modal Logic (2005)
3. Becker, C., Dürr, F.: On Location Models for Ubiquitous Computing. Personal and Ubiquitous Computing 9(1), 20–31 (2005)
4. Beigl, M., Zimmer, T., Decker, C.: A Location Model for Communicating and Processing of Context. Personal and Ubiquitous Computing 6(5-6), 341–357 (2002)
5. Clarke, E.M., Schlingloff, B.H.: Model Checking. Handbook of automated reasoning, pp. 1635–1790. Elsevier Science Publishers B. V, Amsterdam (2001)
6. Emerson, E.A.: Temporal and modal logic. Handbook of Theoretical Computer Science, vol. B, pp. 955–1072. MIT Press, Cambridge (1990)
7. Franceschet, M., Montanari, A., de Rijke, M.: Model Checking for Combined Logics with an Application to Mobile Systems. Automated Software Engineering 11(3), 289–321 (2004)
8. Kripke, S.A.: Semantical Considerations for Modal Logics. The Journal of Symbolic Logic 34(3) (1969)
9. Negri, M., Pelagatti, G., Sbattella, L.: Formal Semantics of SQL Queries. ACM Trans. on Database Systems 16(3), 513–534 (1991)
10. Satoh, I.: A Location Model for Pervasive Computing Environments. In: Proc. 3rd IEEE International Conf. on Pervasive Computing and Communications (PERCOM 2005), pp. 215–224. IEEE Computer Society, Los Alamitos (2005)

Scalable Inter-vehicular Applications

Jonathan J. Davies and Alastair R. Beresford

Computer Laboratory, University of Cambridge,
15 JJ Thomson Avenue, Cambridge, CB3 0FD, UK
{jjd27,arb33}@cam.ac.uk

Abstract. Many pervasive inter-vehicular applications involve the collation, processing and summarisation of sensor data originating from vehicles. When and where such processing takes place is an explicit design-stage decision. Often some processing occurs on vehicles, and some on backend servers, but it is hard for the programmer to optimise this distribution for feasibility or performance. This paper investigates automated task assignment: we define a computational model which captures data aggregation and summarisation explicitly, allowing a compiler to automatically optimise the assignment of processing tasks to particular vehicles and servers. Our model allows a compiler to apply program transformations to data processing, which can further improve task assignment.

Modern motor vehicles contain a plethora of on-board computing equipment. Today's cars have a variety of microprocessors governing diverse aspects of the vehicle's operation. We believe that trends in decreasing power requirements, size, and cost of manufacture mean that in future we can expect vehicles to provide embedded computing platforms supporting the execution of general applications. As cars become increasingly connected—to each other and to the Internet—these applications will evolve beyond disconnected intra-vehicle applications and will help to improve the safety, efficiency and comfort of using transport [1]. This vision of communicating vehicles will enable applications involving multiple participants, such as:

Collection of vehicle position data. Known as *floating car data*, information regarding the locations and velocities of vehicles using the road network can be used to identify levels of road congestion and used as input to journey-time prediction applications [2].

Real-time weather map. Most modern vehicles already contain thermometers. If data from vehicles' onboard weather sensors could be aggregated, a real-time weather map of high resolution could be composed.

Real-time road map updates. Traditional techniques for updating road maps involve manual surveying and data entry. Timely integration of changes to the road network can instead be done automatically based on vehicles' location traces [3].

Road hazard detection. Acceleration data collected from vehicles containing accelerometers can be used to build a map of road hazards by noting points in the road network where many vehicles have been found to swerve or brake sharply [4].

R. Meersman, Z. Tari, P. Herrero et al. (Eds.): OTM 2007 Ws, Part II, LNCS 4806, pp. 876–885, 2007.

Such inter-vehicular applications collate, process and summarise raw sensor data from a large number of vehicles, and typically the greater the number of vehicles involved, the more useful the application becomes. A given application might be written to execute in one of a variety of different architectural configurations. At one end of the spectrum is the fully centralised approach, where all of the source data is transferred to a single processing node. At the other end is the fully decentralised, or peer-to-peer, approach, where there is no infrastructural support. Vehicular applications are not readily-suited to a centralised model of computation because this approach does not scale well and requires ubiquitous network coverage. Fortunately, most applications are inherently parallelisable by partitioning their input data into subsets which can be processed independently. For example, the data set could be decomposed into geographical subsets, with each containing data concerning a particular spatial region.

Today, a programmer writing an inter-vehicular application must manually define where the processing of sensor data takes place. This paper describes an alternative strategy, allowing programmers to write applications which can be distributed *automatically*, therefore potentially at run-time and with a greater degree of optimality because they do not need to rely on design-stage assumptions. More specifically, we devise a computational model which captures the notion of data aggregation or summarisation in an application. This permits the automation of *program transformations*, where the order in which data processing and distribution takes place in an application can be rearranged by a compiler, thereby allowing the exploration of many different software architectures. In addition, we take inspiration from work in parallel and distributed computing and explore how, for a given application with a specific software architecture, the application might be automatically distributed and executed across a heterogeneous network of on-board computers in vehicles and backend servers. We also explore what optimality might mean in the context of vehicular networks, and describe a few metrics on which we can measure and optimise the configuration of an application. Finally, we describe our experience implementing a tool which applies program transformations to an application described in our computational model and automatically simulates the distribution of such an application across many computing resources.

1   Computational Model

It is necessary to be able to define an application and the topology of the network in which it is to be executed at a given instant in time. We model the former (the software) with a *task graph* and the latter (the hardware) with a *computation resource graph*. The task graph is weighted with values indicating the application's requirements, whilst the computation resource graph is weighted with values indicating the hardware's capabilities. We will describe how each is modelled, in turn. We will then turn to the question of how to determine the best computation resource on which to execute each task.

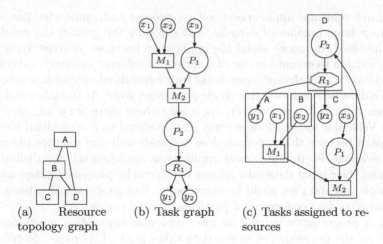

(a)   Resource topology graph

(b) Task graph

(c) Tasks assigned to resources

Fig. 1. Example graphs

1.1 Modelling the Network Topology

At any point in time, the topology of the network of available computation resources is static. It can be modelled by a weighted graph $G_p = (E_p, V_p)$. The set of vertices, V_p, model the processing nodes; the edges, E_p, model direct communication links between nodes. An example is shown in Fig. 1(a). The processing nodes, which have local memory, are not assumed to be homogeneous in their processing power.

If we assume that all the communication links are symmetric, the graph can be undirected. The transitive closure of the graph indicates which nodes can directly or indirectly communicate with any other nodes.

The resource graph's vertices are weighted with values describing their computational characteristics, such as processor speed. The edges are weighted with values characterising the link, such as maximum throughput or latency.

1.2 Modelling the Algorithm

An application's algorithm can be described by a directed weighted graph $G_t = (E_t, V_t)$ called the *task graph*. The set of vertices, V_t, are the tasks and the edges, E_t, indicate the direction of data flow between tasks. A task is a set of instructions which must be executed sequentially on a single processor. An edge (v_1, v_2) indicates that task v_2 receives the output of task v_1, and that v_2 cannot commence execution until the execution of v_1 is complete. An example is shown in Fig. 1(b). We express algorithms in terms of five types of task nodes:

Source nodes are points where data is produced. These can be thought of as functions of type $\text{unit} \rightarrow \alpha$, for some type α.

Sink nodes are points where data is consumed. These can be thought of as functions of type $\alpha \rightarrow \text{unit}$, for some type α.

Processing tasks are functions which transform a tuple of inputs into a tuple of outputs. In general, they have type $\alpha \times \beta \times \ldots \to \gamma \times \delta \times \ldots$.

Merge tasks are a special type of processing task which are commutative, associative, binary functions with type $\alpha \times \alpha \to \alpha$, for some α.

Replication tasks are a special type of processing task which have type $\alpha \to \alpha \times \alpha$ and additionally where the two outputs are each identical to the input. Hence, it is not possible to modify the data in a replication task.

Identifying merge tasks as a special class of processing tasks is particularly important in applications involving large numbers of inputs, such as those which are the focus of this paper. Because of the wealth of input data, it is usually necessary to be able to aggregate data into a significantly smaller amount of information to make their processing computationally feasible.

Some examples of simple merge tasks are set union, addition and maximisation. In order to express an arithmetic mean of multiple values in terms of a binary function, we must keep track of both the numerator and denominator in the calculation, otherwise we will lose track of the number of items which have contributed to the mean. The merge task is therefore an operation on a pair of values of type `real` $\times$ `int` and can be expressed as $M((a_n, a_d), (b_n, b_d)) = (a_n + b_n, a_d + b_d)$. The value of the mean is yielded by a subsequent processing task with type `real` $\times$ `int` $\to$ `real`, $P(a_n, a_d) = \frac{a_n}{a_d}$.

The vertices of the task graph are weighted with values relating to the computational requirements of the application. The graph's edges are weighted with values relating to the characteristics of the data flow between tasks.

2 Program Transformations

For some programs, it is possible to express the graph of tasks in a variety of semantically-equivalent ways. For example, an algorithm to compute the exponential of the sum of three numbers could be equivalently expressed as the product of exponentials: $e^{x+y+z} = e^x e^y e^z$. The formula e^{x+y+z} is an additive *merge* of the three numbers x, y and z followed by an exponentiation *processing* task; a transformation of the formula gives $e^x e^y e^z$ which is three exponentiation *processing* tasks followed by a multiplicative *merge* of the resulting numbers.

We have defined four transformations which can be applied on any sub-graph of a task graph. After the application of one or more of these transformations, the task graph may be better suited to efficient execution by processors in a particular network. Section 3 will describe the ways in which this can be measured.

2.1 Exchange of Merge and Processing Tasks Transformation

The exponentiation example is an instance of a program transformation involving the exchange of merge and processing tasks which preserves semantic equivalence. Rather than merging some inputs and processing the result, we can process each input individually and merge the results. Figure 2(a) depicts this transformation for a unary processing function of type $\alpha \to \beta$.

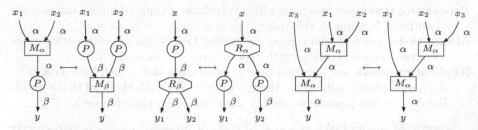

(a) Exchange of merge and processing tasks

(b) Exchange of processing and replication tasks

(c) Associativity of merge tasks

Fig. 2. Three program transformations (task graph edges labelled with types)

Firstly, it is notable that there are more tasks on the right side of the transformation than on the left. This means that the total amount of work required may be different after the transformation is applied, depending on the sizes of tasks M_α and M_β. Moreover, there is one processing task $P : \alpha \to \beta$ on the left and there are two such tasks on the right. Furthermore, the merge task required on the left deals with values of type α and the merge task required on the right deals with values of type β. In the exponentiation example, these functions were addition and multiplication.

Depending on the relative sizes of values of types α and β, the volume of data flow may be affected by this transformation. If values of type β are significantly smaller than values of type α then early processing to β is most favourable, so the volume of data flow is smaller on the right.

2.2 Exchange of Processing and Replication Tasks Transformation

Similarly, rather than performing some processing and then replicating the result, we can replicate the input and process each replica individually. This transformation is depicted in Fig. 2(b).

As above, there is a difference in the number of processing tasks before and after the transformation. On the right, there are two such tasks. As before, depending on the relative sizes of values of types α and β, the volume of data flow may be affected by the transform. If values of type α are significantly smaller than values of type β then late processing to β is most favourable, so the volume of data flow is smaller on the right.

2.3 Merge and Replication Transformations

Two transformations follow directly from the associativity of merge functions and the equivalence of outputs from replication tasks. These transformations are useful to alter how the merging or replication of a large number of values takes place in a distributed manner. Figure 2(c) depicts the transformation for merge tasks; the transformation for replication tasks is analogous.

3   Execution Strategy

Once we have a model of the application in terms of its constituent tasks and flow of data, and a model of the topology of the network of processors that could execute the application, it remains to define where each task is to be executed.

An assignment function $A : V_t \rightarrow V_p$ maps tasks to processing nodes, indicating where in the network each task should be executed. Source and sink vertices in the task graph must be mapped to the particular nodes in the network where data is produced and consumed, respectively. Other tasks can be mapped to network nodes which are reachable from source nodes and from which sink nodes are reachable via communication links. An example of an assignment function is shown pictorially in Fig. 1(c), where M_1 is mapped to processor A, P_1 and M_2 are mapped to processor C, and P_2 and R_1 are mapped to processor D.

The decision about which nodes to use affects the efficacy of the assignment. It impacts on the duration of execution of the algorithm; the privacy of the originators of the data; the amount of network bandwidth consumed; and a variety of other factors. The efficacy of the assignment can be described quantitatively by a *cost function* specific to each application. A cost function $C : G_t \times G_p \times (V_t \rightarrow V_p) \rightarrow \mathbb{R}$ is a function of an assignment function yielding a real number indicating the cost of the assignment. Applications will use a cost function which embodies the trade-offs they desire between relevant metrics. For example, one application may express in its cost function the policy that only a total execution time of less than two minutes is acceptable, and that minimising the use of network bandwidth is the next most important concern. Another application may seek to minimise total execution time at the expense of all other metrics.

Several candidate metrics for evaluating an assignment function are relevant to many applications:

Total execution time. The duration of time elapsed from the start of the algorithm's execution to the result being delivered to the final recipient.

Quality of result. Thus far, we have assumed that any transformations applied will maintain semantic equivalence of the algorithm. In the exponentiation example, we required the merge functions—addition and multiplication—to be appropriate to maintain this equivalence. But, in the case that a merge function for one data type is not an exact analogue of the one used for the other, the exchange of merge and processing tasks transformation (Sect. 2.1) no longer maintains semantic equivalence, and the algorithm's output approximates the true result. For example, a merge task which sums values followed by a processing task which quantises the sum is approximated by summing quantised values. The value of this metric will relate to the accuracy of the approximation.

Privacy. The level to which the privacy of the originators of the input data is respected is important in applications where personally-identifiable data is processed. The value of this metric could relate to an observer's view of the number of individuals who could have a particular identity.

Energy consumption. Another useful metric is the energy consumption caused by the execution of particular tasks. By associating with each processor a value indicating its power consumption, the total energy consumption for a given assignment function can be calculated.

4   Automatic Task Assignment

In a system where the available resources change dynamically, such as when vehicular resources are utilised, the particular resources which are available for an application to use are not known until run-time. Thus, the assignment of tasks to processors cannot take place at the same time that the application is defined. *Automatic* task assignment is thus required, and this implies a need for a programming language and compiler which can perform this when the program is about to be run.

The compiler must be able to split the program into constituent tasks; determine the optimal mapping with respect to the program's cost function (which may involve the use of some program transformations); distribute the tasks to their processors, ensuring that they can communicate with each other and are able to deal with failures.

4.1   Implementation

We have developed a prototype framework to investigate the feasibility of automatic task assignment. As input, it accepts a description of the computation resource graph and an application's task graph in text format, along with a cost function. The weights of the graphs' nodes and edges that are used by the cost function are also specified.

The framework can automatically derive an optimal assignment function using an exhaustive search, or find a near-optimal assignment function in polynomial time using an approach which involves choosing a reasonable initial assignment and then improving it incrementally by changing the assignment of the tasks until no improvement can be found. The transformations described in Sect. 2 are also applied automatically and the cost of assigning the resulting task graph is compared with that of original graph. The framework also allows a user to assign tasks to processors manually through a graphical front-end, depicted in Fig. 3, which uses *Graphviz* graph layout software to visualise the graphs. It does not execute the application, but merely simulates the effect of the chosen assignment function.

The implementation has highlighted the necessity of adopting a sub-optimal search technique: the sheer size of the search space renders the search for an optimal assignment infeasible for even very small task and resource graphs. However, it has also become clear that implementing a suitable heuristic on which to base the search is challenging. It is difficult to model the effect on the cost of either applying a transformation or modifying the assignment since there are so many variables; this means that traditional techniques for solving the global optimisation problem, such as branch-and-bound and simulated annealing, are hard to apply.

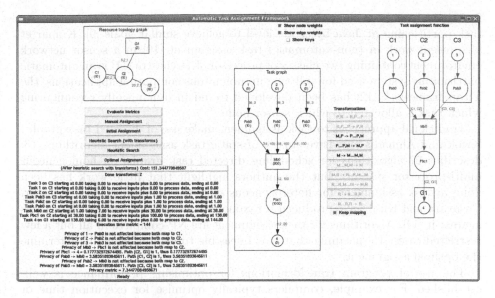

Fig. 3. Prototype framework interface

5   Related Work

In recent years, with the emergence of grid computing, the advent of network processors, and the amount of processing possible in wireless sensor networks, there has been a considerable rise in the level of interest in task assignment. In its usual form, grid computing differs from computation in vehicular networks in several characteristics. For example, it is usually the case that the processors comprising a grid, or the data that is processed, are owned by a single entity, so privacy is of no concern. Furthermore, grids may have different priorities regarding the desire to balance load evenly across available processors. Multi-core network processors can be thought of as distributed systems on a single chip, consisting of a matrix of independent processors, each with local memory, sharing a communication bus. The challenges faced in designing compilers for these systems are similar to those examined in this paper. The problems associated with using a global, shared address space have led to suggestions such as the use of linear types [5]. Ennals et al. have devised a set of program transformations which exploit linear types as the programmer expresses the task assignment function [6]. Major differences between network processors and the large-scale distributed systems considered in this paper are that the processors are arranged systematically, are powered and controlled by a single entity, and have predictable communication links and network behaviour.

Various research has been undertaken into automatically off-loading computation from resource-constrained devices with a view to minimising energy consumption. Several frameworks for off-loading the processing of expensive functions have been devised [7,8]. Ou et al. have implemented an automatic task partitioning at Java bytecode level to achieve similar goals [9]. Kumar et al. present work in (non-automatic) task assignment [10] in a sensor network environment containing two classes of processor. J-Orchestra [11] is an automatic partitioning system used for splitting up ubiquitous computing applications; the Titan framework [12] has been developed to aid in dynamically reconfiguring which task is allocated to which sensor node.

Traditional approaches to task assignment make use of directed task graphs. Kwok and Ahmad's extensive survey of static task assignment algorithms [13] describes 27 algorithms for scheduling directed task graphs on homogeneous multi-processor systems, but the authors highlight that little work has been done in task assignment for heterogeneous systems, the subject of this paper. Casavant and Kuhl have produced a taxonomy and classify various algorithms against it [14]. Algorithms for task assignment are NP-complete in all but a few restricted cases [15] meaning that it is infeasible to computationally determine the optimal assignment.

The use of program transformations for optimising performance is well-established. For example, compilers typically optimise for execution time or memory footprint by performing semantics-preserving transformations. In databases, there are often many different ways of processing data to formulate the result of a query; it is the job of a query optimiser to choose the optimal approach, which involves rewriting the query into a more efficient form [16]. However, to the best of our knowledge, there is no prior work in compiler theory which makes use of transformations which do not preserve semantics.

6   Conclusions

Traditionally, when designing an application to collate, process and summarise sensor data from a large number of vehicles, a programmer must manually define where these tasks will be executed. We have proposed a strategy whereby the assignment of program tasks to processors is done automatically. The expression of a program in terms of merge, process and replication tasks mean that certain program transformations can be applied automatically by a compiler to allow a more optimal assignment.

Whilst this work was motivated by the problems faced in the implementation of applications involving vehicles, it has a broader applicability to other ubiquitous computing scenarios, in particular to wireless sensor networks.

Further work will examine how the movement of vehicles can be represented in the model, perhaps by defining several vehicles in a spatial area as a single processing unit. We will also consider whether the optimisation of task assignment can be effected on a local scale rather than globally. We plan to continue to explore the ideas described in this paper by implementing a task partitioning and assignment engine to distribute and execute applications described in an augmented version of the Java programming language.

Acknowledgments

The authors thank Samuel Kounev, Andrew Rice, David Cottingham, Tom Craig and Ripduman Sohan for their helpful comments and suggestions; and in particular Andy Hopper for his financial support and Alan Mycroft for the exponentiation example.

References

1. Cottingham, D.N., Davies, J.J.: A vision for wireless access on the road network. In: Proc. WIT 2007 Technische Universität Hamburg-Harburg, pp. 25–30 (2007)
2. Day, P., Wu, J., Poulton, N.: Beyond real time. ITS International 12(6), 55–56 (2006)
3. Davies, J.J., Beresford, A.R., Hopper, A.: Scalable, distributed, real-time map generation. IEEE Pervasive Computing 5(4), 47–54 (2006)
4. Gruteser, M., Grunwald, D.: Anonymous usage of location-based services through spatial and temporal cloaking. In: Proc. MobiSys 2003, pp. 31–42. ACM Press, New York (2003)
5. Ennals, R., Sharp, R., Mycroft, A.: Linear types for packet processing. In: Schmidt, D. (ed.) ESOP 2004. LNCS, vol. 2986, pp. 204–218. Springer, Heidelberg (2004)
6. Ennals, R., Sharp, R., Mycroft, A.: Task partitioning for multi-core network processors. In: Bodik, R. (ed.) CC 2005. LNCS, vol. 3443, pp. 76–90. Springer, Heidelberg (2005)
7. Kremer, U., Hicks, J., Rehg, J.H.: A compilation framework for power and energy management on mobile computers. Technical Report DCS-TR-446, Rutgers University (2001)
8. Li, Z., Wang, C., Xu, R.: Computation offloading to save energy on handheld devices: A partition scheme. In: Proc. CASES 2001, pp. 238–246. ACM Press, New York (2001)
9. Ou, S., Yang, K., Liotta, A.: An adaptive multi-constraint partitioning algorithm for offloading in pervasive systems. In: Proc. PERCOM 2006, pp. 116–125 (2006)
10. Kumar, R., Tsiatsis, V., Srivastava, M.B.: Computation hierarchy for in-network processing. In: Proc. WSNA 2003, pp. 68–77. ACM Press, New York (2003)
11. Liogkas, N., MacIntyre, B., Mynatt, E.D., Smaragdakis, Y., Tilevich, E., Voida, S.: Automatic partitioning for prototyping ubiquitous computing applications. IEEE Pervasive Computing 3(3), 40–47 (2004)
12. Lombriser, C., Roggen, D., Stäger, M., Tröster, G.: Titan: A tiny task network for dynamically reconfigurable heterogeneous sensor networks. In: Kommunikation in Verteilten Systemen (KiVS), pp. 127–138. Springer, Heidelberg (2007)
13. Kwok, Y.K., Ahmad, I.: Static scheduling algorithms for allocating directed task graphs to multiprocessors. ACM CSUR 31(4), 406–471 (1999)
14. Casavant, T.L., Kuhl, J.G.: A taxonomy of scheduling in general-purpose distributed computing systems. IEEE Trans. on Soft. Eng. 14(2), 141–154 (1988)
15. Fernández-Baca, D.: Allocating modules to processors in a distributed system. IEEE Transactions on Software Engineering 15(11), 1427–1436 (1989)
16. Ioannidis, Y.E.: Query optimization. ACM CSUR 28(1), 121–123 (1996)

Towards Personal Privacy Control

Susana Alcalde Bagüés[1,2], Andreas Zeidler[1], Carlos Fernandez Valdivielso[2],
and Ignacio R. Matias[2]

[1] Siemens AG, Corporate Research and Technologies
Munich, Germany
[2] Public University of Navarra
Department of Electrical and Electronic Engineering
Navarra, Spain
{susana.alcalde.ext,a.zeidler}@siemens.com
{carlos.fernandez,natxo}@unavarra.es

Abstract. In this paper we address the realization of personal privacy control in pervasive computing. We argue that personal privacy demands differ substantially from those assumed in enterprise privacy control. This is demonstrated by introducing seven requirements specific for personal privacy, which are then used for the definition of our privacy policy language, called SenTry. It is designed to take into account the expected level of privacy from the perspective of the individual when interacting with context-aware services. SenTry serves as the base for implementing personal privacy in our User-centric Privacy Framework for pervasive computing.

1 Introduction

Privacy is a prime concern in today's information society where personal sensitive data has to be revealed in common daily tasks. Thus, laws exist that shall control the collection and processing of sensitive information and should prevent its misuse by enterprises. Individuals, though, often are not really aware of personal privacy issues and mostly make decisions casually or on the move. Even in open settings, like the Internet, users control privacy mostly manually and are limited to acknowledging some prefabricated privacy statements. To our believes, for the upcoming era of so-called *Ambient intelligence* [1], which fosters the deployment of heterogeneous Context-Aware Mobile Services (CAMS), such habitual control of *personal privacy* eventually will fall short. The large number of services alone will make a manual per-use authorization of access to personal data (as required by law) an impossible task.

Personal privacy is a more "intimate" concern than the enterprise's requirement to meet existing legislations. The challenge is to meet the individual's expected level of privacy when information is revealed to third parties. In this paper, we focus on managing personal privacy "offline", e.g., beforehand of actually being in a particular situation. To do so, we have elaborated requirements for personal privacy control and applied them in the design of the *SenTry* language, which allows users to generate appropriate User Privacy Policies to automatically govern all accesses to their sensitive data.

The SenTry language is presented in this paper as a centerpiece of our ongoing work, focused in the development of the User-centric Privacy Framework (UCPF) [2]. The

R. Meersman, Z. Tari, P. Herrero et al. (Eds.): OTM 2007 Ws, Part II, LNCS 4806, pp. 886–895, 2007.

SenTry language allows the specification of fine-grained constraints on the use of personal data to conform to a user's privacy criteria. Thus, UCPF takes the roll of a trusted *privacy enforcement point* and acts as the sentry of its users' personal privacy by supervising the application of policies in the interaction of the user with CAMSs.

The remainder of this document is structured as follows: First, Section 2 provides some background information on privacy policies. After that, in Section 3, requirements for implementing personal privacy are elaborated. Section 4 then details the central features of the SenTry language, which is followed by Section 5 where details of the implementation and simple use cases are presented. Finally, Section 6 indicates the directions of future work and concludes the paper.

2   Background

While there are many languages for access control they are rarely adequate for enforcing privacy policies [3]. Privacy control in general needs expressive methods to delimit accesses to and usage of sensitive information. A privacy policy language should cater for the enforcement of a user privacy criteria as well. Also requirements differ when we consider the needs of an enterprise or those of an individual.

In many countries legislation regulates the collection and use of privacy data. It prevents its misuse and demands that enterprises comply with certain privacy practices (directives 95/46/EC and 2002/58/EC in Europe). Therefore, the main requirement of an enterprise from a privacy policy system is that it allows for automatic enforcement of enterprise privacy practices. Thus, the enterprise reduces the risk of unauthorized disclosure and the risk of misuse of the collected data. A description of enterprise-level privacy requirements is given in [4].

There exist only a few policy languages for the support of an automated analysis of a privacy criteria. Probably the best known are the Platform for Privacy Preferences (P3P) [5], the Enterprise Privacy Authorization Language (EPAL) [6] and the eXtensible Access Control Markup Language (XACML) [7]. P3P is an standard from the World Wide Web Consortium (W3C) that enables websites to express their privacy policies and to compare them with the user's privacy criteria, which in turn can be specified by using A P3P Preference Exchange Language (APPEL) [8]. While APPEL provides a good starting point for expressing user privacy preferences, it cannot support the richness of expressions needed in ambient intelligence scenarios, see Section 3. In [9] user requirements for a privacy policy system are detailed with emphasis in the granularity of constraints users might want to apply to control the distribution of their location information. Here, rules are implemented as system components called *validators* without defining a concrete implementation language, though. Apart from the lack of expression, P3P does not address the problem of enforcing a website's privacy policy [4]. The use of P3P alone does not give assurance about the *actual* privacy practices in the back-end of the website and whether "obligations" implicitly included in the user privacy preferences, such as the purpose of the data collection, are respected.

EPAL and XACML are two platform-independent languages that support the definition and enforcement of privacy policies and obligations. IBM submitted EPAL 1.2 to the W3C in November, 2003, for consideration as a privacy policy language standard (still pending). XACML 2.0 is an XML-based language designed primarily for access control and extended for privacy. It has been accepted as an OASIS standard and is widely deployed. In [3] a comparison of both languages shows that EPAL offers only a subset of the functionalities of XACML. Nevertheless, while XACML has been developed for some time and has reached a high level of standardization, it has only started to take possible privacy constraints on information management into account. It may be enough for enterprise privacy control but it lacks of some important features to enforce personal privacy. In the next Section, we present those requirements that a privacy policy language from the perspective of the individual shall meet.

3   Requirements for Implementing Personal Privacy

A privacy policy system used to define and enforce an individual's privacy criteria should provide a way of describing the elements involved in his/her interaction with a CAMS (Figure 1), together with the environmental conditions in which such interaction could occur or not, and operate accordingly. Next we outline seven requirements, which have guided the design of our policy system and particularly the definition of the SenTry language, and how they were incorporated into our implemented solutions.

Centralized privacy enforcement. In order to offer a controlled distribution of sensitive personal data, the policy system should centralize the collection and distribution of context-related data of its users. Thus, we provide a *trusted privacy enforcement point* with our UCPF, which collects context information of its users from appropriate context providers and controls all accesses to privacy-relevant information. Figure 1 represents the role of the UCPF in our model of protected interaction chain, showing the following entities: i) A *Target* is the tracked individual and the source of any *Resource*(location, calendar, situation, medical data, etc); ii) A *CAMS* compiles the resource and carries out the *Action* (purpose of the interaction); iii) A *Subject* is the user of the service and the final recipient of the data; iv) A *Context Provider* acts as an intermediate entity, responsible for collecting and disseminating context; v) The *UCPF* acts as sink of its users' context and enforces their privacy criteria, providing a protected interaction with the service.

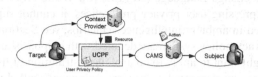

Fig. 1. Our Model of Interaction Chain with a CAMS

User aware. A privacy policy system should be aware of its users needs; avoiding situations in which the target has a passive role limited to accept or reject CAMSs' privacy practices without explicitly specifying his privacy preferences, as is the case of some previous approaches e.g., PawS [10]. In the UCPF individuals have an active role in the specification and enforcement of their privacy criteria.

Context awareness. Policies should be context-aware in the sense that their evaluation involves checking individuals' context against the applicable policy. Additionally to the typical restrictions on the entities of Figure 1, constraints related to a user's context should be possible. Moreover, the peer's context is a rich source of information that also can be employed to restrict when a rule applies. Therefore, our SenTry language incorporates dynamic constraints that take into account individuals' context values and express *context-awareness* rules. For realizing the idea of the UCPF being the sink of its users's context information from potential different context providers, a standard model for context representation is assumed.

Semantic awareness. Obviously, the interaction with pervasive services and user customized applications demands some awareness of the underlying semantics. Therefore, the policy language used for reasoning on context information has to be expressive and aware of the underlying meaning at the same time. This recommends the use of a semantic representation model for privacy policies and context data, which also will provide a common semantic frame for the collaboration of the different entities involved in the interaction chain.

Post-disclosure awareness. Privacy control should not be limited to a pre-disclosure phase typical of access control. To guarantee the expected privacy once the data has been disclosed, is vital to provide mechanisms that delimit the extension of the granted action, e.g. a target may want to limit the time that an action is granted, or to restrict any secondary use on the data transmitted. For that, we use *Obligations* [11], they allow to bind an entity (service, subject) with the obligation of performing a predefined action in a future on an specified object. We include the so-called *Positive Obligation Permission* as a possible result of the *Positive Authorization Rule*, to trigger a negotiation process between the UCPF and a CAMS before any resource is transmitted. As a result the CAMS might agree on holding an obligation with the UCPF's user, allowing post-disclosure control.

Transformations. The user should have means to reveal "fuzzy" information when using a certain service. We introduce the concept of Transformations [12] to allow users to better specify their privacy preferences. Basically, we define Transformations as any process that the tracked user may define on a specific piece of context information to limit the maximum accuracy to be revealed, e.g coordinates accuracy max. 500 m, or filter calendar items labeled as private. We have created the *Positive Transformation Permission* and the *Positive Transformation* and *Positive Obligation Permission*. Both include a Transformation instance added to the positive permission of disclosure, which is performed before transmitting the data.

Constraints on Active interactions. For practical reasons (like fine-grained control of stateful information disclosure; see examples in Section 5 for more details), in the interaction of a user with a CAMS, the active and passive roles must be distinguishable. We classify users interactions as *active* or *passive*. In the active case the user is actively using the service and some action from the service is requested, e.g. where to find the closest-by Italian restaurant. To compile an appropriate answer the CAMS asks the current location in return. The interaction is passive if the user receives a request from the CAMS without previously requesting the service e.g., if a colleague asks the Friend-finder application where a user currently is located, the disclosure of the user's location does not necessarily involve that he gets any benefit in return. We will show that the introduction of constraints to limit a disclosure on whether or not a service request is based on an active interaction, brings much greater privacy control (cf. 5).

4 *SenTry* Language

The SenTry language (SeT) is built on top of the Web Ontology Language (OWL) and the Semantic Web Rules Language (SWRL) as a combination of instances of our user-centric privacy ontology (SeT ontology) and SWRL predicates. The SeT ontology describes the classes and properties associated with the policy domain in OWL DL using a unique XML namespace. OWL provides considerable expressive power for modeling a domain knowledge. However, it also has some limitations, which mainly stem from the fact that is not possible to capture relationships between a certain property and another in the domain [13]. For instance, we cannot define in OWL that if a person *hasName* "Pablo" and *hasGroup* "UPNA" the property *hasAccessMyLocation* should get the value "True". A possible way to overcome this restriction is to extend OWL with a semantic rule language. An important step has been taken with the definition of the Semantic Web Rules Language, based on a combination of OWL DL and OWL Lite sublanguages with RuleML. Due to the mentioned limitations on the OWL language we have included SWRL predicates to reason about the individuals provided by the SeT ontology, primarily in terms of classes and properties.

In common with many other rule languages SWRL rules are written as antecedent-consequent pairs. In SWRL terminology, the antecedent is referred to as the rule body and the consequent as the head. The head and body consist of a conjunction of one or more atoms (SWRL predicates). SWRL does not support more complex logical combinations of predicates than the conjunction, represented with the symbol "∧" in the examples shown in Figure 5 and 6.

The SeT ontology has three main constructors, namely: *Policy*, *Service Request* and *User Request*. Together they model the elements needed to provide a protected interaction of a user with a CAMS. The OWL classes and properties of the Policy constructor are shown in Figure 2, an instance of the class policy represent the privacy criteria of an entity who is identified with the property *onTarget*. In our model, the system holds a unique policy instance per target, which contains a collection of rules associated with the property *hasRule*. A SenTry rule is created as a combination of instances of the class *set:Rule* and of the class *swrl:Imp* (SWRL rules).

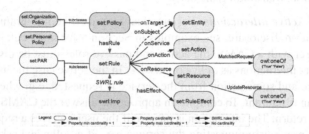

Fig. 2. Classes and Properties of a Policy

A policy is divided into two disjoint subclasses *PersonalPolicy* and *OrganizationPolicy*. Organizational policies covers the need of companies to manage specified context of a predefined group of individuals (employees, clients, etc). A person as part of an organization may adhere, usually based on a contractual relationship, to some privacy policy which might be orthogonal to her own. The personal policy is mainly about how individuals control the use of personal information in every days life. In the following we focus on the personal policy, leaving the organizational policy for future work.

In SenTry, we have included two rule modalities: the Positive Authorization Rule (PAR) and the Negative Authorization Rule (NAR). By using one of these two rule subclasses we explicitly delimit the elements involved in the interaction described in Figure 1 with the properties *onSubject, onService, onAction* and *onResource* and specify the result of such interaction with the property *hasEffect*. The only possible effect of a NAR rule is to deny the disclosure of the requested data. This rule includes an instance of the Negative Authorization Permission (NAP) as result. On the other hand, the PAR rules allow the disclosure of the data by using one out of four positive permissions: the Positive Authorization Permission (PAP), the Positive Transformation Permission (PTP), the Positive Obligation Permission (POP) and the Positive Transformation Obligation Permission (PTOP).

Since a policy might contain multiple rules and since rules might be evaluated to different results for a given request e.g., the NAR "Deny", the PAR "Grant", "Grant with transformation", "Grant with obligation", or "Grant with transformation and Obligation". The system determines potential rule conflicts, see Section 5, before compiling the effect of the policy evaluation process. The final result is assigned to the request class through its property *hasEvaluation* (Figure 3(a)).

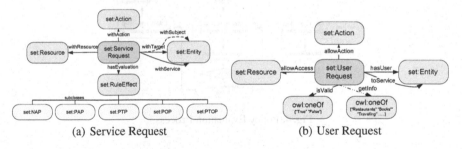

(a) Service Request   (b) User Request

Fig. 3. Classes and Properties

Our second constructor is the Service Request, it holds the information needed to check the applicability of a SenTry rule against a third-party request. The *policy decision point* (PDP) [14] of the UCPF is fired to evaluate an instance of the service request class shown in Figure 3(a). The value given by the OWL property *withTarget* selects a policy among the policies stored in the system. The rest of properties of the service request class, *withSubject, withService, withAction* and *withResource* check whether the rules included in the selected policy match that request. A rule applies if all the values included in the service request are equal to those given in the rule by *onSubject, onService, onAction* and *onResource*.

The last constructor created in SeT is the User Request. This element models *active interactions* between a user and a CAMS, see Section 3. A user identified with the property *hasUser* in Figure 3(b) might actively apply for a particular CAMS, the service instance attached to the property *toService*, and allow the action defined with *allowAction* on the resource contained in the property *allowResource*, of a user request instance. The CAMS uses the resource to compile the expected answer. We explicitly distinguish between rules that will affect an active interaction with a service by evaluating an instance of the class user request against an instance of the service request.

5   Implementing a SenTry Policy

Our PDP component has been developed in JAVA on top of the Java Expert System Shell (Jess). Once the PDP gets an evaluation request message, it triggers the execution of the Jess rule engine and accesses the policy repository to retrieve the policy applicable for the current situation, which is evaluated based on the provided service request. Along with rules in the Jess language, Jess accepts rules formulated in SWRL [15], which are translated into Jess rules and facts with the Java API SWRL factory. Jess implements the well-known *Rete algorithm* to efficiently evaluate rule predicates and facts.

Each time a user defines a new rule, a new instance of the set:Rule is created together with the needed SWRL predicates. In our system rule predicates are first expressed with SWRL and then automatically translated to Jess. The SWRL predicates are classified in three groups, namely: *Filter predicates*, *Static predicates* and *Dynamic predicates*.

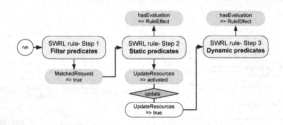

Fig. 4. Policy Evaluation Process

The SWRL predicates included in a SenTry rule are grouped in three SWRL rules and enforced consecutively as shown in Figure 4. In step 1, the Filter predicates are evaluated. They arc a set of common predicates used to filter those instances of the set:Rule class that match a request. There is only a single SWRL "rule- step 1" shared by all the rules. If it is evaluated to true the rule *MatchedRequest* property (see Figure 2) is set to "true" and the SWRL "rule- step 2" is triggered. On the other hand, if it is false for all the set:Rule individuals the process ends without further evaluation.

The evaluation of the Static predicates leads (if true) to two possible results: i) there are not Dynamic predicates associated with the matched set:Rule. Thus, the SWRL "rule- step 2" asserts the appropriate instance of the *RuleEffect* class; ii) there exist Dynamic predicates defined and the evaluation of the Static predicates has the effect of setting the *UpdateResources* property of that set:Rule instance from "false" to "activated". The Static predicates include exclusively constraints on multiple recipients and/or the active interaction constraint. Once this second step is finished, an internal process checks if there are any instance of a set:Rule with the property *UpdateResources* equals to "activated" and then updates the list of context values from the respective context providers. Finally, it sets the *UpdateResources* property to "true".

Step 3 of the evaluation process evaluates Dynamic predicates. They include dynamic constraints that should be updated before further evaluation, such as time constraints, constraints on the target's context, and on a target peer's context. As consequence it always returns a *RuleEffect*. The last part of this process consist of applying a combining algorithm on the *RuleEffect* instances returned in the process described. It combines rule effects generated in the second and third step. The system generates NAR rules per service and subject by default, only when a user explicitly creates a PAR, it is possible to permit a requested action. We use a *grant overrides* combining algorithm to resolve potential conflicts with four variations that allow or disallow to inherit obligations and transformations from the POP, the PTP and PTOP results.

5.1 Examples

We now introduce two simple use cases that show how a SenTry policy is implemented. Use case 1 includes only two SWRL rules (step 1 and 2), while the use case 2 has also Dynamic predicates (step 3). Step 1, Figure 5(b), as mentioned is common to both use cases. It filters the policy with target "Pablo" and the rules within that policy than specify the same resource and action that the given request.

Use case 1: *Pablo uses his mobile phone to call a taxi in the evening, he wants the taxi company to determine his location automatically to ensure a smooth pickup, but he does not want them to be able to trace him once the journey is over.*

In the above use case the target, Pablo, allows being tracked during a limited time slot meanwhile the request for the taxi service is still "active". Thus, the "rule- step 2" (Figure 5(c)) includes only a constraint: the active interaction constraint. Step 2 here checks whether an active instance of a user request exists with the flag *isActive* equals to true, for a given service request and returns a grant permission, defined by the set:Rule (PabloPAR-UC1) in Figure 5(a).

Use case 2: *When Pablo is on a business trip, he likes to meet old colleagues that may also be in the city. He allows to be located by the peer group Colleagues, but only if the potential subject is in the same city and with a maximum accuracy of 500 m.*

```
<set:PersonalPolicy rdf:ID="PabloPolicy">
    <set:hasTarget rdf:resource="#Pablo"/>
    <set:hasRule>
        <set:PAR rdf:ID="PabloPAR_UC1">
            <set:onService rdf:resource="#Taxy-service"/>
            <set:onResource rdf:resource="#Coordinates"/>
            <set:onAction rdf:resource="#Tracking"/>
            <set:hasEvaluation rdf:resource="#PAP"/>
        </set:PAR>
    </set:hasRule>
</set:PersonalPolicy>
```

(a) set:Policy

```
set:ServiceRequest(?varRequest) ∧
set:withAction(?varRequest, ?varAction) ∧
set:withResource(?varRequest, ?varResource) ∧
set:withTarget(?varRequest, ?varTarget) ∧
set:PersonalPolicy(?varPolicy) ∧
set:hasRule(?varPolicy, ?varRule) ∧
set:onTarget(?varPolicy, ?varTarget) ∧
set:onResource(?varRule, ?varResource) ∧
set:onAction(?varRule, ?varAction) ∧
───────▶
set:MatchedRequest(?varRule, true)
```

(b) SWRL rule (step 1)

```
set:MatchedRequest(PabloPAR_UC1, true) ∧
set:withAction(?varRequest, ?varAction) ∧
set:withResource(?varRequest, ?varResource) ∧
set:withTarget(?varRequest, ?varTarget) ∧
set:withService(?varRequest, ?varService) ∧
set:UserRequest(?varUrRequest) ∧
set:hasUser(?varUrRequest, ?varTarget) ∧
set:toService(?varUrRequest, ?varService) ∧
set:allowAction(?varUrRequest, ?varAction) ∧
set:allowResource(?varUrRequest, ?varResource) ∧
set:isActive(?varUrRequest, true) ∧
set:hasEffect(PabloPAR_UC1, ?varEvaluation)
───────▶
set:hasEvaluation(?varRequest, ?varEvaluation)
```

(c) SWRL rule (step 2)

Fig. 5. Pablo's Rule - Use Case 1

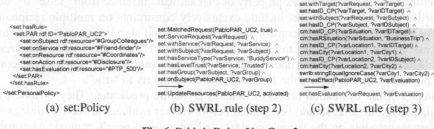

(a) set:Policy   (b) SWRL rule (step 2)   (c) SWRL rule (step 3)

Fig. 6. Pablo's Rule - Use Case 2

Use case 2 requires constraints based on multiple subjects and dynamic constraints. The "rule- step 2", Figure 6(b), checks the type of service as well as the group of the requesting subject. Here, Pablo allows the disclosure of his coordinates with a maximum accuracy of 500m (PTP-500) to the group Colleagues. The "rule- step 3" applies constraints based on the target's situation (Business trip) and limits the disclosure to cases in which target and subject are in the same city, Figure 6(c). There the rule includes predicates expressed over instances of the context model ontology (CM), situation and coordinates. We assume an standard context model defined as an OWL ontology. The CM is related with the SeT ontology through the property *hasID-CP* per person and per context provider as is shown in Figure 6(c).

Assuming that the system holds also two NAR rules created by default to govern the interaction of the taxi-service and the friend-finder application with Pablo, Figure 5 and 6 show all the necessary rules to meet Pablo's demands for granting both services.

6 Conclusions and Outlook

In this paper we introduce seven requirements for Personal Privacy Control and outlines how they are incorporated into our implemented solutions: the User-centric Privacy Framework (UCPF) and the SenTry policy language. In particular, we give a detailed description of our SenTry language, built as a combination of OWL individuals from the SeT ontology and SWRL predicates. It allows for the specification of fine-grained constraints on the use of personal and sensitive data to conform to a user's privacy criteria. The UCPF then provides a *trusted privacy enforcement point* that acts as the *sentry* of its users' personal privacy by supervising the application of policies in the interaction chain of the user with context-aware services. Additionally, this paper presents two examples that show how a *SenTry policy* can be specified and what it looks like.

Beside its application in our prototype of the UCPF, as part of the smart home lab infrastructure, the SenTry language is currently used in the IST project CONNECT. The CONNECT Project is centered around two main scenarios: the "Home CareWork" and the "Mobile Advertising" scenario. For both exist ongoing implementation work in their respective demonstrator applications where SenTry policies are used. This allows us to study the applicability of the policy language in real scenarios and in greater detail.

Further development of the SenTry language obviously involves to provide an intuitive and easy-to-use user interface, which will include three complementary tools, namely: *Privacy Management Tool*, *Privacy Status Tool* and *Deniability Tool*. This will open-up the management and control of privacy to the average user of context-aware services. Another ongoing line of work is the extension of the UCPF with an obligation negotiation protocol, which allows for binding services with respective SenTry obligations. Altogether, in this paper we outlined how a combination of our UCPF and the SenTry language can be leveraged to control and manage the dissemination of personal sensitive data to the outside world.

References

1. IST Advisory Group (ISTAG): Ambient Intelligence: From Vision to Reality. For participation - in society and business. Luxembourg: Office for Official Publications of the European Communities (2003)
2. Bagüés, S.A., Zeidler, A., Valdivielso, C.F., Matias, I.R.: Sentry@home - leveraging the smart home for privacy in pervasive computing. the International Journal of Smart Home (to appear, 2007)
3. Anderson, A.: A Comparison of Two Privacy Policy Languages: EPAL and XACML. Technical report, Sun Microsystems Laboratories, Technical Report TR-2005-147 (2005)
4. Dekker, M., Etalle, S., den Hartog, J.: Security, Privacy and Trust in Modern Data Management. ch. 25: Privacy Policies. Springer, Heidelberg (2007)
5. Cranor, L., Langheinrich, M., Marchiori, M., Reagle, J.: The platform for privacy preferences 1.0 (P3P1.0) specification. W3C Recommendation (2002)
6. Ashley, P., Hada, S., Karjoth, G., Powers, C., Schunter, M.: Enterprise Privacy Authorization Language (EPAL 1.2) Specification (2003), http://www.zurich.ibm.com/security/enterprise-privacy/epal/
7. OASIS standard: eXtensible Access Control Markup Language. Version 2 (2005)
8. Langheinrich, M., Cranor, L., Marchiori, M.: Appel: A P3P Preference Exchange Language. W3C Working Draft (2002)
9. Myles, G., Friday, A., Davies, N.: Preserving privacy in environments with location-based applications. IEEE Pervasive Computing 2(1), 56–64 (2003)
10. Langheinrich, M.: A privacy awareness system for ubiquitous computing environments. In: Borriello, G., Holmquist, L.E. (eds.) UbiComp 2002. LNCS, vol. 2498, pp. 237–245. Springer, Heidelberg (2002)
11. Kagal, L., Finin, T.: Modeling Conversation Policies using Permissions and Obligations. Journal of Autonomous Agents and Multi-Agent Systems (2006)
12. Bagüés, S.A., Zeidler, A., Valdivielso, C.F., Matias, I.R.: A user-centric privacy framework for pervasive environments. In: Meersman, R., Tari, Z., Herrero, P. (eds.) On the Move to Meaningful Internet Systems 2006: OTM 2006 Workshops. LNCS, vol. 4278, pp. 1347–1356. Springer, Heidelberg (2006)
13. Horrocks, I., Patel-Schneider, P.F., Bechhofer, S., Tsarkov, D.: OWL rules: A proposal and prototype implementation. J. of Web Semantics 3(1), 23–40 (2005)
14. Yavatkar, R., Pendarakis, D., Guerin, R.: RFC2753 - A framework for policy-based admission control (January 2000)
15. O'Connor, M., et al.: Supporting Rule System Interoperability on the Semantic Web with SWRL. In: Gil, Y., Motta, E., Benjamins, V.R., Musen, M.A. (eds.) ISWC 2005. LNCS, vol. 3729, Springer, Heidelberg (2005)

Supporting Adaptive Application Mobility

Francis M. David, Bill Donkervoet, Jeffrey C. Carlyle,
Ellick M. Chan, and Roy H. Campbell

University of Illinois at Urbana-Champaign, Urbana IL 61820, USA
{fdavid,donkervo,jcarlyle,emchan,rhc}@uiuc.edu
http://choices.cs.uiuc.edu/

Abstract. Application mobility has the potential to enhance user experience in ubiquitous computing environments by providing a flexible and reusable solution to managing applications across myriad computing devices, especially when applications adapt to the characteristics of individual devices. Using example scenarios, we argue that application mobility is a better solution to the problem of accessing remote applications than schemes like remote desktop which only export displays. Our mobile application framework provides the opportunity for applications to better adapt their user interface to the new environment. This ability is enhanced through the use of the Model-View-Controller design pattern. Our framework also uses discovery mechanisms to find potential migration targets. Lost applications are recovered through a simple scheme called homing. A preliminary implementation for our framework is based on the JADE mobile agent platform.

1   Introduction

We live in a world where we are constantly surrounded by computers. Libraries, airports and coffee-shops have wireless networking or public computer terminals which provide Internet connectivity. Networked mobile computing devices like laptops, PDAs and smart phones are also widely available. With the constant availability of computing devices, it has become desirable to access our data and work from anywhere. We can achieve this to some degree using tools such as remote desktop, VNC [1] and XMOVE [2] which export user interfaces through a network; however, these solutions are not perfect and are deficient in several aspects.

Let us consider the remote desktop solution to accessing applications available on your office PC through your laptop when you are away from the office. This would require that both your office PC and your laptop are connected to the Internet. This requirement holds even if the applications you are accessing don't have any use for a network. Interactivity is also poor because of network latency issues. If your applications were mobile, you could move the applications to your laptop at any time and work with them without sacrificing response times even when you are disconnected from the network.

Let us consider another scenario. You are editing a file on your office PC and would like to continue editing the file on a PDA during your commute back home.

R. Meersman, Z. Tari, P. Herrero et al. (Eds.): OTM 2007 Ws, Part II, LNCS 4806, pp. 896–905, 2007.

Normally, you would close the file on your PC, save the file to your PDA and start editing it using a PDA application. Wouldn't it be nice if the application jumped from your PC to the PDA? And automatically adjusted itself to the smaller screen size? And reduced the use of energy intensive graphical effects to save battery power? After you finish editing the file, you start watching an on-demand movie on the PDA. When you walk into your home, it would be extremely convenient if the movie player migrates to your home computer and starts to use your big plasma screen on the wall after being transformed into a high-bandwidth application. As you are watching the movie, your boss calls your smart phone. You cannot hear the phone ringing over the sound of the movie and miss the call. Ideally, a remote notification application should have been placed on the home computer by your smart phone when you walked in. The incoming phone call would have then triggered this application to take appropriate action like pausing the movie. From these examples, it is easy to see that when applications are mobile, they can present themselves to users on a more convenient device in a form more suited to that device. This enhances user experience and increases productivity.

Virtualization technology presents a different solution to the problem of pervasive application and data access. Products such as VMWare's VMotion [3] and several research projects [4,5] propose the use of virtual machine migration to manage the mobility of data and applications. One drawback of this approach is that this scheme does not work across heterogeneous platforms. Yet another drawback is that migrating a complete virtual machine consumes more time and bandwidth than just selective application migration.

We believe that a framework which supports application mobility provides a flexible and reusable mechanism to manage applications in a world where we are surrounded by computing. We discuss our design for supporting mobility and the rationale behind the choices we make in Section 2 and describe a preliminary implementation in Section 3. We highlight other related projects in Section 4 and conclude in Section 5.

2   Supporting Adaptive Application Mobility

Our design is based on the functionality already provided by frameworks like JADE [6] that support mobile agents. We can leverage existing research and work in this area in order to support application mobility. In this section, we describe modifications and extensions to the basic migration and naming functionality provided by agent frameworks.

2.1   Target Discovery

Discovery is used to continuously check for potential migration destinations. In the case of mobile devices, the framework can discover new targets as the user moves around. In the case of fixed devices, the mobility framework can be initially configured with a fixed set of target hosts and discovery can be used to detect new or mobile devices. Once a set of potential migration targets is compiled, the system can choose to migrate applications automatically based on preset application and user preferences, or it can provide a management interface and leave the decision making to the user. When one or more possible targets have been identified, parts of the application code can be speculatively migrated to them before the final destination is identified and state hand-off is performed. Speculative migration helps reduce hand-off time.

2.2   Adaptive Migration

Process migration is a well researched topic [7]. But application mobility in ubiquitous computing environments has challenges that are not addressed by simple migration. The architectures of the devices between which migration takes place could be very different and they might provide completely different types of I/O resources to the applications running on them. A common virtual instruction set, like Java, Microsoft Intermediate Language (MSIL) or LLVA [8] can be used to enable migration across heterogeneous architectures. However, in order to provide the most utility to the user, applications should adapt their user interfaces, computational requirements, and network usage when they migrate. In other words, applications need to undergo change in order to better fit new devices.

A quick solution to this problem that avoids migration completely and sim-ulates mobility is to have pre-installed device specific applications and only mi-grate portable state information and data. This is the approach taken by current PDA and cell phone synchronization software on PCs. This solution works well when software that can interpret the data is present on all possible devices that a user is expected to interact with; however, when some of the devices are pub-lic terminals this is a requirement that cannot be easily met. This problem is more pronounced when licensed, proprietary or custom applications are involved. Storage constraints also limit the usefulness of this approach.

In our framework, a migrating application is made aware of the change it is experiencing. The application can then take advantage of this information to perform pre- and post-migration tasks which facilitate adaptation to new environments. This has the positive side effect of avoiding the problems usually associated with transparent migration like maintaining network connections, file handles, and other forms of non-serializable application state.

2.3   GUI Decoupling

The GUI for the application can be migrated by serializing the object repre-senting the GUI and reconstructing the object at the destination. However, fine grained adaptation can only be achieved by re-creating the GUI. This might be required if the source and the destination have extremely different input or out-put devices. For example, this would be the case if the migration is between a PC and a PDA. In order to be able to destroy the GUI at the source and re-create it at the destination with adaptation, it needs to be decoupled from the rest of the

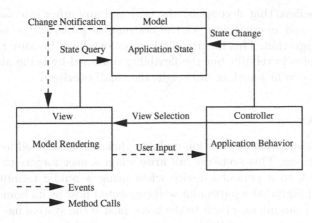

Fig. 1. Model-View-Controller

application. A version of the Model-View-Controller (MVC) pattern [9] can be used to design applications with decoupled GUIs. Figure 1 shows the decoupling of the various components of a GUI application which uses MVC. The model stores all application data and notifies the view when any data changes. The view just displays data in the model. User input is handled by the controller, which can make modifications to both the view and the model.

Multiple GUIs could be programmed for different devices and packaged with the application. The MDAT project [10] and the MAIS project [11] allow developers to specify the GUI in abstract terms and the compiler generates code for different architectures automatically. Automatic GUI code generation combined with a system of GUI repositories can be used to reduce the amount of code transfer during migration. When a new target is discovered, a remote repository of GUIs can be consulted to check for the availability of a GUI for the new target. If there is a GUI available for the new target, then the GUI can be downloaded from the repository. We believe that programming a single GUI object that adapts to different device environments when initialized is a far better approach. This, however, requires careful design by developers.

While automatic code generation lightens the burden on the programmer, pre-compiled GUI code will not work well with newer hardware. We are investigating at the use of ontologies to describe an abstract GUI that can be instantiated at run time in a device dependent manner. In this case, migration will only involve moving the GUI specification. This scheme saves network bandwidth and can work with any new device that can instantiate objects based on the specification. Our preliminary implementation does not include this functionality yet.

There have been other approaches to migrating GUIs as well. GUIevict [12], which works with the X server, saves and moves the entire GUI of an X client as pixmaps. Another approach is to interpose a proxy which maintains the state of the GUI on the display. The proxy can then save and re-create the GUI easily on another machine. In both of these cases, the code is very platform specific and hardware dependent and will not work with heterogeneous devices.

We therefore believe that decoupling the GUI and any other interface using the MVC pattern, and re-creating it at the destination provides the best solution for adaptive migration. This incurs a little more overhead because the interface has to be completely rebuilt, but the flexibility obtained by being able to create a completely different interface outweighs the small overhead.

2.4 Homing

An important issue with application mobility that needs to be addressed is that of lost applications. This scenario can arise when a user forgets to migrate an application back to a personal device when using a public terminal. If every application is designated a particular well connected host as its home, the application and data can migrate back to the home host if the system has determined that the user is no longer available.

Homing is not an issue with remote display since the program and all data are held on the server. However, using mobility, any program data not migrated remains on the host. In the case of public terminals, this equates to lost data and possible security issues. Homing, thus, provides mobile applications with a mechanism to ensure that all program data is moved to a safe, well-connected location. Homing can be initiated using a timeout mechanism or user sensing (e.g. RFID or bluetooth) to ascertain if the application has been abandoned and, if so, migration occurs automatically. In the future, we may investigate methodologies to support speculative homing, where state is saved to the destined location in the background before the timeout.

2.5 Disconnected Operation

Since applications and data dependencies are migrated to the device, rather than run directly off the network, applications that don't require network connectivity can operate without interruption after migration. This is yet another advantage of using application migration as opposed to remote application access mechanisms like VNC [13] and Windows Remote Desktop [14]. If the network connection is lost, or degraded, our applications continue to run with full interactivity on the client. Only the initial and final migration phases require access to a network.

2.6 Data Access

The data needed by applications needs to be available to them after migration. A solution that can avoid the use of a distributed file system is to program the application to migrate the data it needs as a pre-migration task. We are looking at integrating our framework for application mobility with a mobile distributed file system such as Segank [15], which also supports disconnected operation. Segank can also be used to address issues encountered with consistency when copies of data are stored in multiple locations. Open files used by a migrating

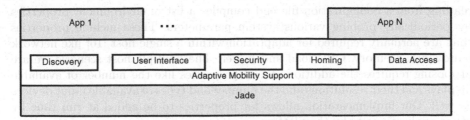

Fig. 2. System Architecture Layers

application are treated similarly to GUIs in that they cannot be migrated but instead their state is saved and they are re-created on the destination system. Open handles to system resources such as communication sockets and audio devices are re-created in a similar fashion.

2.7 Security

Running trusted applications on untrusted hardware is an issue that has a lot of open questions. The data accessed by a mobile application and the input it receives both need to be secured. There are solutions to secure data used by non-interactive programs based on computing on encrypted data. However interactive applications cannot avoid access to unencrypted data. We are currently investigating certificate based authentication of the software on migration targets with run time software signature verification using secure coprocessors. This scheme is still susceptible to various forms of attack at the I/O level when the data is being handled by peripheral devices like touch screens or keypads (e.g. false keypads and card readers have been used to steal debit card information and PIN numbers from users of ATM machines.)

3   Implementation

JADE is a Java-based platform that supports mobile agents across heterogeneous hardware. We use JADE as the base for our preliminary implementation. The JADE platform provides naming, object serialization and migration services. The architectural layers of our implementation are illustrated in figure 2. Services provided by the mobility support layer are utilized by applications for adaptive migration. Though we already have a design for homing, this has not been implemented yet.

In order to support target discovery, we assume that all devices that exist in proximity to each other have at least one network interface in a common subnet. This allows for discovery through simple IP packet broadcasts on a predetermined port at network join and periodically thereafter. In order to support adaptive migration, we added a class called Environment to the JADE code. There is only one instance of this class per host. This object initializes itself by reading from a configuration file and compiles a list of environment properties by periodically probing various system parameters. These include properties that are normally required for adaptation within a single host [16] like network bandwidth, memory, CPU load and power. The multiple host scenario we are proposing requires the addition of I/O properties like the number of available displays and their resolutions and the number and types of available input devices as well. Our implementation allows for properties to be added at run time by other applications or by the platform.

Since we chose JADE as our platform, all applications have to implement the Java interface specified by JADE. They also have to conform to behaviors defined by JADE. This constrains the ways in which applications are programmed. For example, indefinite loops have to be programmed differently. Multi-threaded applications are not supported; concurrency must be attained using multiple agents. In spite of these limitations, the JADE programming model is still powerful enough to write most types of applications.

The JADE agent interface includes notification method calls that are used to convey start and completion of migration. An application can obtain information about its environment through the Environment object. The Environment object can be queried for various properties of the system. An application programmer can use such queries to write code that changes the behavior of the application depending on changes in the system properties.

4   Related Work

Process migration has been studied extensively over the years [17,18,19,20]. Our design builds on existing process migration techniques and advocates support for adaptation to the execution environment. There is a lot of research on applications which continuously adapt to changes in their executing environments at run time. Network multimedia applications constantly track latency and bandwidth and adapt audio or video quality. Power aware applications can track energy usage on portable devices and make changes to prolong battery life. Researchers have also proposed a programming interface for application-aware adaptation for mobile devices [16]. However, adaptation after migration has not been addressed in any of these projects.

The Multi-Device Authoring Technology (MDAT) project [10] at IBM research created a toolkit to reduce the complexity of creating applications that execute on heterogeneous devices. Developers can specify an application's common aspects in a device independent form and the toolkit generates code that supports both design time and load time adaptation. The MAIS project [11] also looked at designing applications that adapt to different environments at program load time. They used an Abstract Interface Unit (AIU) model to specify the interface and decouple it from the rest of the application. Additionally, ICrafter [21], uses service descriptions to generate remotely displayed GUIs on heterogeneous hosts. However, none of these designs allow for run time adaptation.

The 2k project [22] built an architecture-aware OS which uses reflection to modify itself upon changes in its operating environment. This project concentrated on OS adaptation with applications relying on changing OS services. In this system, since applications are bound to OS services, there is implicit adaptation during migration. The GAIA project [23] also uses MVC, but simulates application mobility by only migrating state information. One.world [24] migrates applications by checkpointing the state of a running application. While it supports application adaptation, it lacks a programming model for adaptive migration. Additionally, none of these projects have the concept of homing. A recent publication presents an alternative framework for adaptive migration using ontologies [25] and is closely related to our work. The use of ontologies to govern adaptation is complementary to our design.

More recently, several web-based designs have emerged as a solution to cross-platform and multi-machine application management. Web operating systems such as YouOS [26] and EyeOS [27] allow persistence of a web desktop environment via session management. A user may log out of one machine, and resume the session at another machine. Many web services such as those offered by Google, have some limited adaptation support. For example, the web application is still usable when being viewed on a small PDA screen. In terms of state persistence, these web applications periodically save data, so that state is not lost due to loss of connectivity, or other errors. In contrast to these web-based systems, our design does not require Internet connectivity and works across local networks.

The Internet Suspend/Resume [28] project uses virtualization technology to freeze the state of a complete running system and migrates it over the Internet. In contrast to this approach, our design migrates individual applications and also supports adaptation.

5   Conclusions

In this paper, we show that adaptive application mobility is a useful and viable service in distributed environments. Application mobility can allow users to access their applications and data wherever they are. Unlike other solutions like remote access using VNC or Windows Remote Desktop, adaptive application mobility provides better user interfaces and faster response times and allows for disconnected operation. While our simple implementation is not completely flexible because of the constraints imposed by JADE, it does illustrate the basic concepts very well.

There are several open research issues that we have encountered as part of this project. We have not yet addressed mobility of distributed applications. In this case, a decision needs to be made as to which device each component of the distributed application should migrate to. Ontologies could be used to write a description of the interconnections and capabilities required by a distributed application and this can then be mapped onto the current environment.

The security of code and data during migration is an aspect that has not been addressed yet. We are studying the use of cryptography to create signatures in order to detect tampering of code during migration.

An important limitation of our design is that it requires that applications be written using a new programming model. Existing applications will need to be rewritten in order to make them mobile. While this is an impediment to the adoption of this technology, we believe that it is time to revisit the design of applications in order to exploit the benefits of the ubiquitous computing environments of the future.

Acknowledgments

We would like to thank David Raila and Anand Ranganathan for discussions that led to some of the ideas described in this paper. Part of this research was made possible by a grant from DoCoMo Labs USA.

References

1. Richardson, T., Stafford-Fraser, Q., Wood, K.R., Hopper, A.: Virtual Network Computing. IEEE Internet Computing 2(1), 33–38 (1998)
2. Solomita, E., Kempf, J., Duchamp, D.: XMOVE: A pseudoserver for X window movement. The X Resource 11(1), 143–170 (1994)
3. VMWare: VMotion. Last Checked: September 10, 2007, http://www.vmware.com/products/vi/vc/vmotion.html
4. Cáceres, R., Carter, C., Narayanaswami, C., Raghunath, M.: Reincarnating PCs with Portable SoulPads. In: MobiSys 2005: Proceedings of the 3rd International Conference on Mobile Systems, Applications, and Services, pp. 65–78. ACM Press, New York (2005)
5. Satyanarayananan, M., Kozuch, M.A., Helfrich, C.J., O'Hallaron, D.R.: Towards Seamless Mobility on Pervasive Hardware. In: Pervasive and Mobile Computing, vol. 1, pp. 157–189 (2005)
6. Bellifemine, F., Poggi, A., Rimassa, G.: JADE - A FIPA-Compliant Agent Framework. In: Proceedings of PAAM 1999, pp. 97–108 (1999)
7. Milojicic, D., Douglis, F., Paindaveine, Y., Wheeler, R., Zhou, S.: Process Migration Survey. In: ACM Computing Surveys (2000)
8. Adve, V., Lattner, C., Brukman, M., Shukla, A., Gaeke, B.: LLVA: A Low-level Virtual Instruction Set Architecture. In: Proceedings of the 36th annual ACM/IEEE international symposium on Microarchitecture, San Diego, California (2003)
9. Gamma, E., Helm, R., Johnson, R.: Design Patterns: Elements of Reusable Object-Oriented Software. Addison-Wesley, Reading (1995)
10. Banavar, G., Bergman, L.D., Gaeremynck, Y., Soroker, D., Sussman, J.: Tooling and System Support for Authoring Multi-Device Applications. Journal of Systems and Software 69(3), 227–242 (2004)
11. Ausiello, G., Bertini, E., Calí, A., Catarci, T., Kimani, S., Santucci, G.: Designing Adaptable Multidevice Applications. In: MAIS project report (2003)
12. Zandy, V.C., Miller, B.P.: Checkpoints of GUI-based Applications. In: USENIX Annual Technical Conference, General Track, pp. 155–165 (2003)

13. RealVNC, Ltd.: RealVNC - Open-Source Cross-Platform Remote Control Solution. Last Checked: September 10, 2007 http://www.realvnc.com/what.html
14. Microsoft: Remote Desktop Protocol. Last Checked: September 10, 2007, http://msdn.microsoft.com
15. Sobti, S., Garg, N., Zheng, F., Lai, J., Shao, Y., Zhang, C., Ziskind, E., Krishnamurthy, A., Wang, R.: Segank: A Distributed Mobile Storage System. In: Proceedings of the 3rd Conference on File and Storage Technologies, San Francisco, California (2004)
16. Noble, B.D., Price, M., Satyanarayanan, M.: A Programming Interface for Application-Aware Adaptation in Mobile Computing. Technical Report CS-95-119 (1995)
17. Milojičić, D.S., Douglis, F., Paindaveine, Y., Wheeler, R., Zhou, S.: Process Migration. ACM Computing Surveys 32(3), 241–299 (2000)
18. Steketee, C., Zhu, W., Moseley, P.: Implementation of Process Migration in Amoeba. In: International Conference on Distributed Computing Systems, pp. 194–201 (1994)
19. Theimer, M.M., Lantz, K.A., Cheriton, D.R.: Preemptable Remote Execution Facilities for the V-System. SIGOPS Operating Systems Review 19(5), 2–12 (1985)
20. Douglis, F., Ousterhout, J.K.: Transparent Process Migration: Design Alternatives and the Sprite Implementation. Software - Practice and Experience 21(8), 757–785 (1991)
21. Ponnekanti, S., Lee, B., Fox, A., Hanrahan, P., Winograd, T.: ICrafter: A Service Framework for Ubiquitous Computing Environments. In: Proceedings of the 3rd International Conference on Ubiquitous Computing, London, UK, pp. 56–75. Springer-Verlag, Heidelberg (2001)
22. Kon, F., Singhai, A., Campbell, R.H., Carvalho, D., Moore, R., Ballesteros, F.J.: 2K: A Reflective, Component-Based Operating System for Rapidly Changing Environments. In: Magnusson, B. (ed.) ECOOP 1998 and SCM 1998. LNCS, vol. 1439, Springer, Heidelberg (1998)
23. Roman, M., Campbell, R.H.: Gaia: Enabling Active Spaces. In: Proceedings of the 9th workshop on ACM SIGOPS European workshop, pp. 229–234. ACM Press, New York (2000)
24. Grimm, R., Anderson, T., Bershad, B., Wetherall, D.: A System Architecture for Pervasive Computing. In: Proceedings of the 9th ACM SIGOPS European Workshop, pp. 177–182 (2000)
25. Thant, K.P., Naing, T.T.: A Migration Framework for Ubiquitous Computing Applied in Mobile Applications. In: 6th Asia-Pacific Symposium on Information and Telecommunication Technologies, pp. 213–218 (2005)
26. WebShaka: YouOS. Last Checked: September 10, 2007, http://www.youos.com/
27. eyeOS Core Team: EyeOS. Last Checked: September 10, 2007, http://eyeos.org/
28. Kozuch, M., Satyanarayanan, M.: Internet Suspend/Resume. In: Proceedings of the Workshop on Mobile Computing Systems and Applications, pp. 40–46 (2002)

Workshop on Peer to Peer Networks (PPN)

PPN 2007 PC Co-chairs' Message

Welcome to the proceedings of the 1st International Workshop on Peer-to-Peer Networks (PPN 2007). The workshop was held in conjunction with the On The Move Federated Conference and Workshops (OTM 2007), November 25-30, 2007.

The primary goal of this workshop is to serve as an active forum for researchers in the field of peer-to-peer networking, to meet and exchange background information and current challenges in this area.

For its launch this year, the workshop attracted 19 submissions from Asia, Canada, Europe, Africa, and the USA. Each paper went through a rigorous peer review process, with each submission receiving at least three reviews from members of the Technical Program Committee as well as additional reviewers. These reviews provided higher and detailed comments on the quality of the submitted papers. Based on the referee reports, we selected 11 best quality papers for presentation at the workshop. Other submissions are also good quality, but unfortunately could not be included in the workshop due to the strict requirement of the acceptance rate imposed across all the federated conferences.

We believe that the accepted papers provide interesting and up-to-date results on different areas related to peer-to-peer networking. We hope that you will find the papers interesting.

We would like to thank all the people who made an effort to make this workshop a success. First, we would like to thank all the members of the Technical Program Committee and the invited reviewers for their amazing effort in providing high-quality reviews which greatly helped us in the selection process.

We would also like to thank the OTM 2007 Organizing Committee for their help and support. Special thanks are due to Zahir Tari and Pilar Herrero, respectively, OTM General Co-chair and OTM Workshops General Chair, and Vidura Gamini Abhaya, Workshops Publication Chair, for their help and guidance over the past few months. Finally, we would like to thank all the authors who chose PPN to submit part of their recent work.

August 2007

Farid Naït-Abdesselam
Jiankun Hu
Azzedine Boukerche

Nomad: Virtual Environments on P2P Voronoi Overlays

Laura Ricci and Andrea Salvadori

Department of Computer Science
Largo Bruno Pontecorvo, Pisa, Italy
ricci@di.unipi.it,
asalvad@cli.di.unipi.it
http://www.di.unipi.it/ricci

Abstract. *Nomad* is a support for the development of *distributed virtual environments(DVE)* on P2P networks. In a *DVE* each avatar generally interacts with other ones in its area of interest, i.e. the *DVE* region surrounding its location. *Nomad* exploits a *Voronoi partition* of the *DVE* to define a highly dynamic P2P overlay that connects a peer with those controlling avatars in its area of interest. This paper introduces an accurate definition of neighbours and describes the basic protocol defined by *Nomad*. This protocol is then refined through a set of optimizations. Finally, preliminary experimental results are presented.

1 Introduction

In the last years, several novel applications have been added to first generation *P2P* applications, such as file sharing and instant messaging. A new class of applications which could fully exploit the P2P model is that of *Distributed Virtual Environments (DVE)* which include both distributed simulations developed for military or civil protection purposes and massively multiplayer online games (MMOG), for instance World of Warcraft or Second Life. The definition of a scalable communication support is a basic issue for the wide diffusion of these applications in a P2P environment. Several proposals [4,5,6,7,10] exploit the concept of *Area of Interest (AOI)* to define these supports. The *Area of Interest* of an avatar in a *DVE* is the region of the virtual world surrounding it. Each avatar is generally aware of the avatars and passive objects in its *AOI*. *AOI* may be implemented through multicast groups or through publish subscribe systems but the scalability of these solutions is fairly low. An alternative solution exploits a fully decentralized overlay network, to support direct interactions between peers controlling avatars aware of each others. These overlays are highly dynamic because their structure may change anytime the position of an avatar changes. Direct interactions support a reduction of latency while scalability may be increased by exploiting the *AOI* to minimize the number of interactions. The main problem posed by this approach is to guarantee the connectivity of the P2P overlay, and to inform an avatar when another one enters its *AOI* from the *DVE* portion outside its *AOI*. This requires the definition

R. Meersman, Z. Tari, P. Herrero et al. (Eds.): OTM 2007 Ws, Part II, LNCS 4806, pp. 911–920, 2007.

of a set of 'beacon avatars' to detect new nodes entering the AOI that may be chosen among those in the AOI. However, if the AOI is empty, avatars located outside the AOI have to be chosen in order to guarantee network connectivity. A recent proposal [3,4] suggests the adoption of *Voronoi Diagrams* to partition the DVE and assign each resulting region to a distinct DVE node. In this approach, 'beacon' neighbours of a node n are chosen according to the resulting partitioning.

This paper presents *Nomad*, a P2P support for DVE based on a Voronoi approach. A node may exploit *Nomad* to join a DVE, to notify the update of its position to its neighbours, to discover new neighbours and to leave the overlay. In *Nomad* each avatar has a partial view of the DVE, hence different avatars may have inconsistent views of the DVE. Furthermore, the latency and the reliability of the transport protocol may increase these inconsistencies. As a matter of fact, a reliable connection oriented protocol is not suitable because of the high dinamicity of the overlay and of the low amount of exchanged data. *Nomad* introduces mechanisms to speed up the convergence of the partial views of the different nodes and to reduce the number of messages.

The paper is organized as follows. Sect.2 describes some recent proposals of P2P DVE. Sect.3 presents the approach based upon Voronoi Diagrams. Sect.4 describes the main *Nomad* functionalities, while Sect.5 describes the support in more details. Preliminary experimental results are shown in Sect.6 and some conclusion are discussed in Sect.7.

2   Related Work

Solipsis [5] is a P2P support for DVE that has the goal of scaling to an unbounded number of participants. Each peer implements the entities of the virtual world and perceives its surroundings. Each entity perceives only a part of the virtual world, its *Awareness Area*, inhabited by some entities and it should be aware of all updates to the virtual representations these entities. The DVE should satisfy a *Global Connectivity* property to guarantee that no entity will 'turns its back' to a portion of the world.

Mopar [6] defines an overlay network using both a Pastry DHT and an hybrid P2P architecture. It decomposes the DVE into hexagonal cells, and introduces master, slave and home nodes. Each cell has at most one master node and several slave nodes. If a new node finds that there is a cell with no master node, it registers itself as the master node. Otherwise, it becomes a slave node. Master nodes build direct connections with the neighboring master nodes through home nodes of neighboring cells. Slave nodes query the master node to build direct connections with their neighboring slave nodes. In this way, master nodes have not to be involved in the protocol to notify the accurate positions to the neighboring participants. SimMud [7] is a support for DVE built on top of Pastry, a widely used DHT, and it uses Scribe, a multicast infrastructure built on top of Pastry, to disseminate game state. The approach exploits locality of interest typical of DVE. Von [4] exploits Voronoi Diagrams [2] to assign to each node a distinct DVE region resulting from a proper partition of the virtual world. Von preserves high overlay topology consistency in a bandwidth-efficient manner. [8,9,10] describe a publish subscribe approach to the definition of DVE and introduce a set of mechanisms to guarantee the consistency of the virtual environment. An exhaustive comparison of these approaches is presented in [3].

3   Voronoi Dynamic Overlays

In the following we consider a Voronoi tessellation of the plane because 2D DVE will be taken into account. We suppose that each node of the DVE control a single avatar. As a consequence in the following the term node and avatar will be used in an interchangeable way, according to the context.

Given a set of points S in the plane, a Voronoi tessellation is a partition of the plane which associates a region $V(p)$ with each point p in S so that all points in $V(p)$ are closer to p than to any other point in S. The goal of a Voronoi based approach for DVE is to define a highly dynamic overlay network, where each node n interacts mainly with the neighbours in its AOI. The Voronoi approach supports a straightforward definition of the set B of 'beacon nodes' of a node n which inform n when a new node enters its AOI. B should be chosen in a way that pairs each region surrounding $AOI(n)$ with one node of B. In this way, each node approaching $AOI(n)$ is detected by one beacon node which then informs n. Furthermore each beacon may inform n of new neighbours discovered as a result of its movement. We can notice that the beacon nodes are not necessarily located in the AOI of a node because, if the AOI is empty, the node cannot discover new neighbours as a result of a position update. So the node looses the connectivity with the rest of the network. To solve this problem we must define a new type of neighbours, called enclosing neighbours in order to guarantee the connectivity of the overlay. An accurate definition of AOI and 'beacon' neighbours is required to define the *Nomad* protocol. At first, we define them by considering $Voro_{Glob}$, the Voronoi diagram including all the nodes of the DVE. Then, we update the definitions by considering $Voro_n$, the Voronoi diagram built by considering the partial view of the DVE at node n and prove that the two definitions are consistent. Let n,m be a pair of nodes in $Voro_{Glob}$.

Definition 1. *m is an* enclosing neighbour *of n iff exists a border of $V(m)$ that overlaps a border of $V(n)$.*

Definition 2. *m is an* AOI neighbour *of n iff $m \in AOI(n)$.*

Definition 3. *m is a* boundary neighbour *of n iff at least one of the following conditions holds.*

- *$m \in AOI(n)$ or it is an enclosing neighbour of n and $V(m) \cap AOI(n) \neq V(m)$*
- *$m \in AOI(n)$, $V(m) \cap AOI(n) = V(m)$, $\exists k$, $k \notin AOI(n)$, k enclosing neighbour of m, such that $V(k) \cap AOI(n) \neq 0$.*

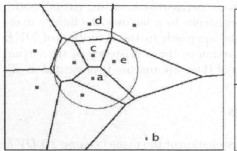

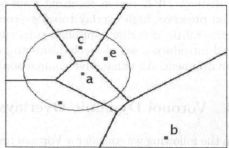

Fig. 1. Neighbour Classifications   **Fig. 2.** Voronoi Diagram at node a

The boundary neighbours correspond to the 'beacon nodes'. According to Def.3 boundary neighbours of n are the nodes whose Voronoi region intersects the borders of the $AOI(n)$ or whose enclosing neighbours are located outside the AOI, but control a region intersecting $AOI(n)$. Enclosing neighbours of n should be considered as well, anytime the Voronoi partition does not assign some regions of the DVE surrounding n to one of the previous neighbours.

Fig. 1 shows distinct neighbours of node a. e is an AOI neighbour, because it belongs to $AOI(a)$. It is also a boundary neighbour because $V(e)$ intersects the border of the AOI. Furthermore e is also an enclosing neighbour. Even if $V(c)$ is a subset of $AOI(a)$, c is a boundary neighbour because its enclosing neighbour d is located outside the AOI, and $V(d)$ intersects the AOI. Finally, even if $V(b)$ does not intersect $AOI(a)$, b is a boundary neighbour because it is an enclosing neighbour that covers a region not assigned to other boundaries.

The final goal of Voronoi based approach is to minimize the number of nodes each node n of the DVE interacts with. Since n has a partial view of the DVE, its Voronoi diagram $Voron_n$ differs from $Voro_{glob}$, because it includes only a node subset. As an example, a partial view of n may include its AOI and enclosing neighbours. In Fig.2, $Voro_a$ differs from $Voro_{glob}$ because it either associates with AOI or with enclosing neighbours of a, some regions that in $Voro_{glob}$ belongs to nodes outside $AOI(a)$.

A simpler condition can be defined to find out boundary neighbours in $Voron_n$ in such a way that each node finds out its boundary nodes by looking at its AOI and enclosing neighbours. This condition requires to check if the Voronoi region of any AOI or enclosing neighbour of n includes at least a point outside $AOI(n)$. The following theorem shows that this condition is equivalent to the one of Def. 3.

Theorem 1. *Let m and n be two nodes, where m is an AOI or an enclosing neighbour of n. If $V(m) \cap AOI(n) \neq V(m)$ holds in $Voron_n$, then m is a boundary neighbour of n in $Voro_{glob}$.*

Proof. Let us consider n and m in $Voron_n$ such that $V(m) \cap AOI(n) \neq V(m)$ holds. Hence, $V(m)$ includes at least a point outside $AOI(n)$. If this condition holds in $Voro_{glob}$ as well, m is trivially a boundary neighbour of n because of the first condition of Def. 3. Let us now suppose that the condition is violated in $Voro_{glob}$, so that a subset of $V(m)$ is assigned to some of its enclosing neighbours which are not visible by n, i.e. are located outside $AOI(n)$. Here, m is a boundary neighbour of n because the second condition of Def.3 is satisfied.

Hence, each node n can detect its boundary neighbour by detecting in its Voronoi diagram any AOI and the enclosing neighbours whose region contains at least a point outside its area of interest. It is worth noticing that our definition of boundary neighbours updates those in [3,4] to guarantee the connectivity of the overlay.

4   The Nomad Protocol

Nomad defines a protocol to manage a Voronoi overlay. The current version of *Nomad* only supports the notification of position updates among avatars, that act as *heartbeats* notifications. We are currently extending the protocol to manage passive DVE objects. Furthermore, in Nomad, the AOI of avatars are circular region centered on the avatar, and any AOI has the same (static) radius.

While [3,4] defines a P2P Voronoi overlay, it does not investigate the definition of a support for the application level protocol on a real transport protocol. Instead, to fully exploit the overlay characteristics, *Nomad* Voronoi overlay is based upon the UDP transport protocol. As a matter of fact, the structure of the overlay is highly dynamic, i.e. it may change at each movement of an avatar, and a low amount of information, i.e. the avatar position and a payload with status information, is exchanged at each interaction. Hence, the choice of TCP is not appropriate, because of the cost of opening and closing dynamically a large amount of new connections. On the other hand, *Nomad* requires a set of mechanism to recover errors due to the low reliability of the underlying protocol as well as mechanisms to increase the consistency of the local views of the DVE which may differ at distinct nodes.

This section introduces the basic *Nomad* mechanisms, while the following one describes those that increase the reliability of the protocol and the consistency of the distributed views. The basic functionalities supported by *Nomad* are:

- **Join**. A new node J joins a *Nomad* overlay by notifying its initial position in the DVE and the IP address of a *bootstrap node*, i.e. any active node on the overlay, to the *Nomad* support which notifies the proper enclosing and AOI neighbours to J.
- **Position Update**. An application exploits this function when the avatar moves or to periodically send a keep alive and application payload messages. If an avatar A changes its position, the support updates the Voronoi diagrams of the node associated with A and of its neighbours.
- **Neighbour Discovery**. *Nomad* defines mechanisms to discover new AOI neighbours or enclosing neighbours of a node in the DVE.

- **Leave.** *Nomad* properly updates the Voronoi diagram of all the neighbours of a node that departs from the Voronoi overlay.

The join request is forwarded from the bootstrap node to the node A such that $V(A)$ includes the initial position of the joining avatar J, by a greedy routing algorithm [4]. That algorithm forwards, at each routing step, the join request to the neighbour closest to the final destination. A inserts J in its Voronoi diagram and exploits this diagram to discover the neighbours of J. These are notified to J, but no notification is sent to J's neighbours that can discover J through the discovery mechanisms associated to position updates messages.

Nomad periodically sends an heartbeat, including the current position of J and an optional application payload, to both the AOI neighbours and the enclosing neighbours of J. The maximal speed of J is bounded in order to guarantee that J can bypass a boundary neighbour B only after B has notified J of unknown neighbours. A heartbeat is sent also to the nodes that are no longer neighbours of J after its movement. In this way, they can discover that J is no longer their neighbour.

On the other way, no explicit request should be done at application level to discover new neighbours because the support detects them by pairing a discovery request with each Nomad heartbeat to the boundary neighbours. This is implemented by tagging the heartbeat message to these neighbours to inform them of a discovery request. When a node receives a discovery request from a node N, it looks for new neighbours of N by exploiting the received position of N and the radius of $AOI(N)$. N may receive duplicate notifications for the same neighbour, because the same neighbour notifies the same node more than one time or the same notification is received by distinct neighbours. The next section describes some strategies to reduce the number of redundant notifications. N discards duplicate notifications by checking if the notified node is already present in its Voronoi diagram and notifies the new nodes to the application. Sect. 5.2 introduce a further mechanism to discover neighbours.

Nomad can manage both voluntary peer disconnections from the overlay and unexpected peer departures. In the former case, a leave message is explicitly sent by the leaving node. Unexpected departures are managed by assigning a TTL to each neighbour, which is periodically decremented by an *aging procedure*. Furthermore, the TTL of n is set to its starting value when receiving a message from n. When the value of the TTL becomes 0, the corresponding node is eliminated from the local Voronoi diagram and it is inserted into a *Blacklist* that also includes nodes that have sent a *Leave Message*. The loss of a leave message is managed as an unexpected departure.

The *Blacklist* is exploited to avoid fuzzy situations, where a node n eliminated from the local Voronoi Diagram is immediately reinserted because some boundary neighbour still notifies the presence of n. This may happen in a fully P2P distributed environment where all the neighbours of a node cannot detect the node disconnection simultaneously. When a neighbour n is notified, it is inserted in the Voronoi Diagram if and only if it is does not belong to the blacklist. A node n can be removed from the blacklist if a notification is received from it.

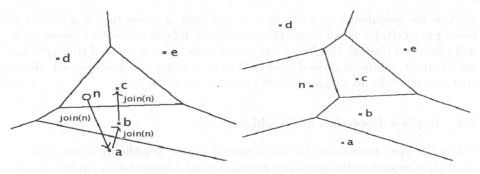

Fig. 3. Joining node n   **Fig. 4.** Voronoi Diagram of node c

Finally, an *aging mechanism* the blacklist nodes is defied as well to limit its size and to recover erroneously detected disconnections.

5   Improving the Protocol

The protocol defined in the previous section may be improved to speed up the convergence of the local views of the DVE at distinct nodes and minimize the number of messages. *Nomad* cannot avoid any inconsistency because of both the latency of the interconnection network and the reliability of the underlying protocol. For instance, latency jitter prevents a simultaneous notification of a position update to all the neighbours of a moving avatar and notifications may be lost or delivered out of order. The number of messages to implement the protocol greatly affects the system scalability. Therefore, a set of optimizations to minimize these messages is a key point of *Nomad*.

5.1   Joining *Nomad*: Improving Initial Neighbours Discovery

In the protocol previously described, the neighbours of a node J joining the overlay are notified by the node A holding the region including the starting position of the avatar controlled by the node. Because of its limited view of the DVE, A may not know some neighbours of J. To speed up both the discovery of the neighbours of the joining node and of the new node by the other ones, *Nomad* involves in the discovery process each node that routes the join message toward its destination. When routing a join message, a node checks if the joining node J is an AOI or an enclosing neighbour. In this case, it adds J to its neighbour list, it discovers J neighbours in the updated diagram and notifies them to J. This increases the number of neighbours discovered during the joining phase and it make the protocol more robust by increasing the probability that a node joins the network even if some messages are lost.

Figure 3 shows the routing path of the join message of node n in $Voro_{Glob}$ if a is the bootstrap node and the initial position of n belongs to $V(c)$. The path includes node a, b and c. Figure 4 shows $Voro_{Glob}$ after the insertion of

n. Now the neighbours of n are a, b, c, d. Figure 3 shows that if a does not belong to $AOI(c)$, c is not aware of a and cannot inform n about its presence. n will discover a through the *Nomad* discovery mechanism associated to heartbeat notifications. In the improved protocol, when routing the message, a realizes that n is its enclosing neighbour and inform n about itself.

5.2   Implicit Detection of Neighbours

While the basic mechanism for the detection of new neighbours is based upon an explicit request to the boundary nodes, *Nomad* defines also an implicit mechanism to detect new neighbours because when receiving an heartbeat from n, a node checks if n belongs to its area of interest and, in this case, it adds n to its neighbour list. This mechanism is sound because all the area of interest have the same radius so that the relation 'belongs to an area of interest' is symmetric, i.e. if A belongs to the $AOI(B)$, B belongs to that of $AOI(A)$.

5.3   Managing Out of Order Notifications

Notifications may be received out of order because of the underlying protocols. Each instance of *Nomad* associates a *logical timestamp* with each message sent on the overlay. The timestamp is a counter incremented when sending a heartbeat. An heartbeat received by node m from n, is out of order if the value of its timestamp is less than the more recent timestamp from n. Such an heartbeat is discarded because it is obsolete. The timestamp is exploited to detect aged replies to discovery request as well. Each node receiving a discovery request with timestamp T, pairs T with the reply message, i.e. the discovery response. The sender of the discovery request can discard replies that are obsolete.

5.4   Reducing Discovery Traffic

The explicit *Nomad* discovery mechanisms sends a discovery request to the boundary neighbours which send a reply message with the new discovered neighbours. Each boundary node B may simply insert in the reply message all the nodes it has detected and that belong to the $AOI(N)$ or are enclosing neighbours of N, where N is the requesting node. This set may include nodes already known by N, because they have been previously notified by B itself or by other neighbours of N. To minimize the number of messages, *Nomad* should include in the discovery reply the nodes not known by N only. Since in a fully distributed environment the corresponding protocol is extremely complex, because of the limited knowledge of each node, *Nomad* implements a simpler strategy. Each boundary node insert in a discovery reply a node only if it has not been previously notified to the same neighbour. For this reason, a boundary node maintains, for each neighbour n, $KnownList(n)$, the list of nodes notified to n. A node may be removed from the knownlist when the neighborhood relations are updated and be re-inserted in that list later.

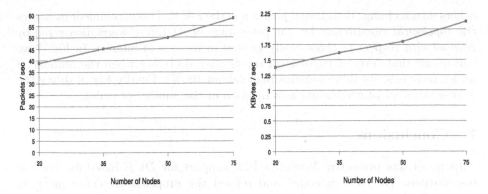

Fig. 5. Network Traffic: Packets   **Fig. 6.** Network Traffic: KBytes

6   Experimental Results

Nomad has been implemented in JAVA [1] and tested on a *LAN*. We have experimented the inefficiency of the standard JAVA serialization mechanism and we have defined an ad hoc serialization strategy reducing the number of bytes needed to code *Nomad* messages. The first set of experiments test the scalability of the support by measuring the *network traffic*, i.e. the number of packets/kbytes exchanged as a function of the number of nodes. In each experiment the network traffic has been measured for 5 minutes, starting from the instant when all the nodes have joined *Nomad*. The application is a simple game where avatars moves randomly. Each avatar chooses a random direction, then moves in that direction and changes its direction when it finds some obstacles or when a fixed period of time has elapsed. Fig.5 and Fig.6 show, respectively, the average number of packets/KBytes transmitted by a node in a second. The experiments show that the increase of sent packets/kBytes is small when the number of nodes increases. In the second set of experiments a single avatar of the *DVE* moves, while the

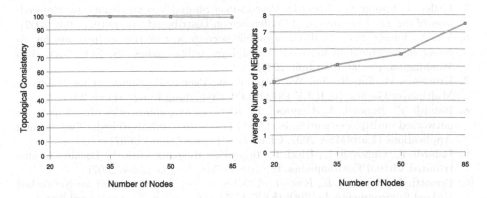

Fig. 7. Topological Consistency   **Fig. 8.** Average Neigh. Number

others do not change their initial position in the DVE. This experiment measures the *topological consistency*, i.e. the percentage of *AOI neighbours* discovered by *Nomad* with respect to those really located in its area of interest. In Fig. 7 we can notice that, even if topological consistency decreases when the number of nodes increases, its value remains high (about 98 %). Finally Fig. 8 shows the average number of neighbours as a function of the number of nodes.

7   Conclusions

This paper has presented *Nomad*, a P2P support for DVE based on *Voronoi* tessellations. We have extended and refined the proposal presented in [4] in several directions. A formal definition of different kind of neighbours has been given which guarantees the connectivity of the overlay. We have developed and evaluated a prototype which speeds up the convergence of the different views of the DVE at different nodes. While the focus of [3,4] is not the choice of the most suitable transport protocol to support DVE applications, we have chosen UDP and defined a set of mechanisms to face its unreliability. We plan to extend *Nomad* to consider the management of DVE passive objects and *crowding scenarios*, i.e. situations where a high number of avatars are located in the same zone of the DVE. Finally, we are porting *Nomad* on a WAN to test the effect of latency jitter and high rates of packet loss on its behaviour.

References

1. Salvadori, A.: Nomade: Overlay Network Dinamiche basate su Diagrammi di Voronoi per Ambienti Virtuali Distribuiti, Grad. Thesis, Univ. of Pisa (April 2007)
2. Aurenhammer, F.: Voronoi Diagrams-A Survey of a Fundamental Geometric Data Structure. ACM Computing Surveys 23(3), 345–405 (1991)
3. Hu, S.: Scalable Peer-to-Peer Networked Virtual Environment, Master's thesis, Tamkang University, Taiwan (January 2005)
4. Hu, S., Chen, J., Chen, T.: VON: A scalable peer-to-peer network for virtual environments. IEEE Network 20(4), 22–31 (2006)
5. Keller, J., Simon, G.: Toward a Peer-to-Peer Shared Virtual Reality. In: Proceedings of the 22nd International Conference on Distributed Computing (July 2002)
6. Yu, A., Vuong, S.T.: MOPAR: a mobile peer-to-peer overlay architecture for interest management of massively multiplayer online games. In: NOSSDAV 2005 Washington, pp. 99–104 (June 2005)
7. Knutsson, B., Lu, H., Xu, W., Hopkins, B.: Peer-to-Peer Support for Massively Multiplayer Games. In: IEEE INFOCOM 2004, Hong Kong (March 2004)
8. Baiardi, F., Bonotti, A., Genovali, L., Ricci, L.: A publish subscribe support for networked multiplayer games. In: IASTED Internet and Multimedia Systems and Applications (EuroIMSA 2007, Chamonix, France (March 2007)
9. Baiardi, F., Genovali, L., Ricci, L.: Improving Responsiveness by Locality in Distributed Virtual Environments. In: 21th ECMS, Prague (June 2007)
10. Bonotti, A., Genovali, L., Ricci, L.: DIVES: A Distributed Support for Networked Virtual Environments. In: 20th IEEE AINA 2006, Wien, Austria (April 2006)

A Novel Overlay Network for a Secure Global Home Agent Dynamic Discovery

Ángel Cuevas, Rubén Cuevas, Manuel Urueña, and Carmen Guerrero

Departamento Ingeniería Telemática, Universidad Carlos III de Madrid,
28911 Leganés (Spain)*
{acrumin,rcuevas,muruenya,guerrero}@it.uc3m.es

Abstract. Mobile IP and Network Mobility are the IETF proposals to obtain mobility. However, both of them have routing limitations, due to the presence of an entity (Home Agent) in the communication path. Those problems have been tried to be solved in different ways. A family of solutions tries to improve the routing by locating closer Home Agents making shorter the communication path. These techniques require a method to discover a close Home Agent from the Mobile Device. We proposed peer-to-peer based solution, *Peer-to-Peer Home Agent Network*, in order to discover a close Home Agent. This paper defines the necessary mechanisms to make this solution secure based on a mechanism named Secure Join Procedure.

1 Introduction

Mobile IP (MIP) is the mechanism proposed by the Internet Engineering Task Force (IETF) to enable host mobility, in IPv4 (MIPv4) [1] and IPv6 (MIPv6) [2]. However, mobility is also required in networks (planes, trains, etc). Hence, support for network mobility is required. The Network Mobility (NEMO) Basic Support Protocol [3] [4] is the IETF proposal to provide network mobility support. The basic solution of Mobile IP and NEMO presents the so-called triangular-routing as the main performance limitation: mobile nodes' communications must pass through an entity, called the Home Agent (HA). It is possible that some communications suffer from higher delays than those required by some kind of applications (e.g. real time applications like voice or video) in order to obtain an acceptable performance. Several solutions have been proposed in order to solve these routing problems. One family of solutions proposes (so as to improve the routing) to reduce the distance between the HA and the mobile devices as much as possible, minimizing the total path length. This paper is based on [9] which proposes the use of an overlay peer to peer network (Peer-to-Peer HA Network), formed by HAs, in order to discover a close HA to a certain mobile device. It is simple, fully global,

---

* This work was supported by the European Commission through NoE CONTENT FP6-CONTENT-038423, the Spanish government through the Project IMPROVISA TSI2005-07384-C03-027 and the Madrid regional government through the Project BIOGRIDNET CAM-S-0505/TIC-0101.

R. Meersman, Z. Tari, P. Herrero et al. (Eds.): OTM 2007 Ws, Part II, LNCS 4806, pp. 921–930, 2007.

dynamic and it can be deployed in IPv4 and IPv6. But [9] does not consider the security aspects, thus, this paper describes the main security mechanisms needed to make the Peer-to-Peer HA Network a secure solution.

Peer-to-Peer Home Agent Network (P2PHAN) is an architecture focused on a structured DHT (Distributed Hash Table) based Peer-to-Peer network. This kind of Peer-to-Peer (p2p) networks have been extensively investigated and several approaches have been defined (e.g. Chord [10] or Kademlia [11])[1]. An important effort has been done in security aspects for p2p networks and the main problems have been identified, specially for file sharing p2p networks [13] [14]. Moreover, security becomes a primary issue when p2p is applied to scenarios as the one considered in this paper, the Home Agent discovery.

This paper focuses on the security of a specific application, the P2PHAN. It has some specific features different from the file sharing scenario but also common problems. Security issues can be solved because of some specific features of this architecture, as verifiable data based on Border Gateway Protocol (BGP) information [15] and a reduced number of peers in comparison with a file sharing p2p networks.

In addition, a mechanism that secures the communications between MRs and HAs (i.e to guarantee the trust between HAs and MRs) must be used. It is IKEv2 [16] and its application to mobile environment can be done as it is proposed in [17].

All this guarantees a practical high security level for the P2PHAN approach. The paper proposes a main mechanism, *The Secure Join Procedure*, and some others associated to this one (as redundancy or parallel queries) to guarantee the security on the P2PHAN. The Secure Join Procedure is based on the use of a central bootstrapping server. The presence of bootstrapping nodes is used in commercial p2p networks since it is an efficient method for the peers to join the network and find other peers (e.g. Emule [12]). The Secure Join Procedure contains a secure Peer-ID assignment based on random assignment. This solves the main cause of possible attacks in the structured DHT p2p Networks which is that peers can choose its own Peer-Id. In addition, the paper evaluates the complexity of the possible attacks concluding that the proposed mechanisms introduce a practical level of security.

The structure of this document is as follows. In section 2, the Peer-to-Peer Home Agent Network will be more accurately defined. Section 3 exposes the security problems of the Peer-to-Peer Home Agent Network and the mechanisms which solve them and Section 4 shows the conclusions extracted from this work and introduces the further work to be developed.

2   Peer-to-Peer Home Agent Network Architecture

Peer-to-Peer Home Agent Network is a structured DHT p2p network with ring structure formed by HAs. It is similar to Chord [10]. In our scheme the search key will be: *hash(AS number)*.

---

[1] Detailed information about peer to peer networks can be found in [21] and [22] which are surveys about this technology.

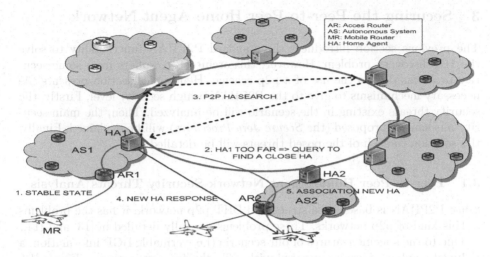

Fig. 1. P2P Home Agent Network Scheme

When one node joins the P2PHAN, it chooses an identifier (*Peer-ID*) from the ids pool. Its position in the ring is determined by the chosen id (it is placed between the two nodes with the closest higher and the closest lower id *Peer-ID* than its own id). Each peer has direct references to its two neighbors and also with other peers (crossing the ring) so as to make faster the routing within P2PHAN. These references are called *fingers*. Each peer uses the fingers to create its P2PHAN routing table.

On the other hand, each peer must store its Autonomous System (AS)[2] number within P2PHAN. The peer obtains a key by computing the *hash(AS number)*. Then, it looks for the peer with the most similar Peer-ID to that key and sends to this peer the key and its IP address. The destination peer stores the pair *<key, IP address>*.

Eventually, an MR connected to a HA_1 detects that the distance to this HA is higher than the desired (e.g. it measures RTT with HA_1 higher than a threshold). Then, it launches the procedure to discover a closer HA. The MR sends its current CoA to HA_1. At this point, HA_1 discovers (using BGP) the CoA's AS number. Afterwards, it computes the *hash(AS number)* which is the *search key*. With this *search key* the HA launches a search within the P2PHAN and obtains the list of the HAs placed at the same AS as the MR. The list is sent to the MR which decides its preferred HA.

Fig. 1 shows the P2PHAN functionality explained above. A more detailed explanation can be found in [9]. It must be noticed that the solution has been explained for NEMO but it also works on MIP.

---

[2] In the Internet, an autonomous system (AS) is a collection of IP networks and routers under the control of one entity (or sometimes more) that presents a common routing policy to the Internet [5].

3   Securing the Peer-to-Peer Home Agent Network

The previous section introduces the standard P2PHAN functionality to solve the HA discovery problem. However, this architecture suffers from some security problems due to the use of a p2p scheme. Hence, this section presents the necessary mechanisms to give to the P2PHAN a high security level. Firstly, the security threats existing in the scenario will be analyzed. Then, the main security mechanism proposed (the *Secure Join Procedure*) will be explained. Finally, the solution to each of the posed threats will be detailed.

3.1   Peer-to-Peer Home Agent Network Security Threats Analysis

Since P2PHAN is based on a structured DHT p2p network, it has the problems of this kind of p2p networks. These problems are fully detailed in [13] and [14].

Due to the specific features of our scenario (i.e. verifiable BGP information, a reduced number of users compared with a file-sharing application as Emule [12] or Kademlia [11]), it allows to define a number of security and trust mechanisms which offer a high security level. It must be also considered that when we refer to a malicious node it can be both a node of a malicious user or a non-malicious node with a bad behaviour (e.g. due to an hardware or software failure). HAN approach.

Following, the security problems affecting the P2PHAN are described:

1. *Starvation Attack:* A malicious node does not answer (or return false information) to the queries which are soliciting information that this node is storing. Then, the nodes which have registered their information in this malicious node cannot be contacted. In the P2PHAN if a malicious peer selects its Peer-ID as close as possible to a given key = hash(AS), it becomes the closest HA to that key and thus, the responsible of storing the information of the given AS. In this situation, this malicious peer can starve the given AS by not answering (or by giving false answers) to the queries soliciting the information of this AS. We call this attack *targeted starvation attack*. There is a less sophisticated version of this attack. It occurs if the malicious peer does not have any specific target AS, thus, it obtains any Peer-ID and if this Peer-ID has associated any AS key (i.e. hash(AS number)) it could starves that AS. Finally, the attack can be performed by one single attacker or by several attackers. When several attackers works together, the attack is called *Collaborative Attack* and usually it is more harmful because the resources to perform the attack increase with the number of attackers collaborating.

2. *Routing Attack:* A malicious node does not route the messages or select bad routes for the queries. If we focus on the P2PHAN, this attack can have different objectives. The first one is affect the performance of the P2PHAN without any other more specific objective, this can be interpreted as a non-targeted routing attack. On the other hand, the objective of the routing attack could be starve a victim node. That is, the malicious peer selects the Peer-IDs so as to obtain all the fingers of the victim node. In this situation,

all the queries sent by the victim node must be routed by the malicious peer(s) which does not route (or selects bad routes for) the queries of the victim node. We call to this attack the *targeted routing attack* and it is a good example where the collaborative attack is more effective.

3. *High Rate of Joins and Leaves:* A malicious node joins and leaves the P2PHAN continuously in order to make the topology unstable and generate a huge amount of signalling traffic. It could be also a collaborative attack.

4. *Register False Information:* A malicious node registers a false AS different from the AS where it is located.

5. *Multiple Registers:* A malicious node joins the network several times with the same IP Address in order to obtain as many Peer-ID as possible.

3.2   Secure Join Procedure

In order to define the *Secure Join Procedure* (SJP) in the proposed scenario the re-use of a Bootstrapping Server as security point is proposed. The main security function of this Bootstrapping Server is to assign a random identifier for the new HA which wants to become a member within the P2PHAN. However, this bootstrapping server can not guarantee the Secure Join Procedure itself. Therefore, the next method will be applied in order to get a secure access to the P2P network.

First of all, if an organisation managing an AS wants to introduce HAs in the P2PHAN, it has to create a pair public key-private key (AS_pu_key - AS_pr_key). Therefore, if a HA wants to register itself within the P2PHAN, it must own the AS_pr_key to be able to register its information within the P2PHAN. The list of HAs of an AS is stored for a node in the P2PHAN (which is another HA). This node is called *Responsible HA*. When a HA tries to register its information, its *Responsible HA* will use the AS_pu_key as it will be described later to check that the new HA trying to join the P2PHAN knows the AS_pr_key. This implies that the new HA is an authorised node of that AS. The *Responsible HA* can obtain the AS_pu_key from a repository or it could be included in the registration message of the first HA of an AS which is registered within the P2PHAN. Following the SJP is described.

A new HA which wants to join the P2PHAN sends a *Join Request* to the Bootstrapping Server (See step 1 in fig. 2) with its IP address, the AS number and a cheksum of all this information ciphered with the private key (AS_pr_key) of the AS where it is located, that is, its signature.

After that, the Bootstrapping Server generates a random peer-ID for the new HA and launches a search in the p2p network to find the HA which has the most similar ID to the peer-ID generated, which is the Responsible HA (See step 2 in fig. 2). Then, the bootstrapping server forwards the message received from the new HA adding the peer-ID generated to the Responsible HA (See step 3 in fig. 2).

The function of this *Responsible HA* is to make several tests in order to check if the new HA is a malicious node. If one of this test is not successful the *Responsible HA* returns a *Check Failure* to the *Bootstrapping Server*. Otherwise, the new HA information is stored by the *Responsible HA*. The following three tests are executed (See step 4 in fig. 2):

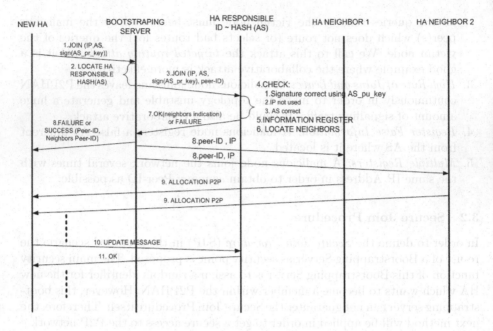

Fig. 2. Secure Join Procedure Message Exchange

1. The *Responsible HA* uses the AS_pu_key which has stored in order to check if the checksum obtained is the same than the checksum ciphered with the AS_pr_key for the new HA. If it is not, it returns a *Check Failure*, otherwise it runs the second test.

2. The *Responsible HA* checks whether it has information stored for the IP which appears in the *Check Request* or not. If it has information stored for that IP, it returns a *Check Failure*, otherwise it runs the third test.

3. The *Responsible HA* checks if the IP within the *Check Request* belongs to the AS present in the request. In order to get this information the *Responsible HA* obtains the AS path for the IP address using BGP. Then, it checks if the AS number given in the *Check Request* matches with the last AS number returned in the AS path.

If some of these tests are not successful the *Responsible HA* returns a *Check Failure*. Otherwise, since all test were successful it registers the new HA information (See step 5 in fig. 2) and sends a query to the P2PHAN in order to find the neighbors for the new HA, i.e. the the two nodes with the closest higher and the closest lower Peer-ID (See step 6 in fig. 2). After locating the neighbors, the *Responsible HA* sends to the Bootstrapping Server a *Check Success* adding the neighbors IP addresses and peer-IDs (See step 7 in fig. 2). Next, the Bootstrapping Server sends the neighbors peer-IDs and the random peer ID to the new HA. In parallel the Bootstrapping server sends the peer-ID and the IP of the new HA to the neighbors (See step 8 in fig. 2). When the neighbors receive the message from the Bootstrapping Server, they allocate the new HA using standard P2P techniques (See step 9 in fig. 2).

From this moment, the new HA will send periodically update messages in order to indicate that it is alive to the *Responsible HA*. If the Responsble HA does not receive these update messages during a pre-configured time out, it removes the entry for that HA (See steps 10 and 11 in fig.2).

It must be noticed that a P2PHAN member has to sign the update messages sent to the *Responsible HA* with its AS_pr_key.

3.3 Security Problems Resolution

The SJP is the basic mechanism used to solve the problems introduced in Section 3.1. In this section we analyze how the proposed solution solve those problems. For this purpose we divide the attacks into three categories: targeted attacks, non-targeted attacks and other attacks.

Targeted Attacks. The key point in our solution is that the Peer-ID assignment is performed by the network instead of each HA can choose its position in the P2PHAN. Hence, in order to perform any of the targeted attacks defined in 3.1 the attacker should solicit as many Peer-ID as necessary until obtain one valid Peer-ID for its purpose.

Based on the study developed in [24], table 1 shows a realistic example of the results that would be obtained in a targeted *Starvation Attack* with a probability of success equal to 0.7. This scenario has 10000 HAs where the time spent to obtain one Peer-ID is 1 second. It is shown the number of Peer-IDs and time needed for 1, 5 and 10 replicas available within the P2PHAN.

Table 1. Example Scenario for a Targeted Starvation Attack with $P_s = 0.7$

Replicas	Peer-IDs Required	Time (hr) ($)
1	23333	6.48
5	135420	37.62
10	275540	76.5

The analysis in [24] focuses on the targeted starvation attack, thus, it can be used as a method to analyze the complexity of this kind of attack introduced in 3.1. Therefore, it seems that the attack is quite feasible for one attacker and it becomes easier in the case of a collaborative attack. Thus, an attacker must obtain 23333 Peer-IDs in order to perform and attack with P_s equal to 0,7 and only one replica. That is, the attacker should receive more than the double of the Peer-IDs in the P2PHAN, thus, the Bootstrapping server can detect easily the attack by evaluating the rate of solicited Peer-IDs. This rate would be in a normal situation 10000 Peer-IDs in the time of live of the P2PHAN, whereas in an attack scenario this rate would be hundred or even thousand of Peer-IDs per day. Therefore, it would be really easy to detect attack attempts.

In the targeted routing attacks, again a high number Peer-IDs is needed by the attacker. In this case the attacker needs to become all the fingers of the target HA. This implies the solicitation of many peer-IDs ,based on the analysis made in [24]. Therefore, it would be easy to detect attack attempts due to the high rate of Peer-IDs solicitation received in the Bootstrapping Server.

In a nutshell, a targeted attack against the P2PHAN could be viable in terms of time, but it is easily detectable by applying an access control policy based on the high rate of joins attempts.

Non-Targeted Attacks. In this subsection the non-targeted starvation and routing attacks are analyzed. The non-targeted starvation attack is solved with the use of replicas. That is, in an scenario where r replicas are being used in order to store the information of each AS, the malicious node would be responsible of 1 of r replicas. Then, the victim AS would never be starved. In this situation, if the peers only send one query in order to obtain the information about the desired AS, the malicious peer would affect to the $\frac{1}{N} * 100\%$ of the queries for the victim AS, because 1 of each r queries would arrive to the malicious node. In order to avoid this loss of performance the peers sends in paralell at less three queries for different paths (i.e. using different fingers). Statistically, each query arrives to different peers storing different replicas. By doing so, the correct result would be the most common among the responses. Obviously, this mechanism is more efficient with higher r and number of paralell replicas.

On the other hand, the non-targeted routing attack is characterized by a node which does not route the queries or it uses bad routes for them. In this case, the number of replicas is not a critical point. However, the solution is also the utilization of parallel queries in order to reach the destination. Again, the correct result is the most common among the responses and the mechanism is also more efficient with a high number of parallel queries.

Furthermore, it must be highlighted that all these proposed security methods are also useful if non-malicious nodes have non-standard behaviour which may affect the performance of the P2PHAN.

Other Attacks. The rest of the security problems described in Section 3.1 are solved by the SJP as is following described:

- *High Rate of Joins and Leaves:* The Bootstrapping Server will have a list with the IP addresses of the recent joins. Based on this list it is easy to check if a node is continuously joining the network. In this case, this node will be introduced in a *Black List* and its join requests will not be accepted during an established time. The fact of preventing continuous joins inherently avoids continuous leaves. This mechanism prevents the inconsistent behaviour of non-malicious nodes with an unstable network connection.
- *Register False Information:* It is checked by the *Responsible(s) Node(s)* during the SJP in the second and third performed tests.
- *Multiple Registers:* The SJP prevents that one node joins the network twice with the same IP address twice with the same IP address by using the second test performed by the *Responsible(s) Node(s)*.

4   Conclusion and Further Work

This paper is focused on adding security to the Peer-to-Peer Home Agent Network [9]. This architecture is used to discover HAs geographically distributed in a simple, dynamic, fully global and distributed way. Besides, it works over IPv4 and IPv6. The main security mechanism proposed is the *Secure Join Procedure*. This method reuse the Bootstrapping Server (present in commercial p2p networks) so as to assure a secure Peer-ID assignment procedure. The conclusion after the analysis of the SJP is that an attacker must have the following characteristics to perform an attack to the P2PHAN:

- The attacker must know a private key which is controlled for the organisation which manages the AS.
- BGP capabilities, it is hard because it is not necessary only supports BGP but have any relationship with other AS in order to obtain BGP information.
- HA capabilities, it is feasible.
- Thousand of IP addresses, if the P2PHAN is formed by thousand of HAs.

In addition, if the attacker fulfils all the previous requirements, the attack can be easily detected in case of targeted attacks due to the increment in the Peer-ID solicitation rate, and easily avoidable in the case of non-targeted attacks due to the utilization of replication and multiples parallel queries.

Therefore, the security solution presented in the paper has a practical security level which makes any attempt of attack against the P2PHAN unaffordable.

The future work will be the implementation of the P2PHAN with all the security framework proposed in this paper for both scenarios: simulation environment and real testbed.

References

1. Perkins, C.: Mobility Support for IPv4. RFC 3344 (August 2002)
2. Johnson, D., Perkins, C., Arkko, J.: Mobility Support in IPv6.,RFC 3775 (June 2004)
3. Devarapalli, V., Wakikawa, R., Petrescu, A., Thubert, P.: Network Mobility (NEMO) Basic Support Protocol. RFC 3963 (January 2005)
4. Leung, K., Dommety, G., Narayanan, V., Petrescu, A.: IPv4 Network Mobility (NEMO) Basic Support Protocol. IETF Draft (June 2006)
5. Hawkinson, J., Bates, T.: Guidelines for creation, selection, and registration of an Autonomous System (AS)., RFC 1930 (March 1996)
6. Chowdhury, K., Khalil, M., Akhtar, H.: Home Subnet Prefix or the Home Agent discovery for Mobile IPv6. IETF Draft (April 2004)
7. Jang, H.J., Yegin, A., Choi, J.: DHCP Option for Home Agent Discovery in MIPv6. IETF Draft (February 2006)
8. Yen, Y.S., Hsu, C.C., Chao, H.C.: Global dynamic home agent discovery on mobile IPv6. Wireless Communications and Mobile Computing 6(5), 617–628
9. Cuevas, R., Guerrero, C., Cuevas, A., Caldern, M., Bernardos, C.J.: P2P Based Architecture for Global Home Agent Dynamic Discovery in IP Mobility. In: Proc. IEEE 65th Vehicular Technology Conference (April 2007)

10. Stoica, I., Morris, R., Karger, D., Kaashoek, M.F., Balakrishnan, H.: Chord: A scalable peer-to-peer lookup service for internet applications. In: Proc. ACM SIG-COMM 2001 (2001)
11. Maymounkov, P., Maziers, D.: Kademlia: A peer-to-peer information system based on the xor metric. In: LNCS, pp. 53–65 (2002)
12. Emule. [Online]. Available: http://www.emule-project.net
13. Wallach, D.S.: A survey of peer-to-peer security issues. In: Lecture Notes in Computer Science, pp. 42–57 (2003)
14. Sit, E., Morris, R.: Security considerations for peer-to-peer distributed hash tables. In: Druschel, P., Kaashoek, M.F., Rowstron, A. (eds.) IPTPS 2002. LNCS, vol. 2429, Springer, Heidelberg (2002)
15. Rekhter, Y., Li, T., Hares, S.: A Border Gateway Protocol 4 (BGP-4). RFC 4271 (January 2006)
16. Kaufman, C.: Internet Key Exchange (IKEv2) Protocol. RFC 4306 (December 2005)
17. Devarapalli, V., Dupont, F.: Mobile IPv6 Operation with IKEv2 and the revised IPsec Architecture. IETF Draft (December 2006)
18. Perera, E., Sivaraman, V., Seneviratne, A.: Survey on network mobility support. Mobile Computing and Communications Review 8(2) (2004)
19. Thubert, P., Wakikawa, R., Devarapalli, V.: Global HA to HA protocol. IETF Draft (October 2005)
20. Bagnulo, M., García Martínez, A., Bernardos, C.J., Azcorra, A.: Scalable Support for Globally Moving Networks. In: ISWCS 2006 (September 2006)
21. Risson, J., Moors, T.: Survey of Research towards Robust Peer-to-Peer Networks: Search Methods. Internet Draft, draft-irtf-p2prg-survey-search-00.txt
22. Lua, E.K., Crowcroft, J., Pias, M., Sharma, R., Lim, S.: A Survey and Comparison of Peer-to-Peer Overlay Networks Schemes. IEEE Communications Surveys 7(2) 2nd Quarter
23. Castro, M., Drushel, P., Ganesh, A., Rowstron, A., Wallach, D.: Secure routing for structured peer-to-peer overlay networks. In: Proc. of OSDI 2002, Boston, MA (2002)
24. Cuevas, R., Cuevas, A., Urueña, M., Banchs, A., Guerrero, C.: Analysis of the Full Starvation Attack in structured p2p systems. Available: http://www.it.uc3m.es/rcuevas/tech_report/p2p_full_starvation.pdf

Recursive Replication: A Survival Solution for Structured P2P Information Systems to Denial of Service Attacks*

Xavier Bonnaire[1] and Olivier Marin[2]

[1] Departamento de Informática
Universidad Técnica Federico Santa María, Valparaíso, Chile
xavier.bonnaire@inf.utfsm.cl
[2] Laboratoire d'Informatique de Paris 6 - INRIA
Université Pierre et Marie Curie, Paris, France
olivier.marin@lip6.fr

Abstract. Structured Peer to Peer overlays have shown to be a very good solution for building very large scale distributed information systems. Most of them are based on Distributed Hash Tables (DHTs) that provide an easy way to manage replicas, thus facilitating high availability of data as well as fault tolerance. However, DHTs can also be affected by some well known Distributed Denial of Services attacks that can lead to almost complete unavailability of the stored objects. Very few powerful solutions exist for this kind of security weakness, and increasing the number of replicas for a given object seems to be the best known one. In this paper, we show how a recursive replicating schema can provide a good solution for this kind of attack.

1 Introduction

Peer to Peer systems have shown to be a very good solution for building very large scale distributed information systems [1,11]. Their self organization as well as their ability to provide a high level of fault tolerance make it much easier to build complex systems with millions of nodes [12], without the need of a central administration. Distributed Hash Table (DHTs) overlays like PASTRY [14], CHORD [17] or TAPESTRY [19] are especially interesting as they provide a very good way to implement data replication, which leads to a very high availability of stored objects. In the following, we will take the example of PASTRY to illustrate our D.D.O.S. (Distributed Denial Of Service) [4] attack. In the same manner, this kind of attack can be applied to other DHT overlays.

DHTs have a lot of security weaknesses [15]. Distributed Denial Of Service (D.D.O.S.) attacks [9] like the well-known Sybil Attack [6] allow an attacker to control a considerable number of nodeIDs. Controlling a large number of nodeIDs makes it possible for a malicious node to control an important part of the system,

---

* This work is part of a CNRS/CONICYT international cooperation project between France and Chile.

R. Meersman, Z. Tari, P. Herrero et al. (Eds.): OTM 2007 Ws, Part II, LNCS 4806, pp. 931–940, 2007.
© Springer-Verlag Berlin Heidelberg 2007

and therefore an important number of replicas. In this paper, we will show how a Recursive Replication schema can allow a DHT to survive a strong IP based D.D.O.S. attack. We will illustrate our solution using the PASTRY [14] overlay, and will make a short discussion on how Recursive Replication can be made with other DHTs.

Section 2 presents the PASTRY overlay that will be used to illustrate our attack. Section 3 explains the different types of possible D.D.O.S attacks , and Section 4 an efficient way to build such an attack in PASTRY. In Section 5, we give some solutions and drawbacks, and Section 6 describes the Recursive Replication mechanism and gives simulation results. Finally, we present some conclusions and future work in Section 7.

2   The PASTRY P2P Overlay

In PASTRY, nodes are organized in a ring where nodeIDs are 128 bits unsigned integers computed using the SHA [13, 8, 16] strong hash table function. The application can choose how to really compute the nodeID, but a good way to obtain unique nodeIDs is to use the node's IP address ($SHA(@IP)$). In PASTRY, each node in the ring maintains a set of L nodes numerically closest in the ring, called its Leaf Set. For node X, the Leaf Set consists of nodes whose nodeIDs are the closest numerically to that of X; the $\frac{L}{2}$ nodes with a numerically smaller nodeID and $\frac{L}{2}$ nodes numerically greater one. One of the properties of the SHA function is to ensure that, for two elements $E1$ and $E2$ that are numerically close, $SHA(E1)$ and $SHA(E2)$ are very different. In other words, two nodes with very similar IP addresses (for example in the same sub-net) will have two very different nodeIDs, and then will be numerically far in the ring. Therefore, two nodes numerically close in the ring cannot be physically near (in term of their IP addresses). Since all the nodes belonging to the same leaf set are physically far on the Internet, in case of a network partition the probability that all the nodes fail or that all the nodes will be on the same side tends towards zero. The leaf set in PASTRY has been designed to be the perfect place for an application to make replicas.

As an example, the PAST [7] application implemented above PASTRY uses the leaf set in the following way. It computes key $k = SHA(Object1)$ to store $Object1$. Then node Y, whose nodeID is numerically closest to key k, becomes the root for $Object1$, and the Leaf Set of node Y is used to make L replicas of $Object1$.

The routing algorithm in PASTRY is a prefix based algorithm. The function $Route(M, k)$ routes message M to node Y: the numerically closest node to key k. Each node on the route tries to find in its routing table a node that shares a prefix with at least one more digit than the current node's ID. The algorithm has two important properties. The first one is that the number of hops required to arrive at destination is around $Log_b(N)$ where b is the numerical base used for nodeIDs representation, and N is the number of nodes in the overlay. This routing algorithm makes the overlay very scalable as the number of hops required to arrive at a destination increases much slower than N does. The second property of the PASTRY routing algorithm is that it ensures that message M will at least arrive at a node in the Leaf Set of node X, if node X is the root for key k.

3 Distributed Denial of Service Attacks

There are several well-known ways to build a Distributed Denial Of Service attack above PASTRY or any other DHT. The key idea is to make as many replicas as possible unavailable. A brief survey of existing D.D.O.S. attacks is required to further explain our solution.

1. **The Sybil Attack** [6]: It consists for a node in obtaining a large number of nodeIDs in order to be able to control an important part of the ID space. This means that the node can control a great number of replicas in the ring. Potentially, the node can control all the replicas for a given key k by taking over all the leaf set corresponding to key k, as well as its root nodeID.

   A good solution for this problem has been proposed by P. Druschel and M. Castro in [2] by using certificates. To enter the ring, a node must obtain a certificate, and the cost for obtaining a certificate is determined to deter nodes from trying to acquire a large number of nodeIDs. The cost can be a financial one or it can be a cost in computational time, where the node must solve a cryptographic puzzle. This is not a pure Peer to Peer solution, as it uses centralized servers. Yet it is important to note that a pure P2P solution has fundamental limitations, as proven in [6] .

2. **The Overlay Maintenance Attack:** An overlay D.D.O.S attack consists in trying to modify the content of the structures used to manage the overlay. A typical way to do this for an attacker is to control a set of nodes that will provide false information during the *join()* procedure for a new node, or that will not forward messages correctly. Providing false information during a *join()* procedure will build routing tables with bad nodes, and the routing of a message could need more hops to reach a given object's leaf set. For a given object *Object1*, it can be relatively easy for the attacker to control a set of nodes near the leaf set of *Object1*. As every route to the leaf set of *Object1* has a very high probability to go through one of these nodes, it is very likely the attacker will be able to make an important part of the replicas unavailable if the controlled nodes do not forward the messages to *Object1*'s leaf set.

3. **The Internal Message Flooding attack:** It occurs when the attacker floods the leaf set of an object using the internal routing function of the overlay. The advantage for the attacker is to be able to also flood a set of nodes near the leaf set, as normal routes for a given leaf set have a very high probability to go through nodes very close of the target leaf set. However, to be able to build an efficient flooding, the attacker will need to control several nodes in the overlay.

4. **The IP Flooding Attack:** It is among simplest and also among the most efficient attacks. The idea is to use an IP packet flooding against a node, so that this node – or its corresponding router – will not be able to answer legitimate requests. The overflow of incoming packets forces the node to drop a lot of packets, thus also dropping normal requests. Attacked nodes quickly become unable to reply to lookup queries.

In the case of PASTRY, the ideal target set to attack replicas of *Object1* is the leaf set of node Y which is the root node of key $k = SHA(Object1)$. Finding caches of Object1 is also possible using Diversity Routing [2]. Diversity Routing consists in using diverse routes to arrive to the leaf set of *Object1* in order to go through different nodes before entering into the leaf set. As cache for *Object1* is usually located on nodes numerically close to the leaf set, it is highly probable to find several ones using diverse routes. The attacker can then build an IP flooding to attack both the leaf set and caches for *Object1* from computers that do not necessarily belong to the overlay. In the following section, we describe a simple way to efficiently build such an attack.

4   Building a Distributed IP Flooding Attack in PASTRY

In this paper, we focus on IP based D.D.O.S. attacks. The main advantage of the IP flooding attack is that it only requires the attacker to control one node of the overlay. The attack basically comprises the two following steps:

4.1   First Step : Targets Localization

Let us suppose that Z is a malicious node. Our attack only requires from Z to be a member of the overlay, that is Z has a single valid nodeID. As we only need one nodeID for Z, it can easily be obtained at a very limited cost, even if the overlay uses a financial cost or a cryptographic cost to deliver a nodeID.

The target for a given object is the set of nodes corresponding to the object replicas in the overlay, including the object's root itself. If the object is known, then Z can compute the associated key with $k = SHA(Object1)$. If the object is unknown, Z can choose a random key k. Using the routing mechanism of PASTRY, Z can find the root node R corresponding to the key k. At this point, there is no difference if the root node is obtained using the normal PASTRY routing table or the secure (constrained) one. Because of network partitions or node failures, node R might not be the root node but a node belonging to the leaf set of the root node.

If R is the root node, then Z can obtain its target leaf set from R. The target set for Z is then $T = R, L_0, ..., L_q$ where L_i represents the leaf set of R.

If R is not the root node but a node from the leaf set of the root node, then Z must add a few nodes on both sides of the leaf set of R (if R is smaller or greater than key k). This will build a target set a little larger than a normal leaf set, but will ensure that the target set will include all the nodes of the root node's leaf set, that is all the nodes that hold a replica for key k.

4.2 Second Step : The D.D.O.S. Attack

The Distributed Denial of Service Attack (D.D.O.S.) [9,3] itself consists in making unavailable all nodes that hold a replica for a given object *Object1*. The attacker who has a set H of hacked computers on the Internet can use this set H to make an IP flooding of all hosts from set T built by node Z. Ideally these computers should be well distributed on the Internet, that is geographically far (for example, with very different IP addresses). The number of hacked hosts for the attacker, and their network link capacities, directly determine the maximum number of replicas that the attacker can make unavailable. The attacker must assign each host of set H to a specific target in set T in order to minimize the network latency between a hacked computer and its target. This allows the attacker to maximize the flooding rate to this target. Figure 1 shows a typical IP flooding attack.

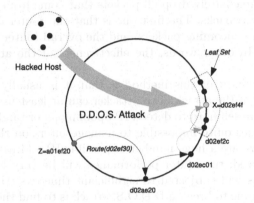

Fig. 1. D.D.O.S. Attack

4.3 Attack Consequences

The first consequence of any type of the above mentioned Distributed Denial Of Service attacks in a structured P2P overlay is to try to make all the replicas unavailable for a given object *Object1*. This basically means that a normal message routing to key k, such that $k = SHA(Object1)$ will have a very low probability to either enter the leaf set corresponding to k, or to go through a node near the leaf set containing a cache for *Object1*.

The maintenance operations for the overlay will have the following consequences:

– Every node under attack will loose the ability to communicate in the overlay, and will get out of every close nodes' leaf set,
– Every node under attack will disappear from the other nodes' routing table when the routing algorithm will consider them improper for message forwarding.

All nodes under attack will then slowly but completely disappear from the overlay.

5   Solutions and Drawbacks

In this section, we show that in practice the traditional solution for IP based D.D.O.S. attacks are both inefficient and hardly feasible. We also show how increasing the number of replicas is quite the only acceptable way to protect the DHT from this kind of attack.

5.1   Stopping the Attack

For the external D.D.O.S. attack, the solution usually consists in modifications on the router corresponding to the sub-net, or on the computers themselves [10, 18]. It previously requires to be able to detect that the sub-net is under attack, and what the origin of the attack is. When the source of the attack is identified [18], the router or computer modification consists in adding some filtering rules to immediately drop all packets that come from this source. This solution has two drawbacks. The first one is that the router could spend a lot of time dropping the incoming packets, and the overall router performance will strongly decrease. In other words, the effectiveness of the attack is still very good.

The second drawback of this method is that it is usually very difficult to identify the source of the attack. The attacker can at least use two well-known methods to make himself hard to detect: IP Spoofing [5] or random IP addresses. Therefore, it becomes quite impossible to manage filters on the router to drop the packets, as there are a huge number of possible IP addresses to filter. Even with filtering enforced, the router performance will be very low, and the P2P overlay applications will subjected to significant timeouts. Unfortunately, the only efficient technique to break a D.D.O.S. attack is to find the multiple origins of the attack, which makes it quite impossible to stop the attack in a reasonably short time.

Stopping an internal message flooding is easier than stopping an external one, as the flooding nodes belong to the overlay. Each node can easily monitor the message forwarding requests coming from a given node Z, and if the rate is greater than a given threshold, the node can reject the request from node Z. Node Z can then try to route its messages using Diversity routing.

5.2   Increasing the Number of Replicas

As it is very difficult to stop an IP flooding D.D.O.S. attack, a well-known solution consists in increasing the number of replicas to make it very difficult for the attacker to reach them all at the same time. However, increasing the replicas is not sufficient, as it also requires that the replicas that are not under attack will be kept available for legitimate nodes. A traditional way to increase the number of replicas in PASTRY is to increase the size of the leaf set to ensure that it will be much more difficult for an attacker to make all the replicas unavailable. However, in PASTRY, the size of the leaf set directly influences the maintenance of the structure of the overlay. In order to maintain its leaf set, each node of the overlay sends a KEEP_ALIVE message to each node of its leaf set every period of time p. Hence if the overlay contains N nodes, and the size of the leaf set is L, the leaf set maintenance procedure will globally generate $Mess_{LeafSet} = p \times N$ messages. Then increasing the number of replicas of r will produce an amount of $p \times r$ extra messages.

Previous D.D.O.S attacks made against well known organizations have shown that an attacker can easily obtain hundreds of hosts to complete his attack. This means that increasing the leaf set size to be able to handle hundreds of replicas is not a scalable solution as it will generate millions of maintenance messages in the whole overlay.

6   Recursive Replication

Recursive replication replicates the object using various leaf sets. We define the function $SHA^m(k)$ as the following:

$$SHA^m(k) = SHA(SHA(SHA(...(SHA(k)))))),\quad m\,times \tag{1}$$

that is m recursive calls to function $SHA()$ from the initial key k. Then the replication of *Object1* is made using m small sized leaf sets. If L is the size of the leaf set, the total number of replicas is $R = m \times L$. For a standard Pastry overlay, the value of L is 16 or 32. The procedure is shown in figure 2.

Replicate(Object obj, Key k, level m
)
{   Key k1;
    if (m == 0) exit;
    if (k==0) k1 = SHA(obj);
      else k1 = SHA(k);
    Route(k1, REPLICATE, obj);
    Replicate(obj, k1, m-1);
}

Fig. 2. Recursive Replicate Procedure

When receiving a REPLICATE message, a node replicates the object *obj* in all nodes of its leaf set. After m levels of replication, m leaf sets are populated with replicas of object *obj*.

When a node wants to find a given replica, it does a maximum of T_r attempts. The number of attempts are equally distributed over all the leaf sets and the lookup terminates at the first replica found. The number of attempts required for a hit is a geometric probabilistic distribution.

The lookup procedure in Recursive Replication uses $\frac{T_r}{G}$ attempts per leaf set. The leaf set order l is randomly chosen from 1 to m and a lookup message is sent to the key $SHA^l(Object1)$. Diversity Routing is used for lookups in the same

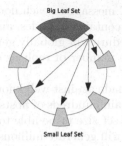

Fig. 3. Small Leaf Sets versus Big Leaf Set

leaf set to increase the probability to arrive to a different node of the leaf set, in case of a previous failure (unavailable replica). A Tabu List is used in order to avoid an immediate cycle (the random selected leaf set is the same as the previous one), yet allows a new visit after a given number of attempts (in our case the leaf set size).

The cost in number of messages of the Recursive Replication procedure is $M_R = G \times L$, where L is the size of the leaf set. If we consider that we make T_r attempts to find a valid replica, then the lookup cost in terms of messages is exactly the same for both replication approaches, as in the traditional one, a node will have to use Diversity Routing to try to reach a valid node of the leaf set. However, the maintenance cost of the overlay will be exactly the same as with the traditional method using a large, unique leaf set.

Compared to the traditional leaf set based replication, the recursive approach does not have the dramatic leaf set management cost increase due the size of the leaf set.

Recursive Replication in CHORD

Implementing Recursive replication in the CHORD overlay can be easily made using the set of S successors that manage every CHORD node instead of the PASTRY leaf set. Recursive Replication can be used in every DHT overlay where nodes manage a set of numerically closest nodes.

6.1   Results of Simulation

We have built a simulation to evaluate the performance of both Traditional Replication and Recursive Replication. The simulation runs were made with the following parameters:

- Number of replicas: R
- Number of unavailable replicas: U
- Maximum number of attempts to find a replica: T
- Number of groups of replicas: G
- Small leaf set size: L

We have evaluated the percentage of replicas found as a function of the maximum number of attempts, given a number of 100000 lookup for a given object.

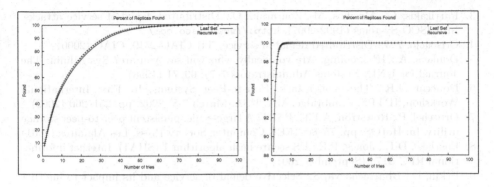

Fig. 4. Simulation Results - Respectively Strong and Weak Attack

We analyzed two cases. A strong attack, where less than 10% of the replicas are still available in the overlay, and a weak attack where more than 87% of the replicas are available. The parameter values used in the previous figures are R=867, U=800, G=51, L=17, T={1 2 3 4 5 6 7 8 9 10 15 20 25 30 35 40 45 50 55 60 70 80 100} for the strong attack, and R=867, U=100, G=51, L=17 and T={1 2 3 4 5 6 7 8 9 10 15 20 25 30 35 40 45 50 55 60 70 80 100}. The simulation results (figure 4) show that both replication approaches have the same performance in term of percentage of hints in the best (very small number of attacked replicas) and worst cases. For the strong attack, only 35 attempts are required in average to have a replica hit of 95%, while 10 attempts give a hit ratio greater than 56%. For the weak attack, only 6 attempts are needed for both approaches to have a hit of 100% , as can be easily expected.

7   Conclusion

Distributed Denial Of Service attacks are very efficient ones that can lead to complete unavailability of replicas in DHTs. Increasing the number of replicas is a good solution, but increasing the number of nodes in the PASTRY leaf set has a very high management cost. We have shown that our approach using Recursive Replication has the same performance as the traditional replication schema in term of hints, but at a dramatically lower cost in term of the overlay management. However, there is still some important work to do about replicas management in structured P2P overlays, especially about how to hide a part of the replicas from a potential attacker.

References

1. Androutsellis-Theotokis, Spinellis: A survey of peer-to-peer content distribution technologies. CSURV: Computing Surveys 36 (2004)
2. Castro, M., Drushel, P., Ganesh, A., Rowstron, A., Wallach, D.: Secure routing for structured peer-to-peer overlay networks. In: Operating System Design and Implementation, OSDI 2002, Boston, MA (2002)

3. Patrikakis, C., Masikos, M., Zouraraki, O.: Distributed denial of service attacks. In: CISCO Systems (1992-2007), http://www.cisco.com/
4. Criscuolo, P.J.: Distributed denial of service. TR CIAC-2319, CIAC (2000)
5. Donkers, A.: IP spoofing: Are you really who you say you are? Sys Admin: The Journal for UNIX Systems Administrators 7(7), 69–71 (1998)
6. Douceur, J.R.: The sybil attack. Peer-to-Peer Systems. In: First International Workshop, IPTPS, Cambridge, MA, USA, March 7-8, 2002, pp. 251–260 (2002)
7. Druschel, P., Rowstron, A.I.T.: PAST: A large-scale, persistent peer-to-peer storage utility. In: HotOS, pp. 75–80. IEEE Computer Society Press, Los Alamitos (2001)
8. Eastlake, D.E., Jones, P.E.: US secure hash algorithm 1 (SHA1). Internet informational RFC 3174 (September 2001)
9. Etkin, D., Bhattacharya, S.: Selective denial of service and its impact to internet based information systems (May 08, 2000)
10. Di Francesco, P., Bianchi, G., Fabio, G., Oriti, N.: A new distributed defense to distributed denial of service attacks. Miami, Florida, USA, WSEAS (2004)
11. Lua, K., Crowcroft, J., Pias, M., Sharma, R., Lim, S.: A survey and comparison of peer-to-peer overlay network schemes. Communications Surveys & Tutorials, IEEE, 72–93 (2005)
12. Mislove, A., Post, A., Reis, C., Willmann, P., Druschel, P., Wallach, D.S., Bonnaire, X., Sens, P., Busca, J.-M., Arantes, L.B.: Post: A secure, resilient, cooperative messaging system. In: Jones, M.B. (ed.) HotOS, pp. 61–66. USENIX (2003)
13. Preneel, B., Bosselaers, A., Govaerts, R., Vandewalle, J.: Collision-free hashfunctions based on blockcipher algorithms (IEEE catalog number 89CH2774-8). In: Proceedings 1989 International Carnahan Conference on Security Technology, Zurich, Switzerland, Oct 3–5 1989, pp. 203–210. IEEE Computer Society Press, Los Alamitos (1989)
14. Rowstron, A., Druschel, P.: Pastry: Scalable, decentralized object location, and routing for large-scale peer-to-peer systems. In: Guerraoui, R. (ed.) Middleware 2001. LNCS, vol. 2218, pp. 329–339. Springer, Heidelberg (2001)
15. Sit, E., Morris, R.: Security considerations for peer-to-peer distributed hash tables. In: Druschel, P., Kaashoek, M.F., Rowstron, A. (eds.) IPTPS 2002. LNCS, vol. 2429, pp. 261–269. Springer, Heidelberg (2002)
16. Stallings, W.: SHA: the Secure Hash Algorithm. Dr. Dobb's Journal of Software Tools 19(4), 32–34 (1994)
17. Stoica, I., Morris, R., Karger, D.R., Kaashoek, M.F., Balakrishnan, H.: Chord: A scalable peer-to-peer lookup service for internet applications. In: SIGCOMM, pp. 149–160 (2001)
18. Sung, M., Xu, J.: IP traceback-based intelligent packet filtering: A novel technique for defending against internet DDoS attacks. IEEE Transactions on Parallel and Distributed Systems PDS-14(9), 861–872 (2003)
19. Zhao, B.Y., Huang, L., Stribling, J., Rhea, S.C., Joseph, A.D, Kubiatowicz, J.D.: Tapestry: A resilient global-scale overlay for service deployment. IEEE Journal on Selected Areas in Communications 22(1), 41–53 (2004)

Adaptive Expression Based Routing Protocol for P2P Systems

Imran Rao[1], Aaron Harwood[2], and Shanika Karunasekera[3]

NICTA Victoria Labs,
CSSE, The University of Melbourne Australia
imran@csse.unimelb.edu.au, aharwood@csse.unimelb.edu.au,
shanika@csse.unimelb.edu.au

Abstract. There is an emerging trend of using P2P systems for computational
and data intensive tasks, such as online collaborations, distributed database applications and message passing interface (MPI) algorithms. For better resource utilization in these and similar applications, it is necessary to discover the resource
capabilities of the collaborating peers. Another requirement of these applications
is to locate the optimal resource based on a search criteria, e.g., to seek a resource
with earliest execution time. We define this kind of routing as *expression based
routing*. However, in the absence of the centralized controlling node, tracking the
capabilities of the participating peers is very difficult. Moreover peers join and
leave the system dynamically and, thus, makes the discovery of a desired resource
even more complex. In this paper we investigate a novel algorithm for expression
based routing in a P2P system. It evaluates peer u such that l_u (the value of u)
is minimum where e.g. l_u is load of the peer u. Our contribution includes a detailed algorithm to search for the least loaded peer in the system. We analyze the
accuracy and cost of our proposed algorithm through detailed simulations.

1 Introduction

Some of the benefits of a P2P approach include: improving scalability by avoiding
dependency on centralized points; eliminating the need for costly infrastructure by enabling direct communication among clients; and enabling resource aggregation. However, most of the popular publicly available peer-to-peer systems such as Bittorent [1]
focus on sharing of multimedia data and files. There is an emerging trend of using
P2P systems for computational and data intensive tasks, such as online collaborations,
distributed database applications and message passing interface (MPI) algorithms. For
better resource utilization in these and similar applications, it is necessary to discover
the resource capabilities of the collaborating peers. Another requirement of these applications is to locate the optimal resource based on a search criteria, e.g., to seek a
resource with earliest execution time. Other similar routing parameters are to route to
the least or most loaded peer, cheapest resource in terms of service utilization cost,
peer with highest network bandwidth, etc. We define this kind of routing as *expression
based routing*. Some of the motivating applications for expression-based routing are
load-balancing, performance optimization and fault tolerance.

Existing P2P routing techniques allow users to search for a resource that meets the
minimum job requirements, e.g., to locate a resource which can complete a task within

R. Meersman, Z. Tari, P. Herrero et al. (Eds.): OTM 2007 Ws, Part II, LNCS 4806, pp. 941–949, 2007.

given deadline T. Since these techniques simply search resources matching a given value or range, we refer to these routing techniques as *value based routing*. It is important to realize the difference between value and expression based routing. The value based routing is considered to be successful as soon as it discovers a resource matching the search criteria, whereas expression based routing demands the traversal of the whole system as it seeks unknown extremes or aggregates. In the absence of the centralized controlling node, tracking the capabilities of the participating peers is very difficult. Moreover peers join and leave the system dynamically and, thus, makes the expression based routing even more complex.

In this paper we propose a novel algorithm for expression based routing in a P2P system. Our algorithm finds the peer u with minimum value for the expression in the system. To put it formally, our proposed routing protocol searches a peer u such that $l_u = min_{i \in S}\{l_i\}$. That is, route towards the node u with its minimum value, l_u; where e.g. l_u is load of the peer u. Our routing algorithm can also evaluate other similar expressions, e.g. least loaded, most loaded, cheapest, etc. Our contribution includes a decentralized and semi-structured routing algorithm to search the least loaded peer in the system. Our proposed algorithm falls in the category of semi-structured P2P routing protocols, that is, it learns the structure as the routing proceeds.

The rest of the paper is organized as follows: Section 2 summarizes the related works; Section 3 describes our proposed adaptive least-loaded peer search algorithm; Section 4 present the simulation results and their analysis and Section 5 concludes the paper along with outlines for the future directions.

2   Background

In peer-to-peer systems nodes are connected through a logical topological structure, termed as an overlay. To locate a resource is one of the most challenging issues in decentralized P2P systems. To achieve high success rate, the P2P routing algorithms employ different file placement and routing topologies, which can be classified as *structured* and *unstructured* [4,15]. Ref. [12] presents a thorough categorization and description of many approaches.

The unstructured P2P routing protocols offer arbitrary network topology and file placement. These systems do not put any constraints on placement of data items on peers and how peers maintain their network connections. On the other hand, the structured routing algorithms provide strict rules for file placement and object discovery. In these systems, the overlay topology is tightly controlled and files are placed at specified locations that will make subsequent search easy to satisfy. In *highly structured* systems, both network topology and and the placement of files are precisely determined. Structured P2P lookup systems including Chord [6], CAN [7], Pastry [8] and Tapestry [9] are primarily based on Distributed Hash Tables (DHTs). In the DHTs, as data can be looked up within a logarithmic overlay routing hops, they perform very efficient searches. In spite of their advantages, structured P2P networks incur big overheads during peer join/leave operations. High join/leave rate is one of the essential characteristic of a P2P system, and is known as *churn rate* and with a large churn rate it may be impossible to index file locations, making some files inaccessible even though peers have them available.

To mitigate these issues a variation of structured overlays, called *semi structured*, is proposed. In semi-structured P2P networks, there is limited control on the placement of files; Freenet [5] is an example of such systems. Mercury [18] do not apply randomizing hash functions for organizing data items and nodes. It rather organizes nodes into a circular overlay and places data contiguously on this ring. As Mercury does not apply hash functions, data partitioning among nodes is non-uniform. Hence it requires an explicit load-balancing scheme. Besides this, a gossiping and one-hop routing approach has also been proposed for maintaining the routing overlay [19].

As far as searching in P2P overlays is concerned, flooding is one of the most primitive approach used in Gnutella-like P2P systems [2]. Ref. [13] proposes a variation of the flooding scheme with peers randomly choosing only a ratio of their neighbors to forward the query to. The same procedure is followed in [14] by nodes with no information about the location of a file. If an object is found, the query takes the reverse path to the requester, storing the document location at those peers. Nodes with location information contact the specific node directly. In the Random Walks method [15], the requesting node sends out query messages to K randomly chosen neighbors. These neighbors further forward this query to K randomly chosen neighbors. Each of these queries follows its own path and is known as walker. At times user is also interested in to search a peer which meets a range of selection criteria rather than a specific value. Such routing protocols are termed as *range based* routing protocols and is topic of interest of many researchers [16,18,17].

However the goal of all of the existing routing algorithms is to seek a peer which matches a 'given' search criteria and they fail to perform an expression-based routing. In this paper, we propose a novel expression based routing protocol for P2P systems. Our algorithm searches the peer u with least its value l_u minimum in the system, where e.g. l_u is load of the peer u. This proposed routing algorithm can also evaluate other similar expressions, e.g. least loaded, most loaded, cheapest, etc. Our algorithm falls in the category of semi-structured decentralized P2P routing protocols. Some of the important applications of such a routing protocol are that it can help users to estimate the capabilities of the system and better utilize and manage these P2P resources. Other indirect benefits of our proposed work are fault tolerance, load balancing, load optimization, resource scheduling and management.

3 Proposed Solution

3.1 System Model

In this section we give a formal description of our P2P system on which our subsequent discussion is based. The complete notation of our proposed model is listed in Table 1. We assume that there is a set of N peers in a P2P system $S = \{1, 2, 3, ..., N\}$. Each peer $u \in S$ has a value l_u, for example the workload, associated with it. We define workload of a peer as an indicator of its availability to perform a certain task. Workload can be as simple as the number of jobs in the queue to be processed by that peer or a complex function of job queue length, available memory, storage capacity and CPU processing power. Let l_u change by a value δ_u at the rate of α_u. If α_u follows a Poison distribution,

Table 1. Notation of proposed system model

N	NUMBER OF PEERS IN THE SYSTEM	K	SIZE OF THE ROUTING TABLE MAINTAINED BY A PEER
C	NUMBER OF PEERS TO BE UPDATED, A SUBSET OF K	B	NUMBER OF PEERS TO INITIALIZE THE ROUTING TABLE, A SUBSET OF K
H	MAXIMUM NUMBER OF HOPS A QUERY CAN TRAVERSE IN THE SYSTEM	λ_u	THE RATE WITH WHICH A PEER u COMES ONLINE
δ_u	CHANGE IN WORKLOAD OF A PEER u	α_u	WORKLOAD CHANGE RATE OF A PEER u
μ_u	THE RATE WITH WHICH A PEER u GOES OFF-LINE	ρ_u	QUERY ARRIVAL RATE AT PEER u

```
1  PROCEDURE: join(peer u, bootstrapPeer b)
2  begin
3      Vector bList := b.receiveList()
4      for each peer p ∈ bList
5      begin
6          | updateRTable(u, tuple(p, l_p))
7      end
8      status_u := online
9      updateRTable(u, tuple(u, l_u))
10     for each peer x ∈ T_ui i < C
11     begin
12         | updateRTable(x, tuple(u, l_u))
13     end
14 end
```

Fig. 1. New Peer Join - Algorithm

the time t at which next event for workload change will be triggered can be calculated as $-ln(1 - P_{change}(u))/\alpha_u$. Let queries arrive at u at a rate ρ_u with the arrival rate following a Poison distribution. Let u join and leave the system with frequencies λ_u and μ_u respectively. If λ_u and μ_u follow Poison distribution, the probability $P_{online}(u)$ of u being online at time t can be calculated as $1 - e^{-t\lambda_u}$. Similarly the probability $P_{offline}(u)$ of u being off-line at time t can be calculated as $1 - e^{-t\mu_u}$. We propose that peer u maintain a query routing table T_u of size less then or equal to K where $K \leq N$. Routing Table T_u is an array of tuples (u, l_u).

3.2 Proposed Routing Protocol

Initially when a new peer u joins the system, it gets an address of the *bootstrap* peer b. Peer u gets from peer b at most B least loaded peers and updates its routing table T_u where $0 \leq B \leq K$. If after this operation size of T_u increases the routing table size

```
1 PROCEDURE: query(peer u)
2 begin
3 |   result := null if (status == online)
4 |   begin
5 |   |   result := search(u, u, u)
6 |   end
7 end
```

Fig. 2. Query processing - Algorithm

```
1 PROCEDURE: search(originator u, sender v, receiver x)
2 begin
3 |   result := null
4 |   if ( v != x )
5 |   begin
6 |   |   updateRTable(v, tuple(x, l_x))
7 |   end
8 |   peer p := find least loaded peer in T_x
9 |   loop
10 |   begin
11 |   |   if (status of p == offline)
12 |   |   begin
13 |   |   |   remove p from T_x
14 |   |   |   p := find least loaded peer in T_x
15 |   |   end
16 |   end
17 |   until (status of p == online)
18 |   if ( p is same as x )
19 |   begin
20 |   |   result := x
21 |   end
22 |   else if (query traversed less then H hops)
23 |   begin
24 |   |   result := search(u, x, p)
25 |   end
26 |   return result
27 end
```

Fig. 3. Least Loaded Peer Search - Algorithm

limit K, the most loaded peer are removed from the T_u to bring ti equal to K. In the next step, u will send its current workload to update C least loaded peers in l_u where $0 \leq C \leq K$. When a peer again comes online from its off-line state, it is treated as a newly joined peer and its routing table is updated in the same way as is for a newly joined peer.

```
 1  PROCEDURE: updateRTable(peer u, tuple(v, lᵥ))
 2  begin
 3  |   if(v exists in Tᵤ)
 4  |   begin
 5  |   |   update tuple(v, lᵥ) in Tᵤ
 6  |   end
 7  |   else if(size of(Tᵤ < K))
 8  |   begin
 9  |   |   insert tuple(v, lᵥ) in Tᵤ
10  |   end
11  end
```

Fig. 4. Peer Routing Table Update - Algorithm

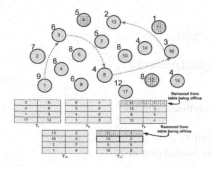

Fig. 5. Overview of the proposed routing algorithm

The goal of our search algorithm is to find u such that; $l_u = \min_{i \in S}\{li\}$. A search request can start at any peer in the system. The query is discarded if the peer is not online at the time of the query arrival. If u is online at that time, it will search the least loaded node in the system as explained in the text to follow and return the result to user. To avoid query being routed infinitely in the system, we define a maximum hop counter H a query can route in the system.

Peer u forwards a search query to the least loaded peer v in T_u. In response v sends its current workload l_v to u and forwards the query to the least loaded peer w in T_v. If u does not receive a response from v in a reasonable time, it assumes v to be off-line. Peer u removes the entry for v from T_u and seeks the next least loaded peer in T_u. The query propagates in the network, until it reaches to peer x which finds its self as the least loaded peer in T_x. Peer x not only sends its current workload l_x to the peer from where the query routed to it (represented by *sender* in the algorithm), but also sends its address to the source of the query (represented by *originator* in the algorithm). The time to wait for the acknowledgement can be adjusted by the system administrators and should be big enough to accommodate the network latency.

Fig.5 elaborates this algorithm with an example. Peer 1 will pick the least loaded node 3 from its routing table T_1 and invoke $search(1,1,3)$ method. Peer 3 routes query $search(1,3,8)$ to peer 8 and sends its new load to peer 1. Peer 8 finds peer 11 as the least loaded peer in its table T_8 and routes query $search(1,8,11)$ to it. Peer 8 does not receive the load updated message from peer 11 in the expected time. Assuming peer 11 dead, peer 8 removes it from T_8 and seeks next least loaded peer in its routing table. Similarly, query $search(1,16,13)$ reaches to peer 13 from peer 16. Peer 13 finds itself as the least loaded peer in T_{13} and sends its address to peer 1 and updates peer 16 with its current workload.

4   Simulation and Results

In this paper, we focus on the search success rate and search cost of our proposed protocol. We define the search success rate as a ratio of successful queries to the total number of queries processed by the system. Search cost is calculated as number of messages exchanged for a successful query. We use α, λ, μ and ρ as input parameters to measure the accuracy and cost of our proposed algorithm.

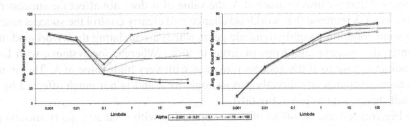

Fig. 6. Evaluating (a) Avg. success percentage and (b) Avg. message count per query; with varying α and λ and keeping $N = 1000, B = 20, K = 20, C = 20, H = 20, \mu = 1.0, \rho = 1.0$

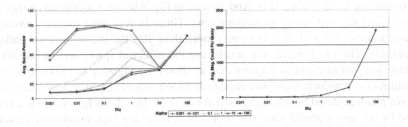

Fig. 7. Evaluating (a) Avg. success percentage, (b) Avg. message count per query; with varying α and μ; and keeping $N = 1000, B = 20, K = 20, C = 20, H = 20, \lambda = 1.0, \rho = 1.0$

We tested the accuracy and cost of our proposed protocol using a single-threaded, discrete event simulator. All the simulation run from the same machine. In this work, we did not simulated the network delays and solely concentrated on the accuracy and

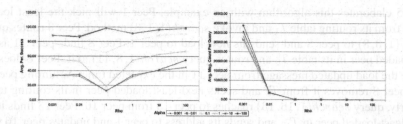

Fig. 8. (Evaluating avg. success percentage, avg. message count per query with varying α and ρ; with $N = 1000, B = 20, K = 20, C = 20, H = 20, \lambda = 1.0, \mu = 1.0$

overhead of the protocol. To start off, the status of all peers is set to off-line. The join the system at the rate of λ. Moreover, the workload of all the peers N=1000, B=20, K=20, C=20 and H=20 and vary the values of α, λ, μ, ρ for one by one. We ran each simulation test 10 times and then took their average. The Fig.6(a) shows the accuracy of our proposed protocol with varying workload change rate and peer online arrival rate. For lower λ, as most of the time a peer stays off-line, their are less peers online in the system to search from and hence we get high success rate. It is also observed from Fig.6(a) that for lower values of λ, the value of α does not affect the success rate. When value of λ increase 0.1, workload change rate clearly control the success rate. So for a system containing highly dynamic peers which are updating their workload more frequently, the success rate drops to drops to 28%. Whereas for systems, with lower α, we achieve close to 100% success rate even with very high value of λ. The number of messages exchange are directly proportional to λ and is not much affected by the workload change rate as see in Fig.6(b).

The Fig.7(a) reflects the affect of μ in conjunction with α on avg. query success rate and avg. query cost. When the peers off-line arrival rate is low, we find a great variation in the query success rate. For peers with high workload change rate and low off-line arrival rate, the success rate is merely 10%. And for low α and high μ, the success rate is around 90%. For α greater then 10.0 has no affect on the query success rate and is solely controlled by the value of μ. And as seen in Fig.7(b), peer workload change rate does not affect the avg. query processing cost. There is very little message exchange for lesser values of μ and as it increases more than 10., the message count increase exponentially. The most interesting results are the affect of query arrival rate on the avg. query success rate and avg. query cost as shown by Fig.8(a) and (b) respectively. Low values of α results in high success rate and high values results in low success rate and is not much affected by the query arrival rate ρ. However, the query cost is entirely affected by the query arrival rate only. For low ρ, there are high number of messages exchanged to search a least-loaded peer in the system as seen in Fig.8(b).

5   Conclusion and Future Work

In this paper, we propose a new class of P2P routing, called expression based routing. After presenting the motivation of this work, we give a detailed explained of our routing proposed semi-structured routing protocol. Through extensive, simulation results, we show that our proposed routing protocol gives close to 100% success rate for lower workload-change rate and on average the query cost is around 10 for very high query arrival rate. In the future, we intend to emulate existing DHT based structured P2P routing protocol to solve expression based queries and compare results with our proposed protocol. We also intend to solve more complex expressions in the future, e.g. Average loaded peer, K^{th} loaded peer, etc.

References

1. Pouwelse, J.A., Garbacki, P., Epema, D.H.J., Sips, H.J.: The Bittorrent P2P File-sharing System: Measurements and Analysis. In: Castro, M., van Renesse, R. (eds.) IPTPS 2005. LNCS, vol. 3640, Springer, Heidelberg (2005)
2. Clip2: The Gnutella Protocol Specification (2001), http://www.clip2.com
3. Napster (1999), http://www.napster.com
4. Milojicic, D.S., Kalogeraki, V., et al.: Peer-to-Peer Computing. HP Lab (2002)
5. Clarke, I., Sandberg, O., Wiley, B., Hong, T.W.: Freenet: A Distributed Anonymous Information Storage and Retrieval System. In: Federrath, H. (ed.) Designing Privacy Enhancing Technologies. LNCS, vol. 2009, pp. 46–66. Springer, Heidelberg (2001)
6. Stoica, I., Morris, R., Karger, D., Kaashoek, F., Balakrishnan, H.: Chord: A Scalable Peer-To-Peer Lookup Service for Internet Applications. In: SIGCOMM 2001, pp. 149–160 (2001)
7. Ratnasamy, S., Francis, P., Handley, M., Karp, R., Shenker, S.: A scalable content-addressable network. In: SIGCOMM 2001, CA, United States, pp. 161–172 (2001)
8. Rowstron, A., Druschel, P.: Pastry: Scalable, Decentralized Object Location, and Routing for Large-Scale Peer-to-Peer Systems. In: Guerraoui, R. (ed.) Middleware 2001. LNCS, vol. 2218, Springer, Heidelberg (2001)
9. Zhao, B.Y., Kubiatowicz, J.D., Joseph, A.D.: Tapestry: An Infrastructure for Fault-tolerant Wide-area Location and Routing. Tech Report: UCB/CSD-01-1141 (April 2001)
10. Cormen, T.H., Leiserson, C.E., Rivest, R.L., Stein, C.: Introduction to Algorithms, 2nd edn. The MIT Press, Cambridge (2001)
11. Berman, K.A., Paul, J.: Fundamentals of Sequential and Parallel Algorithms. PWS Publishing Co. (1996)
12. Tsoumakos, D., Roussopoulos, N.: A Comparison of Peer-to-Peer Search Methods. In: Processdings of WebDB (2003)
13. Kalogeraki, V., Gunopulos, D., Zeinalipour-Yazti, D.: A local search mechanism for peer-to-peer networks. In: Proceedings of the 11th International Conference on Information and Knowledge Management, pp. 300–307 (2002)
14. Menasce, D., Kanchanapalli, L.: Probabilistic Scalable P2P Resource Location Services. ACM Sigmetrics Performance Evaluation Review 30(2) (2002)
15. Lv, C., Cao, P., Cohen, E., Li, K., Shenker, S.: Search and replication in unstructured peer-to-peer networks, Reprint (2001)
16. Andrzejak, A., Xu, Z.: Scalable, Efficient Range Queries for Grid Information Services. In: Proceedings of P2P, vol. 00, IEEE Computer Society, Los Alamitos (2002)
17. Schmidt, C., Parashar, M.: Flexible Information Discovery in Decentralized Distributed Systems. In: Proceedings of the 12th IEEE International Symposium on HPDC (2003)
18. Bharambe, A.R., Agrawal, M., Seshan, S.: Mercury: supporting scalable multi-attribute range queries. In: SIGCOMM 2004, pp. 353–366 (2004)
19. Spence, D., Crowcroft, J., Hand, S., Harris, T.: Location based placement of whole distributed systems. In: Proceedings of the ACM conference on Emerging network experiment and technology, pp. 124–134 (2005)

CAP: A Context-Aware Peer-to-Peer System

Marguerite Fayçal[1,2] and Ahmed Serrhrouchni[1]

[1] GET/ENST – Telecom Paris, 46 rue Barrault, 75013 Paris, France
[2] Orange Labs R&D, 2 avenue Pierre Marzin, 22307 Lannion cedex, France
marguerite.faycal@orange-ftgroup.com, ahmed@enst.fr

Abstract. Despite various hurdles, as copyright issues and security concerns, peer-to-peer (P2P) networks and systems still grow and gain in popularity, and the research community continues to develop interesting applications of P2P technology, together with new platforms for application development. Because the overall performance of these platforms and applications also depends on the performance of the background routing protocol, new systems are mostly based on distributed hash tables (DHT), which are algorithms that provide efficient mechanisms for resource location. However, DHTs assume the system is uniform in available resources and that every node participating in the DHT is within the same transport domain. A number of earlier works has gone on topology-awareness, as a possibility to improve the system's lookup performance. In this paper, we go on to a configurable context-aware system and propose CAP, a novel scheme, which promises contextual data retrieval, alleviated message latency, and enhanced lookup time.

Keywords: P2P, DHT, HMAC, HKey, VDHT.

1 Introduction

In P2P computer networks, nodes are dynamic and can be heterogeneous (routers, servers or any terminal client), but they all provide resources, including bandwidth, storage space, and computing power. Thus, as nodes arrive and demand on the system increases, the total capacity of the system also increases. Besides resources, peers may provide also services (e.g. printing). P2P systems reduce thus the infrastructure and operation costs of a network operator or service provider.

P2P networks are commonly classified into three main categories, according to their degree of centralization. Decentralized architectures can be structured or unstructured (pure P2P systems). Most of the structured decentralized architectures are based on DHTs, which are distributed algorithms that can efficiently route messages to the unique owner of any given key. They are typically designed to scale to large numbers of nodes and to handle continual node arrivals and failures. This paper will concern only the DHT-based structured decentralized architectures.

In P2P networks, users should be able to share, search, and access resources in a distributed and efficient manner; but the overall performances of P2P applications greatly depend on the routing mechanism. In current DHT based routing algorithms, a common assumption is that every node participating in the DHT is within the same

R. Meersman, Z. Tari, P. Herrero et al. (Eds.): OTM 2007 Ws, Part II, LNCS 4806, pp. 950–959, 2007.

transport domain; however in DHTs, routing tasks are distributed across all system peers in an autonomous and spontaneous way and they do not take the IP network into account, and even in an overlay network, the delay in searching information and transferring data depends necessarily on the IP network topology, since a single overlay hop is likely to involve multiple IP routing hops, and the physical path so travelled is often less than optimal. Furthermore, DHTs assume the system is uniform in resources; whereas available resources (bandwidth, CPU, storage capacity, mobility, etc.) differ from one peer to another. Moreover, P2P use cases are various, e.g.: in a multimedia streaming context, a P2P system should be able to support various applications and file formats; in group communications, like in a secure context, a P2P system should present an access control; etc.

To achieve these goals, it is particularly important to address the challenges in a context-aware architectural design or P2P routing protocol that takes different parameters into account, e.g. administrative domains, available resources, security parameters, network access policies, etc. Some solutions for topology-awareness already exist, e.g. Plethora [1], Hieras [2], Bypass [3], etc., but they take only the topology proximity into account. So, we propose CAP, a generic configurable context-aware system, to address the above mentioned problems.

In DHT-based systems, each node is responsible for objects; they are identified respectively by *nodeID* and *objectID*. These numerical identifiers are generated by a cryptographic hash function, applied to the node's address IP or the data item name. To meet our target, we need to use a context-aware cryptographic hash function and propose HMAC (*Hash based Message Authentication Codes*) [4], a keyed-hashing function, originally aimed for message authentication.

In order to avoid confusion with the key designing the matter of a query request in the P2P system, we call HKey, the key of the HMAC function. The HKey will be a configurable system parameter, that we will identify our context. Thus, when a HMAC function is used in the system, identifiers will be computed on a HKey, and resulting DHTs based on it. We call then such DHTs, VDHT (for *Virtual DHT*).

The remainder of this paper is organised as follows. In Section II, we give some background information about DHT and HMAC. Section III presents our concepts of HKey and VDHT, and the system design and its mechanisms. We discuss our system, called CAP, in Section IV, and Section V discusses related work. Finally, we conclude this paper and introduce our future work in Section VI.

Note the use of following abbreviations in the whole paper: P2P, DHT (and VDHT), HMAC, ISP for Internet Service Provider, AS for Autonomous System, ARD for Autonomous Routing Domain, which represents the whole ASs under a same authority or responsibility, and QoS for Quality of Service. Note also that in this paper, the words *peer* and *node* will be used interchangeably.

2 Background

2.1 Distributed Hash Tables

Mathematically, a DHT is a distributed injective hash function that associates keys with values, both in the same logical key space K. In DHT-based structured P2P systems, the ownership of this key space is split among the active peers of the network. A unique identifier is thus assigned to every node and object: *nodeID* and *objectID* respectively. DHT algorithms mapped then *objectID*s to the node responsible for them. And in order to locate an object, a node uses the DHT to map its *objectID* to a *nodeID* responsible for that object. The responsible node then either supplies the object directly or indicates where (or how) it can be acquired.

So, given a hash function h and a peer p, with an IP address @IP, that wants to share a resource identified by r, then the *objectID* and the *nodeID* are computed respectively as $h(r)$ and $h(@IP)$, and both are elements of K. Then, in the DHT, each *objectID* is mapped to the next numerically closest *nodeID*, and the resource r, or a link towards it, is stored on the peer identified by this *nodeID*. Consequently, each key lookup is resolved by iteration, in multiple steps, resulting in a multi-hop path at the P2P layer.

The key feature of DHTs used to build P2P architectures is that they use *consistent hashing*, which means that changes to the location where data items are stored are minimal as peers enter and leave the network.

All DHT-based P2P architectures share these four main properties [5]:

1. *Low degree*: each peer keeps only a small number of active connections.
2. *Low diameter*: any peer can be reached in a small maximal number of hops.
3. *Greedy routing*: peers independently calculate a short path to the destination.
4. *Robustness*: even if links or peers fail, a path to the destination can be found.

DHTs can thus effectively route messages to the unique owner of any given key and are typically designed to scale to large numbers of nodes. But, they assume the system is uniform in available resources and all peers are in a same transport domain.

2.2 Keyed-Hash Message Authentication Code

HMAC [4] aims to guarantee integrity between a sender and a receiver. It enables two parties holding a common secret key to generate, with a standard hash function, hash values that are completely different from the outcome of the standard hash function.

$$HMAC\,(h, k, m) = h(k \oplus opad\,, h(k \oplus ipad\,, m))$$

where h is a commonly used cryptographic hash function, like the ones used for building DHTs; k is a value known by all the nodes (it will be the HKey we will define in the next section); m represents the text to be hashed (it will be either the object name or the peer's address IP); *ipad* and *opad* are the bytes 0x36 and 0x5C respectively, each repeated 64 times; $\oplus$ denotes the bitwise XOR operation, and the comma denotes the concatenation of two bit strings.

3 System Design and Mechanisms

Using HMAC does neither affect key spaces nor DHT's principle. In fact, since the output of HMAC is nothing else than the output of the hash function, key spaces are of same nature and length. And obviously, both routing procedure (for key location) and maintenance procedures (as nodes join/leave the network) remain unchanged.

The HKey is not a fixed value for the whole system: the P2P network will then be a logical heterogeneous network of different node groups, known as zones, each characterized by a HKey used to compute there *nodeIDs* and *objectIDs*.

3.1 Semantic Definition of a HKey

A HKey could be *simple* or *compound*. We define a *simple HKey* as based on a single parameter or criterion, and a *compound HKey* as based on the combination of two characteristics (parameters or criteria) or more. They will be computed as the output of the cryptographic hash function, used by HMAC, applied to the concatenation of those characteristics; e.g., if we are interested in searching a resource according to three metrics m_1, m_2, and m_3, then the *compound HKey* $= h(m_1, m_2, m_3)$.

A *compound HKey* could be derived in two or more, simple or compound HKeys: these *derived HKeys* are then typically based on one or more different characteristics from those the initial compound HKey is based on. *Derived HKeys* could also be defined as the combination of two or more *simple HKeys*.

The type of the HKey will be defined in a global system profile, and in case of a *derived HKey*, the profile will define which combination of two or more values or characteristics to take into account, and eventually in which order. In fact, the profile defines a set of priority criteria and identifies the parameters to be taken into account in the routing protocol, defining thus the HKey structure. These criteria and parameters are quantified as the peer arrives in the system, according to an external procedure that does not affect the routing mechanism and that could be based on some database reports. In this paper, we do not discuss those external mechanisms.

Some examples of what a *simple HKey* could be are: the AS identifier, the administrative domain name or the group name or identifier of a specific group communication or a specific secure group, a QoS parameter (e.g. minimum available bandwidth, minimum battery power for mobile systems, etc.), a type of shared file (e.g. movie file, audio file, etc.), a secret key, etc.

3.2 Virtual Distributed Hash Tables and System Architecture

In order to be efficient, a context-aware routing has to be performed according to the HKey. Consequently, the P2P overlay network will be logically layered in a set of different uniform overlays, each one characterized by a specific HKey, based on which identifiers are computed there. This HKey will be then considered as a label for both the new overlay and its corresponding DHT. We call this new DHT, a VDHT.

Each node resides at least in the global P2P overlay and participates in one or more secondary ones, called local overlay, depending on its properties (available resources, locality, group membership, etc.) and according to the semantic of the HKey that is taken into account. In case the system profile does not define *derived HKeys*, there will be only one secondary overlay; else, there will be L more local overlays, where L will be equal to the number of the different possible defined *derived HKeys* per peer.

Hereafter, unless explicitly mentioned, we will neither discuss *derived HKeys*, nor distinguish between simple and compound HKeys. Thus, besides the global overlay, each node may participate in only one local overlay.

Each peer will have a *nodeID* in the key space of each overlay it resides in: it will be computed in each key space with HMAC based on the corresponding HKey. (An analogy could be done with a terminal with two active network interfaces.) For simplicity, we do not require uniqueness across key spaces, and we suppose each key space (so each overlay) use the same DHT algorithm (e.g. Chord [6], Pastry [7], etc.). And to avoid collisions at the global overlay between the *nodeIDs* of a heterogeneous population of peers building their identifier with the HMAC method, we propose either to apply our method to the different local overlays only, or to fix the HKey to a given value (e.g. 0x00 repeated 64 times) for the global overlay. In the latter possibility, new joining nodes will have their *nodeID* also computed with HMAC at the global overlay, and as fast as the network is dynamic, all the *nodeIDs* will be uniform, but since, in some cases, this may seem somewhat complicated and may still result in collisions, we propose to introduce HKeys, and its subsequence on overlay and DHT, only at the secondary layers of local overlays.

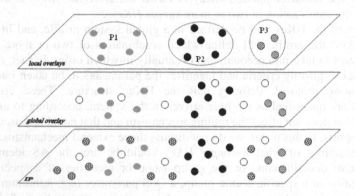

Fig. 1. Overview of the system with no *derived HKeys*

A simple illustration of our so defined system is shown in Fig. 1. P1, P2, and P3 are subsets of the global overlay, each of them characterized by a different value of a same parameter, which is the defined HKey (e.g., if we aim searching by ASs, the HKey at the local overlays layer will be the AS ID, and each P*i* will represent a different AS). In Fig. 1, gray dots represent the nodes of P1, black dots represent the nodes of P2, striped dots represent the nodes of P3, and white dots represent nodes of the global overlay that reside in no local overlay; dotted dots at the IP level represent active nodes of the internet that do not participate in our P2P system.

3.3 Routing Mechanism Using HKeys and VDHT

The routing mechanism is the same at each level according to the P2P routing protocol in use in the system (defined at the global overlay); the only difference is the HKey characterising the level and thus the usage of the consequent different *nodeIDs*, *objectIDs*, and DHT (or VDHTs). Note that at the secondary levels, *objectIDs* are, like *nodeIDs*, computed using HMAC with the corresponding HKey. So, a same data item may be stored at each level under a differently computed *objectID*. (It is absolutely like an object has several aliases.)

When a peer needs a data item, it first searches the local overlay it resides in (using the corresponding HKey-based *nodeID* and *objectID*, and the corresponding VDHT), and so, the query is forwarded only to the local successors of the data item requester. If the query fails, the peer will then searches the global overlay for the data item it needs, using then its original *nodeID* and the DHT defined at the global layer.

Note that when *derived HKeys* are defined, before searching the global overlay, a peer searches each overlay it resides in (using corresponding identifiers and VDHTs) until the needed data item is retrieved. Searching each of those overlays is done either simultaneously or consecutively in accordance with the order defined by the system policy in the global system profile.

To improve the data retrieval performance in the system, we propose to automatically cache a data item retrieved from the global overlay in the requester's local overlay (and of course, its *objectID* will be then computed using HMAC and based on the HKey corresponding to that local overlay).

3.4 Node Operations

The joining and failure/departure operations are the same at each level according to the routing protocol implemented in the system; but after a node joins the system (at the global overlay) and before it joins any local overlay at any secondary level, some operations must be taken.

When a new node arrives in the system, it needs to build its tables and inform other nodes of its presence. It first joins the global overlay and initializes its tables according to the implemented DHT algorithm, the same way in any P2P system based on that algorithm (e.g., with Chord [6], by sending a join message to a nearby member node of the system). But afterwards other operations must be taken in order to join the secondary overlay if it exists or to build it if possible. These operations allow the new node identify another node already member of the overlay it wants to join.

Therefore, the new node computes first the identifier of the local overlay it should joins, i.e. *h*(HKey), where HKey is the label of that overlay, according to the global system profile. (Note that the way the system profile is loaded in the system does not affect node operations or routing mechanism; e.g. it could be loaded on a kind of bootstrap node managed by the system operator or service provider.)

Otherwise, the local overlay identifier is stored at the global overlay in a data table, we call *zone table*. The structure of this table is shown in Table 1, where *zone ID* is *h*(HKey), the *zone label* is a textual string of the HKey, and the reminder data in the zone table are *nodeID*s in the global overlay of nodes already participating in the local overlay (if less than four nodes resides in the local overlay, then the remainder cells are set to NULL). This table is considered at the global overlay as an object and is stored on a rendezvous point: the node whose *nodeID* is the numerically closest to its *zoneID*. (The *zone label* is particularly useful for conflict avoidance in case derived HKeys are defined in the system.) The rendezvous node periodically checks the status of the nodes in the zone table, and in case of a no response (i.e. node failure or departure), a new routing procedure is performed to update the table. Note that if the rendezvous node leaves the system, the responsibility of the zone table is given to another node, like any other key object. But for fault tolerance, a zone table could also be duplicated on several nodes; in particular, the node of largest *nodeID* could take

Table 1. Zone Table Structure

Zone ID	Zone label	Largest nodeID	Second largest nodeID	Second smallest nodeID	Smallest nodeID

the responsibility of first storing the zone table, then, for the system consistency to put it on the local overlay (so on the node of the local overlay, whose local *nodeID* is the numerically closest to the *zoneID*).

So, when the new node has computed the *zone ID*, it sends (at the global overlay) a 'zone table request' message the same way it sends any request message. If the concerned zone table exists, it will be stored at the rendezvous node. Now, two situations may arise depending on the response of the rendezvous node.

In the first case, the zone table exists and the rendezvous point returns it. The new node knows then up to four nodes of the local overlay it wants to join, chooses one, computes its own *nodeID* based on the HKey of the overlay it wants to join, and sends a join message to the existing node it has just known. With this join message, the new node sends its local *nodeID* and the HKey to let the responding node knows that it has to respond with its corresponding local *nodeID*. Afterwards, the procedure continues as a simple join procedure in the global overlay. Note that just after a successful join to the local overlay the newly added node has to compare its global *nodeID* with those in the zone table in order to replace one of them if its *nodeID* is larger than the second largest *nodeID* or smaller than the second smallest one. And if a replace is necessary, messages are sent for an update.

In the second case, the new node is the first node to wants to join its local overlay. In this case, either the rendezvous node returns a message informing that the required object (as it happens, it is the zone table) does not exist, or the new node deduces that the zone table does not exists as a timeout expires. So, the new node creates a zone table with the *zone ID* and zone label of the local overlay it wanted to join, its global *nodeID* as the largest *nodeID*, and the remainder *nodeID*s fields set to NULL. The new node stores then the created zone table at the rendezvous node and starts its local overlay. Note that in a well populated network, the latter case is unlikely, particularly if the HKey is not so particular (e.g. if HKey stands for AS, a typical available bandwidth in some network, a popular file type, etc.)

Note also that in case derived HKeys are defined in the system, the procedure of joining a local overlay is repeated for each layer (with its corresponding HKeys).

In P2P systems, a node may leave the system or fail silently. These cases are treated the same way at each level (according to the implemented DHT algorithm at the global overlay). But they are treated independently at each level as soon as they are noticed (however, in general, they are noticed simultaneously by a few seconds).

4 Discussion

The use of HMAC in computing node and object identifiers enables context-awareness in a DHT-based P2P system through the HKey, without affecting the DHT algorithm, but allowing creation of virtual contextual secondary layer(s) with a corresponding VDHT for contextual data searching and management. Searching CAP for a resource or data item is thus context oriented, beginning at the contextual (local) overlay layer first, and since in VDHTs all identifiers are context-aware, computed specially for the concerned context, CAP ensures then that the result of a context-oriented query is context-oriented. In fact, at each secondary layer, when an overlay is populated, all nodes and objects have their identifiers based on the HKey characterising the overlay; so if a query initiated by a node residing in the overlay does not fail, that means that the response is present in this overlay. And even if the response is a pointer to the requested object, the object is necessary in the overlay, since every data presented or represented in a secondary overlay has its identifier necessarily computed based on the HKey characterising that overlay.

Consequently, CAP guarantees that a data item will be retrieved from the zone where it has been found; this must alleviate message latency and enhance lookup time. Note also that the caching mechanism (at the local overlay), we defined as part of the routing procedure, favours the popularity of the different objects (whether there are popular are not) in the system. These consequences are of great importance for network operators (particularly if HKey stands for local domain, or AS, or a typical available bandwidth, etc.) and service providers (if HKey stands also for a specific file type for example, or for a specific service characteristic).

In particular, the benefits of CAP are important for ISPs since their interconnection strategy is a financial matter. In fact, despite peering agreements between ISPs, in which context they share their network capacities for free, they are still led to buy and sell these capacities, because such agreements cannot be signed unless each ISP exchange with the other a same amount of traffic, and each ISP has enough clients in order to have a great income. So, ISPs still buy and sell each other their network capacities, and their income is conditioned by the terms of the agreement they sign. The terms of such agreements are about the volume of traffic exchanged inside and outside each ISP's ARD (both ways are charged), the maximum network bandwidth, and its cost, the geographical point of interconnection, the security policies, the dynamic configuration, the inter-domain routing policies, etc. So, defining and negotiating such agreements is a permanent burden for ISPs, and having a system that favours local and contextual searching, and guarantees a local or contextual data retrieval (of course, if the data item is found) is of great interest for ISPs, especially that major part of aggregate worldwide P2P traffic is a traffic of media formats, and compared to most web pages, a streaming media file is well huge, and moreover today network bandwidth cost (due to the above mentioned agreements) leads to poor quality media. So, since CAP is able to take advantage of an ISP network, it is of great interest for ISPs and their clients.

CAP can also be used as a core routing system for any application or service with a suitable interface allowing the user to determine or pick (from a predefined list) the zone he wants to search for his desired resource (data item or service).

5 Related Work

To the best of our knowledge, CAP is the first system that uses HMAC for other than its intended purpose. In fact, HMAC was defined for message authentication and the guarantee of integrity between a sender and a receiver. However, using it for building node and/or object identifiers in P2P systems is not a novel idea, but an idea that remained till now for security purposes, e.g., in smaller networks, where a shared secret between all nodes is used as a seed together with the IP address to generate *nodeID*s using an HMAC-SHA-1 scheme, or like in [8], where HMAC is used to guarantee a secured and unique *nodeID* for nodes that use prepaid P2P services.

Furthermore, as far as we know, CAP is the first configurable context-aware DHT-based P2P system, when many papers have gone on topology-aware P2P systems. (Note that this is a particular case of CAP, e.g. when HKey is the AS ID.) Some of existing systems, like Bypass [3], reserve the secondary layer only to high-capability peers. However, two systems, among other existing ones, inspired CAP design and mechanisms: Plethora [1], by its global and local overlays and its caching mechanism, and Hieras [2], by its P2P ring table. In both systems, a peer has a unique identifier in the whole system, generated by the consistent hash function of the underlay DHT algorithm.

Plethora [1] is a two-layer wide area read-write repository, where peers are expected to be semi-static (partially persistent network state) and to have good Internet connectivity in terms of bandwidth. The lower layer is the global overlay, which contains all the peers, and on the top of which all the peers are organized, at a same layer, into different local overlays according to ASs and associated proximity. Local overlays serve as locality-aware caches for the global overlay, and their size is defined by system parameters and managed by specific distributed algorithm.

Hieras [2] is a hierarchical system to relieve the problem of distributed routing tasks across all system peers without awareness of underlay network link latency. It is organized in several levels: the highest one contains all the peers, and at the lower ones, topologically adjacent peers are grouped into different P2P rings in such a strategy that the average link latency between two peers in lower level rings is much smaller than in higher level rings. Information of different P2P rings are maintained in *ring tables*, which structure is very similar to our *zone table*, and that serve also for a new joining node.

Otherwise, in MDHT [9], a multi-space DHT for multiple transport domains, a node can be a bridge between multiple key spaces (each using the same DHT algorithm) and have thus a *nodeID* in each one. But MDHT looks more like a solution to map requests through a mesh network of independent key spaces and DHTs, thanks to the bridge nodes. Furthermore MDHT does not deal with the system context, although it is a solution to the problem resulting from the common assumption that every node that participates in the DHT is within the same transport domain.

6 Conclusion and Future Work

This paper illustrates our ongoing work on a novel scheme, called CAP, a context-aware DHT-based P2P system. To the best of our knowledge, this is the first paper raising the issue of a configurable and extensible context-aware P2P system. To do so, we propose to generate object and node identifiers with HMAC instead of a consistent hash function, and we take the context information we are interested in into account through the key of the HMAC function, that we called HKey. Consequently, our system presents virtual secondary overlays that are context-oriented, but still based on the underlying DHT algorithm. However, key spaces of each overlay are independent and those of virtual overlays are managed in a virtual DHT, we called VDHT.

Despite its simplicity, we believe in the potential of our concept. Its main advantages are contextual data retrieval (when there are available in the defined contextual zone), alleviated message latency, and enhanced lookup time.

Our future work begins with simulation experiments to evaluate CAP routing performance. Afterwards, it continues with a quantitative analysis of CAP overheads. Looking further ahead, we also intend to implement CAP as a routing core of a promising scalable and robust P2P application.

References

1. Ferreira, R.A., Grama, A., Jagannathan, S.: Plethora: an efficient wide-area storage system. In: Bougé, L., Prasanna, V.K. (eds.) HiPC 2004. LNCS, vol. 3296, pp. 252–261. Springer, Heidelberg (2004)
2. Xu, Z., Min, R., Hu, Y.: Hieras: a DHT based hierarchical P2P routing algorithm. In: 32nd International Conference on Parallel Processing, pp. 187–194. IEEE Press, New York (2003)
3. Kwon, G., Ryu, K.D.: Bypass: topology-aware lookup overlay for DHT-based P2P file locating services. In: 10th International Conference on Parallel and Distributed Systems, pp. 297–304. IEEE Press, New York (2004)
4. Krawczyk, H., Bellare, M., Canetti, R.: HMAC: keyed-hashing for message authentication. RFC 2104 (1997)
5. Risson, J., Moors, T.: Survey of research towards robust peer-to-peer networks: search methods. Technical Report (2004)
6. Stoica, I., Morris, R., Liben-Nowell, D., Karger, D.R., Kaashoek, M.F., Dabek, F., Balakrishnan, H.: Chord: A Scalable Peer-to-peer Lookup Protocol for Internet Applications. In: SIGCOMM 2001, pp. 149–160. ACM Press, New York (2001)
7. Rowstron, A., Druschel, P.: Pastry: Scalable, decentralized object location and routing for large-scale peer-to-peer systems. In: Guerraoui, R. (ed.) Middleware 2001. LNCS, vol. 2218, pp. 329–350. Springer, Heidelberg (2001)
8. Emmert, B., Jorns, O.: $P^2 2P$ - Prepaid Peer-to-Peer Services. In: 6th International Conference on P2P Computing, pp. 223–224. IEEE Press, New York (2006)
9. Harwood, A., Truong, M.: Multi-space distributed hash tables for multiple transport domains. In: 11th IEEE International Conference on Networks, pp. 283–287. IEEE Press, New York (2003)

A Socially Inspired Peer-to-Peer Resource Discovery Service for Delay Tolerant Networks

Tuan Dung Nguyen and Siegfried Rouvrais

GET / ENST Bretagne, France
{td.nguyen,siegfried.rouvrais}@enst-bretagne.fr

Abstract. The increasing popularity of wireless computing devices has promised a vision for mobile resource sharing applications. The scalability of such environments and their intermittent connection characteristics raise new challenges for both network protocols and system design. This paper proposes an overlay-based resource discovery service with a socially inspired peer-to-peer lookup algorithm for delay tolerant networks. Several simulation scenarios have been carried out to evaluate the algorithm's efficiency and scalability in comparison to classical approaches.

1 Introduction

The recent years have seen a remarkable diffusion of mobile appliances into our daily life. Nowadays, people often go around with their favorite handheld computing devices such as smart phones, PDAs or music players. Apart from their increasing processing and storage capacity, these devices can communicate via mobile ad hoc networks (MANETs) created on-the-fly using any available wireless interfaces. Many spontaneous applications [1] can be imagined that enable resource sharing among users anytime and anywhere (e.g. in the street, on the campus, at the airport).

However, spontaneous environments are often characterized by a large-scale population and intermittent connections between nodes due to their frequent mobility. Moreover, users can temporarily switch off their devices due to battery shortage. The end-to-end connectivity assumption in traditional MANETs does not hold all the time as nodes may be temporarily located in different network partitions. Delay tolerant [2] paradigms have been recently proposed to cope with the above-mentioned problems in which network protocols and system services should exploit node's physical mobility for message transmission with an acceptable delay.

So far, research works have primarily focused on routing issues to find a way to send a message from one node to another in these challenged networks [3]. But to the best of our knowledge, few relevant middleware services (e.g. naming, resource discovery) exist today to facilitate the construction of applications in these environments. For instance, the aim of resource discovery is to find the location of available resources (e.g. document, data) on the network. In traditional distributed systems, this service is often based on centralized directories

R. Meersman, Z. Tari, P. Herrero et al. (Eds.): OTM 2007 Ws, Part II, LNCS 4806, pp. 960–969, 2007.

and does not perform well in decentralized environment. Consequently, it is of interest to investigate peer-to-peer resource discovery techniques [4].

A scalable peer-to-peer resource lookup technique should avoid blind flooding in the network as it generates a lot of redundant messages leading to network congestion and battery wasting [5]. Furthermore, coping with intermittent connections is crucial to successfully provide relevant resources to their requesters. Our contribution is a socially-inspired resource lookup and delivery service that answers the aforementioned requirements. To the best of our knowledge, this is the first work composing several overlays for resource lookup in intermittently connected networks.

The rest of this paper is organized as follows. First, section 2 presents existing resource lookup and delivery approaches. Section 3 proposes our algorithm relying on two overlays of resource interest and human mobility. Section 4 evaluates and compares the performance and scalability of our approach with classical algorithms through simulations. Section 5 reviews the related work. Finally, section 6 concludes the paper and gives some perspectives for future work.

2   Background

We review in this section existing solutions for resource discovery in general and then clarify challenges related to inherent characteristics of delay tolerant networks. We discuss the benefits as well as issues relating to the composing of two overlays based on resource interest and human mobility.

2.1   Peer-to-Peer Resource Lookup and Delivery

Nodes are connected by logical links in an overlay network on top of underlying physical networks. A structured lookup technique (e.g. Chord, Pastry) imposes strict constraints with the overlay formation while an unstructured lookup technique (e.g. Gnutella) lets nodes self-organize to fulfill this task [4]. The former can achieve a more efficient resource lookup but also incur costly overhead due to the structured overlay's maintenance. The latter is believed to be better suited for coping with frequent topology changes and intermittent connections [6]. Additionally, unstructured lookup techniques can support better keyword searches.

Existing resource lookup mechanisms can be classified into three main categories: push-based, pull-based and combined push/pull [4]. Push-based solutions proactively broadcast resource advertisements in the networks. Nodes cache this information for their local lookup operations. This approach can also be used to disseminate small size spatial temporal resources. Pull-based solutions broadcast requests for resources on demand. Nodes having the requested resources answer with a reply message. This message can also be cached for later use to reduce the communication overhead. Other hybrid solutions combine the two previous push/pull approaches by broadcasting resource requests as well as advertisements. Push-based and push/pull based solutions are not relevant to non spatial temporal resources due to their generated communication overhead. On the other hand, pull-based solutions generate queries on demand and exploit the information from their propagation to send back reply message, e.g. using source routing.

Benefits of a resource interest overlay. Empirical studies of popular file-sharing systems (e.g. eDonkey) [7] have clearly demonstrated the interest proximity of resources offered by peers. Peers already having some common resources are likely to have others in the same category. Interestingly, this observation holds not only for popular but also for rare resources. Some works [7] have exploited this to significantly improve the lookup performance of Internet peer-to-peer file-sharing systems. A resource interest overlay can be obtained either by using peer's explicit profiles (peer's interest, sharing resource's categories) or implicitly investigating the lookup history. A peer tries to contact with his resource interest neighbors before soliciting other peers in the conventional overlay network. However, in networks with frequent mobility, a node may be unable to communicate directly with its corresponding resource interest neighbors.

2.2 Issues with Delay Tolerant Networks

For discovery service, a simple broadcast can fail to reach relevant nodes due to mobility and intermittent connectivity. To cope with this issue, periodic broadcast or gossip-based (epidemic-style) [8] approaches should be taken into consideration. The former repeatedly diffuses messages to every 1-hop neighbors. The latter stores received messages in a buffer and forward them later a defined number of times t. Each time, a message is sent to a defined number f of randomly selected nodes. Additionally, frequent mobility often leads to the invalid reverse path. The return of reply message using source routing results in high latency or even failed delivery. On the other hand, a simple use of delay tolerant routing protocols (e.g. epidemic-routing) can be costly in term of high communication overhead.

Benefits of a human contact overlay. Experimental results [9] have demonstrated that human mobility is not totally arbitrary. The *small world* phenomena [10] suggested that two random U.S. citizens could be connected by hand-passing a letter by a short chain of six acquaintances. In reality, human movements are solicited by various social relations (e.g. family, workplace) [11]. Mobility-aided techniques, where messages can be transmitted thanks to node physical movements, have exploited this result to resolve routing problems in intermittently connected networks [3]. Likewise, these nodes can also be considered as good forwarding candidates in resource discovery mechanisms. A human contact overlay can be built by monitoring the meeting frequency with node's physical neighbors.

2.3 Composing Two Overlays

Keeping an updated global view of these overlays at each mobile nodes is an unfeasible task due to limited resources and frequent mobility. Therefore, desired

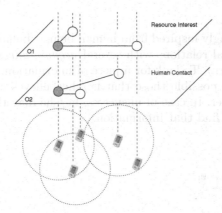

Fig. 1. Two overlays vs. physical topology

global behaviors should be obtained from aggregated local information using appropriate rules without a complete system view. In our proposal, a node only keeps information about its neighbors in each overlay and obtains information about other nodes during lookup operations. Moreover, as depicted in Figure 1, nodes appearing as neighbors in one overlay may not be neighbors in the other. The inconsistency between two overlays O_1 and O_2 and node's physical neighbors raises difficulties for lookup operations and requires an algorithm capable of working on these two superposed overlays.

3   Socially-Inspired Resource Discovery

3.1   System Model

The network is modeled by a graph $G = (V, E)$, where $V = p_1, p_2, \ldots, p_n$ is the set of nodes and E is the set of direct connections among them. The network dynamism implies that V and E evolve over time as nodes can be switched off or out of transmission range. A node has a list of physical neighbors currently in its transmission range. For our two selected overlays, we propose that each node maintains two semantic neighbor lists defined as follows:

1. **Interest Neighbor.** Resources are classified into a limited number of categories (e.g. music, events). Node keeps a list of nodes sharing the same interest, namely *interestNeighbor*, which are likely to provide resources in a category. The Least Recently Used (LRU) strategy [7] is used to maintain these lists with nodes having already replied to previous queries in the corresponding category.
2. **Contact Neighbor.** A node keeps a list of neighbor nodes in its social networks, namely *contactNeighbor*, i.e. they are more likely to be met in the future. Items in this list are in the form of $\{(p_i, timer)\}$ where *timer* represents the meeting frequency with p_i. The value *timer* is initially set to 0 and incremented by 1 at each meeting time.

3.2 Algorithm

Our algorithm is strongly inspired from human being's behaviors. An individual usually has some social relations with several *acquaintances* (e.g. family members, friends, colleagues). To look for a piece of information, one often starts by contacting as soon as possible those that he/she believes to be more likely to have a favorable answer. In case of unsuccess, he/she may then lean on his/her *acquaintances* to help find that information.

```
 1  initialisation:
 2  begin
 3  |    interestNeighbor ← ∅;
 4  |    contactNeighbor ← ∅;
 5  |    msgBuffer ← ∅;
 6  end

 7  while true do
 8  |    wait until event;
 9  |    switch event do
10  |       case LookupEvent
11  |          msg = CreateReqMessage ();
12  |          for p ∈ physicalNeighbor do  send (p, msg);
13  |       case RequestReceivedEvent (msg)
14  |          if IsResourceFound (msg) then
15  |             SendResponse ();
16  |          else if msg.src ∈ contactNeighbor then
17  |             msg.interestNeighbor ← msg.interestNeighbor ∪ interestNeighbor;
18  |             msgBuffer ← msgBuffer ∪ msg;
19  |       case GossipTimerEvent
20  |          CGossip ();
21  |          Update (contactNeighbor);
22  |       case ResponseReceivedEvent (msg)
23  |          Update (interestNeighbor);
24
```

Fig. 2. Main Lookup Algorithm

The main algorithm (cf. Figure 2) remains idle until it is triggered by one of these four principal events:

- LOOKUPEVENT: event generated by applications when they want to look for a resource;
- REQUESTRECEIVEDEVENT: event generated by the communication layer to inform its higher layers about a request message arrival;
- GOSSIPTIMEREVENT: event generated periodically to trigger a gossip round;
- RESPONSERECEIVEDEVENT: event generated by the communication layer to inform its higher layers about a response message arrival.

For the sake of clarity, Figure 2 illustrates the pseudo code of the main algorithm for one resource category. The query propagation works as follows. The generated query messages are first sent to any available corresponding interest neighbors. If this step fails, query messages are sent to every nodes in the transmission range and the message is augmented with the source node's interest neighbors (cf. line 16). This allows nodes to discover gradually other neighbors in the resource interest overlay from their initial local knowledge. Node receiving a query and having the required resource will send back a reply message to the query's origin. Otherwise, it saves a copy of the query in the message buffer and gossips it using the previously described mechanism. The gossiping algorithm is depicted in Figure 3. Furthermore, the lookup process for distant resources is carried out through a stable path consisting of social-related neighbors. The return of reply messages is realized using source routing to exploit the social traces left by the propagation of query messages.

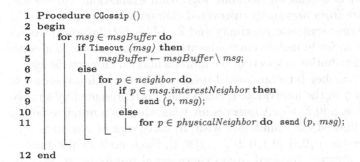

```
1  Procedure CGossip ()
2  begin
3     for msg ∈ msgBuffer do
4        if Timeout (msg) then
5           |  msgBuffer ← msgBuffer \ msg;
6        else
7           for p ∈ neighbor do
8              if p ∈ msg.interestNeighbor then
9                 |  send (p, msg);
10             else
11                |  for p ∈ physicalNeighbor do  send (p, msg);

12 end
```

Fig. 3. Controlled Gossip Algorithm

4   Evaluation

4.1   Simulation Environment

We realized our experimental study using OMNet++ [12], a modular simulation environment. We focused on the sent messages overhead of different decentralized lookup algorithms. We did not consider message loss nor low layer's detail in this study. Messages are supposed to be transmitted between nodes in transmission range with a random delay uniformly selected in the range $[0.1, 0.4]$ (s). The main simulation parameters are presented in Figure 4. During the simulations, nodes move following two mobility models: Random Way Point (RWP) and Community-based (CMM). The former is a classic model in which each node chooses a random target location, moves to the destination with a random speed, then waits for a random period of time before repeating this process. The latter

Parameters	Values	Parameters	Values
Simulation time	1800 s	Transmission delay	Uniform in $[0.1, 0.4]$ s
Simulation area	2000 m × 2000 m	Node speed	Uniform in $[1, 6]$ m/s
Transmission range	50 m	Request interval	120 s

Fig. 4. Simulation parameters

is a recently proposed model that exploits social networks to generate more realistic mobility traces [11].

We evaluated our proposal by comparing it with two classical algorithms: *periodic broadcast* and *epidemic-based*. In the former one, node keeps its own queries in a buffer to periodically broadcast them to its neighbors. In the latter one, node does not only keep its own queries but also received queries in a buffer to periodically send them to a random subset of its current neighbors.

4.2   Resources and Queries Distribution Model

A realistic workload is crucial for accurate algorithm evaluation. We assume that resources inspire from previously discovered characteristics of peer-to-peer file sharing applications: semantic proximity and Zipf-like distribution[7]. As no real workload exists so far in mobile environments, a synthetic model is used to generate resource distribution respecting the aforementioned characteristics.

We adopted the Number Intervals model used in [13] in which a resource is represented by a point in the interval $[0, 1]$ and a query is represented by a range within that interval, e.g. $[0.2, 0.5]$. A query is hit when there is a resource falling into its range. We limited our simulation with 10 categories of resources represented by 10 intervals $[0, 0.1], [0.1, 0.2], \ldots, [0.9, 1]$. Each node is also limited with one semantic category. Inspired from experimental results in [7], we proposed that a resource is generated in its corresponding semantic nodes with a probability equals to 0.7.

As in [13], a resource query is also represented by a range $(center, range)$ where $range$ takes a random value according to a normal distribution with mean 0.05 and variance 0.002 whereas $center$ is selected according to a Zipf-like distribution. The probability that $center$ falls into a category i is $\dfrac{(\frac{1}{i})}{\sum_{j=1}^{10}(\frac{1}{j})}$.

4.3   Simulation Results and Analysis

The presented algorithms are evaluated using the three following criteria: *success ratio* (the number of successful resource deliveries/the number of resource requests) and *total sent messages*.

The hit ratio is depicted in Fig. 5 (*a*) and the message overhead is illustrated in Fig. 5 (*b*) with the message overhead in log scale using the CMM mobility model. The periodic broadcast and epidemic algorithms trade off message overhead for hit ratio. The simulation results show that our algorithm achieved a hit ratio as good as other algorithms while reducing message overhead by order of magnitude.

Impact of node density. We ran our simulations with 50, 100, 150, 200, 250 nodes to verify the impact of node density to the algorithm's performance and scalability. A smaller number of nodes results in a sparser network that is closer to our application's environment. The simulation results clearly show that our proposed algorithm is really appropriate for intermittently connected networks.

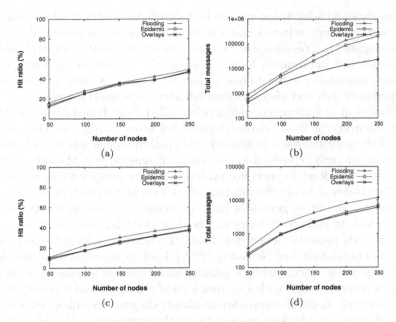

Fig. 5. Hit ratio vs. message overhead: (a), (b) with CMM model; (c), (d) with RWP model

Impact of mobility model. The simulation results in Fig. 5 (*a*) and Fig. 5 (*b*) show that our algorithm works considerably better than other algorithms with the CMM model, i.e. with social-related human mobility. As depicted in Fig. 5 (*c*) and Fig. 5 (*d*), with the RWP model where a node's movement is completely random, the gain is less significant but its performance is not worse than the other algorithms.

5 Related Work

Several works have been proposed for lookup in highly dynamic environments using push, pull or push/pull approaches and caching mechanisms. None of them have taken into account the coexistence of multiple overlays for representing non-functional contexts.

Lindemann et al. [14] proposed an epidemic-based peer-to-peer lookup service, namely Passive Distributed Indexing (PDI), to cope with intermittent connectivity and high mobility. Query and response messages are transmitted using local broadcast and query results are cached in participating nodes for later use. Most queries could be resolved locally thanks to the implicit dissemination of index entries in the network using node mobility. This work uses a pull-based approach as ours but does not take into account content semantic proximity and social-related mobility.

Motani et al. [15] presented the PeopleNet architecture for searching information in a wireless virtual social network. Information queries which represent both requests and advertisements are first directed via infrastructure-based networks to k randomly selected nodes in non overlapping geographical areas and then propagated in a peer-to-peer manner. The key idea is that a better probability of match (i.e. information found) is achieved when related matching queries are closely placed in a defined area. Unlike our work, this work uses a combined push/pull approach and also needs an infrastructure-based network.

Wolfson et al. [13] proposed an algorithm called Rank-Based Broadcast (RBB) for the discovery of local spatio-temporal resources in high mobility environments. This work follows a combined push/pull approach where both resource information (namely reports) and queries are disseminated in the network by periodic 1-hop broadcast Reports are ranked by their relevancy to queries received from other nodes and only the top-ranked ones are sent in each broadcast round. However, this algorithm generates high communication overhead with large size resources and its lookup efficiency can be degraded with rare resources.

Hui et al.[16] presented a new communication scheme, namely Osmosis, for file sharing in Pocket Switched Networks (PSN). Lookup messages are disseminated using an epidemic scheme while a penalty-based osmosis scheme is proposed to send back reply messages with a certain level of reliability and without overloading the network. Reply messages are implicitly directed to lookup senders using traces left during the lookup process due to the propagation of query messages. This work is inspired from a biological phenomenon and does not exploit social and resource semantic as in our work.

6 Conclusion

We have proposed a socially-inspired algorithm for resource lookup and delivery in delay tolerant networks. We have demonstrated through simulations that our algorithm can achieve as good success ratio as existing solutions while considerably reducing the communication overhead and latency. We plan to carry out further evaluations with better metrics to determine more accurate interest and contact neighbors as well as with a a real resource workload. Future work will also consist of proposing a middleware to take into account different contexts (e.g. battery, reputation, application's non-functional requirements) and to generalize our multiple overlays approach.

References

1. Feeney, L.M., Ahlgren, B., Westerlund, A.: Spontaneous Networking: An Application-Oriented Approach to Ad Hoc Networking. IEEE Communications Magazine 39(6), 176–181 (2001)
2. Burleigh, S., Hooke, A., Togerson, L., Fall, K., Cerf, V., Durst, B., Scott, K., Weiss, H.: Delay-Tolerant Networking: An Approach to Interplanetary Internet. IEEE Communications Magazine 41(6), 128–136 (2003)
3. Pelusi, L., Passarella, A., Conti, M.: Opportunistic Networking: Data Forwarding in Disconnected Mobile Ad Hoc Networks. IEEE Comm. Mag. 44(11) (2006)

4. Li, X., Wu, J.: Searching Techiques in Peer-to-Peer Networks. In: Wu, J. (ed.) Handbook of Theoretical and Algorithmic Aspects of Ad Hoc, Sensor, and Peer-to-Peer Networks, pp. 613–642. Auerbach Publications, Boston, MA, USA (2005)
5. Ni, S.Y., Tseng, Y.C., Chen, Y.S., Sheu, J.P.: The Broadcast Storm Problem in a Mobile Ad Hoc Network. In: Proc. of MobiCom 1999, Seattle, Washington, USA (1999)
6. Oliveira, L.B., Siqueira, I.G., Macedo, D.F., Loureiro, A.A.F., Wong, H.C., Nogueira, J.M.: Evaluation of Peer-to-Peer Network Content Discovery Techniques over Mobile Ad Hoc Networks. In: Proc. 6th IEEE Int. Symp. on a World of Wireless, Mobile and Multimedia Networks (WoWMoM 2005), Taormina, Italy (2005)
7. Handurukande, S.B., Kermarrec, A.M., Fessant, F.L., Massoulié, L., Patarin, S.: Peer Sharing Behaviour in the eDonkey Network, and Implications for the Design of Server-less File Sharing Systems. In: Proc. 1st European Conf. on Computer Systems (EuroSys 2006), Leuven, Belgium, ACM Press, New York (2006)
8. Eugster, P.T., Guerraoui, R., Kermarrec, A.M., Massoulié, L.: Epidemic Information Dissemination in Distributed Systems. IEEE Computer 37(5), 60–67 (2004)
9. Chaintreau, A., Hui, P., Crowcroft, J., Diot, C., Gass, R., Scott, J.: Impact of Human Mobility on the Design of Opportunistic Forwarding Algorithms. In: Proc. of INFOCOM 2006, Barcelona, Spain (2006)
10. Milgram, S.: The Small-World Problem. Psychology Today, 61–67 (1967)
11. Musolesi, M., Mascolo, C.: A Community Based Mobility Model for Ad Hoc Network Research. In: Proc. 2nd ACM/SIGMOBILE Int. Workshop on Multi-hop Ad Hoc Networks: from theory to reality (REALMAN 2006), Florence, Italy, pp. 31–38 (2006)
12. OMNeT++ (2006), http://www.omnetpp.org
13. Wolfson, O., Xu, B., Yin, H., Cao, H.: Search-and-Discover in Mobile P2P Network Databases. In: Proc. 26th IEEE Int. Conf. on Distributed Computing Systems (ICDCS 2006), Lisboa, Portugal, IEEE CS Press, Los Alamitos (2006)
14. Lindemann, C., Waldhorst, O.P.: A Distributed Search Service for Peer-to-Peer File Sharing in Mobile Applications. In: Proc. 2nd IEEE Int. Conf. on Peer-to-Peer Computing (2002), Linköping, Sweden (2002)
15. Motani, M., Srinivasan, V., Nuggehalli, P.S.: PeopleNet: Engineering A Wireless Virtual Social Network. In: Proc. 11th ACM Int. Conf. on Mobile Computing and Networking (MOBICOM 2005), Cologne, Germany, pp. 243–257. ACM Press, New York (2005)
16. Hui, P., Leguay, J., Crowcroft, J., Scott, J., Friedman, T., Conan, V.: Osmosis in Pocket Switched Networks. In: Proc. 1st Int. Conf. on Communications and Networking in China (CHINACOM 2006), Beijing, China (2006)

In-Network Event Processing in a Peer to Peer Broker Network for the Internet of Things

Holger Ziekow

Humboldt-Universität zu Berlin,
Institute of Information Systems, Spandauer Str. 1, 10178 Berlin, Germany
ziekow@wiwi.hu-berlin.de

Abstract. With the rise of RFID technology, the so called internet of things for managing supply chain events has gained increasing attention. This paper proposes a peer to peer broker network on top of existing specifications for this infrastructure. With the broker network, event based communication is enabled for the internet of things. However, the main contribution of this paper is a mechanism to optimize event queries for in-network processing of supply chain events. Here, models of supply chain processes are exploited for optimizing event queries and enabling improved mapping of query operators to the broker network. Experiments are presented which show that the proposed mechanism reduces network load compared to existing approaches.

Keywords: In-network processing, complex event processing, internet of things.

1 Introduction

In the past few years new technologies have begun to change how companies monitor their supply chains. In particular the RFID technology enabled capturing more logistic events in greater detail than it was possible with barcodes [12]. These changes have caused the demand for new internet services to exchange RFID data between companies in the value chain. This network infrastructure is frequently referred to as the internet of things [4]. In this paper a broker service is proposed on top of existing specifications for this infrastructure. This service is designed in a distributed fashion as a peer to peer broker network for exchanging supply chain events using an event based communication scheme. This design accounts for requirements such as decentralization and flexibility that are crucial in the application context of supply chain management. The proposed service uses in-network processing techniques for event queries to reduce network load. An important novelty is that the service exploits knowledge of the supply chain structure to optimize event queries.

The remainder of the paper is structured as follows. In section 2 the distributed broker service is presented. This is done along with a description of core services in the internet of things and a discussion of the need for the purposed extension. Section 3 focuses on the issue of in-network event processing.

R. Meersman, Z. Tari, P. Herrero et al. (Eds.): OTM 2007 Ws, Part II, LNCS 4806, pp. 970–979, 2007.

This covers a description of the role of complex events in supply chain management and techniques for distributed complex event processing. The main part of section 3 is on a new method to optimize in-network processing that exploits process knowledge for query optimization. Section 4 presents experiments that demonstrate the benefits of the proposed approach. Related work is discussed on section 5 and the paper is concluded in section 6.

2   Peer to Peer Broker Services for the Internet of Things

The internet of things is an infrastructure for exchanging information about physical objects that are tagged with RFID chips. The so far dominating application domain is supply chain management [12]. International effort was made to specify core services and data standards for the internet of things. In particular the global standardization consortium GS1 played a major role in the standard development. Currently specified core services are the Object Name Service (ONS) and the EPC Information Services (EPCIS) [8]. For this paper, only the EPCIS is relevant. The EPCIS provides network interfaces for accessing object related information. These are read events of the RFID tag which corresponds to the respective object. Such read events may be enriched with some context information like the process step during which the event was captured.

EPCISs provide interfaces for ad-hoc queries and a callback interface for standing queries. By implementing a callback interface EPCISs serve the event driven nature of RFID data well. However, the direct addressing of callback interfaces limits flexibility [11]. Event queries would need to be adopted whenever the supply chain changes. This is a contradiction to the increasing demand for flexibility in modern supply chains [7]. To solve this problem this paper proposes to use an event based communication scheme and broker services on top of EPCISs. This combines server initiated communication with indirect addressing [11].

In the proposed setup, event consumers register event queries at a broker service instead of directly addressing an EPCIS. The broker service routes events from EPCISs to the consumers which registered to a corresponding query. This way, changes in the supply chain structure do not affect the event consumers because the broker service is in charge of the adoption. A simple way of implementing a broker service would be to establish a central instance that maintains all registered event queries and is connected to all EPCISs. The central instance could register standing queries at the EPCISs according to the active subscriptions of event consumers. Whenever events match a registered query the broker service would forward these events to the corresponding consumer. This implementation follows the definition of event based systems as defined by Carzaniga et al. [5]. However, despite simple implementation, a centralized solution is infeasible for a number of reasons. Major technical reasons are that a centralized solution does not scale, creates a single point of failure, and accounts for inefficient event routing that results in large communication overhead. Furthermore, political concerns exist against the establishment of centralized services in the internet of things [13]. That is, decentralized solutions are favored to prevent single parties from gaining control over the infrastructure.

2.1 Distributed Event Management on Top of EPCISs

From the above discussion a number of requirements result for a broker service that communicates events in the internet of things. The most important requirements are:

1. No single party is allowed to control the whole service.
2. The service must adapt to changes in the supply chain structure.
3. The service must operate on top of the EPCIS specification, because changes of the EPCIS standard are practically infeasible.
4. The service must support event queries which are typical to the supply chain domain.
5. The service must be scalable and provide means for handling large numbers of events.

This paper proposes to use an overlay network of distributed broker nodes where users can register event queries. The broker network communicates with EPCISs on behalf of the users. Subsequently it is discussed how this solution takes the above listed requirements into account. In order to meet requirement 1 this paper proposes to decentralize the functionality of the broker service in a peer to peer network. The purposed services build upon technologies of distributed event based systems [11] and their combination with peer to peer technologies [17,15]. In event based systems, event producers and event consumers are registered as clients to a broker service. Brokers provide methods $sub(query)/unsub(query)$ that consumers use for registering/unregistering to certain queries. Furthermore, brokers provide a method $pub(event)$ that is called by event producers to notify about a captured event. The broker service delivers event messages that are published by pub to clients that subscribed to a matching query. In a distributed event based system the task of routing event messages from producers to consumers is split up between several broker nodes. This naturally avoids the problem that a single instance may control the whole network.

To meet requirement 2, message routes need to adapt to changes of event sources. That is because changes in the supply chain structure can impact in which EPCIS certain events are captured. In distributed event based systems, brokers use routing tables to direct event messages. Each node maintains a routing table that denotes which incoming event messages must be forwarded to which neighbor. Each table entry comprises a filter that specifies events of interest and the neighbor broker or local client to which matching events must be forwarded. Updates to the networks routing tables are triggered by calls of sub or $unsub$. Several approaches like covering based routing or merging based routing exist to reduce the network traffic for updates of routing tables [10]. These techniques allow the broker services to adapt to changes and meeting requirement 2.

The third requirement (not to change existing specifications) prevents the application of an event based system in its pure form. That is because event sources need to actively call the *pub* method of a broker service in order to realize the provider initiated communication paradigm in event driven systems. However, implementing this behavior in the EPCISs would require changes in the specification. This paper proposes broker services that work on top of the EPCIS standard instead of changing it. In order to realize the producer initiated event communication the broker services use the callback interface of the EPCIS. For each filter in its routing table a broker service registers a standing query at all EPCISs that are connected to it.

Another challenge is to meet requirement 4 (support for supply chain specific queries). The supply chain domain adds complexity to event queries. Event queries in this domain will mainly be for complex events that occur due to a set of events. Examples are, trace queries that comprise several supply chain stations. This poses the need for supporting complex event patterns for event queries [9]. Details on how this is realized in the proposed broker network are discussed in section 3.1.

Finally, most challenging and most important is requirement 5 (handling large numbers of events). It is expected that millions of RFID tagged items will create a large number of event messages. The challenge is to reduce the resulting traffic by communicating relevant events only. Existing RFID middleware solutions use low pass filters to suppress duplicates and false events due to read errors (e.g. [3]). However, relevance of an event can only fully be determined by evaluation the query patterns of registered queries. Techniques of in-network query processing that push this evaluation close to the event sources can help to significantly reduce the network traffic [16,14]. This has especially been used in the sensor network community (e.g. [18]). To meet requirement 5, this paper proposes to adapt such technologies in the presented broker services. The broker services use a novel optimization technique that is tailored to the internet of things. Details of this mechanism are presented in section 3.2.

3   In-Network Event Processing for the Internet of Things

In this section it is described how the proposed broker services realize in-network event processing for reducing network load. Therefore section 3.1 discusses complex event patterns and related processing technologies. Section 3.2 presents a novel optimization technique that improves operator mapping for in-network query processing. This optimization technique makes the proposed broker services particular efficient in the application context of supply chain management.

3.1   Complex Events in the Internet of Things

Complex events are defined as events that occur because a set of other events has occurred [9]. For instance, a complex event C may be that the primitive event A is followed by the primitive event B within 5 minutes. Event queries in the internet of things will often be for complex event patterns rather than for isolated primitive events. Examples are trace queries in supply chains or monitoring queries. E.g. the event of a correctly finished process is if an item has passed several process steps of a certain kind in a certain order and maybe within a certain timeframe.

Several high level languages have been proposed for the definition of complex event queries (e.g. [6]). Expressions in these languages follow the same general structure. Expressions (often referred to as rule) comprise an action, an event pattern, and constraints. Actions are used to describe the systems reaction to the respective complex event, e.g. by issuing a corresponding event message. Event patterns describe relations between primitive events (e.g. "followed by" or "AND") that must hold for the complex event to happen. Each pattern can be translated to and detected by a state machine [11]. Constraints are defined on attributes of events in the pattern. Especially equality checks on the corresponding item IDs of events are of major importance in the RFID domain. This is because events matching monitoring and trace queries must correspond to the same item. Consequently equality checks are highly selective and evaluated within the state machine for high speed query processing on RFID events [19].

Pushing query operators into the network is as a efficient mechanism to reduce network load [16,14]. For the internet of things this means operators for complex event processing must be pushed to the proposed broker network. In particular, broker services must support pattern matching and equality checks in a distributed manner. Distributing query operators comprises two parts. First, the query must be disassembled in isolated operators. This step must be done with the network structure in mind to find the optimal solution [14]. The second step is optimized mapping of operators to physical network nodes. The most common target function for optimization is to minimize the network load. For application in the internet of things this paper proposes a novel technique to transform query patterns based on process knowledge. This impacts the first part of distributing query operators and enables improved mapping of operators to the broker network. The following subsection explains this approach in detail.

3.2   Query Distribution Using Process Knowledge

This section presents a novel mechanism for distributing query operators for complex event queries. The mechanism is implemented by a mapping component that realizes registrations and unregistrations of event queries. A core principle of the proposed approach is to independently maintain a set of generic event queries Q_G and a set of process models P, which both serve as input for the distribution mechanism. Queries in Q_G can be generic queries that are defined on the level of event types and may be reusable across different companies or supply chains. Process models in P describe the currently implemented instances of business processes. In the supply chain context this is the current structure of the monitored supply chain. For each process step the model denotes what events are issued when a certain item passes this step. E.g. the outbound of a distribution center issues events of outgoing shipments for items that pass.

Existing approaches operate on a set of queries Q_G and a model of the overlay network N for creating a mapping of query operators to network nodes. The distribution mechanism proposed in this paper adds an additional step. It transforms queries in Q_G to query instances Q_I based on the process models in P. The queries in Q_I are than mapped to the network, based on existing mapping heuristics (e.g. [14]). Note that in modern supply chains P is frequently changed when organizations leave or enter the supply network. Separating P and Q_G has the advantages that the business logic encoded in Q_G must not adapt when the supply chain structure changes. Instead the distribution mechanism automatically ensures that queries in Q_G are optimized for the current processes in P. The distribution mechanism is triggered whenever P or Q_G changes. Changes in Q_G cause that query operators are distributed to the network or revoked. Changes in P cause a new creation of Q_I based on the changed P. This may cause a remapping and redistribution of query operators to the network.

The proposed mechanism for query distribution creates Q_I from Q_G and P by analyzing which event sequences that can be created by processes in P would actually match queries in Q_G. Therefore the mechanism determines for each process $p \in P$ the set $eventsequences(p)$, which holds all event sequences that the process p can create. This is done by a depth search from starting steps in p to ending steps in p and extracting all events that can be created in the respective steps. Running all queries $q_G \in Q_G$ on $\bigcup_{p \in P} eventsequences(p)$ results in the set $match(P, Q_G)$ which holds all event sequences that processes in P can create and that match queries in Q_G. The proposed mapping component than uses the sequences in $match(P, Q_G)$ to create state machines which match these sequences. These machines are simply sequences of states that are connected by transitions reflecting the event sequences in $match(P, Q_G)$. The subsequent step transforms resulting state machines by merging identical parts of the machines.

For illustration consider the following example. Let "$A \rightarrow B \rightarrow C$" be the only query in Q_G ($\rightarrow$ denotes "followed by"). Let "$A1 \mapsto B1; B1 \mapsto C; A2 \mapsto B2; B2 \mapsto C$" be the model of the only supply chain in P ($\mapsto$ denotes "delivers to"). The supply chain model describes a distribution center C that receives goods from two first tier supplier ($B1, B2$) which receive goods from corresponding second tier suppliers ($A1, A2$). The query in Q_G is a trace query that monitors product flows in this supply chain.

Existing approaches would register a query operator for "$A \rightarrow B$" at the event sources for $A1$, $A2$, $B1$, and $B2$. The mapping mechanism proposed in this paper transforms the query "$A \rightarrow B \rightarrow C$" to "$((A1 \rightarrow B1)\|(A2 \rightarrow B2)) \rightarrow C$". Consequently an operator for "$A1 \rightarrow B1$" is registered only at sources for $A1$, $B1$ and an operator for "$A2 \rightarrow B2$" is registered accordingly. This operator splitting allows for better optimization of the operator mapping because the operators can be positioned closer to the event sources. Figure 1 visualizes the effect for this example. The depicted mapping is based on the optimization method proposed by Pietzuch et al. [14].This method positions network nodes in a virtual cost space where the distance between nodes reflects the communication latency between them. A query plan is treated as a network of springs that is connected by the query operators. A mapping that minimizes network traffic is at positions that correspond to a relaxed state of the springs. Figure 1 illustrates in which

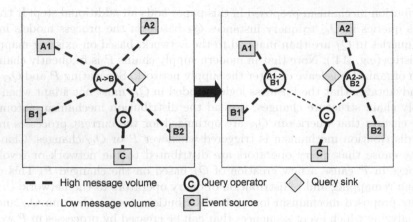

Fig. 1. Mapping without (left) and with the proposed optimization (right)

way the proposed mapping mechanism enables improved operator placement. The next section presents experiments to study the resulting gain.

4 Experiments

This section describes experiments which target the benefits of the proposed mechanism for mapping queries into the network. The investigations addressed the impact of the network size on the achieved gain. Furthermore, the experiments analyzed how the size of the monitored supply chain influences the results.

4.1 Experimental Setup

The experiments tested mappings of a sample query to network models of various sizes. The mapper used the query $"A \to B \to C"$ from section 3.1 in its unmodified form and after optimization with the purposed approach. The nodes of each used network model were uniformly arranged on a square plain. Coordinates on the plain correspond to coordinates in the cost space that represents communication latency [14]. The assumed supply chain structure for the experiments is the one used in example in section 3.1. That is, the supply chain has a number of two step routes that lead to a distribution center. The number of such inbound routes changes throughout the experiments to test the impact of the supply chain size.

The quality of each mapping is given by a cost function that evaluates the overall resulting network traffic. That is the summed up product of the latency of each communication link and the consumed bandwidth along this link [14]. The value for the bandwidth of the event stream at each event source was 1. The experiments assumed a product flow such that the selectivity of each query operator $" \to "$ was 10%. Further assumptions were that event sources are randomly distributed and that no network congestion exists. Note that these are worst case assumptions for the proposed approach. In reality most event sources of one supply route are most likely correlated in their locations. This correlation favors the mapping of optimized plans even more. Furthermore, the proposed approach is likely to be even more beneficial if network congestion is considered. This is because the split up of query operators supports distribution of the load. However, the conducted experiments just focus on effects of overall traffic reduction in the network.

4.2   Experimental Results

Figure 2 shows results of the conducted experiments. The chart on the left side depicts how much network traffic is saved dependent on the network size for different sizes of supply chains. Each point in the chart represents the average of 400 test runs. The y-axis denotes the percentage of traffic which can be saved due to the proposed query optimization. The x-axis denotes the network size in terms of network nodes in a logarithmic scale. The chart shows that in the tested setting generally a high proportion of the network traffic is saved by the proposed approach. An exception is the case where the supply chain has only one inbound route. This result is expected since in this case the optimized query equals the original one. For all other situations the chart shows a generally high amount of saved traffic. However, the experiments show no effect on the percental saved traffic in dependence of the network size.

The right hand side of figure 2 depicts results of experiments regarding the influence of the supply chain size. In these experiments the network size was fixed at 25 nodes and supply chains with up to 256 inbound routes were tested. The chart is the result of 100 test runs for each data point and reveals that the gain stabilizes at a high level. The previous decline can be explained looking at the operator for processing events form the distribution center C. With a growing number of inbound routs the amount of network traffic from C increases. This causes that the operator for matching with C is eventually mapped to the same

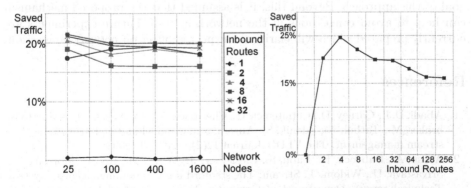

Fig. 2. Experimental results for saved network traffic

network node for both, optimized and unoptimized query plans. From this point on, network traffic can only be saved by better operator placement for matching with events A, B. This is reflected in the displayed saturation effect for large scale supply chains. Overall the experiments show that the proposed approach enables to save a significant amount of traffic. The gain changes with the supply chain size but stabilizes at a high level for large scale supply chains.

5   Related Work

Much of the related work has already been discussed throughout this paper. This section adds a discussion of relevant approaches for query mapping, in-network processing and event processing. Much work exists on technologies for processing event streams. Prominent examples are for instance the projects Stream [2] and Auroa [1]. Wu et al. particularly addressed processing of RFID related event queries [19]. However, these projects rather focus on processing technologies than on distribution heuristics for query operators. Distribution heuristics of query operators are mainly targeted in the domain of sensor networks because of the special need to prevent energy consuming communication (e.g. [18]). Widom et al. addressed operator distribution among heterogeneous devices [16] and Pietzuch et al. presented distribution heuristics for general overlay networks [14]. However, unlike the solution presented in this paper, none of these approaches considers process knowledge for query optimization.

6   Conclusions and Future Work

This paper describes the design of a service for event processing that meets the requirements of the internet of things. It identifies suitable technologies for implementing the service and describes how the new service can be situated among existing ones. However, the main contribution of this paper is a novel mechanism for optimizing in-network query processing. Experiments show the benefits of this approach regarding the absolute network load, which is the main goal of this approach. Beyond this, it is assumed that the proposed mechanism can help to avoid congestions in the network as well. Future experiments are currently prepared which will clarify the expected effect on network congestion.

References

1. Abadi, D.J., Carney, D., Cetintemel, U., Cherniack, M., Convey, C., Lee, S., Stonebraker, M., Tatbul, N., Zdonik, S.: Aurora: a new model and architecture for data stream management. The VLDB Journal 12(2), 120–139 (2003)
2. Arasu, A., Babcock, B., Babu, S., Cieslewicz, J., Datar, M., Ito, K., Motwani, R., Srivastava, U., Widom, J.: Stream: The stanford data stream management system. Technical report, Department of Computer Science, Stanford University (2004), http://dbpubs.stanford.edu:8090/pub/2004-20

3. Bornhövd, C., Lin, T., Haller, S., Schaper, J.: Integrating automatic data acquisition with business processes - experiences with sap's auto-id infrastructure. In: VLDB, pp. 1182–1188 (2004)
4. Bose, I., Pal, R.: Auto-id: managing anything, anywhere, anytime in the supply chain. Commun. ACM 48(8), 100–106 (2005)
5. Carzaniga, A., Di Nitto, E., Rosenblum, D.S., Wolf, A.L.: Issues in supporting event-based architectural styles. In: ISAW 1998, pp. 17–20. ACM Press, New York (1998)
6. Chakravarthy, S., Krishnaprasad, V., Anwar, E., Kim, S.-K.: Composite events for active databases: Semantics, contexts and detection. In: VLDB 1994, Morgan Kaufmann Publishers Inc. San Francisco, CA, USA (1994)
7. Christopher, M.: The agile supply chain: Competing in volatile markets. Industrial Marketing Management 29(1) (2000)
8. EPCglobal. Epc information services (epcis) version 1.0 specification. (last accessed 7/7/2007) (2007) http://www.epcglobalinc.org/standards/epcis/EPCIS_1_0-StandardRatified-20070412.pdf
9. Luckham, D.C.: The Power of Events: An Introduction to Complex Event Processing in Distributed Enterprise Systems. Addison-Wesley Longman Publishing Co., Inc., Boston, MA, USA (2001)
10. Mühl, G., Fiege, L., Gärtner, F.C., Buchmann, A.: Evaluating advanced routing algorithms for content-based publish/subscribe systems. In: MASCOTS 2002, p. 167. IEEE Computer Society, Washington, DC, USA (2002)
11. Mühl, G., Fiege, L., Pietzuch, P.: Distributed Event-Based Systems, New York, Inc., Secaucus, NJ, USA. Springer, New York (2006)
12. Niederman, F., Mathieu, R.G., Morley, R., Kwon, I.-W.: Examining rfid applications in supply chain management. Commun. ACM 50(7), 92–101 (2007)
13. German Ministery of Exonomics and Technology. European policy outlook rfid: draft version. (last accessed 7/7/2007) (2007), http://www.rfid-outlook.de
14. Pietzuch, P., Ledlie, J., Shneidman, J., Roussopoulos, M., Welsh, M., Seltzer, M.: Network-aware operator placement for stream-processing systems. In: ICDE 2006, p. 49. IEEE Computer Society, Washington, DC, USA (2006)
15. Pietzuch, P.R., Bacon, J.: Peer-to-peer overlay broker networks in an event-based middleware. In: DEBS 2003: Proceedings of the 2nd international workshop on Distributed event-based systems, pp. 1–8. ACM Press, New York (2003)
16. Srivastava, U., Munagala, K., Widom, J.: Operator placement for in-network stream query processing. In: PODS 2005, pp. 250–258. ACM Press, New York (2005)
17. Terpstra, W.W., Behnel, S., Fiege, L., Zeidler, A., Buchmann, A.P.: A peer-to-peer approach to content-based publish/subscribe. In: DEBS 2003: Proceedings of the 2nd international workshop on Distributed event-based systems, pp. 1–8. ACM Press, New York (2003)
18. Villanueva, F.J., Villa, D., Moya, F., Barba, J., Rincón, F., López, J.C.: Lightweight middleware for seamless hw-sw interoperability, with application to wireless sensor networks. In: DATE 2007, pp. 1042–1047. ACM Press, New York (2007)
19. Wu, E., Diao, Y., Rizvi, S.: High-performance complex event processing over streams. In: SIGMOD 2006, pp. 407–418. ACM Press, New York (2006)

Structured Peer-to-Peer Network for Live-Content Distribution*

Sergio Machado, Javier Ozón, and Xavier Hesselbach

Department of Telematic Engineering, Technical University of Catalonia,
Avinguda del Canal Olímpic, 15, 08860 Castelldefels
smachado@entel.upc.edu

Abstract. In this paper we present the mathematical analysis of a structured peer-to-peer network designed for the distribution of live-content such as television channels, which is modeled using media distribution graphs. As far as we know, the use of a structured network is a novel proposal in the area of live-content distribution and allows a peer to schedule consecutives data chunks produced by the media source using an algorithm that we provide. In our analysis we obtain a bound for the maximum delay between the media source and any of the peers of the network and the buffer size needed to allow consecutive scheduling. We study a type of media distribution graph called minimum distribution graph obtaining its order and diameter, which means that under estable conditions we can ensure certain quality-of-service parameters.

1 Introduction

Peer-to-Peer (P2P) live content streaming is an emerging Internet application expected to be the next disruptive IP communication technology [1]. There are several systems providing this service, such as CoolStreaming [2], PPLive [3], and SopCast [4]. The reader can found an evaluation of the performance of SopCast and PPLive in [5]. A peer shares its buffer content and issues unsorted requests for unavailable data chunks to other peers.

Scheduling is the process of requesting parts of the media transmission to other peers. As far as we know, all the proposals and implementations of P2P live content applications, such as [2], [3], [4], and others, have similar engines for downloading video. For example, in CoolStreaming, video stream is fragmented in data chunks of 1 second length. Scheduling uses a *buffer map* which represents data chunks availability on a peer. Peers exchange periodically or upon request their own buffer maps. A buffer map of approximately 120 seconds is enough to ensure that any other peer is not interested in the data chunks out of this window. The buffer map is a table of 120 bits and 2 bytes more which correspond to the sequence number of the first data chunk represented in the buffer map. An heuristic algorithm is used for scheduling. It first computes the potential

* Work supported by the Spanish Ministerio de Ciencia y Tecnología (MCYT) and FEDER under projects TSI2005-06092 and TSI2006-12507-C03-03.

R. Meersman, Z. Tari, P. Herrero et al. (Eds.): OTM 2007 Ws, Part II, LNCS 4806, pp. 980–990, 2007.
© Springer-Verlag Berlin Heidelberg 2007

number of suppliers for each one of the data chunks represented in the buffer map and then sort them in ascending order according to the number of potential suppliers. For each data chunk, the selected supplier is the one having higher bandwidth and whose data chunk expires later in its buffer map. Received data chunks are reassembled in order and buffered forming a local streaming file in memory. When the streaming file length crosses a predefined threshold, media playing begins.

All those systems are unstructured in the sense that the peers form logical connections to download chunk according to the chunk availability on other peers which are known by the peer through a gossip protocol. The main contribution of this paper is the proposal of a structured P2P network for live-content distribution. The main advantages of this kind of network is that we can define a consecutive scheduling algorithm, that is, we can request chunks in order, and besides we can establish some important quality-of-service parameters under stable conditions of the network as the worst-case chunk delay, the buffer size needed and the number of peers that can be connected to the network receiving the media stream at the same generation rate that it was generated.

The rest of this paper is organized as follows. Sect.2 presents the analytical model that we propose for a structured P2P live-content distribution network, providing general definitions and presenting the media distribution graphs, particularly, the graph called minimum distribution graph that is the main focus of analysis. Sect.3 presents the proposed scheduling algorithm, the distribution time of a single media chunk and the buffer size of peers. The analysis of the minimum distribution graph to obtain the diameter and the order of the graph is presented in Sect.4. Finally, we conclude in Sect.5.

2   Mathematical Model

2.1   Preeliminaries

We denote by r the live-content media source, also called *root*, which generates chunks at a rate of λ chunks/sec. We discretize time to the duration of a chunk, 1 t.u. $= 1/\lambda$, so the resulting generation rate becomes 1 chunk/t.u. We model the chunks distribution network by means of the directed multigraph $G = (V, E)$, where $r \in V$. A vertex $v \in V - \{r\}$ models a peer that receives chunks from any other peer, including the root, and which can forward the received chunks to any other peer, except the root. An edge $e = (u, v) \in E$ models a one-way transport layer unicast connection between peers u and v with a bandwidth of $(1/K)$ chunks/t.u., $K \geq 2, K \in \mathbb{Z}$. Graph edges correspond to logic connections in the context of the underlaying network, such as TCP or UDP sockets in TCP/IP networks. The bandwidth of $(1/K)$ implies that the transmission time of a chunk through any edge is K t.u. We neglect the effects of network latency and approximate the transmission delay between any two peers to K t.u. Note that K must have a minimum value of 2 in order to allow the connection of low bandwidth peers. Moreover, a peer should receive chunks at unitary rate, so it must establish K download connections of bandwidth $(1/K)$ in order to achieve the unitary download rate. Note that the model considers the usual Internet assymetric access networks.

From the point of view of u, the edge $e = (u, v)$ models an uploading connection of bandwidth $(1/K)$, that is, u sends chunks to v, whereas from the point of view of v it is a downloading connection, e.g., v receives chunks from u. If there is enough bandwidth between u and v they can establish up to K connections between them, each one modeled by K different graph edges. We define the *maximum out-degree* of u, denoted by $d_{up,\max}(u)$, as the maximum number of upload connections that u can establish with any other peer $v \in V - \{r\}$, with $d_{up,\max}(u) \geq 1$, $\forall u \in V$. Then each peer must have an upload bandwidth of at least $(1/K)$ to be admitted in the network. Denoting by $U_{bw}(u)$ the uploading bandwidth of peer u, we have $d_{up,\max}(u) = K \cdot U_{bw}(u)$. By definition the root has bandwidth α and then $d_{up,\max}(r) = \alpha \cdot K$, which means that the root can serve directly to α peers, which corresponds to the client-server paradigm. We define the *in-degree* of a peer $v \in V - \{r\}$, denoted by $d_{down}(v)$ as the number of downloading connections that it has established. Since each peer must receive at unitary rate, $d_{down}(v) = K$, $\forall v \in V - \{r\}$, and $d_{down}(r) = 0$.

2.2 Distribution Graphs

A distribution graph $G = (V, E)$ models the network which has the maximum number of peers connected. We define the *distribution graph for α, K, and u* denoted by $G_{\alpha,K,u} = (V, E)$ as the maximum graph such that $d_{up,\max}(v) = u$, $1 \leq u < K$, $\forall v \in V - \{r\}$. Note that u is strictly lower than K because for $u \geq K$ we have the problem of multicast overlay which is not the subject of this paper. We define the *minimum distribution graph* $G_{\alpha,K,1} = (V, E)$ *for α, and K* as the distribution graph where $u = 1$, that is, the maximum graph where all the peers, except the root, have the minimum upload bandwidth: $U_{bw}(v) = (1/K)$ $\forall v \in V - \{r\}$. For the aim of simplicity we use the notation $G_{\alpha,K} = (V, E)$ indistinctly for both distribution and minimum distribution graphs. Note that the order of the distribution graph is finite because when a peer joins the network, it consumes K units of the total upload bandwidth, whereas it only contributes with (u/K) units, $u < K$, resulting a negative balance of $(u/K) - 1$ units.

Consider the existence of a joining algorithm such that beginning with root gives the general distribution graph $G_{\alpha,K} = (V, E)$ after $|V| - 1$ joints. We define the *set of suppliers of the peer v*, denoted by $\mathbb{S}_e(v)$, as the set of input edges incidents at v, that is, if $s_e \in \mathbb{S}_e(v)$ there exists at least one vertex v such that $s_e = (u, v) \in E$. As a peer needs K downloading connections, the cardinality of the set $\mathbb{S}_e(v)$ is $|\mathbb{S}_e(v)| = K$. We also define the *set $\mathbb{S}(v)$ of supplier peers or parents of the peer v* as the set of peers that send chunks to peer v. That is, if $u \in \mathbb{S}(v)$, there exists the edge $e = (u, v) \in \mathbb{S}_e(v)$. We define the *set $\mathbb{C}_e(v)$ of children of the peer u* as the set of edges which join u to any other peer of the graph. We also define the *set $\mathbb{C}(u)$ of child peers of the peer u* as the set of peers which receive chunks from u, that is, if $v \in \mathbb{C}(u)$ there exists $c_e = (u, v) \in \mathbb{C}_e(u)$.

We denote by $\mathbb{P}(v)$ the set of paths which goes from root r to peer v. Let $p_{rv} \in \mathbb{P}(v)$ be one of these paths and $|p_{rv}|$ the number of edges of the path p_{rv}, we define the *level of peer* $v \in V - \{r\}$ at the distribution graph $G = (V, E)$ as $l(v) = \max\{|p_{rv}|, \forall p_{rv} \in \mathbb{P}(v)\} - 1$. Note that a peer v such as $\mathbb{S}(v) = \{r\}$ will have level $l(v) = 0$. Then we calculate the *diameter* of the distribution graph $G = (V, E)$, denoted by H, as $H = \max\{l(v), \forall v \in V\} + 1$.

3   Scheduling

The main aim of a structured P2P network for live-content distribution is to allow consecutive scheduling of chunks. As any peer in the network has K downloading connections, a execution of the scheduling algorithm results in K chunks to be requested, one for each of the K connections. In this case, the parent buffers of a peer must have the K chunks of the resulting request.

Let $\mathbb{O}$ be the set of all the chunks generated by the root r, where each $o_i \in \mathbb{O}$ represents the chunk generated by the root at time t_i. Let $B(v)$ the set of chunks stored at buffer of peer v. In our model we consider that if o_i has been requested by the peer v and it is being received during the interval $[t, t+K)$, then $o_i \in B(v)$. Although physically o_i is not yet in the buffer of peer v it could be requested to v during the interval $[t, t + K)$ and v could forward it after time $(t + K)$.

3.1   Scheduling Algorithm

Now we describe a scheduling algorithm. Denoting by $o_{\text{last}}(s_i)$ the last chunk in the buffer $B(s_i)$ of peer s_i, the algorithm computes the minimum index q for all $o_{\text{last}}(s_i)$ with $s_i \in \mathbb{S}(v)$. The set $\mathbb{S}_e(v)$ of supplier edges is supposed to be sorted in decreasing order as function of $o_{\text{last}}(s_i)$ of the supplier peer of the edge. The for loop of lines 3-6 demand each o_i of the K chunks of the request, each o_i for one s_j of the K supplier edges. Let o_q be the chunk with an index such as $q = \text{minindex}\{o_{\text{last}}(s_i) \ \forall s_i \in \mathbb{S}(v)\}$, for the scheduling algorithm correctness we need:

$$\{o_{q-(K-1)}, \ldots, o_q\} \subseteq \bigcap_{s_i \in \mathbb{S}(v)} B(s_i) \tag{1}$$

The condition is necessary because in the worst case we request K different chunks to supplier peers and all the chunks must be buffered in $B(s_i)$. And it is sufficient because the requester is not interested in previous chunks.

Algorithm 1. SCHEDULING

1: $q \longleftarrow \text{minindex}\{o_{\text{last}}(s_i) \ \forall s_i \in \mathbb{S}(v)\}$
2: $j \longleftarrow 0$
3: **for** $i = q - (K - 1)$ to q **do**
4:    REQUEST o_i TO s_j, where $s_j \in \mathbb{S}_e(v)$ and $o_i \in \mathbb{O}$
5:    $j \longleftarrow j + 1$
6: **end for**

3.2 Total Distribution Time and Chunk Buffer Size

In this section we calculate the delay and the required buffer size for the correct distribution of the chunks. We define the *total distribution time* of a segment of K chunks as the time elapsed between the time that the segment is available to be forwarded by the root r and the time it is received by all the peers.

Lemma 1. *Let $G_{\alpha,K} = (V, E)$ a distribution graph with diameter H. The total distribution time of a segment of K chunks is at most $\Delta_o = K \cdot H$*

Proof. A segment delays K time units when it is transmitted from supplier u to peer v. As a segment has to be retransmitted H times at most, then it results a maximum delay of $\Delta_o = K \cdot H$. □

Moreover, the chunk buffer size must be enough large to verify (1).

Theorem 1. *Let $G_{\alpha,K} = (V, E)$ a distribution graph with diameter H. If the buffer size of all peers is of $B \geq K \cdot (H + 1)$ chunks, then (1) is accomplished.*

Proof. The worst case for (1) occurs when the root is a supplier of a gain child v at the last level $H - 1$, and v has also a parent u at the previous level $H - 2$. Without loss of generality, consider that r begins to generate and buffer at t_0 the first chunk segment $[o_0, \ldots, o_{K-1}]$. Then, u receives all the segment at time $t_0 + \Delta_o$, which counts the time generation K plus the transmission delay $K \cdot (H - 1)$. At this time, r has just generated the chunk segment $[o_{\Delta_o-K}, \ldots, o_{\Delta_o-1}]$, since the root generates 1 chunk/t.u. Then, r and u sends the first chunk segment to v during $[t_0 + \Delta_o, t_0 + \Delta_o + K]$. Therefore, the root must buffer the chunk segment $[o_0, \ldots, o_{K-1}]$ since t_0 until $t_0 + \Delta_o + K$. During this interval of duration $\Delta_o + K$ the root has generated $\Delta_o + K = K \cdot H + K$ chunks which have to be stored in $B(r)$, giving us the maximum buffer size of the distribution graph. □

According to the argumentation of the latter proof, the maximum buffer size of a peer at level l has to be of $B \geq K \cdot (H + 1 - (l + 1)) = K \cdot (H - l)$.

4 Minimum Distribution Graph

In this section we analyze the minimum distribution graph (MDG) for α and K, $G_{\alpha,K} = (V, E)$, and we calculate its order and diameter. Note that in this case $u = 1$, that is, each peer except the root has one upload connection of bandwidth $1/K$. Thus, the MDG provides us with bounds for the general distribution graph $G_{\alpha,K,u} = (V, E)$ where $d_{up,\max}(v) = u$, $1 \leq u < K$, $\forall v \in V - \{r\}$.

We consider a joining algorithm that proceeds as follows: initially, the algorithm assigns to level 0 the first α peers, which are root children. Once the level 0 has been completed, the next level 1 is formed grouping the peers of level 0 in groups of K peers, and assigning to each group a child at level 1. All these peers are called *direct children*. Note also that if $(\alpha \mod K)$ is not equal to 0, then some peers of level 0 will remain without children. For the generic level $l > 1$ the algorithm proceeds in the same way: it forms groups of K peers that will be the parents of a direct child at level $(l + 1)$. Only when at a level l is not possible to form one more group of K peers, the algorithm can recover the remaining peers from the previous levels which do not have children, to form a new child at level $(l + 1)$. In this case, we say that the new child is a *gain child*.

Then, the next time that we have accumulated at level l' a total of $(K - 1)$ direct peers without children, we group them together with the last gain peer to form a new gain child at level $(l' + 1)$. Therefore, we call *gain child* a peer whose parents belong to different levels or also a peer who has as a parent a previous gain peer. Note that, in this latter case, all the parents, including the gain one, can belong to the same level. According to this, we say that a peer is a *direct child* if it is not a gain child, that is, if all its parents belong to the same level, and all the parents of its parents belong to the same previous level, and so forth (and then we can say that a direct child has been formed "directly", without the participation of "remaining" peers). Thus, by definition, all the parents of a direct child belong to the same level and are also direct children. The level with the α root children is numbered level 0.

Lemma 2. *If there is a gain child at level l, then it is the only gain child at this level. Besides, if v is the first gain child of MDG and it belongs to level l, at most $(K - 1)$ of its parents belong to level $(l - 1)$ and at least another parent belongs to a level previous to $(l - 1)$.*

Proof. First we prove the second part of the lemma. If v is the first gain child of MDG then all its parents are direct children. In this case, at level $(l - 1)$ the graph can have up to $(K - 1)$ of these parents (otherwise, they would supply a direct child), and up to $(K - 1)$ at the rest of the previous levels for the same reason, resulting a total of $2 \cdot (K - 1)$ potential parents at levels previous to l. Thus, in this case we can have at level l only one gain child (for each child we need K suppliers) with at most $(K - 1)$ parents from level $(l - 1)$ and therefore with at least one parent at a level previous to $(l - 1)$.

Then, suppose that the second gain child appears at level l'. In this case, we can have at level $(l' - 1)$ at most $(K - 1)$ potential parents which are direct children, and from previous levels we can have at most $(K - 1)$ parents without children, according to the argumentation of the latter paragraph. We could also find the first gain child at level $(l' - 1)$. In this case we count at maximum $2 \cdot (K - 1) + 1 = 2K - 1$ potential parents for the second gain child and therefore at level l' we will have at maximum one gain child. Then, we recursively repeat the argumentation for the following gain children considering that at the previous level we can only have one gain child. □

4.1   Diameter Bounds

In MDG the root has α children and the rest of peers one or none, and all the peers, except the root, have K supplier edges. Then, denoting $n(l)$ the number of peers at level l, we have:

$$n(l) \geq \lfloor \alpha/K^l \rfloor \quad \text{for } 0 \leq l \leq H_t - 1 \qquad (2)$$

where $H_t = \max\{h \in \mathbb{Z} : \alpha/K^{(h-1)} \geq 1\} = 1 + \lfloor \log_K(\alpha) \rfloor$. Note that (2) does not consider all peers at level l since it does not count the gain children. Note also that initially we consider levels from 0 to $H_t - 1$.

Lemma 3. *Let $G_{\alpha,K} = (V, E)$ be the minimum distribution graph for α and K, and let H be the diameter of $G_{\alpha,K} = (V, E)$, then:*

$$H_t \leq H \leq H_t + 1, \quad \text{with } H_t = 1 + \lfloor \log_K(\alpha) \rfloor \tag{3}$$

Proof. If we consider exceptionally that the graph does not have gain children, then each level l will have exactly $\lfloor \alpha/K^l \rfloor$ peers, and we can add a new level until this quantity is higher or equal to one. According to the definition of H_t, this gives us a last level $H_t - 1$ and a diameter of $H = H_t$ since the first level of the root children is numbered with level 0. Otherwise, if we accumulate at least K potential suppliers without direct children at level $H_t - 1$, we can add one additional level resulting $H = H_t + 1$. Note also that, according to Lemma 2, we can only add one new gain child per level, and thus only one new additional level from $H_t - 1$, resulting $H_t \leq H \leq H_t + 1$. □

4.2   Order of the Minimum Distribution Graph

If we do not take into account the gain children to obtain the MDG order $|G_{\min}|$ we obtain a first bound:

$$|G_{\min}| \geq 1 + \sum_{i=0}^{H_t-1} \lfloor \frac{\alpha}{K^i} \rfloor$$

Anyway, it is possible to obtain a tighter bound $N_{min,c}$ of $|G_{\min}|$ according to the next lemma.

Lemma 4. *Let $G_{\alpha,K} = (V, E)$ be the minimum distribution graph for α and K, then:*

$$|G_{min}| \geq N_{min,c} = 1 + \left\lfloor \sum_{i=0}^{H_t-1} \frac{\alpha}{K^i} \right\rfloor = 1 + \left\lfloor \alpha \cdot \frac{1 - (\frac{1}{K})^{H_t}}{1 - \frac{1}{K}} \right\rfloor \tag{4}$$

Proof. We define the summatory S:

$$S = \sum_{i=0}^{H_t-1} \frac{\alpha}{K^i} = \alpha \cdot \frac{1 - \left(\frac{1}{K}\right)^{H_t}}{1 - \frac{1}{K}} \tag{5}$$

Then we show that when $1 + S$ is higher than $|G_{min}|$, the difference is lower than 1, which proves the lemma since we have rounded down $1 + S$. We can see that $1 + S$ is higher than $|G_{min}|$ in the case that a peer has no direct child in next level, since in this case S counts a fraction of $1/K$ children for the next level, $1/K^2$ for the next, and so forth. Thus, if the peer without children is at distance

L to the last level $H_t - 1$, it contributes to the summatory of S with a number of children of:

$$\sum_{i=1}^{L} \frac{1}{K^i} < \sum_{i=1}^{\infty} \frac{1}{K^i} = \frac{1}{K-1} \leq 1 \quad for\ K \geq 2$$

Let $\epsilon = (1 + S) - |G_{\min}|$ and let N be the number of parents accumulated at level $H_t - 2$ without direct children. If $N \leq K - 1$, then:

$$\epsilon < N \cdot \frac{1}{K-1} \leq \frac{K-1}{K-1} = 1$$

When $N \geq K$ the graph could use the first K parents to supply a gain child that should be discounted in ϵ, so for this K parents we have:

$$\epsilon < K \cdot \frac{1}{(K-1)} - 1 = \frac{1}{(K-1)} \leq 1$$

Consequently, we have $N' = (N - K)$ peers without a direct children. If $N' \leq K - 2$ and we count the inherited excess $\epsilon < 1/(K-1)$, it results:

$$\epsilon < \frac{1}{(K-1)} + (K-2) \cdot \frac{1}{(K-1)} = 1$$

Moreover, for $N' \geq K - 1$ we can form new gain children. In this case, we would need only $(K-1)$ unoccupied peers plus the last gain child to supply the next gain child. Thus, we have to count the last $\epsilon < 1/(K-1)$ plus the excess from the new $(K-1)$ parents minus the new gain child:

$$\epsilon < \frac{1}{(K-1)} + (K-1) \cdot \frac{1}{(K-1)} - 1 = \frac{1}{(K-1)} \leq 1$$

Now the argumentation applies recursively until we have no peers enough to supply a new gain child.   □

Anyway the order of MDG can be obtained with independence of H_t according to next theorem.

Theorem 2. *The order of MDG for α and K is:*

$$|G_{\min}| = 2 + \left\lfloor \frac{\alpha - 1}{1 - \frac{1}{K}} \right\rfloor \tag{6}$$

Proof. When a peer joins the network it takes K units of bandwidth and offers $(1/K)$. Let i be the number of peers connected to the network at any time, the available bandwidth is $BW(i) = \alpha + (i-1) \cdot (1/K - 1)$. Then, we can add peers to the graph as long as $BW(i) \geq 1$ and therefore $|G_{\min}|$ is the minimum integer i such that $BW(i) < 1$ resulting the expression of theorem.   □

4.3   Diameter of the Minimum Distribution Graph

In previous sections we have proved that either $H = H_t$ or $H = H_t + 1$ with $H_t = 1 + \lfloor \log_K(\alpha) \rfloor$. Now we find the condition that makes H to take one of these two values.

Theorem 3. *Let* $G_{\alpha,K} = (V, E)$ *be the MDG for* α *and* K *with order* $|G_{\min}|$ *which is lower bounded by* $N_{\min,c}$ *as seen in* (4), *then:*

$$N_{min,c} < |G_{min}| \iff H = H_t + 1 = 2 + \lfloor \log_K(\alpha) \rfloor \tag{7}$$

Proof. At proof of Lemma 4 is shown that until last level $H_t - 1$ we have $\epsilon < 1$ and now we prove that $\epsilon \geq 0$, where $\epsilon = (1 + S) - |G_{\min}|$. As shown in that proof, at a level where the graph has not already a gain child we have $\epsilon \geq 0$ (note also that we have $\epsilon = 0$ only when $(1 + S) = |G_{\min}|$, that is, in the case that we have only direct children). When the first gain child is added at level l, we know from Lemma 2 that at most $K - 1$ parents come from level $l - 1$ and at least one parent comes from a previous level. Therefore, the $K - 1$ suppliers have added, at least, $1/K$ to S and, at least one of them, would have added $1/K + 1/K^2$. Then, discounting the contribution of the gain child:

$$\epsilon \geq K \cdot \frac{1}{K} + \frac{1}{K^2} - 1 = \frac{1}{K^2} > 0$$

For the second gain children, the worst case contribution in S would occur when the two gain children are at consecutive levels, and we have:

$$\epsilon \geq \left[K \cdot \left(\frac{1}{K} + \frac{1}{K^2} \right) + \frac{1}{K^3} \right] + (K - 1) \cdot \frac{1}{K} - 2 = \frac{1}{K^3} > 0$$

Generalizing this result to have a total number of i consecutive gain children, we have $\epsilon \geq 1/K^{i+1} > 0$. Therefore, we have proved that $0 \leq \epsilon < 1$ until last original level $H_t - 1$, which is the last step that takes into account the summatory of S. Therefore, only if we can connect another gain peer at new level H_t we can have $\epsilon < 0$. Moreover, as in this case there exists at least one gain children, at level $H_t - 1$ we have $0 < \epsilon$ and consequently $0 < \epsilon < 1$. Now we discount to the excess $0 < \epsilon < 1$ the last gain child at level H_t, resulting $-1 < \epsilon < 0$ and then $N_{\min,c} < |G_{\min}|$. In this case, since we have a new last level H_t we also have that $H = H_t + 1$, as we wanted to prove. □

The consequence of Theorem 3 is that the error resulting of computing the order with $N_{\min,c}$ is due to the additional level H_t. Now, we show that in this case $|G_{\min}| = N_{\min,c} + 1$.

Theorem 4. *Let* $G_{\alpha,K} = (V, E)$ *be the minimum distribution graph for* α *and* K *with order* $|G_{\min}|$ *lower bounded by* $N_{\min,c}$, *and with graph diameter* $H = H_t + 1 = 2 + \lfloor \log_K(\alpha) \rfloor$, *then* $|G_{\min}| = N_{\min,c} + 1$

Proof. We know that the difference between H and H_t is due to the admission of a gain child at level H_t that is not counted when calculating $N_{\min,c}$. From

Lemma 2 we know that we have only one gain child at this level. Moreover, from proof of Theorem 3 we know that when $H = H_t + 1$ we have $0 < \epsilon < 1$ at level $H_t - 1$, where $\epsilon = (1 + S) - |G_{\min}|$. Therefore, we have to discount to the excess $0 < \epsilon < 1$ the last gain child at level H_t, resulting $-1 < \epsilon < 0$. This is the difference between $1 + S$ and $|G_{\min}|$. Finally, as we obtain $N_{\min,c}$ rounding down $1 + S$ it follows $N_{\min,c} = |G_{\min}| - 1$. □

Moreover, for the case $H = H_t$ we have that $0 \leq \epsilon < 1$ and then, since we round down $1 + S$ to obtain $N_{\min,c}$ we have $|G_{\min}| = N_{\min,c}$, resulting:

$$|G_{\min}| = \begin{cases} 1 + \left\lfloor \alpha \cdot \frac{1 - (\frac{1}{K})^{H_t}}{1 - (\frac{1}{K})} \right\rfloor & \text{if } H = H_t \\[2mm] 1 + \left\lceil \alpha \cdot \frac{1 - (\frac{1}{K})^{H_t}}{1 - (\frac{1}{K})} \right\rceil = 1 + \left\lfloor \alpha \cdot \frac{1 - (\frac{1}{K})^{H_t+1}}{1 - (\frac{1}{K})} \right\rfloor & \text{if } H = H_t + 1 \end{cases} \tag{8}$$

Note that for the case $H = H_t + 1$, since $-1 < \epsilon < 0$ we can round up the expression from $N_{\min,c}$ or alternatively we can go on one step more in the summatory of S until a level H_t and then round down. Therefore, we can compare (8) to (6) to determine whether $H = H_t$ or $H = H_t + 1$.

4.4   Gain Children Distribution in MDG

Finally, we analyze the distribution of the gain children and then we recalculate H and $|G_{\min}|$. We denote by α_i the number of peers without direct children at level i, $\alpha_i = \alpha / K^i \bmod K$. Note that $\alpha_i < K$ is the i-th digit of α in base K, that is $\alpha = \alpha_0 + \alpha_1 \cdot K + \alpha_2 \cdot K^2 + \ldots + \alpha_{H_t - 1} \cdot K^{H_t - 1}$. We denote by $R(l)$ the number of peers that have not formed direct children until level l:

$$R(l) = \sum_{i=0}^{l} \alpha_i \text{ for } 0 \leq l \leq H_t - 1$$

We also denote by R the number of peers without direct children until level $l = H_t - 1$ inclusive, that is $R \triangleq R(H_t - 1)$. Note also that when in MDG there are g gain children we need $K + (g - 1) \cdot (K - 1)$ parents without direct children to admit them, K parents for the first gain child and $K - 1$ for the rest since they take as a parent the previous gain child. We define $g(l)$ as the number of gain children from parents up to level l, that is, the number of gain children which have been admitted until level $(l + 1)$, $g(l) = \max\{g \in \mathbb{Z} : K + (g - 1) \cdot (K - 1) = K \cdot g - (g - 1) \leq R(l)\}$. Then:

$$g(l) = \begin{cases} 0 & \text{if } R(l) = 0 \\[2mm] \left\lfloor \frac{R(l) - 1}{K - 1} \right\rfloor & \text{if } R(l) > 0 \end{cases} \quad \forall \, 2 \leq l \leq H_t - 1 \tag{9}$$

Note that (9) is defined for $2 \leq l \leq H_t - 1$ because neither level 0 nor level 1 would ever have gain children.

Theorem 5. *The diameter for MDG for α and K is:*

$$H = \begin{cases} H_t + 1 & \text{if } g(H_t - 1) > g(H_t - 2) \\ H_t & \text{if } g(H_t - 1) = g(H_t - 2) \end{cases} \tag{10}$$

Proof. When $g(H_t - 1) > g(H_t - 2)$ the difference corresponds to the gain child at additional level H_t and thus $H = H_t + 1$. Equally, when $g(H_t - 1) = g(H_t - 2)$ there is no additional gain child and no additional level H_t and then $H = H_t$. □

Finally, we can apply $g(l)$ to calculate a new expression for the number $|G_{\min}|$ of peers of MDG, considering the root, the α root children at level 0, the direct children at the rest of the levels, and all the gain children, and resulting

$$|G_{\min}| = 1 + \alpha + \sum_{i=1}^{H_t - 1} \left\lfloor \frac{\alpha}{K^i} \right\rfloor + g(H_t - 1) = 1 + \sum_{i=0}^{H_t - 1} \left\lfloor \frac{\alpha}{K^i} \right\rfloor + \left\lfloor \frac{R-1}{K-1} \right\rfloor$$

5   Conclusions

In this paper we have presented a proposal for a structured P2P network for live-content media distribution. We have modeled the network as a family of media distribution graphs and we have analyzed a particular kind of these graphs called the minimum distribution graph, achieving relevant measures of the network quality-of-service as a function of graph parameters. Due to space constraints we can not present other important results for the minimum distribution graph that we plan to state in an extended version of this paper together with experimental results from real live-content transmissions.

References

1. Cherry, S.: The battle for broadband [Internet Protocol television]. IEEE Spectrum 42(1), 24–29 (2005)
2. Zhang, X., Liu, J.C., Li, B., Yum, P.: CoolStreaming/DONet: A data-driven overlay network for efficient live media streaming. In: Proc. IEEE INFOCOM, Miami, FL, USA, March 2005, vol. 3, pp. 2102–2111 (2005)
3. PPLive: http://www.pplive.com
4. SopCast: http://www.sopcast.org
5. Ali, S., Mathur, A., Zhang, H.: Measurement of commercial peer-to-peer live video streaming. In: Proc. of ICST Workshop on Recent Advances in Peer-to-Peer Streaming, Waterloo, Ontario, Canada (August 2006)

Bittella: A Novel Content Distribution Overlay Based on Bittorrent and Social Groups

Rubén Cuevas, Carmen Guerrero, Isaías Martinez-Yelmo , Ángel Cuevas,
and Carlos Navarro

Departamento Ingeniería Telemática. Universidad Carlos III de Madrid,
28911 Leganés (Spain)*
{rcuevas,guerrero,imyelmo,acrumin,cnavarro}@it.uc3m.es

Abstract. This paper presents *Bittella*: a new social network for content distribution based on Peer-to-Peer technologies. It exploits the common interests of the users in order to create social groups based on an algorithm called *Ranking Algorithm*. On the other hand, *Bittella* is deployed over a semantic-search based and unstructured p2p network, in spite of this it uses Bittorrent-like download techniques in order to improve the download time. For this purpose, a new Bittorrent trackerless scheme is proposed.

1 Introduction

Peer-to-Peer (p2p) systems have become one of the most successful technologies in the Internet during the last years supported by file-sharing applications like Gnutella [6], Bittorrent [1] or Kademlia [8]. Two main categories of p2p systems have been defined so far [7]: unstructured p2p systems (e.g. Gnutella [6]) and structured p2p systems (e.g. Kademlia[8]). The main problem of unstructured p2p systems is the generation of massive traffic in the search procedure which is usually based on flooding. On the other hand, there is not control on the data placement, thus, these systems are resilient in dynamic environments. The structured p2p systems are based on Distributed Hash Table (DHT). Therefore, there is control on data placement and the search procedure generates less traffic than in the p2p unstructured systems. However, this control on data placement produces a very high cost to maintain a consistent distributed structure in the typical dynamic environment of file-sharing applications. Moreover, there is a third type of p2p network model which provides a better download rate: Bittorrent [1]. It is based on a web servers infrastructure in order to perform the searching procedure of the file. The web servers store the *.torrent* file which indicates the IP address of the *Tracker*. This is the central entity responsible of the management of the sharing process within the Bittorrent swarm.

* This work was supported by the European Commission through the NoE CONTENT FP6-CONTENT-038423, the Spanish government through the Project IMPROVISA TSI2005-07384-C03-027 and the Madrid regional government through the Project BIOGRIDNET CAM-S-0505/TIC-0101.

R. Meersman, Z. Tari, P. Herrero et al. (Eds.): OTM 2007 Ws, Part II, LNCS 4806, pp. 991–1000, 2007.

On the other hand, the Social Networking is a novel phenomenon which is growing exponentially in the Internet during the last years. This concept was firstly studied on the Social Sciences [10], and latter it was adopted by the Internet community in different applications (e.g. Skype or MSN). However, if we consider the content distribution applications based on p2p systems, and more concretely the file-sharing applications, the concept of social relationship is not used. Hence, the advantages of the social relationships (defined by the common interests among the peers in the content distribution applications) are not exploited on the current p2p systems.

This paper proposes *Bittella*, a novel *Social Network for Content Distribution*. *Bittella* creates the social groups with common interests transparently to the final user. For this purpose, it is based on a semantic-search based p2p network. This type of search permits a better exploitation of the common interests and thus, the formation of a more robust and reliable Social Network. Due to the semantic search is much easier to be performed on the unstructured p2p networks, we decided to implement *Bittella* over this type of p2p networks. In order to solve the problem of the overloaded traffic generated in the searching process in unstructured p2p, *Bittella* exploits the social relationships and the common interests among peers. Thus, a peer queries firstly the peers sharing its interests. These peers have the highest probability of storing the desired content. By doing so, the flooding is drastically reduced.

On the other hand, since it has been demonstrated that Bittorrent is the most effective p2p technology in terms of download rate and fairness, the decision was to use it for the file-sharing in our scheme. This implies to utilize Bittorrent over an unstructured and semantic-search based p2p network. For this purpose, a novel trackerless Bittorrent model for *Bittella* is defined in this paper.

The *Bittella* architectural framework is composed by three different layers: the lowest layer is the unstructured p2p network (e.g. Gnutella); the medium layer is formed by the Bittorrent-like swarms for file-sharing; the upper layer is the social layer where we can find the social groups with common interests.

Finally, the paper introduces the *Ranking Algorithm*, a novel procedure which leads to the creation of groups of peers with common interests. It ranks the known peers considering those in the first positions of the ranking as partners on the social group.

2   Basic Functionality of *Bittella* Protocol

This section explains the functionality of *Bittella*. Firstly, we present the procedure to create the Bittorrent-like swarm to share a given content. Indeed, this is the description of our trackerless Bittorrent. After that, the section presents the functionality of the *Bittella* searching protocol. Then, the section describes the *Ranking Algorithm* and the creation of the social groups. Finally, we introduce the concept of the *Secure Permanent Peer ID* which leads to obtain a reliable and robust social structure along the time.

2.1 *Bittella* Swarm Creation

This section assumes that there is an unstructured p2p network already deployed. Indeed, Gnutella will be considered during the remainder of the section, but it could be any other unstructured p2p. In this scenario, when a peer has a new content to share, it operates similarly as a file provider in Bittorrent: it divides the content into chunks, computes the hashes of each chunk and the complete file, and creates a *.bittella* file with the number of each chunk and the hash associated to it and the hash of the complete file. At this point, there is a seed (named *Bittella Seed*) for the given content and this is available on the p2p network. Eventually, a peer (e.g. *Peer A*) solicits the content as will be explained in section 2.2. When the query from *Peer A* reaches the Bittella Seed, this one answers including in the response the *.bittella* file. In the instant when the *Peer A* receives the response, it can starts to download the content by soliciting chunks to the seed. Afterwards, another peer, e.g. *Peer B*, could send a query looking for the same content. This query can reach to: only the seed, only the *Peer A* or both the seed and the *Peer A*. If we suppose that *Peer A* is the one which answers, it delivers to *Peer B* the following data: **(i)** the *.bittella* file; **(ii)** the list of chunks of the content owned by itself; **(iii)** a list of seeds and peers known in the swarm[1] (in this example this list is only composed by the seed). Therefore, the swarm will be growing while new peers solicit the content.

In the traditional Bittorrent, the *.torrent* file is obtained from a Web Server and the management of the sharing process is performed by the Tracker. Some trackerless schemes based on structured p2p networks have been proposed so far. The main objective of these proposal is to remove the single point of failure represented by the Tracker. However, the mechanism of swarm creation introduced in this section can be intended as a novel trackerless Bittorrent scheme. For the best of our knowledge, this is the first proposal where the search of the *.bittorrent* file is done by using an unstructured p2p network.

2.2 *Bittella* Searching Protocol

In this section, we suppose a stable *Bittella* network with several swarms enough populated. In this scenario, if a peer, e.g. *Peer C*, wants to obtain a content, e.g. *Content X*: Firstly, the peer checks if it has stored local information related to *Content X* (The procedure to create the local information is explained in Section 2.3). If it has, then, it sends the query directly to the peer(s) which have high probability to answer the query. This(ese) peer(s) answers to *Peer C* including in the response the following information: **(i)** the *.bittella* file; **(ii)** the list of *Content X* chunks owned by itself (themselves); **(iii)** a list with the IP addresses of seeds and other peers within the swarm. Then, *Peer C* selects other peers from the obtained list and asks them about their list of chunks. At this point, *Peer C* starts the downloading procedure by soliciting chunks to other peers. In addition, if it is necessary the peers can perform a gossiping protocol

---

[1] It can include all the peers and seeds known by *Peer A* or maybe only a random selection of them.

in order to identify more members within the swarm. This mechanism is named *Peer Exchange* [2].

On the other hand, if *Peer C* has not local information about the *Content X*, it queries those peers present in its same social group. These are the peers sharing its interests and the ones with a higher probability of know the location of the content. However, in some cases the peer can ask for content that differs from its interests, then if it only queries within its social group these unexpected queries could be unsuccessful in many occasion. In order to avoid this, the queries will be also sent to some of the underlying neighbors[2] in order to increase the probability of success of the unexpected queries.

It must be noted that the described procedure must be applied for both type of queries, those generated by the peer and those to be routed by the peer.

Finally, as it occurs in Bittorrent, when a peer finishes the download of a given content it becomes a seed for this content.

2.3   *Bittella* Ranking Algorithm

Firstly, it must be highlighted that we consider a social group as a group of peers with common interests within the p2p network. The *Ranking Algorithm* is the defined procedure in order to find out which peers have common interests, and therefore, belong to the same social group. The Ranking Algorithm runs on each individual peer and is a passive procedure. That is, it does not create any kind of message and only uses the messages of the protocol to evaluate which peers have common interests. This is a great advantage because the *Ranking Algorithm* does not produce any kind of overhead. In addition, the *Ranking Algorithm* allows the creation of social groups in a transparent manner to the user and maintaining the anonymity. These are two of the most important features that the users require to the content distribution applications: *transparency* means simplicity from the user point of view; that is, the application optimizes the searches of contents without any kind of configuration or waste of time from the user, apart from selecting the content to be downloaded. On the other hand, *anonymity* is another required feature since nobody wishes that others could identify what kind of contents he/she is downloading. Therefore, with these three features in mind (no overhead, transparency and anonymity) the *Ranking Algorithm* is defined as follows.

Each peer generates a ranking of the other peers in the *Bittella* network. The top one is the peer with most similarity. In order to rank the other peers, *Bittella* uses two different mechanisms. The first one is the number of swarms where the peer has been met (e.g. a peer receives one point per each swarm where it is found out). It is an intuitive mechanism, if *Peer A* and *Peer B* have common interests they will meet each other in many swarms and they will rank each other in a high position. The second mechanism considers the queries to be routed. That is, if a peer has to route a query which matches with some of its last queries (e.g. the last 20 queries), it gives some points to the peer which

---

[2] Neighbors in the underlay p2p network.

generated this query. Due to Bittella uses semantic queries, the query can fully match or partially match any of my previous queries, then the points assigned to that peer may vary: for instance, from 0 when there is not match to 1 when the query fully match[3], assigning intermediate values between 0 and 1 when there is a partial match. Here, we can observe the importance of the semantic search in the case of dealing with social groups. It leads to an accurate evaluation of the common interests among the peers. This is the main reason to reject the use of DHT-p2p systems in our purpose, these systems do not permit the use of semantic searches reducing the power of the *Ranking Algorithm* in the creation of the social groups.

Equation 1 includes the formula in order to obtain the Rank of a Peer. The $SwarmMatch_i$ is equal to 1 when the peer has been met on the swarm of the content i and 0 if it has not been met. $QueryMatch_k$ indicates the matching between the kth routed query and the previous generate queries, thus it varies from 0 to 1. The factor β adjust the query matching depending on if the query matched is an old or a recent one, it varies from 0 (oldest) to 1 (the most recent). Finally, α is the factor which adjusts the importance of each one of the mechanism (swarm matching or query matching) and varies from 0 (query matching more important) to 1 (swarm matching more important).

$$PeerRank = \alpha * \sum_i SwarmMatch_i + (1 - \alpha) * \sum_k \beta * QueryMatch_k \quad (1)$$

When a peer has elaborated the ranking it learns the information about the peers placed in the highest positions (e.g., the top 100; this number depends on the host capacity). This procedure, named *Bittella Learning Procedure*, is performed by directly requiring to the top peers the following information: (i) the contents that they have already downloaded; (ii) the contents that they are currently downloading; (iii) the list of the top peers on their rankings. This is performed every time that a top peer is found in a swarm. That is, when a peer find out one of its top peers in a swarm the former requires from the latter, apart from the list of chunks, the information described above. By doing so, the result of the *Bittella Learning Procedure* is that each peer has the updated information about the available contents in its social group. These are the most likely contents to be solicited by the peer.

Another important feature of the *Ranking Algorithm* is that it is self-adaptive. If the interests of a given node change, the algorithm brings this peer to its new social group due to the peers on this new social group will start to receive points and occupy the first positions in the ranking. In order to make the algorithm more adaptive, peers which are not found in any swarm or whose queries have not been received during a period of time decrease their rank.

2.4   Secure Permanent Peer ID

If we analyze the behaviour of p2p users, they are available on the network intermittently. In the study developed by Pouwelse et al. [9] we find that only

---

[3] The complete matching is equivalent to find the peer on a swarm.

the 17% of the peers stay more than 1 hour on the Bittorrent network after they finished the download. Then, the peers leave and come back continuously to the network. Furthermore, the most of the users have a dynamic IP address. Hence, when an user leaves the network and rejoins it the next time, the connection will be established (with high probability) with a different IP address. If we consider these factors, a peer only belongs to the social group during the time that it is connected to the *Bittella* network. If it leaves and rejoins again it is considered as a new user and therefore the procedure to discover its social group must be done again. It is not a desirable feature. Thus, in order to avoid this, but keeping the anonymity and transparency initially defined, the following procedure was defined. The peer obtains a Permanent ID when firstly connects to the network. This ID is the public key of a *Public/Private Key Pair*. Then, it discovers its social group. At a certain moment, finally, it leaves the *Bittella* network. Then, all the peers which have this one on their top ranking discover that the peer left and freeze its entry, put it out of the ranking and assigns a timeout to this entry. In order to discover top peers which have left each peer sends periodically keep alive messages to the top nodes in its ranking. If the peer does not return to the *Bittella* network before the time out expires, the entry is removed. Otherwise, if the peer rejoins the network before the timeout expiration (in the worst case with a new IP áddress) it informs the peers in its top ranking about its presence in the network and its new IP address. Besides, it can be done in a secure way since the others can challenge it with a nonce (encoding something with its ID, that is the *Public Key*) and it can answer the challenge, demonstrating that it is the right node. The peers which receive the message storing the new IP address, recover the entry and put the peer again on the ranking. Therefore, by applying this mechanism the peers are able to leave and join the networks without losing the information about its social group and using a secure process. Moreover, in order to improve the anonymity the peers can change their Permanent Peer-ID. For this purpose, the node generates a new *Public/Private Key Pair* and sends the new Permanent Peer-ID (that is the Public Key) signed with the previous private key to its top ranking nodes. The top ranking nodes can decipher the new Permanent ID and store it.

3   *Bittella* Three Layer Architecture

This section describes *Bittella* from an architectural point of view. *Bittella* is a three layer architecture where the lower layer is called *Underlay Layer* and it is basically the underlying p2p network; the medium layer is the *Swarm Layer* and is formed by the Bittorrent-like swarms; finally, the higher layer is the *Social Layer* which is formed by the social groups based on common interests. Figure 1 represents this architecture.

- **Underlay Layer:** This is the fully distributed and unstructured p2p network (e.g. Gnutella). In this layer each node has some neighbors which are called *Underlay Neighbors*. Basically, when *Bittella* uses this layer, it uses the search mechanism defined on it. Usually, it is flooding.

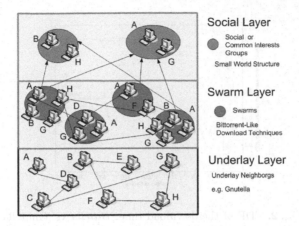

Social Layer

● Social or
Common Interests
Groups

Small World Structure

Swarm Layer

● Swarms

Bittorrent-Like
Download Techniques

Underlay Layer

Underlay Neighborgs

e.g. Gnutella

Fig. 1. Three Layer *Bittella* Architecture

- **Swarm Layer:** This layer is formed by different Bittorrent-like swarms. In this paper we focus on the file-sharing, but as future work we will add the live streaming and the VoD distribution. Therefore, there would be different type of swarms. In addition, the swarm layer is basic for the *Ranking Algorithm*.
- **Social Layer:** This layer is formed by different *Social Groups* based on common interests. From the topological point of view, this layer has a *Small-World* structure, that is, a loosely connected graph of highly connected subgraphs [11]. In *Bittella*, the highly connected subgraphs are the *Social Groups* and they are loosely connected by the links among the *Underlay Neighbors*. The *Small-World* structure on p2p networks has been previously analyzed [5] [3] [4].Therefore, all the demonstrated advantages of the Small World structure can be considered *Bittella* advantages as well.

4   An Initial Simulation Analysis

Bittella was simulated in an small scenario with 1000 nodes. Therefore, each node is going to include in its ranking all the other known nodes due to the reduced size of the network. The performed experiments analyzes if the *Bittella Learning Procedure* proposed in this paper is useful. If the results are positive, we can consider the *Ranking Algorithm* as an extension which should be applied in larger networks with a higher number of nodes.

The *Bittella Simulator* was implemented in Java, and it is a discrete time simulator: the time is divided into cycles and each node finishes all the pending events on each cycle. The simulations were deployed with the following parameters: N (the number of nodes forming the Gnutella network) equal to 1000; n (the number of neighbors that each node has in the *Underlay Network*) equal to 5; R (the number of contents offered in the p2p system) equal to 62; C (the number of chunks per content) equal to 100; Q (the number of queries -i.e. content solicitation- generated during the simulation) equal to 3000; S (the number

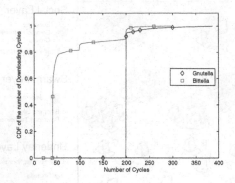

Fig. 2. CDF of the download time: *Bittella* vs. Gnutella

of initial seeds per content) equal to 2; TTL (the radius of the flooding queries) equal to 5; BW (the number of chunks which can be simultaneously downloaded or uploaded) equal to 4. The experiments were deployed in a static situation, where nodes do not join or leave the network and they compare the performance of Gnutella and *Bittella* in terms of *Bandwidth Consumption* and *Download Time*.

The first metric analyzed is the CDF (Cumulative Distribution Function) of the number of simulation cycles spent in the download process. The results are presented in Figure 2. The Figure shows that in a time equal to 50 cycles the 80% of the files have been downloaded using *Bittella* whereas Gnutella downloads never takes less than 200 cycles. Therefore, in the 80% of the download processes *Bittella* reduces the time spent in a factor 4. And in the 100% of the cases it improves the download time of Gnutella. Basically, despite of Gnutella and *Bittella* nodes are configured with the same bandwidth, *Bittella* uses it more efficiently. This behaviour is due to the use of Bittorrent-like download techniques.

The second experiment focuses on evaluating the reduction of bandwidth generation during the searching procedure obtained by *Bittella* in front of Gnutella. In this experiment, we measure the total number of queries generated every 20 simulation cycles. That is, the original queries but also the replication of they forwarded due to the flooding algorithm. In addition, the local hit rate offered by *Bittella* is measured. That is, the ratio of queries which do not need to be flooded (because the peer has local information to solve the query) in front of the total number of the generated queries. Again, it is measured in periods of 20 simulation cycles. Figures 3 and 4 show the results of the experiment. The upper graphic in Figure 3 shows the *Bandwidth Consumption* of Gnutella and *Bittella* in terms of relative bandwidth units. We can see that at the beginning of the simulation, *Bittella* and Gnutella present the same *Bandwidth Consumption*, but according with the simulation advance, Gnutella keeps the same *Bandwidth Consumption*, around $2.5 * 10^5$ bandwidth units, whereas *Bittella* reduces it. At the middle of the simulation (around the cycle 500) *Bittella* offers a *Bandwidth Consumption* around $1.75 * 10^5$ that represents the 30% of reduction compared

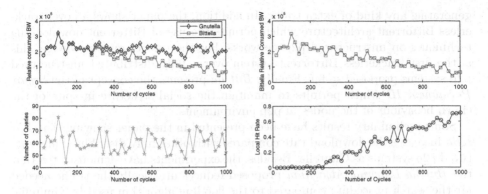

Fig. 3. BW consumption & Generated Queries : Gnutella vs. *Bittella*

Fig. 4. *Bittella*: BW Consumption & Local Hit Rate

with Gnutella. Even more, at the end of the simulation, *Bittella* shows a *Bandwidth Consumption* below $1 * 10^5$ bandwidth units which means a reduction of the 150% (1.5 times) compared with Gnutella. This reduction in the *Bandwidth Consumption* is produced by the learning procedure described in this paper. The lower graphic on Figure 3 represents the number of queries generated each 20 cycles. If we compare it with the graphic above, it is easy to check that those periods which present a higher number of queries result in more bandwidth consumption, because more queries mean more flooding. However, we can check that the *Bittella Learning Procedure* mitigates this effect. When the nodes have a high level of knowledge (at the end of the simulation) the variation on the number of the queries generated affects in a minor degree because the flooding is drastically reduced.

Finally, Figure 4 shows in the upper graphic the BW consumption of *Bittella* and in the lower graphic the *Local Hit Rate*. This figure demonstrate the behaviour of the *Bittella Learning Procedure*. Along the advance of the simulation, the nodes learn more and more, thus, the local hit rate increases up to reach values near to the 80% at the end of the simulation. The high local hit rate produces the reduction of the number of queries to be flooded and therefore, the drastic reduction of the *Bandwidth Consumption*. Hence, this experiment proves that with the use of *Bittella* the total traffic generated on unstructured p2p networks is reduced.

5   Conclusion and Further Work

This paper presents a novel architecture of social network for content distribution, *Bittella*. It is based on an unstructured and semantic-search based p2p network and exploits the common interests of the users in order to create the *Social Groups*. The paper introduces a new mechanism, the *Ranking Algorithm* which allows the creation of the *Social Groups* in an easy way and without generating any kind of extra traffic. In addition, the paper shows a novel trackerless Bittorrent architecture which permits the use of Bittorrent downloading techniques on unstructured p2p networks. For the best of our knowledge, this is the first trackerless Bittorrent system over a fully distributed unstructured p2p systems proposed so far. Finally, *Bittella* presents the concept of the *Secure Permanent ID* which permits to maintain the social structure in spite of the churn behaviour of the nodes in p2p environments.

Some preliminary results have been presented in the paper showing that *Bittella* increases the download rate compared with other fully distributed unstructured p2p systems as Gnutella. Besides, the experiments have demonstrated that the *Bittella Learning Algorithm* proposed reduces drastically the traffic needed for the search procedure compared to the flooding algorithm used by Gnutella. This is due to the high local hit rate obtained by the application of the learning algorithm.

Further work will focus on the deep analysis of *Bittella* protocol and the *Ranking Algorithm* by means of simulation. In addition, the distribution of VoD and Live Streaming on *Bittella* will be intensively studied.

References

1. Bittorrent, http://www.bittorrent.org
2. Peer exchange, http://en.wikipedia.org/wiki/Peer_exchange
3. Cohen, E., Fiat, A., Kaplan, H.: Associative search in peer to peer networks: harnessing latent semantics. In: Proceeedings of INFOCOM 2003 (2003)
4. Hui, K., Lui, J., Yau, D.: Small world overlay P2P networks. In: Proceedings of IWQOS 2004 (2004)
5. Iamnitchi, A., Foster, I.: On Exploiting Small-World Usage Patterns in File-Sharing Communities
6. Klingberg, T., Manfredi, R.: Gnutella 0.6. Network Working Group (June 2002)
7. Lua, K., Crowcroft, J., Pias, M., Sharma, R., Lim, S.: A survey and comparison of peer-to-peer overlay network schemes. Communications Surveys & Tutorials, IEEE (2005)
8. Maymounkov, P., Mazieres, D.: Kademlia: A peer-to-peer information system based on the XOR metric. In: Druschel, P., Kaashoek, M.F., Rowstron, A. (eds.) IPTPS 2002. LNCS, vol. 2429, Springer, Heidelberg (2002)
9. Pouwelse, J., Garbacki, P., Epema, D., Sips, H.: The Bittorrent P2P File-sharing System: Measurements and Analysis. In: Castro, M., van Renesse, R. (eds.) IPTPS 2005. LNCS, vol. 3640, Springer, Heidelberg (2005)
10. Wasserman, S., Faust, K.: Social network analysis. Cambridge Univ. Press, Cambridge (1995)
11. Watts, D., Strogatz, S.: Collective dynamics of'small-world'networks. Nature (1998)

Managing Difference-Based Objects with Sub-networks in Peer-to-Peer Environments

Daisuke Fukuchi[1], Yuichi Sei[1], and Shinichi Honiden[1,2]

[1] The University of Tokyo
[2] National Institute of Informatics

Abstract. P2P systems are currently being used all over the world. However, existing P2P technology is not able to modify shared objects efficiently. Naive approaches to support modification result in large amounts of traffic and load concentrations. In our study, this issue is being addressed by representing shared objects by their differences. These are stored in a sub-network that is generated for each object. The object is accessed by retrieving and adding the differences stored in the sub-network. A simple simulation demonstrates the effectiveness of this technique in regard to traffic amount and load balancing properties.

1 Introduction

P2P systems are being used all over the world. Especially, the popularity of P2P file sharing software such as Napster [1], Gnutella [2], BitTorrent [3], and Freenet [4] cannot be ignored. However, current P2P systems are not able to support efficient modification of shared objects. Actually, P2P technology is not quite as popular in applications which need modification such as the Web, BBS, or Wikis.

A lot of P2P systems use replicas of shared objects. The burden to host and provide an object is distributed to the multiple nodes holding the replicated instances. Their purpose is to maintain availability of the object by replication, so that node failures will not result in the loss of the object. In fact, replication mechanism is a basic cornerstone for high performance in file sharing software.

However, these beneficial effects of replication rely on the fact that modification and synchronization of shared objects are not supported. When those concerns are of interest, three types of shared objects need to be distinguished. Objects that are unmodifiable, those that are modifiable, but need no synchronization, and object that are both modifiable and require synchronization. In file sharing, a file is represented by a shared object, and once registered, that object is never modified. Furthermore, a list of nodes holding a target file is maintained as a shared object. The content of the list changes according to changes in nodes holding the file. However, it is only necessary that at least one entry on the list is correct. Therefore, lists of different replicas may vary in their content.

In contrast to the above scenario, synchronization of replicas is a fundamental issue for applications in which objects must reflect modifications accurately. If

R. Meersman, Z. Tari, P. Herrero et al. (Eds.): OTM 2007 Ws, Part II, LNCS 4806, pp. 1001–1010, 2007.
© Springer-Verlag Berlin Heidelberg 2007

not taken into consideration properly, maintaining such objects will result in increasing network traffic and load concentrations. For instance, if updates are propagated by flooding, redundant messages are generated every time an object is modified. This is a problem because nodes waiting updates are much less than all nodes on average. Also, it is not sufficient to create a manager node for each object and entrust it with the management of the replica. It will result in an extreme bias of the workload according to a bias of object popularity [5].

In this paper, we suggest a new method to share modifiable objects efficiently. In our approach, a shared object is represented by differences caused by modifications. Sub-networks are established for each object, and the object is shared by acquisition and addition of differences stored in the sub-network. This method helps to limit traffic and balance the load on the network nodes evenly.

The remainder of this paper is structured as follows. Related work is described in Section 2. The difference-based object is introduced in Section 3. Section 4 describes the method to share difference-based objects efficiently by using sub-networks. A performance evaluation is described in Section 5, and the conclusion is given in Section 6.

2   Related Work

The synchronization problem in regard to distributed objects is not new in studies on distributed file systems (DFS) [6,7,8]. However, existing DFS solutions cannot be applied to a P2P scenario, because P2P environments differ from DFS in a number of crucial points.

1. P2P environments consist of a very high number of nodes.
2. P2P environments have a slow transmission rate on average.
3. P2P environments usually have unstable and weak nodes.

A solution based on so called supernodes has been put forward to overcome these problems and provide a technique similar to those in DFS solutions to P2P applications. In this approach, the overall system performance is maintained by tasks being concentrated on supernodes which are assumed to provide relatively stronger computational power [9]. Turning a regular node into a supernode obviously results in an increased workload for that node, which however in many cases is not desirable or acceptable. For instance, it is said that people refrain from using Skype [10] (which uses this supernode technique) because they are concerned about the risk of their machine becoming a supernode, effectively using up all their network bandwidth.

A simple approach to support modification is to create a manager node for each object and force it to manage all replicas. This mechanism can be realized by using a distributed hash table (DHT) such as Chord [11], Pastry [12], Tapestry [13], or CAN [14]. DHTs are in structured P2P environments that have structured overlay topologies (older P2P networks are unstructured). Of course an additional cost is incurred for maintaining this kind of structure in a P2P environment, as nodes leave and join frequently. On the other hand, certain advantages that unstructured P2P networks do not have can be gained. DHTs have the ability to create a unique charge node for an arbitrary key. Therefore, the management of a value associated with the key can be assigned to the charge node of that key. In the following, a node that is assigned a management of a given value V is called V manager. Neighbors of a manager back the value up, so that in case the manager disappears, the system can still recover automatically and locally. Moreover, messages can be delivered from an arbitrary node to the arbitrary key's charge node in a few hops. For example, $O(\log n_{all})$ hops in Chord and Pastry (n_{all} is the number of all nodes).

Although CFS [15] and PAST [16] define themselves as P2P storage systems, they do not support object modification.

OceanStore [17] is a write-enabled P2P storage. Modifications to an object are first sent to a manager group of the object, and are synchronized among them. Subsequently the modifications are applied to all replicas through a replica tree constructed at the application level. Load concentration occurs on nodes of the manager group. Also, there is no method to efficiently construct the replica tree.

IVY [18] is a P2P distributed file system. Modifications to the file system are kept in a log that consists of differences. However, as the logs are shared on the whole network, this approach generates a lot of retrieval traffic.

Reference [19] suggests a method to keep replicas up to date by flooding update messages from the manager and synchronization requests to the manager. The obvious drawback is that flooding causes a lot of traffic and the central manager becomes a bottleneck for requests.

Reference [20] suggests a method for P2P cooperative editing, whereby editing actions yield the same results independently of the order in which they are applied. This means that replicas can synchronize in unstable P2P networks as long as all actions arrive at each replica eventually. However, this work does not elaborate on how these editing commands can be spread efficiently.

SCOPE [21] sends update notifications by multicast, using a replica-partition-tree (RPT), a structure to manage multicast groups. The maintenance cost of an RPT is smaller than that of a sub-network in this study as it specializes in multicast. However, since it is based on a single network, the RPT contains nodes that do not hold the target replica. Therefore, redundant update messages are generated. Also, our method performs better in situations in which the acquisition of sets of differences works well.

3   Difference-Based Objects

3.1   Definition

A difference-based object is one whose updates or modifications are represented in the form of differences. Every modifiable object can be represented in a difference-based form. For example, each posting on a BBS can be considered as a difference, and modification histories saved in a Wiki are just differences. [1]

---

[1] On the other hand, when modifications mainly consist of rewrites and deletions, the different-based form is unsuitable because such modifications are stacked and cause unnecessary increases in the object volume.

A difference-based object A is represented by a set of differences in which the initial state is the state that holds the first difference only. A added n-th difference is denoted by A_n.

$$A_n = \{d_1^A, d_2^A, \ldots, d_n^A\} \quad (d_i^A \text{ is the } i\text{-th difference of } A)$$

3.2   Naive Difference-Based System

The representation of an object as differences has the following advantages.

1. Replication is efficient as differences are not rewritten.
2. Management loads for an object can easily be distributed.

These advantages become clear if we consider the following naive method to share difference-based objects.

Acquire A_n. Access $d_1^A, d_2^A, \ldots, d_n^A$ managers and acquire each difference one by one.

Modify A_n (Add a difference to A_n). Send an addition request together with a new difference to the d_{n+1}^A manager.

This system can be constructed by using DHT. Set a key associated to d_i^A to

$$\text{name}(A) \parallel i \quad (\parallel \text{ is concatenation}),$$

and each difference manager is uniquely set. Moreover, the manager can be accessed in $O(\log n_{\text{all}})$ hops.

In this difference-based system, management loads of A_n concentrate most on d_{n+1}^A manager because the acquisition procedure includes a confirmation that d_{n+1}^A is not registered. Therefore, the risk of a manager is small as n changes after the addition of a difference. Furthermore, using replicas of differences reduces acquisition loads. The reason is that differences already held as replicas don't have to be acquired again. It is also possible to acquire differences at once, if a node holding replicated differences that are not held in the acquirer is on a route for acquisition.

The drawback of this method is that it causes large amount of traffic as each difference manager is retrieved in the entire system. Moreover, there is a tradeoff between decreasing the retrieval hops by using replica and spreading of useless replicas.

4   Sub-networks

Section 3.2 showed how objects can be distributed as differences. However, this mechanism alone is not efficient because it has to retrieve differences in the whole network in order to access a target object. Instead, to reduce the distribution area of the difference-based object, we can define a sub-network N_A in which a difference-based object A is shared.

4.1 System

The entire system is divided into the top level root network N_{root} and lower sub-networks (Figure 1). In N_{root}, the shared object is a pointer to a corresponding sub-network. It consists of a list of nodes in the sub-network. An unique pointer is generated for each sub-network by using a DHT and defining a (key, value) pair in it by

$$(\text{name}(A), \quad \text{pointer to } N_A).$$

A sub-network is constructed in the same way as the difference-based DHT in Section 3.2 and consists of nodes to access A. In it, (key, value) is

$$(\text{name}(A) \parallel i, \quad d_i^A).$$

The charge node for $\text{key}_A = \text{name}(A)$ in N_{root} and N_A is called N_A gateway and N_A representative respectively.

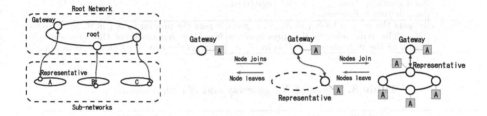

Fig. 1. Entire system   **Fig. 2.** Network transitions

When an acquirer P is in N_A, acquisition of A is achieved by retrieving the differences in N_A. Not in N_A, first, P retrieves a pointer to N_A in N_{root}. Then, P joins N_A. After that, acquisition of differences is started.

The modification of A_n consists of the addition of the difference d_{n+1}^A. When a modifier P is in N_A, this is achieved by sending a request message to d_{n+1}^A manager in N_A. n is known as a result of acquisition of A. When P is not in N_A, first, P retrieves a pointer to N_A in N_{root}. After P joins N_A, a request message is send to d_{n+1}^A manager.

The pointer to N_A should be modified following a change of nodes in N_A. However, it is not necessary to keep an accurate node list as long as one item of the list is accurate. To keep at least one node accurate, the N_A representative regularly reports known nodes to the N_A gateway. The gateway also returns the previous node list, from which the divisions of N_As can be detected.

Because the content of A is managed in N_A, it will be lost if N_A disappears due to node failure. As a result, the content of A would change if a new N_A is generated. To prevent this from happening, a replica of A is held in N_{root} as a backup. Consequently, (key, value) in N_{root} is changed to

$$(\text{name}(A), \quad (\text{pointer to } N_A) \text{ and } A)$$

This backup is updated by reports from the representative whenever it changes and acquires new differences.

Tables 1, 2, and 3 summarize the above points.

Table 1. Root network N_{root} and sub-network N_A

	(key, value)	constitutive nodes	charge node for key$_A$
N_{root}	(name(A), (pointer to N_A) and A)	all nodes	gateway
N_A	(name(A) $\parallel i$, d_i^A)	nodes to access A	representative

Table 2. Acquisition and modification of A by node P

	acquisition	modification
in N_A	P accesses managers for differences not held in P one by one. P acquires new differences or replicas discovered during these accesses. When P accesses d_{n+1}^A manager and confirms that d_{n+1}^A is not registered, acquisition is finished.	P acquires A_n. After that, P sends a difference addition request to d_{n+1}^A manager.
not in N_A	P gets the pointer to N_A in N_{root}. P joins N_A. The remainder of this procedure is same as the procedure that P is in N_A.	P gets the pointer to N_A in N_{root}. P joins N_A. The remainder of this procedure is same as the procedure that P is in N_A.

Table 3. Works of N_A gateway and N_A representative

gateway	representative
- holds pointer to N_A in N_{root}. - receives node list from the representative and returns the previous list. - receives differences from the representative and holds it as backup of A. - provides the backup of A when N_A is generated.	- regularly sends the known node list to the gateway. - sends new differences to the gateway when it is created and acquires new differences.

4.2 Advantages

Routing to retrieve the manager in N_A is shorter than in N_{root} because N_A is smaller than N_{root}.

Moreover, replication efficiency is greatly increased. Because nodes in N_A should hold the replica of A, it makes sense to generate replicas of differences in all nodes on a route. Indeed, the number of messages to acquire a registered difference is $O(1)$ per node. Moreover, the number can be even lower than $O(1)$ because sets of differences can occasionally be acquired at once.

4.3 Maintenance Cost and Auto Leaving

In using a DHT as a sub-network, the maintenance cost increases according to the number of joined sub-networks.

This cost can be reduced. As mentioned earlier, a problem of sharing an object without the difference-based form is a concentrated load on managers.

This factor is negligible as long as the access frequency remains low. Thus, using a sub-network only if access frequency is high can reduce the overall maintenance costs; Hence when the access frequency is low, the object should rather be shared in N_{root}.

This is realized by introducing a mechanism for auto leaving a sub-network. For example, if a node doesn't treat any access message to A within a fixed time, the node leaves N_A. N_A disappears when all nodes leave as a result of sufficiently low popularity of A (Figure 2). Note that even after it disappears, a backup of A is held in the N_A gateway. If node P later wants to access A, it accesses the N_A gateway and acquires the new differences of A. Next, P becomes the first node of N_A. If P has not accessed A after a while, it leaves N_A automatically and N_A disappears again. During this period, P has a link to the N_A gateway because it is N_A representative. However, the maintenance cost of N_A is not generated because there are no other nodes in N_A. If a new node joins N_A, difference-based sharing in N_A is engaged.

5   Evaluation

The output of the simulation conducted is the number of messages required for acquiring and modifying a difference-based object A. If their sum is sufficiently small, the conclusion is that traffic has been suppressed. If the maximum among nodes is small, loads are balanced. The following three methods are compared.

client-manager. Theoretical values when the A manager is forced to be a server of A in the top level network.
one network. Simulation results when A is shared in the top level network by using the difference-based method of Section 3.2.
sub-network. Simulation results when the sub-network in Section 4.1 is used.

The parameters and their default values are listed in Table 4. The DHT used for the simulation is Chord. Node leaves and joins are not considered (a static network). A manager in the **client-manager** does not change in a static network. Therefore, clients directly access the manager without retrievals except for the first time. The whole network constitutes a perfect Chord network. In the **sub-network**, only access nodes constitute a Chord network. A simulation begins from a state in which A has no difference. The simulation finishes when the n_{diff}-th difference is added to A and all access nodes acquire all differences. During this period, additions and acquisitions of differences occur randomly according to r_{diff}.

The results are shown in Figures 3, 4, and 5. It is clear from the average values that loads in the **sub-network** are balanced to about $1/10$ of a manager in the **client-manager**. And the maximum values in the **sub-network** are balanced to a fraction. Moreover, the average values indicate that traffic in the **sub-network** is a fraction of that in the **one network**.

Result of varying the number of access nodes n (Figure 3): As n increases, loads on nodes are distributed in the **sub-network** whereas loads are increased

Table 4. Denotation, default value, and explanation of parameters

denotation	default	explanation
n	10^2	The number of nodes that access A.
n_{all}	10^4	The number of nodes that constitute the entire system.
n_{diff}	10^3	The number of differences of A when the simulation finishes.
r_{diff}	0.05	The rate of difference additions per access. Given a default 0.05, one of 20 accesses is an addition and the rest are acquisitions on average.

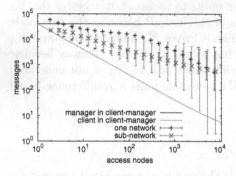 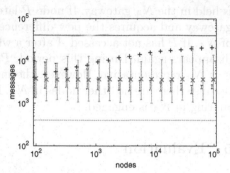

Fig. 3. The number of messages when n is varied

Fig. 4. The number of messages when n_{all} is varied

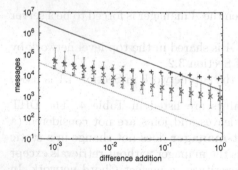

common parts of Figures 3, 4, and 5.

line : theoretical value
point : the number of messages / access node
bar : max and min value among access nodes

Parameters except for x are the same as in Table 4

Fig. 5. The number of messages when r_{diff} is varied

in the **client-manager**. This shows that sub-networks are more efficient for popular objects.

Result of varying the number of all nodes n_{all} (Figure 4): the number of messages in the **sub-network** does not increase when n_{all} is increased, because the sub-network is constituted of constant access nodes.

Result of varying the rate of difference additions r_{diff} (Figure 5): A lower r_{diff} has more advantages for the **sub-network** than for the **client-manager** but not more than for the **one network**.

6   Conclusion

We proposed a P2P system using sub-networks. We have shown that it has a potential to share modifiable objects efficiently. For that, modifiable objects are represented by a difference-based form. The consequence is that synchronization of registered data (differences) and its replicas becomes unnecessary. In addition, by creating each difference manager, the load of object management is distributed. On the other hand, a traffic increase of manager retrievals could be observed as the managers are spread throughout the whole network. To deal with this issue, sub-networks were enacted for each object, limiting manager and access node locations to sub-networks corresponding to the target object. We could show by simple simulations that the sub-networks greatly reduce traffic, and at the same time avoid load concentrations.

There are still a number of tasks left that will help to further advance this research. One of them is the proper verification in dynamic environments. Generally, DHTs in a dynamic environment perform worse than in a static one [11,12]. Measuring that extent will be crucial. In a sub-network, we think that the effect will be greater because node leaves and joins are more frequent. However, we should also consider the fact that a node can respond to the case that the node is in the root network but not in the sub-network. A verification with multiple objects and with realistic access rates following Zipf's law [22,5], a verification of the costs regarding the gateway and representative, and a verification of access overhead from nodes inside and outside the sub-network are also relevant future tasks.

By putting information on the root network and sub-networks in one node, it would be possible to integrate overlapping maintenance processes and detect network divisions. The auto leaving process can be refined. Differences of an object not suitable for a difference-based form could be integrated into a non-difference-based form in suitable contexts. Finally, it may also be interesting to introduce a network tree by recursively enlarging sub-network systems.

References

1. Napster, http://www.napster.com
2. Gnutella, http://www.gnutella.com
3. BitTorrent, http://www.bittorrent.com
4. Clarke, I., Sandberg, O., Wiley, B., Hong, T.W.: Freenet: A distributed anonymous information storage and retrieval system. In: Federrath, H. (ed.) Designing Privacy Enhancing Technologies. LNCS, vol. 2009, p. 46. Springer, Heidelberg (2001)
5. Nielsen, J.: Zipf curves and website popularity, http://www.useit.com/alertbox/zipf.html
6. Satyanarayanan, M., Kistler, J.J., Kumar, P., Okasaki, M.E., Siegel, E.H., Steere, D.C.: Coda: A highly available file system for a distributed workstation environment. IEEE Transactions on Computers 39(4), 447–459 (1990)

7. Anderson, T., Dahlin, M., Neefe, J., Pat-terson, D., Roselli, D., Wang, R.: Serverless network file systems. In: Proceedings of the 15th Symposium on Operating System Principles, ACM, Copper Mountain Resort, Colorado, December 1995, pp. 109–126 (1995)

8. Ghemawat, S., Gobioff, H., Leung, S.: The google file system (2003)

9. Xu, Z., Hu, Y.: Sbarc: A supernode based peer-to-peer file sharing system. iscc 0, 1053 (2003)

10. Skype, http://www.skype.com

11. Stoica, I., Morris, R., Karger, D., Kaashoek, F., Balakrishnan, H.: Chord: A scalable Peer-To-Peer lookup service for internet applications. In: Proceedings of the 2001 ACM SIGCOMM Conference pp. 149–160 (2001)

12. Rowstron, A., Druschel, P.: Pastry: Scalable, decentralized object location, and routing for large-scale peer-to-peer systems. In: Guerraoui, R. (ed.) Middleware 2001. LNCS, vol. 2218, pp. 329–350. Springer, Heidelberg (2001)

13. Hildrum, K., Kubiatowicz, J., Rao, S., Zhao, B.: Distributed object location in a dynamic network (2002)

14. Ratnasamy, S., Francis, P., Handley, M., Karp, R., Shenker, S.: A scalable content addressable network. Technical Report TR-00-010, Berkeley, CA (2000)

15. Dabek, F., Kaashoek, M.F., Karger, D., Morris, R., Stoica, I.: Wide-area cooperative storage with CFS. In: Proceedings of the 18th ACM Symposium on Operating Systems Principles (SOSP 2001), Chateau Lake Louise, Banff, Canada (October 2001)

16. Rowstron, A., Druschel, P.: Storage management and caching in past, a large-scale, persistent peer-to-peer storage utility (2001)

17. Kubiatowicz, J., Bindel, D., Chen, Y., Eaton, P., Geels, D., Gummadi, R., Rhea, S., Weatherspoon, H., Weimer, W., Wells, C., Zhao, B.: Oceanstore: An architecture for global-scale persistent storage. In: Proceedings of ACM ASPLOS, ACM, New York (2000)

18. Muthitacharoen, A., Morris, R., Gil, T.M., Chen, B.: Ivy: A read/write peer-to-peer file system. In: Proceedings of 5th Symposium on Operating Systems Design and Implementation (2002)

19. Liu, X., Lan, J., Shenoy, P., Ramaritham, K.: Consistency maintenance in dynamic peer-to-peer overlay networks. Comput. Networks 50(6), 859–876 (2006)

20. Oster, G., Urso, P., Molli, P., Imine, A.: Data consistency for p2p collaborative editing. In: CSCW 2006: Proceedings of the 2006 20th anniversary conference on Computer supported cooperative work, pp. 259–268. ACM Press, New York (2006)

21. Chen, X., Ren, S., Wang, H., Zhang, X.: Scope: scalable consistency maintenance in structured p2p systems. In: INFOCOM, pp. 1502–1513. IEEE, Los Alamitos (2005)

22. Zipf, G.K.: Human Behavior and the Principle of Least Effort. Addison-Wesley, Reading (1949)

Peer Enterprises: Possibilities, Challenges and Some Ideas Towards Their Realization

Ankur Gupta[1] and Lalit K. Awasthi[2]

[1] Model Institute of Engineering and Technology
Camp Office: B.C Road, Jammu, J&K, India – 180001
ankur_g1@yahoo.com
[2] National Institute of Technology
Hamirpur, Himachal Pradesh, India - 177005
lalitdec@yahoo.com

Abstract. P2P computing has opened up exciting new possibilities by extending the boundaries of scale, creating virtual supercomputers enabling computing in the large, providing gigantic storage clusters and specialized search engines based on shared content and information. However, the P2P revolution has left the large enterprises, small and medium businesses and other organizations largely untouched. These organizations also require access to specialized computing infrastructure, secure scalable storage and to collaborate with other organizations on joint projects or share mutually beneficial information. This research paper proposes to extend the power of P2P networks/computing across enterprises, indicating some possible applications and their benefits, providing an insight into some challenges which need addressing and finally presenting a viable framework which shall enable the realization of the Peer Enterprises concept.

Keywords: Peer-to-Peer Networks/Computing, Peer Enterprises, Fault-tolerance, Security, Resource Location and Management.

1 Introduction

Peer-to-Peer (P2P) networks are currently one of the hottest focus areas amongst the computer science and information technology research community, simply because of the huge untapped potential of the P2P concept – extending the boundaries of scale and decentralization beyond the limits imposed by traditional distributed systems, besides enabling entities to interact, collaborate, share and utilize resources offered by one another. Moreover, P2P architectures are characterized by their ability to adapt to failures and dynamically changing network topology with a transient population of nodes/devices, while ensuring acceptable connectivity and performance. Thus, P2P systems exhibit a high degree of self-organization and fault tolerance.

The P2P paradigm has so far been primarily focused on information sharing by exploiting the resources available through millions of computers connected to the internet in an anonymous manner and without any centralized control. However, organizations ranging from large enterprises to small and medium businesses have

R. Meersman, Z. Tari, P. Herrero et al. (Eds.): OTM 2007 Ws, Part II, LNCS 4806, pp. 1011–1020, 2007.

been largely untouched by the P2P phenomenon. These organizations have the same information/data sharing requirements, scalable storage and resource requirements and solutions which can adapt to dynamically changing environments and are extremely fault tolerant and resilient to security attacks. There is therefore no limitation why the P2P concept cannot be applied to share information, storage and computing resources across organizations. However, the requirements of an organization are somewhat different than that of an anonymous peer connected to the internet. Organizations need to be completely convinced about the security of their data and confidential information, which can be potentially compromised if it is shared with an external agency. Moreover, organizations do not like to expose their resources to the outside world for fear of security attacks which might lead to loss of revenue. Organizations would also not like to interact with an anonymous external agency.

Given these limitations, this research paper examines the potential benefits of P2P networks/computation across enterprises, the issues involved and what it would take to realize such interactions. It finally proposes a viable model to implement such a framework. The proposed architecture is suited to meet the above stated needs of the organizations while addressing issues like resource location/management, fault tolerance and security for the framework under consideration.

Several schemes for harvesting idle CPU cycles have been proposed. SETI@HOME [1], Condor [2] and Avaki [3] are examples of such systems. Lichun and Deters [4] have focused on issues involving the use of distributed P2P applications in an enterprise environment and have proposed a framework to enable seamless execution of distributed P2P applications. Shudo et.al [5] have proposed a middleware which enables aggregation of computational resources and their exchange between peers. A scheme for harvesting idle CPU cycles on the internet has been proposed by Virginia Lo et.al [6] which focuses on application specific scheduling strategies. However, none of the above mentioned approaches focus on peer interactions across organizations.

The Peer Enterprises framework, proposed by the authors shall enable:

   a. Organized P2P interactions between peers residing in different enterprises, organizations or geographical sites, allowing organizations to harness the computing potential of P2P networks.
   b. Aggregation of individual peer resources at the enterprise level providing ease of resource location and substantial savings in terms of search overheads as compared to flooding based P2P systems.
   c. Shall enable Small and Medium businesses which do not have abundant resources to utilize the resources of other peer sites, while offering some resources in return, leading to greater collaboration, resource sharing and utilization.
   d. Shall lead to the evolution of new business/economic models related to pay per use of compute resources.
   e. Shall lead to the creation of new P2P applications promoting collaboration between enterprises.

The rest of the paper is organized as follows: Section 2 outlines the System Model for the Peer Enterprises Framework, while Section 3 presents possible business/

economic models enabled by the Peer Enterprises framework. Section 4 concludes the paper and outlines the future work in this direction.

2 System Model

The proposed system model is depicted in Figure 1. Peer Enterprises interact through designated Edge Peers whose main responsibility is to aggregate the peer resources for a particular organization and try and negotiate the best possible contract with another edge peer. A single edge peer can enter into contracts with multiple other edge peers, leading to the creation of a hierarchical, multi–level P2P network as shown in Fig. 2. Hierarchical P2P networks [7] have been proposed earlier primarily to reduce the lookup times for P2P queries. The context in which the hierarchical P2P network is employed in the Peer Sites framework is novel and represents a new application domain.

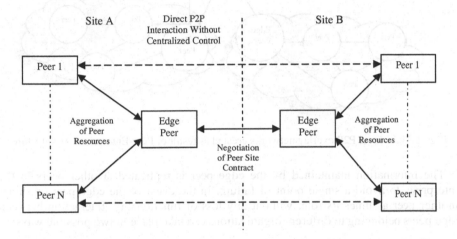

Fig. 1. Conceptual representation of the Peer Enterprises Framework

Broadly, the edge peer is responsible for the following:

- <u>Registration of peers</u> in its site. Peers advertise the resources they offer – information, storage or empty CPU cycles for remote work, which in turn is offered by the edge peer to the edge peer of another enterprise in the negotiation process, while retaining the anonymity of peers in its site and offering assurance to the other site of good behavior by its peers.
- <u>Negotiation of contract</u> with other edge peers on the services provided by peers of its site and the services expected. Once the contract has been agreed upon, peers from both the sites interact independently as in a traditional P2P system.
- <u>Management of Peers</u> – The edge peer monitors the peers in terms of their availability, services provided and ensuring compliance with QoS parameters as negotiated in the contract.

- <u>Trust Management</u> between peers of the two sites. It informs the counterpart edge peer of any malicious peer behavior from its site and takes steps to block the malicious peer or downgrade its trust levels. This information is maintained by the edge peer for future interactions with the same peer enterprise.

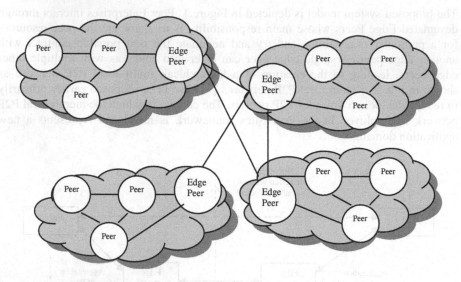

Fig. 2. Multi-level P2P overlay networks created as result of Peer-Enterprises Architecture

The information maintained by the edge peer is replicated at other peers in the enterprise, to avoid a single point of failure. In the event of the edge peer crashing, another peer assumes its role, without any loss of functionality. Interaction between edge peers belonging to different organizations can take place in two possible ways:

a. Trusted Interactions: In such a scenario two organizations enter into an agreement on sharing of resources of different peers in their respective sites. This kind of arrangement leads to interactions between trusted peers and issues such as security and trust computation and management become less relevant.

b. Non-Trusted Interactions: In this scenario a previously unknown edge peer can approach another edge peer offering its services (based on the aggregated resources of the peers in its site) and the services it expects in return. If the offer is found acceptable, the edge peers enter into a contract with pre-defined service level agreements. In this scenario, the edge peer cannot take the offer at its face value and the issues of security and trust as in existing P2P systems are very much applicable.

2.1 Resource Location

Enterprises which wish to join the Peer-Enterprises P2P network must create a P2P network within the enterprise and designate an edge-peer. All other systems send an advertisement of their resources – CPU usage, disk space usage, average uptime, any time constraints affecting availability, available hardware/software platforms etc. to the edge peer, which consolidates the resources available within the enterprise in a single advertisement. Fig.3 depicts a sample advertisement for aggregated resources. Moreover, the edge-peer also accepts requests for specific resources from the peers in its local P2P network which it also consolidates. The edge peer participates in a second P2P network, comprising the edge-peers from all enterprises participating in the Peer-Enterprises P2P network. The edge peer then either responds to requests for specific resources or broadcasts the requests from its local P2P network onto the Peer-Enterprises P2P network. In either case, once the request-response matching has taken place, a contract is negotiated between the two edge peers, regarding service provisioning and QoS (Quality of Service) parameters. Organizations depending upon the available resources may or may not be in a position to commit resources or guarantee a minimum level of service, for instance during peak usage hours. Depending upon the best deals available, the edge-peer enters into a contract with another edge-peer.

Once an "expression of interest" is received by the advertisement peers, other messages which provide further details of computing resources and information content on offer are exchanged along with any pricing information (depending on the type of resource) if applicable. Depending on the requirements of an enterprise only specific resources may be offered or requested.

```
<PE: Aggregated Resource Advertisement>
    <Resources description="Description of Aggregated Resources" >
        <Num_Peers type="Uint32" description="Total no. of peers offering
resources" />
        <Peer_Types type ="List" description=" Various hardware/software
platforms on offer –windows VISTA/NT/XP, linux, solaris, HP-UX etc."/>
        <CPU_Cycles type="Uint32" description="Aggregated number of empty
CPU cylces available" />
        <Storage type="Uint32" description="Total Storage Available" />
        <Content_Size type="Uint32" description="Total content size available" />
        <Content_Classification            type="String"            description="
Video/Audio/Multimedia/Text" />
        <Interaction_Type type="String" description="Interested in exchange of
resources /only a Service Provider/only a consumer" />
        <Service_Level type="String" description="Offering guaranteed fault-
tolerance/task migration or not" />
        <Time_Zone type="String" description="Time Zone in which the
organization resides" />
        <Avg_Uptime type="Uint32" description="The Average time for which
peers participate in the organization's local P2P network" />
    </Resources Description>
</ PE: Aggregated Resource Advertisement >
```

Fig. 3. A sample advertisement depicting aggregated resources of an enterprise

2.2 Fault-Tolerance

Within the Peer-Enterprises framework, ensuring fault-tolerance of deployed P2P applications is a major requirement. This problem has been addressed by earlier work of the authors in which the P4P (Peers-for-Peer) [7] approach for achieving fault-tolerance was proposed. It involves redundantly storing checkpointing data for P2P applications, while taking into account the P2P topology and certain peer characteristics in selecting the peers to hold redundant checkpointing data. By selecting multiple backup peers which are in physical proximity to the primary node executing the remote work, the message exchange overheads are significantly reduced. The selection is based on the ***Neighbour Ranking and Selection Algorithm*** [7], which selects peers based on parameters like avg. CPU utilization, free disk space, avg. uptime and their physical hop distances from the peer submitting the remote work. The primary peer executing the remote work relays the messages that it receives from other components in the P2P application to the backup peers, which then take local checkpoints after receiving a fixed number of messages. Moreover, in case of node transience/failure, the task is migrated to the backup peers, which already have the checkpointing data, thereby alleviating the need for costly retrieval/reconstruction of checkpointing data over the network, as is the case with traditional approaches. Hence, the overall impact on application performance is reduced. Fig. 4 provides a schematic of the P4P strategy.

2.3 Security

Security in traditional P2P systems has been provided based on trust computations [8, 9, 10] and attempting to isolate malicious users/nodes in the P2P network.

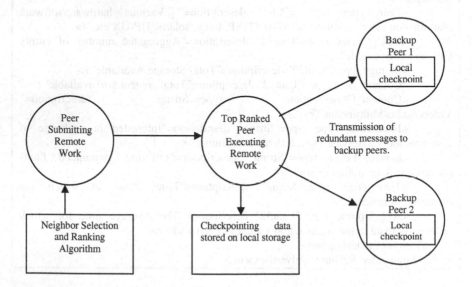

Fig. 4. Schematic representation of the P4P strategy, with the top ranked peer executing remote work and backup peers providing fault tolerance for the P2P application

However, trust computation and distribution of computed trust values for large P2P topologies is less than efficient, warranting the need for developing alternative strategies. Within the Peer-Enterprises framework, security assumes paramount importance, since shared organizational resources cannot be compromised at any cost. To achieve this the authors propose to use a containment based security model, which shall rely on creating specialized security compartments along with fine-grained privileges to restrict the damage that a malicious remote work process/application can cause. The concept of compartments is prevalent in the industry with organizations like HP [11] and Sun [12], having solutions built around compartmentalization and containment. Linux SE [13] is another example of such security frameworks.

Security within the Peer Enterprises framework depends on securing individual computer systems in a P2P network, rather than relying on computing trust values and disseminating these values throughout the P2P network, greatly reducing the overheads of traditional trust management schemes. For each compartment, rules which control access to files, execution of certain commands and access to network resources are defined. Specific compartments can then be created for remote work submitted as part of a larger P2P application, without impacting the rest of the system. The pre-defined rules ensure that the scope of the remote work is not exceeded and any un-authorized operation is not carried out, thereby securing the system. Current work by the authors is focusing on using the fine-grained privileges and access control lists features provided by Linux SE [13] to secure individual peers by defining specific rules.

3   Possible Business/Economic Models

Several studies have assessed the high-level of resource idleness in networked computers on parameters such as CPU [14, 15], but also memory [16] and storage [17]. With such low utilization levels prevailing within organizations, contribution of resources to the Peer Enterprises P2P network is definitely feasible, especially if time-zone differences are factored in. Moreover, organizations could have some special requirements to access specialized infrastructure, which they might not have the resources to invest in. Participating in the Peer Enterprises network would allow them to access such resources on pay-per-use basis or by bartering their resources. The following kinds of business/economic models are envisaged within the Peer Enterprises framework:

3.1   Barter System for Resource Trading

The simplest model of P2P interaction enabled by the Peer Enterprises framework is bartering of resources between two organizations without any financial payments. However, this depends on whether both organizations can offer the required resources to each other through negotiation of a "service contract" and contribution of resources [18], which specifies the services to be provided/expected and any QoS parameters to be enforced.

3.2 Pay-per-Use

The pay-per-use concept for computing infrastructure is not new. Hewlett-Packard's now defunct Utility Data Centre (UDC) [19], allowed consumers to utilize specialized computing infrastructure and services for a fee. Existing research such as [20, 21] focus on computation of fees according to usage of peer resources. The Peer Enterprises Framework goes a step ahead and shall calculate the due payment based on the overall Quality of Service offered as agreed in the service contract, in addition to resource usage.

3.3 Collaborative Contracts

The other business model enabled by the Peer Enterprises framework is "collaborative contracts" where two organizations decide to collaborate on joint products, projects or research. In such cases, the "trusted" mode of interaction is assumed and specific resources required for the collaborative project are opened up to the partner organization. The Peer Enterprises framework shall enable the development of collaborative P2P environments which provide shared resources and information in a secure manner.

3.4 Specialized Service Providers

It is possible for specialized service providers to offer information, storage and specialized computing resources to users at a price, by allowing independent resources to be aggregated. To leverage the millions of computers connected to the internet, a service provider could sub-contract the individually participating peers and offer these resources to interested organizations. This scenario would introduce additional complexities in terms of service and contract management, security, managing micro-payments and revenue sharing between the service provider and the individual resource provider, which need to be addressed before this model can become viable.

The Payment scheme in the Peer Enterprises framework shall depend upon type/category of resources used, service guarantee provided, time of usage of resources, fault-tolerance/task migration features, assurance of data integrity and additional security features.

4 Conclusions and Future Work

We have presented a viable model for Peer Enterprises, which we hope shall allow the power of P2P networks/computing to be extended to organizations, enabling them to harness the specialized computing resources of other organizations, while increasing the utilization of their own resources. The proposed framework which is a work in progress shall also enable new business/economic models, allowing organizations to generate revenue from idle compute resources. Lastly, the framework enables geographically distributed organizations to collaborate on a variety of projects or share information.

As the next step we propose to use the JXTA framework to completely implement the Peer Enterprises framework. Several building blocks of the framework, for instance provisions for fault-tolerance has already been researched and worked out, while a new model for security is proposed. A custom simulator to evaluate the effectiveness of the Peer Enterprises framework in terms of ensuring QoS parameters is also in the offing.

References

1. SETI@Home Website http://setiathome.berkeley.edu
2. Condor Project, http://www.wisc.edu/condor
3. Avaki, http://www.awaki.com
4. Lichun, J., Deters, R.: Coordination and Enterprise-wide P2P Computing. In: IEEE Conference on Services Computing (2005)
5. Shudo, K., Tanaka, Y., Sekiguchi, S.: P3: P2P-based Middleware Enabling Transfer and Aggregation of Computational Resources. In: IEEE Conference on Cluster Computing and the Grid (2005)
6. Lo, V., Zhou, D., Zappala, D., Liu, Y., Zhao, S.: Cluster Computing on the Fly: P2P Scheduling of Idle Cycles in the Internet. In: Voelker, G.M., Shenker, S. (eds.) IPTPS 2004. LNCS, vol. 3279, Springer, Heidelberg (2005)
7. Gupta, A., Awasthi, L.K.: P4P: Ensuring Fault Tolerance for Cycle-Stealing P2P Applications. In: International Conference on Grid Computing and Applications, Las Vegas, USA, June 25-28 (2007)
8. Marti, S., Garcia-Molina, H.: Taxonomy of Trust: Categorizing P2P Reputation Systems. COMNET Special Issue on Trust and Reputation in Peer-to-Peer Systems (2005)
9. Kamwar, S.D, Schlosser, M.T., Garcia-Molina, H.: The EigenTrust Algorithm for Reputation Management in P2P Networks. In: Proc. WWW 2003 (2003)
10. Singh, A., Liu, L.: TrustMe: Anonymous Management of Trust Relationships in Decentralized P2P Systems. In: Proceedings of the Third International Conference on Peer-to-Peer Computing (September 2003)
11. HP-UX 11i Security Containment White Paper http://h20338.www2.hp.com/hpux11i/downloads/SecurityContainmentExecutiveWhitepaper1.pdf
12. Sun Trusted Solaris http://www.sun.com/software/solaris/trustedsolaris/documentation/index.xml
13. Linux SE http://www.nsa.gov/selinux/info/docs.cfm
14. Arpaci, R., Dusseau, A., et al.: The Interaction of Parallel and sequential Workloads on a Network of Workstations, Presented at ACM SIGMETRICS. In: International Conference on Measurement and Modeling of Computer Systems, Ottawa, Ontario, Canada (1995)
15. Acharya, A., Edjlali, G., Saltz, J.: The Utility of Exploiting Idle Workstations for Parallel Computation, Presented at ACM SIGMETRICS. In: International Conference on Measurement and Modeling of Computer Systems Seattle, Washington, United States (1997)
16. Acharya, A., Setia, S.: Availability and Utility of Idle Memory in Workstation Clusters, Presented at ACM SIGMETRICS. In: International Conference on Measurement and Modeling of Computer Systems, Atlanta, Georgia, United States (1999)

17. Bolosky, W., Douceur, J., Ely, D., Theimer, M.: Feasibility of a Serverless Distributed File-System Deployed on Existing Set of Desktop PCs, Presented at ACM SIGMETRICS. In: International Conference on Measurement and Modeling of Computer Systems, Santa Clara, California, United States (2000)

18. Khorshadi, B., Liu, X., Ghosal, D.: Determining the Peer Resource Contributions in a P2P Contract. In: Second International Workshop on Hot Topics in Peer-to-Peer Systems (HOT-P2P 2005), San Diego, California (July 2005)

19. HP Utility Data Centre Technical Report http://hpl.hp.com/techreports/2003/HPL-2003-53.pdf

20. Emmert, B., Jorns, O.: P^2 2P: Prepaid Peer-to-Peer Services. In: Presented at the Sixth International Conference on Peer-to-Peer Computing (P2P 2006), Rhodes Island, Greece (2006)

21. Liebau, N., et al.: Charging in Peer-to-Peer Systems Based on a Token Accounting System. In: Stiller, B., Reichl, P., Tuffin, B. (eds.) ICQT 2006. LNCS, vol. 4033, Springer, Heidelberg (2006)

Workshop on Reliability in Decentralized Distributed Systems (RDDS)

RDDS 2007 PC Co-chairs' Message

Middleware has become a popular technology for building distributed systems from sensor networks to large scale peer-to-peer (P2P) networks. Support such as asynchronous and multipoint communication is well suited for constructing reactive distributed computing applications over wired and wireless networks environments. While the middleware infrastructures exhibit attractive features from an application development perspective (e.g., portability, interoperability, adaptability), they are often lacking in robustness and reliability.

This workshop focuses on reliable decentralized distributed systems. While decentralized architectures are gaining popularity in most application domains, there is still some reluctance in deploying them in systems with high dependability requirements. Due to their increasing size and complexity, such systems compound many reliability problems that necessitate different strategies and solutions. This has led, over the past few years, to several academic and industrial research efforts aimed at correcting this deficiency. The aim of RDDS Workshop is to bring researchers and practitioners together, to further our insights into reliable decentralized architectures and to investigate collectively the challenges that remain.

The program for RDDS 2007 consists of eight research papers of high quality, covering diverse topics. Each paper has been reviewed by a least 3 reviewers. We are very grateful to the members of the RDDS 2007 Technical Program Committee for helping us to assemble such an outstanding program. We would like to express our deep appreciation to the authors for submitting publications of such high quality, and for sharing the results of their research work with the rest of the community.

August 2007   Eiko Yoneki
                                                               Pascal Felber

Improving on Version Stamps

Paulo Sérgio Almeida, Carlos Baquero, and Victor Fonte

Departamento de Informática, Universidade do Minho
Largo do Paço, 4709 Braga Codex

Abstract. Optimistic distributed systems often rely on version vectors or their variants in order to track updates on replicated objects. Some of these mechanisms rely on some form of global configuration or distributed naming protocol in order to assign unique identifiers to each replica. These approaches are incompatible with replica creation under arbitrary partitions, a typical operation mode in mobile or poorly connected environments. Other mechanisms assign unique identifiers relying on statistical correctness. In previous work we have introduced an update tracking mechanism that overcomes these limitations. This paper presents results from recent experimentation, that brought to surface a particular pattern of operation that results in an unforeseen, unlimited growth in space consumption. We also describe informally a new update tracking mechanism that does not exhibit this pathological growth while providing guaranteed unique identifiers for a dynamic number of replicas under arbitrary partitions and the same functionality of version vectors.

1 Introduction

Tracking update dependencies on optimistic replication systems often resorts to the use of version vectors [5] or some of its variants [7]. These mechanisms have been devised and have been successfully supporting traditional scenarios with a fixed number of replicas. Extensions to version vectors have been proposed [6] in order to accommodate a variable number of replicas. However, when trying to cope with the problem of replica identification there is an implicit assumption of global configuration or of a well connected environment in which a distributed naming protocol can be run and replica retirement detected. These assumptions are incompatible with replica creation (and retirement) on distributed systems subject to arbitrary partitions, a typical mode of operation of mobile or poorly connected environments. Other approaches tackle this identification problem relying on statistical correctness [4]. These approaches, not only may lead to occasional errors (which, even very rare, may be unacceptable), but also lead to large identifiers.

Previous work of the present authors [2] introduced the Version Stamp mechanism, a form of decentralized version vector overcoming these limitations. It enables autonomous identity management, update tracking and comparison, solely relying in local or pair-wise knowledge. This confinement is possible because: its structure allows a local management of the identity namespace (both local

R. Meersman, Z. Tari, P. Herrero et al. (Eds.): OTM 2007 Ws, Part II, LNCS 4806, pp. 1025–1031, 2007.

generation of identifiers when forking replicas and merging of identifiers when joining replicas); and the information about updates to replicas is based on the identity namespace in such a way as to allow global comparisons. This structure was devised in such a way that it should naturally grow and collapse as replicas are created and merged in the system.

Version Stamps have been employed in the Panasync [1] decentralized file replication system. Recent experimentation, however, brought to surface a particular pattern of operation that, when repeatedly applied, leads to an unnecessary unbounded growth of its structure.

This paper identifies this particular pattern of operation and illustrates how the version stamp structure degenerates under its occurrence. It also proposes a new version tracking mechanism – inspired by recent insights on autonomous identity management – that does not exhibit this unnecessary structural growth.

2   Version Stamps

Version stamps were devised in order to track update dependencies across a set of replicas in a mobile or poorly connected environment. In this setting, autonomous creation of replicas and pair-wise reasoning over update dependencies are crucial requirements. Operations on version stamps cannot depend on a global view of the system and thus they rely exclusively on local knowledge of each replica.

The structure of a version stamp is made of an *identity* and an *update* component. The identity component distinguishes each replica from all coexisting ones, in any possible configuration. It is also used as an available namespace from which new identities can be generated. This identity generated is achieved by namespace division as described below. The update component records "when" (in which state) changes were applied to a replica. It consists of a single identity-like value collected from the identity of its ancestor when the update was performed.

Three operations are provided: a **fork** operation supports the creation of new replicas whose state is cloned from the original; a **join** operation supports the merging of two replicas keeping one and retiring the other; and an **update** operation accounts for changes on the state of a replica.

An update operation simply copies the *identity* to the *update* component. This means that after an update, subsequent ones do not affect a version stamp. This is an example of the goal, in the design of version stamps, to discard information that is irrelevant to the comparison of coexisting elements in a configuration.

At a fork operation the *identity* of the resulting version stamps is recursively constructed by appending either '0' or '1' to the right of each component of the ancestor *identity*. A fork does not modify the *update* component as it does not introduce any update event (the ones tracked by the mechanism): it simply copies the *update* component to the new version stamps. Regarding identity management, the transformation applied to the identity component can be perceived as the subdivision of the namespace available to a particular replica. Although a local operation, the resulting namespaces are guaranteed to be globally unique and thus distinguish the two new replicas from all the others in the current configuration.

When a join between two elements occur the resulting *identity* is built by merging the two ancestors *identity* components. The *update* component is built likewise, merging the two ancestor *update* components; this reflects the combined knowledge of past updates. Upon the join operation, the resulting identity namespace can be perceived as the union of the two ancestors namespaces. This resulting namespace can then be recursively collapsed each time sibling identity namespaces are present (namespaces that have been previously split upon a fork operation). Since the *identity* component always *dominates* the *update* component (which records information regarding past state changes), this simplification propagates to the *update* component. Ultimately, joining every coexisting replica leads to a completely collapsed version stamp, bringing its structure to its initial value, that is, two empty sets. This division and collapsing feature is an intended design goal of the version stamp mechanism.

Figure 1 shows an example of the version stamp mechanism in action on a replicated system. In this example a version stamp is represented by an [*update* | *identity*] pair, the ϵ denotes an empty set and the δ denotes a local event of state change. Though not shown in this example, as stated above, joining the two remaining replicas would completely collapse the resulting version stamp. A detailed and formal description of Version Stamps including its proof of correctness can be found in [2].

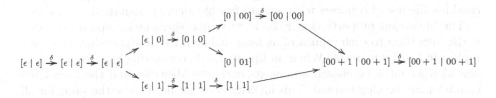

Fig. 1. A set of partially ordered events with version stamps

2.1 Pollution of the Namespace

Exercising the version stamp mechanism, we have observed an undesired growth of version stamps under a simple pattern of operation. This problem is illustrated in Figure 2, which shows the identity component in a scenario where three replicas are created and then we repeat a pattern in which two of them are joined and forked again, while alternating replicas.

Although we end up with only three replicas, the identity components are much more complex than in the configuration after the first two forks (with the same number of replicas). In this scenario, the Version Stamp mechanism leads to a overly refined namespace which cannot be simplified upon these interleaving joins.

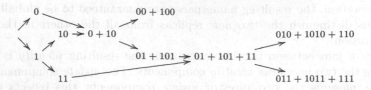

Fig. 2. Pollution in the identity component of version stamps

This growth gets worse every time this operation pattern occurs and recent experimentation does indicate that this can be a fairly common case in practical usage scenarios. Furthermore, when an update occurs, the identity component is copied to the update component, thus aggravating this problem.

This degeneration of the namespace does not imply that the version stamp mechanism is incorrect but that it may consume an unreasonable amount of space. As a result, this growth pattern of version vectors can severely affect its practical application.

3   Dynamic Map Clocks

Dynamic Map Clocks imports from version stamps the basic features that support autonomous identity management. Noticing that the lack of counters, on version stamps, impose important restrictions on identity management that ultimately contributed to the identified growth problem, dynamic map clocks will combine the use of counters with a more flexible identity management scheme.

The important property that rules identity management for update tracking is the allocation to each replica of at least one identity that is exclusive to that replica in a given moment. When an update needs registering in a given replica, one identity must be chosen among its exclusive identities and the associated counter must be incremented. This identity does not need to be the same for all updates in that replica and replicas can handover identities to other replicas.

As a consequence of these insights it is easy to conceive a scheme where replicas can fork by either specializing a binary identity (forking 010 would derive 0100 and 0101) or by partitioning controlled identities that were obtained upon joins (forking 010+10+111 could derive 010+111 and 10). This is the basic mechanism that supports dynamic map clocks and Figure 3 shows a run that illustrates this. More complex rules are enforced when handling joins and setting counters upon joins.

3.1   Non-pollution of the Namespace

Considering the namespace pollution problem that was present on version stamps, it is easy to verify that dynamic map clocks are much more flexible on the handling of names. Figure 4 shows how the run that depicted a name pollution pattern in version stamps is easily handled by this mechanism.

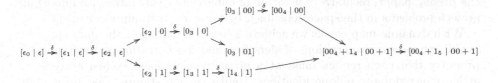

Fig. 3. A set of partially ordered events with dynamic map clocks

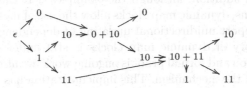

Fig. 4. Non-pollution of the identity component in the dynamic map clock mechanism

In some way, dynamic map clocks try to ally the innovative management of identities that stems from version stamps with the benefits of classical counters and their synthetic encoding of updates.

4 Discussion

Handling replication in highly decentralized systems and large scale settings – in number of nodes, geographical distance or communication latency between nodes– often implies the use of optimistic techniques in order to improve availability. In these scenarios, replicas are allowed to diverge from a consistent global state but reconciliation procedures and update propagation strategies are put in place so that consistency can eventually be restored. All these operations must rely on some dependency tracking mechanism in order to infer the causal relations between replica states. As mentioned before, the standard version vector mechanism assumes a consistent management of replicas names.

In decentralized distributed systems that face partitions, large membership changes under churn and disconnected operation, one can no longer rely on the assumption that globally unique names are available. A way of approaching these settings is to avoid determinism altogether and rely on probabilistic approaches such as generating random replica names, and assume some risk of name collisions [3], or using sets of hashes of replica state to detect updates [4] once again assuming some risk of collision. If reliability cannot be compromised, as is often the case, only a deterministic approach is appropriate. Determinism can only be obtained by recursive generation of names, the approach developed and formalized in our previous work on version stamps. However, as we have shown in the present paper, recursive generation of names can easily introduce important growth problems in the space consumed by the version stamp mechanism.

With dynamic map clocks we achieve a better handling of the data space by avoiding unnecessary partitions of identifiers and concentrating on the important property that each replica must at a given moment have exclusive access to at least one globally unique identifier. Unlike version stamps, that do not use counters, dynamic map clocks are, in a sense, a hybrid mechanism that also relies on counters for registering updates. This usage of counters leads to important savings in size. In addition, although the examples here have only shown runs with join operations, dynamic map clocks allow the use of messages when sending metadata and support unidirectional updating of dependency information.

While the theory of dynamic map clocks is still under development and a proper formalization and formal proof is ongoing work, we already have a running implementation of the mechanism. This implementation has been tested on long random runs under various numbers of replicas and always evaluated as correct. This evaluation is done by contrasting the causal pre-order that the mechanism derives with the equivalent pre-order derived by causal histories. Causal histories are implemented by assuming global knowledge and adding unique update events to a set of events in each replica and relating this sets by set inclusion (see [2,8] for more details on causal histories). We can comment, from our experience, that incorrect mechanisms typically fail these checks after a small number of steps and do not stay correct in long random runs.

Another important property of dynamic map clocks, not present in version stamps, is their potential use as substitutes for vector clocks in autonomous decentralized settings. Vector clocks, that are at the core of causal message delivery protocols and distributed debugging, also rely on globally unique names thus facing the same problems of version vectors.

References

1. Almeida, P.S., Baquero, C., Fonte, V.: Panasync: Dependency tracking among file copies. In: Guedes, P. (ed.) Ninth ACM SIGOPS European Workshop, DIKU - University of Copenhagen, pp. 7–12 (2000)
2. Almeida, P.S., Baquero, C., Fonte, V.: Version stamps – decentralized version vectors. In: Proceedings of the 22nd International Conference on Distributed Computing Systems (ICDCS), pp. 544–551. IEEE Computer Society, Los Alamitos (2002)
3. Heuer, G.A.: Estimation in a certain probability problem. The American Mathematical Monthly 66(8), 704–706 (1959)
4. Kang, B.B., Wilensky, R., Kubiatowicz, J.: The hash history approach for reconciling mutual inconsistency. In: Proceedings of the 23nd International Conference on Distributed Computing Systems (ICDCS), pp. 670–677. IEEE Computer Society, Los Alamitos (2003)
5. Parker, D.S., Popek, G., Rudisin, G., Stoughton, A., Walker, B., Walton, E., Chow, J., Edwards, D., Kiser, S., Kline, C.: Detection of mutual inconsistency in distributed systems. Transactions on Software Engineering 9(3), 240–246 (1983)

6. Ratner, D., Reiher, P., Popek, G.: Dynamic version vector maintenance. Technical Report CSD-970022, Department of Computer Science, University of California, Los Angeles (1997)
7. Saito, Y., Shapiro, M.: Optimistic replication. Technical Report MSR-TR-2003-60, Microsoft Research (2003)
8. Schwarz, R., Mattern, F.: Detecting causal relationships in distributed computations: In search of the holy grail. Distributed Computing 3(7), 149–174 (1994)

Self-healing in Binomial Graph Networks

Thara Angskun[1], George Bosilca[1], and Jack Dongarra[1,2,3]

[1] Department of Computer Science, The University of Tennessee, Knoxville, USA
[2] Computer Science and Mathematics Division, Oak Ridge National Laboratory, USA
[3] Computer Science and Mathematics Schools, The University of Manchester, UK
{angskun,bosilca,dongarra}@eecs.utk.edu

Abstract. The number of processors embedded in high performance computing platforms is growing daily to solve larger and more complex problems. However, as the number of components increases, so does the probability of failure. The logical network topologies must also support the fault-tolerant capability in such dynamic environments. This paper presents a self-healing mechanism to improve the fault-tolerant capability of a Binomial graph (BMG) network. The self-healing mechanism protects BMG from network bisection and helps maintain optimal routing even in failure circumstances. The experimental results show that self-healing with an adaptive method significantly reduces the overhead from reconstructing the networks.

1   Introduction

Recently, several high performance computing platforms have been installed with more than 10,000 CPUs, such as Blue-Gene/L at LLNL, BGW at IBM and Columbia at NASA [1]. However, as the number of components increases, so does the probability of failure. To satisfy the requirements of such a dynamic environment (where the available number of resources is fluctuating), a scalable and fault-tolerant communication framework is needed. The communication framework is important for both runtime environments of MPI libraries and the MPI libraries themselves. In general, the communication framework is based on a logical network topology.

There are several existing logical network topologies that can be used in high performance computing (HPC). Whereas a fully connected topology is good in terms of fault-tolerance and point-to-point performance, it does not exhibit any scalable properties due to its high degree. The bidirectional ring topology is more scalable, but it is hardly fault-tolerant. Hypercube [2] and its variants [3,4,5,6,7,8,9,10], FPCN [11], de Bruijn [12] and its variants [13,14], Kautz [15] and ShuffleNet [16] have a number of node restrictions. They are either not scalable or not fault-tolerant. The Manhattan Street Network (2D Torus) [17] is more flexible (no restriction in number of nodes) than Hypercube-like topologies. However, it has a much higher average hop-distance. Variants of k-ary tree, such as Hierarchical Clique (HiC) [18] and k-ary sibling tree (Hypertree [19]) used in SFTP [20,21], are scalable and fault-tolerant. They are good

R. Meersman, Z. Tari, P. Herrero et al. (Eds.): OTM 2007 Ws, Part II, LNCS 4806, pp. 1032–1041, 2007.
© Springer-Verlag Berlin Heidelberg 2007

for both unicast and broadcast messages. However, all nodes in their topologies are not equal (i.e. the resulting graph is not regular). Topologies, used in structured, peer-to-peer networking based on distributed hash tables such as CAN [22], Chord [23], SkipNet [24], Kademlia [25], Viceroy [26], Pastry [27] and Tapestry [28], are also scalable and fault-tolerant. They were designed for resource discovery in highly dynamic environments. Hence, they may not be efficiently used in HPC owing to the overhead for managing highly dynamic applications.

Binomial graph (BMG) [29] provides desirable topological properties in terms of both scalability and fault-tolerance for high performance computing such as regular graph (every node has the same degree), low diameter, low cost factor, low message traffic density, low fault-diameter, strongly resilient and good optimal probability in failure cases.

BMG is an undirected graph $G:=(V,E)$ where V is a set of nodes (vertices); $|V| = N$; and E is a set of links (edges). Each node i, where $i \in V$ and $i=0,1,...,N-1$, has links to a set of nodes U, where $U=\{i\pm1,i\pm2,...,\pm2^k|2^k \leq N\}$ in circular space, i.e., node i has links to a set of clockwise (CW) nodes $\{(i+1) \bmod N, (i+2) \bmod N,..., (i+2^k) \bmod N \mid 2^k \leq N\}$ and a set of counterclockwise (CCW) nodes $\{(N+i-1) \bmod N, (N+i-2) \bmod N,..., (N+i-2^k) \bmod N \mid 2^k \leq N\}$. The structure of BMG can also be classified in the Circulant graph family[1]. A Circulant graph with N nodes and jumps $j_1, j_2, ..., j_m$ is a graph in which each node i, $0 \leq i \leq n - 1$, is adjacent to all the vertices $i \pm j_k \bmod N$, where $1 \leq k \leq m$. BMG is a Circulant graph where j_k is the power of 2 that $\leq N$. For a BMG size N (having N nodes), each node has a degree δ (the number of neighbors) as shown in Equation (1).

$$\delta = \begin{cases} (2 \times \lceil \log_2 N \rceil) - 1 & \text{For } N = 2^k, \text{where } k \in \mathbb{N} \\ (2 \times \lceil \log_2 N \rceil) - 2 & \text{For } N = 2^k + 2^j, \text{where } k,j \in \mathbb{N} \land k \neq j \\ 2 \times \lceil \log_2 N \rceil & \text{Otherwise} \end{cases} \quad (1)$$

Fig. 1(a) illustrates an example of a 12-node binomial graph. The lines represent all connections in the network. The other way to look at the binomial graph is that it is a topology, which is constructed from merging all necessary links in order to create binomial trees from each node in the graph. Fig. 1(b) shows an example of a binomial tree when node 0 is the root node. The arrows point in the direction of the leaf nodes.

This paper presents a reliability analysis and self-healing capability of the Binomial graph (BMG) networks. The reliability analysis is done using a discrete event simulation. The result indicates a potential of network bisection when the number of failed nodes is more than or equal to the degree δ (i.e. BMG is $\delta - 1$ node fault-tolerance). The self-healing BMG, introduced in this paper, helps protect BMG from network bisection and maintain optimal routing even in failure circumstances. The structure of this paper is as follows: Section 2 describes the reliability analysis of the BMG. The self-healing BMG algorithm

---

[1] The family of Circulant graphs includes fully connected, ring, Recursive Circulants [30] and Midimew [31].

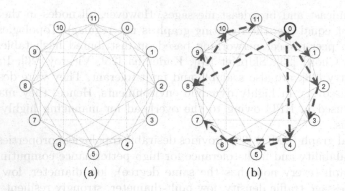

Fig. 1. Binomial graph structure. (a) 12-node BMG. (b) Binomial tree from node 0.

is discussed in section 3. Section 4 presents the experimental evaluations, followed by conclusions and future work in section 5.

2   Reliability Analysis of BMG

The reliability analysis is based on a discrete event simulation. This section presents the description of simulation as well as results from the simulation.

2.1   BMG Reliability Simulation

The reliability of BMG is defined as its ability to maintain an operation over a period of time t, i.e., the reliability $R(t)= Pr$(the network is operational in $[0,t]$). The BMG is *"operational"*, if it can successfully deliver messages from any source to any alive destination even in the case where some intermediary nodes have failed. Due to the fact that multicast and broadcast messages in failure circumstances rely on unicast messages [29], this simulation focuses on reliability of the unicast routing.

The unicast messages in BMG size N are simulated by sending messages from all possible sources (S) to all possible destinations (D), where $S \neq D$. Fortunately, BMG is a vertex symmetric graph (i.e. a graph which looks the same viewed from any node). Thus, the simulation cases for normal circumstances can be reduced from $N \times (N-1)$ to $(N-1)$ cases. During the failure circumstance, the F failed nodes are obtained from combinations of all possible N nodes, i.e., $\binom{N}{F}$ where the source and destination nodes are not one of the failed nodes. Hence, there are $\binom{N-2}{F}$ simulation cases for each unicast transmission.

The total number of simulation cases of unicast message transmission (T) for N nodes of the BMG with F failed nodes is given by

$$T = (N-1) \times \binom{N-2}{F} = \frac{(N-1)!}{(N-F-2)!F!}.$$

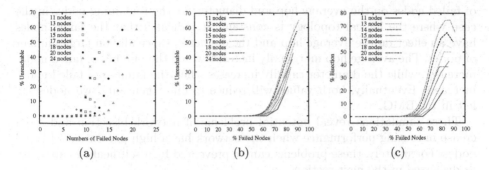

Fig. 2. Percentage of unreachable and bisection in failure circumstances

The transmission of unicast messages is considered successful if the messages can reach the destination. This means that the network can deliver messages even in the presence of failures in the routing path. If there are U unreachable cases due to network bisection, the percentage of unreachable cases (P) is defined by

$$P = (\frac{U}{T}) \times 100.$$

2.2   Simulation Results and Analysis

The results were obtained by simulating all possible cases as described in the previous section. The simulation results were obtained from BMG networks of size 11, 13, 14, 15, 17, 18, 20 and 24. All these BMG topologies have the same degree $(\delta = 8)$. Fig. 2(a) illustrates that destinations become unreachable when the number of failed nodes is more than or equal to eight. However, a percentage of unreachable cases is significant when there are more than 50% of failed nodes as shown in Fig. 2(b) because the percentage of network bisection rapidly increases when the number of failed nodes is more than 50% as shown in Fig. 2(c). Not only did the failed nodes affect the reliability of the BMG, but they also affect the average hop and the diameter of the BMG. Fig. 3 illustrates the effect

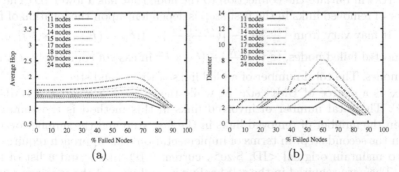

Fig. 3. Average hop and diameter in failure circumstances. (a) Average hop. (b) Diameter.

of failed nodes on the average hop and diameter for the remaining nodes in the case where the logical topology is constant. It indicates that the failed nodes have an effect on the average hop and the diameter, especially on large number of nodes. The average hop marginally increases when the number of failed nodes increases, while the diameter rapidly increases when the number of failed nodes increases. Eventually, both values will reduce to one when only two nodes are left in the BMG.

These simulations reveal potential problems of network bisection and a decrease in routing performance when the network has a high percentage of failed nodes. Fortunately, these problems can be prevented by a self-healing capability as discussed in the next section.

3   Self-healing Capability of BMG

This section presents the self-healing capabilities of BMG as a solution to prevent potential problems of network bisection and a decrease in routing performance when the network has a high percentage of failed nodes. Section 3.1 describes methods that are used to recover the BMG, while section 3.2 discusses the appropriate time to perform the recovery.

3.1   Self-healing Methods

There are two methods presented in this section. The first approach is called the *naive* method. This method destroys the original network and reconstructs the BMG with the remaining nodes. The second method is called *adaptive* method. It only destroys and reconstructs the links that are different between the original BMG and the BMG after excluding of all failed nodes.

Naive Method. This is the simplest method to reconstruct the BMG topology. Suppose there are F nodes in BMG size N. There are two steps involved in this method. The first step removes all existing links. The second step establishes all connections of BMG size $N - F$. For each link in the BMG, a node that has a higher ID will initiate the connection to the node that has a lower ID. The total number of removed links in the first step is dependent upon the location of failed nodes. It may vary from $\left[\frac{\delta_N \times N}{2}\right] - \left[\left(\frac{(F-1) \times F}{2}\right) + \left((\delta_N - (F-1)) \times F\right)\right]$ in case of connected failed nodes, to $\left(\frac{\delta_N \times N}{2}\right) - (\delta_N \times F)$ in case of completely separated failed nodes. The total number of added links of the second step is $\frac{\delta_{N-F} \times (N-F)}{2}$. The δ_N is a degree of BMG size N, while the δ_{N-F} is a degree of BMG size $N - F$. The total number of involved links in this method is the summation between the number of removed links in the first step and the number of added links in the second step. In terms of implementation, this approach requires each node to maintain original <ID, Size>, current <ID, Size> and a list of failed nodes. They are required in the self-healing procedure and the message routing.

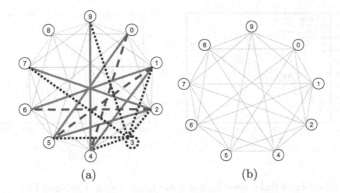

(a)   (b)

Fig. 4. Self-healing BMG using adaptive method. (a) BMG when node ID 3 failed. (b) BMG after the self-healing procedures.

Adaptive Method. This method only removes and adds the links that are different between BMG size N and BMG size $N - F$. Fig. 4(a) illustrates a self-healing procedure of a 10-node BMG using the adaptive method when node 3 failed. When the node 3 failed, all connections that associate with node 3 (represented with dot lines) will be disconnected. A neighbor of node 3 may start the self-healing procedure (dependent upon the frequency as discussed in section 3.2) by broadcasting messages [2] to all nodes. Then each node calculates the different links between BMG size N and BMG size $N-F$. For each connection that needs to be added (represented with thick solid lines), the higher IDs initiate the connections to the lower IDs. All the unnecessary links (represented with dash lines) will be removed. Fig. 4(b) illustrates the BMG topology after the self-healing procedure is completed. All node IDs represented in this figure are IDs according to the original BMG. This method also requires each node to maintain original <ID, Size>, current <ID, Size>, and the list of failed nodes. They are used in the self-healing procedure and the message routing.

A function for calculating added links and removed links of the adaptive method is shown in Algorithm 1. The input variable of this function consists of **N**, **myID$_{org}$**, **dead** and **N$_{dead}$**. The **N** is the size of the original BMG. The **myID$_{org}$** is a node ID in the original BMG. The **dead** is a list of dead nodes. The size of the dead node list is equal to **N$_{dead}$**. There are six steps in this function. The first step is to calculate a new node ID from **myID$_{org}$**. The new node ID is an ID in the BMG that excludes all failed nodes. All node IDs after excluding all failed nodes are continuous. The second step is calculating all neighbor IDs of **myID$_{org}$** in BMG size **N**. The third step is calculating all neighbor IDs of the new node ID (outputs of the first step) in BMG size **n − n$_{dead}$**. The fourth step is to convert outputs of the third step into IDs according to the original BMG.

---

[2] In case of failure, a broadcast message is encapsulated into a multicast message, and then the message is sent from a parent of the failed node to its children in the binomial spanning tree. The children will de-capsulate the multicast message and continue to forward the initial broadcast message.

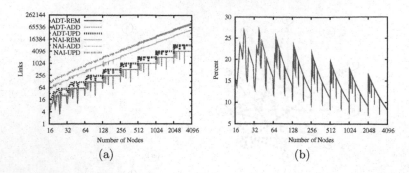

(a)   (b)

Fig. 5. (a) Number of links used in both methods. (b) % Updated link ratio between adaptive and naive methods.

The fifth step is to calculate the **added** links. All these links are linked to all outputs of the fourth step that do not exist in outputs of the second step. On the other hand, the last step is to calculate **removed** links from outputs of the second step that do not exist in the fourth step. The added and removed links are results of this function.

Algorithm 1. Find added and removed links when some nodes failed.

Require: $N \in \mathbb{N}, 1 \leq \mathbf{myID_{org}} \leq N, \{\mathbf{dead}|\forall d \in dead.1 \leq d \leq N\}, \mathbf{N_{dead}} = |dead|$

1: $myID_{new} \Leftarrow$ Convert my original IDs to my new IDs.
2: $NeighborID_{org} \Leftarrow$ Get neighbor IDs of $myID_{org}$ in BMG size N.
3: $NeighborID_{new} \Leftarrow$ Get neighbor IDs of $myID_{new}$ in BMG size $N\text{-}N_{dead}$.
4: Convert all IDs in $NeighborID_{new}$ to IDs in the original BMG.
5: **Added** $\Leftarrow \{\forall a \in Added | a \in NeighborID_{new} \wedge a \notin NeighborID_{org}\}$
6: **Removed** $\Leftarrow \{\forall r \in Removed | r \in NeighborID_{org} \wedge r \notin NeighborID_{new}\}$

Fig. 5(a) illustrates the number of links involved in the self-healing procedures of both methods when there is a failed node. The numbers of added and removed links of each method are roughly the same. The numbers of updated links required by the adaptive method are 10%-30% of the naive method (as shown in Fig. 5(b)), i.e., the adaptive method reduces the overhead from reconstructing the networks up to 90%, especially in large and very large networks.

3.2 Self-healing Frequency

The frequency of recovering the network may vary from recovering when a node in the BMG failed to recovering when a node had $\delta - 1$ failed neighbors. The frequency is a trade-off between recovery time and an overhead from non-optimal routing. The BMG may be configured such that a node initiates the self-healing procedure when the overhead of routing (in terms of hop number) is more than a threshold. Both the number of failed nodes and the number of hops may also

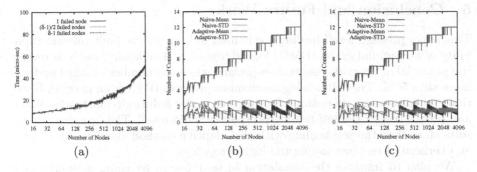

Fig. 6. Experimental Results. (a) Performance of updated link calculation algorithm. (b) Number of added connection per node. (c) Number of removed connection per node.

be used as a threshold to start the recovery procedures. Independent on the threshold, the self-healing procedure must be started before a node had δ failed neighbors. Otherwise the BMG network will become bisectional.

4   Experimental Results

This section presents the evaluation of the algorithm that calculates the added and removed links for the adaptive method (as described in section 3.1). The number of connections reconstruction that are required by naive and adaptive methods are also evaluated.

The updated-link calculation algorithm has been evaluated with several numbers of failed nodes in BMG sized between 16 and 4096 nodes. The experiments have been conducted on an AMD Athlon™ 64 Processor 3500+ 2.2 GHz machine with 1 GB of main memory, running on Linux kernel 2.6.15. Elapsed time of the calling function, that calculates the added and removed links, has been measured. Fig. 6(a) illustrates that the algorithm scales quite well to the number of nodes. The number of failed nodes has marginally affected to the performance of this algorithm.

The number of added and removed connections affects the performance of self-healing procedure. The performance is also dependent on other platform-dependent factors such as time to establish a connection (e.g. three-way handshake in TCP). Fig. 6(b) and Fig. 6(c) present the number of added and removed connections per node in both naive and adaptive methods. The average (mean) number of connections per node for the naive method is significantly higher than for the adaptive method because the number of links that have to be updated by the naive method is larger. The standard deviation (STD) of both graphs indicates that the adaptive method has more load balance than the naive method. Due to the fact that all nodes establish the connection simultaneously, the adaptive method benefit from more parallelism than the naive approach.

5   Conclusion and Future Work

This paper presents a self-healing mechanism to improve the fault-tolerant capability of a Binomial graph (BMG) logical topology. A reliability analysis reveals the potential of network bisection, especially when the number of failed nodes is more than 50%. The self-healing mechanism protects BMG from network bisection and helps maintain optimal routing even under failure circumstances. When and how to trigger the self-healing procedures is discussed. The experimental results show that the self-healing with the adaptive method significantly reduces an overhead of re-constructing the logical topology.

We plan to improve the simulation in near future by using approximation methods (e.g. Monte Carlo) to simulate larger size of BMG. Over the longer term, we hope that BMG will become the default logical topology of the runtime environments within the FT-MPI and Open MPI libraries.

References

1. Dongarra, J.J., Meuer, H., Strohmaier, E.: TOP500 supercomputer sites. Supercomputer 13, 89–120 (1997)
2. Saad, Y., Schultz, M.H.: Topological properties of hypercubes. IEEE Transactions on Computers 37, 867–872 (1988)
3. Banerjee, S., Sarkar, D.: Hypercube connected rings: A scalable and fault-tolerant logical topology for optical networks 24, 1060–1079 (2001)
4. Malluhi, Q., Bayoumi, M.: The hierarchical hypercube: A new interconnection topology for massively parallel systems. IEEE Transactions on Parallel and Distributed Systems 05, 17–30 (1994)
5. El-Amawy, A., Latifi, S.: Properties and performance of folded hypercubes. IEEE Transactions on Parallel and Distributed Systems 2, 31–42 (1991)
6. Kumar, J.M., Patnaik, L.M.: Extended hypercube: A hierarchical interconnection network of hypercubes. IEEE Transactions on Parallel and Distributed Systems 3, 45–57 (1992)
7. Tzeng, N.F., Wei, S.: Enhanced hypercubes. IEEE Transactions on Computers 40, 284–294 (1991)
8. Preparata, F.P., Vuillemin, J.: The cube-connected cycles: a versatile network for parallel computation. Commun. ACM 24, 300–309 (1981)
9. Louri, A., Neocleous, C.: A spanning bus connected hypercube: A new scalable optical interconnection network for multiprocessors and massively parallel systems. IEEE/OSA Journal of Lightwave Technology 15, 1241–1252 (1997)
10. Louri, A., Sung, H.: An optical multi-mesh hypercube: A scalable optical interconnection network for massively parallel computing. Journal of Lightware Technology 12, 704–716 (1994)
11. Ohring, S., Das, S.K.: Folded petersen cube networks: New competitors for the hypercubes. IEEE Transactions on Parallel and Distributed Systems 7, 151–168 (1996)
12. Sivarajan, K.N., Ramaswami, R.: Lightwave networks based on de bruijn graphs. IEEE/ACM Trans. Netw. 2, 70–79 (1994)
13. Ganesan, E., Pradhan, D.K.: The hyper-debruijn networks: Scalable versatile architecture. IEEE Transactions on Parallel and Distributed Systems 04, 962–978 (1993)

14. Chen, C., Agrawal, D.P., Burke, J.R.: dbcube: A new class of hierarchical multi-processor interconnection networks with area efficient layout. IEEE Trans. Parallel Distrib. Syst. 4, 1332–1344 (1993)
15. Panchapakesan, G., Sengupta, A.: On a lightwave network topology using kautz digraphs. IEEE Transactions on Computers 48, 1131–1138 (1999)
16. Karol, M.J.: Optical interconnection using shufflenet multihop networks in multi-connected ring topologies. In: SIGCOMM 1988: Symposium proceedings on Communications architectures and protocols, pp. 25–34. ACM Press, New York (1988)
17. Maxemchuck, N.F.: Regular mesh topologies in local and metropolitan area networks. AT&T Technical Journal 64, 1659–1685 (1985)
18. Campbell, S., Kumar, M., Olariu, S.: The hierarchical cliques interconnection network. Journal of Parallel and Distributed Computing 64, 16–28 (2004)
19. Goodman, J.R., Sequin, C.H.: Hypertree: A multiprocessor interconnection topology. IEEE Transactions on Computers 30, 923–933 (1981)
20. Angskun, T., Fagg, G.E., Bosilca, G., Pješivac-Grbović, J., Dongarra, J.: Scalable fault tolerant protocol for parallel runtime environments. In: Mohr, B., Träff, J.L., Worringen, J., Dongarra, J. (eds.) Recent Advances in Parallel Virtual Machine and Message Passing Interface. LNCS, vol. 4192, pp. 141–149. Springer, Heidelberg (2006)
21. Angskun, T., Fagg, G.E., Bosilca, G., Pješivac-Grbović, J., Dongarra, J.J.: Self-healing network for scalable fault tolerant runtime environments. In: Proceedings of 6th Austrian-Hungarian workshop on distributed and parallel systems, Innsbruck, Austria, Springer, Heidelberg (2006)
22. Ratnasamy, S., Francis, P., Handley, M., Karp, R., Shenker, S.: A scalable content addressable network. Technical Report TR-00-010, Berkeley, CA (2000)
23. Stoica, I., Morris, R., Karger, D., Kaashoek, F., Balakrishnan, H.: Chord: A scalable Peer-To-Peer lookup service for internet applications. In: Proceedings of the 2001 ACM SIGCOMM Conference, pp. 149–160 (2001)
24. Harvey, N.J.A., Jones, M.B., Marvin Theimer, S.S., Wolman, A.: Skipnet: A scalable overlay network with practical locality properties. In: USENIX Symposium on Internet Technologies and Systems. proceedings of the 4th USENIX Symposium on Internet Technol ogies and Systems (USITS 2003), Seattle, WA, USA, pp. 113–126 (2003)
25. Maymounkov, P., Mazieres, D.: Kademlia: A peer-to-peer information system based on the xor metric. In: Druschel, P., Kaashoek, M.F., Rowstron, A. (eds.) IPTPS 2002. LNCS, vol. 2429, Springer, Heidelberg (2002)
26. Malkhi, D., Naor, M., Ratajczak, D.R.: Viceroy: A scalable and dynamic emulation of the butterfly. In: Proceedings of the 21st ACM Symposium on Principles of Distributed Comput ing, pp. 183–192. ACM Press, New York (2002)
27. Rowstron, A., Druschel, P.: Pastry: Scalable, decentralized object location, and routing for large-scale peer-to-peer systems. In: Guerraoui, R. (ed.) Middleware 2001. LNCS, vol. 2218, pp. 329–350. Springer, Heidelberg (2001)
28. Zhao, B.Y., Kubiatowicz, J.D., Joseph, A.D.: Tapestry: An infrastructure for fault-tolerant wide-area location and routing. Technical Report UCB/CSD-01-1141, UC Berkeley (2001)
29. Angskun, T., Bosilca, G., Dongarra, J.: Binomial graph: A scalable and fault-tolerant logical network topology. In: ISPA 2007. LNCS, pp. 471–482. Springer, Heidelberg (2007)
30. Bermond, J.C., Comellas, F., Hsu, D.F.: Distributed loop computer networks: A survey. Journal of Parallel and Distributed Computing 24, 2–10 (1995)
31. Beivide, R., Herrada, E., Balcázar, J.L., Arruabarrena, A.: Optimal distance networks of low degree for parallel computers. IEEE Trans. Comput. 40, 1109–1124 (1991)

A Self-stabilizing Distributed Algorithm for Resolving Conflicts

Brahim Hamid, Mohamed Mosbah, and Akka Zemmari

LaBRI-University of Bordeaux-1
351, cours de la libération
Talence, 33405, France
{hamid,mosbah,zemmari}@labri.fr

Abstract. In this work, we investigate the problem of resolving conflicts in a distributed environment using only local knowledge. The contribution of this paper is twofold. First, we present a self-stabilizing algorithm to deal with this problem. Self-stabilizing algorithms protect against transient failures. The second result gives a particular implementation and analysis based on probabilistic procedures. Thus, the stabilization time is computed in terms of computation steps, then approximated according to the needed synchronizations.

1 Introduction

Resolving conflicts problems in distributed systems has been an active area for more than three decades. In fact, when a group of processes (or processors), composing the system, needs access or uses some resource that cannot be used simultaneously by more than one processor, for example some output device, the system must ensure some properties to manage such conflicts between processors. Here, we study this problem in a distributed environment. A distributed system is represented as a connected, undirected graph G denoted by (V, E). Nodes V represent processes and edges E represent bidirectional communication links. Resources subject to conflicts are called *critical resources* and by extension the use of such resources is named *critical section*.

The Problem. We consider the case when a process needs to have access to *all* its peripheral resources to do any computation. This problem is named *drinking philosophers problem* and was introduced by Chandy and Misra in [1] as a generalization of the *dining philosophers problem* [2], one of the famous paradigms in distributed computing. In a generalization of the drinking philosophers problem, a philosopher can choose one or many bottles and not only all of them [3,4]. In this paper, we consider a network of processes sharing a set of resources. The resources are placed on the edges of the underlying graph. Any algorithm which solves this problem has to ensure the following properties :

- no shared resource is accessed by two processes at the same time, that is, the algorithm ensures the *mutual exclusion* on shared resources,
- if two processes p_1 and p_2 do not share a resource, and hence are not adjacent in the underlying graph, then they can access their resources independently,

R. Meersman, Z. Tari, P. Herrero et al. (Eds.): OTM 2007 Ws, Part II, LNCS 4806, pp. 1042–1051, 2007.
© Springer-Verlag Berlin Heidelberg 2007

and possibly at the same time, that is, the algorithm ensures the *concurrency* property,

- if a resource is asked by two processes p_1 and p_2 and if p_1 formulates its request before p_2 then it must enter the resource before p_2, this is called the *ordering* property,
- if a process p asks to have access to all the resources it needs, p eventually obtain this access, this is called *liveness* property.

Moreover, since fault-tolerance is an important field in distributed computing systems it is important to design robust algorithms working in unreliable networks. In this work we consider transients failures, in the sense that they occur only in the beginning and their duration is limited. As self-stabilization is a suitable approach to deal with such a system, we deal with the problem of designing self-stabilizing algorithm to resolve conflicts. In fact, self-stabilizing [5] algorithms ensure that starting from any configurations, including such with inconsistent data, the system will automatically recover to reach a correct configuration in a finite time.

Previous Works. Several papers investigate the mutual exclusion problem. In [6], the author has given a simple solution to this problem in spite of failures in a complete graph. That is, a validating and checking approach. The probabilistic (or randomized) approach had been introduced in [7] for resolving the dining philosophers problem. This algorithm is the base of many studies such as that of [8]. Randomization is used to cope with the case of asynchronous and anonymous networks [9]. The work is closest to our is [10]. The main difference between our work and [10] is: Here we present a self-stabilizing [5] solution. In [11], we have proposed a formal framework to deal with self-stabilization using means of local computations. This method deals with a predefined set of inconsistencies. Thus, it may capture some of the possible illegal configurations of the system.

Our Contribution. The algorithm we present is based on graph relabeling system and local computations. Any vertex of the graph is labeled by a label which represents its state, the algorithm is encoded by a set of rules: According to its state and to those of its neighbors, a vertex changes its label. The paper has two major contributions. First, we propose a self-stabilizing algorithm to deal with resolving conflicts in a distributed environment. Our solution is local since each computation step involves only local knowledge. Our algorithm stabilizes in $O(N)$ time and uses $O(d(G) + dia(G))$ bits per node where $d(G)$ is the maximum degree of the graph G modeling the network, $dia(G)$ is its diameter and N is the number of its nodes. The second result shows an implementation of the algorithm using synchronization based on randomized procedures.

The rest of the paper is organized as follows. Section 2 presents some preliminaries needed to study the problem. In Section 3, we present our distributed self-stabilizing algorithm to deal with the conflicts. Section 4 presents an implementation of our algorithm using randomized synchronizations. We end the paper with Section 5 which gives conclusions and perspectives.

2 Preliminaries

2.1 Graphs

In each class of graphs we can use the following structure (V, E) to design a
graph G, where V is the set of nodes and $E \subseteq V^2$ is the set of edges. We use
sometimes the notations V_G, E_G to denote respectively V, E where $G = (V, E)$.
Let $(u, v) \in E$, we say that u is a neighbor of v. A path p in G is a sequence
$(v_0, ..., v_l)$ of nodes such that for each $0 \leq i < l$, $(v_i, v_{i+1}) \in E$. The integer l is
called the length of p. The set of neighbors of node u is denoted by $N(u) = \{v \in
V / (u, v) \in E\}$. The degree of a node v, denoted $d(v)$, is the number of neighbors
of v. The degree of G is $d(G) = max\{d(v)$ such that $v \in V\}$. A ball centered on
u with radius l is the set $B_l(u) = \{u\} \cup \{v_j \in V /$ there exists a path $(v_0, ..., v_j)$
in G with $v_0 = u$ and $j \leq l\}$. For simplicity, when $l = 1$ we write $B(v)$ to denote
$B_1(v)$. We use $dist(u, v)$ to denote the shortest path in G between u and v.
Then, the diameter of a graph G, denoted by $dia(G)$, is the longest distance
among all pairs of nodes in G. Formally, $dia(G) = max_{u,v \in V_G} dist(u, v)$. We say
that the graph $G' = (V', E')$ is a subgraph of the graph $G = (V, E)$, if $V' \subseteq V$
and $E' \subseteq E$.

2.2 Graph Relabeling Systems

Let $\mathcal{L}$ be a set whose elements are called *labels*. An $\mathcal{L}$-*labeled graph* is a pair
(G, λ) where $G = (V, E)$ is a connected graph and λ a mapping from $V \cup E$ to
$\mathcal{L}$. The labeled graph (H, λ') is a sub-labeled graph of (G, λ), if H is a subgraph
of G and λ' is the restriction of λ to H. A *graph relabeling rule* is a triplet
$R = (G_R, \lambda_R, \lambda'_R)$ such that (G_R, λ_R) and (G_R, λ'_R) are two labeled graphs.

A *graph relabeling system* (*GRS* for short) is a triplet $\mathcal{R} = (\mathcal{L}, \mathcal{I}, P)$ where
$\mathcal{L}$ is a set of labels, $\mathcal{I}$ a subset of $\mathcal{L}$ called the *initial labels* and P a finite set
of relabeling rules. An $\mathcal{R}$-*relabeling step* is a 5-tuple $(G, \lambda, R, \phi, \lambda')$ such that
R is a relabeling rule in P and ϕ is both an occurrence of (G_R, λ_R) in (G, λ)
and an occurrence of (G_R, λ'_R) in (G, λ'). Such a relabeling step is denoted by
$(G, \lambda) \rightarrow_{R, \phi} (G, \lambda')$. We use the notation $(G, \lambda_0) \xrightarrow[\mathcal{R}]{p} (G, \lambda_n)$ to say that
there exists a sequence of rule applications of length p starting from (G, λ_0) to
(G, λ_n). If the number p is not know, we will use $(G, \lambda_0) \xrightarrow[\mathcal{R}]{*} (G, \lambda_n)$.

Therefore, if an algorithm $\mathcal{A}$ is encoded by a relabeling system denoted $\mathcal{R}$,
an $\mathcal{R}$-*relabeling step* corresponds to an $\mathcal{A}$-*computing step*. A computation in G
is finished if no relabeling rule can be applied on the current labeling. In this
case, the corresponding labeled graph (G, λ') is a result of the algorithm.

In this work, we consider only relabeling in balls of radius 1. That is, only the
center of the ball is able to change its label according to rules depending only on
labels of all nodes composing this ball. We assume that each node distinguishes
its neighbors and knows their labels. In the sequel we use the set $N(v)$ to denote
the set of (locally) ordered immediate neighbors of v which is an input data.

2.3   Self-stabilization

Transient faults cause processes to change their states yielding abnormal local configurations. A self-stabilizing system will be able to repair such a fault by eventually stabilizing to a correct global configuration without restarting the system. We have studied self-stabilizing systems using graph relabeling systems in [11]. To encode a self-stabilizing system we use a graph relabeling system $\mathcal{SSR}$ denoted by a triplet $(L, \mathcal{P}, \mathcal{F})$ where L is a set of labels, $\mathcal{P}$ a finite set of relabeling rules and $\mathcal{F}$ is a set of abnormal local configurations. Let (G, λ) be a labeled graph. We denote by $\mathcal{G}_\mathcal{L}$ the set of all labeled graphs. Let $AB : \mathcal{G}_\mathcal{L} \times \mathcal{F} \longrightarrow \mathbb{N}$ be an application associating to each labeled graph (G, λ), the number of its abnormal configurations in the set $\mathcal{F}$. Let (G, λ') be a labeled graph such that: $(G, \lambda) \xrightarrow[\mathcal{SSR}]{*} (G, \lambda')$. Then the graph (G, λ') satisfies the following properties:

- **Closure**: A computation beginning in a correct state remains correct until the terminal state. Formally, if $AB((G, \lambda), \mathcal{F}) = 0$ then $AB((G, \lambda'), \mathcal{F}) = 0$.
- **Convergence**: The ability of the relabeling system to recover automatically within a finite time (finite sequence of relabeling steps). Formally, if $AB((G, \lambda), \mathcal{F}) > 0$ then there is a labeled graph (G, λ_s) such that $(G, \lambda) \xrightarrow[\mathcal{SSR}]{k} (G, \lambda_s)$ and $AB((G, \lambda_s), \mathcal{F}) = 0$.

3   Self-stabilizing Resolving Conflicts

We consider a set of processors sharing a set of resources. In our graph the resources are placed on the edges, node u is a neighbor of node v if u and v can share a resource. A process can perform its computation if all its needed peripheral resources are allocated to it. Each node v is labeled $\lambda(v) = (St(v), Ord(v))$, where $St(v)$ denotes the current state of v : *tranquil, thirsty* or *drinking*. These states will be respectively encoded by the labels T, Th or D. The label $Ord(v)$ is an integer to manage the order of process requests.

3.1   Description of the Algorithm in a Normal Behavior

Initially all the vertices of the graph are tranquil (this is encoded by the label $(T, -1)$). At each step of the computation, an $(T, -1)$-labeled vertex u may ask to enter the "CS", instead of critical section, which means that it becomes thirsty. In this case u changes its label to (Th, i) where $i = max\{Ord(v)|v \in N(u)\} + 1$. So, the order of u is the maximum of its neighbors. In a complete graph, this order can be seen as a universal time since only one node can change its label to Th at the same time.

If a vertex u, with a label (Th, i), has no neighbor in the critical section (labeled $(D, -1)$) and no neighbor with a label (Th, j) where $j < i$ (it has the lowest rank of its neighbors), the vertex u can enter the critical section. That is, u will have the label $(D, -1)$.

Once, the vertex in the critical section had terminated its work, it changes its label to $(T, -1)$.

3.2   Our Algorithm

Clearly, all the computations starting from configurations where all nodes are $(T, -1)$ using a scheme as the normal behavior are without abnormal configurations. Since self-stabilization deals with computations starting from any configuration, we will consider in our design the fact that some abnormal configurations appear in the started configuration. The goal of the self-stabilization is to ensure that after a finite time, such configurations will disappear. We consider the following abnormal configurations:

- When a node u labeled $St(u) = Th$ has a neighbor v labeled $St(v) = St(u)$ and $Ord(u) = Ord(v)$. We denote this configuration by f_1.
- When a node u labeled $St(u) = D$ has a neighbor v labeled $St(v) = St(u)$. We denote this configuration by f_2.

The algorithm may be encoded by the following graph relabeling system. In the sequel, it is referred to as the $\mathcal{SRC}$ algorithm.

Self-stabilizing Algorithm For Resolving Conflicts : $\mathcal{SRC}$

- **Labels:**
  - $N(v)$: the set of (locally) ordered immediate neighbors of v which is an initial data.
  - $St(v)$: the current state of v. It takes values in $\{T, Th, D\}$.
  - $Ord(v)$: the number of the request, this is an integer in $[-1..\infty]$.
- **Initialization:** Any
- **Rules:**

RC1 : **The node u becomes thirsty, it forms its request to enter in the "CS"**
Precondition :
 * $St(u) = T$
Relabeling :
 * $St(u) := Th$
 * $Ord(u) := Max\{Ord(v), \forall v \in N(u)\} + 1$

RC2 : **Node u is elected in its local ball, so it enters in the "CS"**
Precondition :
 * $St(u) = Th$
 * $\neg \exists v \in N(u)$ such that $St(v) = D$
 * $Ord(u) = Min\{Ord(v), \forall v \in N(u)\}$
Relabeling :
 * $St(u) := D$
 * $Ord(u) := -1$

RC3 : **Node u terminates its use of the "CR", then it leaves the critical section**
Precondition :
 * $St(u) = D$
Relabeling :
 * $St(u) := T$

RC4 : **Thirsty node u finds one of its neighbor also thirsty with the same order, so it changes its order**
Precondition :
 * $St(u) = Th$
 * $\exists v \in N(u)$ such that $St(v) = Th$ and $Ord(u) = Ord(v)$
Relabeling :
 * $Ord(u) := Max\{Ord(w), \forall w \in N(u)\} + 1$

RC5 : **Node u in the "CS" finds one of its neighbor also in the "CS"**
Precondition :
 * $St(u) = D$
 * $\exists v \in N(u)$ such that $St(v) = D$
Relabeling :
 * $St(u) := Th$
 * $Ord(u) := Max\{Ord(w), \forall w \in N(u)\} + 1$

3.3   Proof of Correctness and Analysis

To show the correctness of the $\mathcal{SRC}$ algorithm we propose the two following Lemmas. The proofs of these Lemmas are given using induction on the sequences of relabeling steps [12] and assuming that the correction rules ($RC4$, $RC5$) have higher priority than those of $RC1$, $RC2$, $RC3$.

Lemma 1. *Starting from a configuration (G, λ) without abnormal configuration, all the configurations reached during an $\mathcal{SRC}$-computation remains without abnormal configurations.*

Lemma 2. *Starting from a configuration (G, λ) with abnormal configurations, there is a finite set of the $\mathcal{SRC}$-computation steps after which all the reached configurations are without abnormal configurations.*

The following result shows our first contribution.

Theorem 1. *The $\mathcal{SRC}$ algorithm encodes a distributed self-stabilizing algorithm for resolving conflicts. When the $\mathcal{SRC}$ algorithm is applied on a graph $G = (V, E)$, its time complexity is in $O(\#V)$ [1] steps and its space complexity is in $O(d(G) \times \log d(G) + \log \mathcal{N})$ bits per node. For some upper bound of the possible value of the order $\mathcal{N}$.*

4   Implementation Using Randomized Synchronizations

To implement the GRS, and hence the algorithm it encodes, we need a procedure to ensure that at any step, no two adjacent vertices apply one of the rules at the same time. To do so, we use a randomized procedure studied in [13] and used to implement local computations: In a round, a vertex chooses a random real value and then sends it to all its neighbors. When a vertex v receives all the messages, it compares its value to those sent by its neighbors. If v has the maximum value, then it knows that it is elected in the ball of center v and of radius 1 (and then can perform a step of the GRS). Otherwise, it keeps standing until the next round where it will try again to be locally elected, this local election is called LE_1. In [13], the authors show that if d is the degree of v then the expected waiting time for v to be locally elected is $d + 1$.

Before proving the correctness of the algorithm, let us recall some properties of the graph relabeling system we described in the last section.

Properties 1. *Let (G, λ) be a labeled graph. Let (G, λ') be a labeled graph such that: $(G, \lambda) \xrightarrow[\mathcal{SRC}]{*} (G, \lambda')$. Then (G, λ') satisfies the following:*

1. *Each node u with label $(T, -1)$, which wants to enter the critical sections (CS), changes its label to $(Th, n+1)$, where $n = max\{Ord(v) | v \in N(u)$ and $\lambda'(v) = (X, Ord(v)), X \in \{T, Th, D\}\}$.*
2. *To change its label to $(D, -1)$, a (Th, i)-labeled node must not have a neighbor with the label $(D, -1)$ or with the label (Th, j) where $j < i$.*
3. *Each thirsty node u with label (Th, i) connected with a node v labeled also (Th, i), updates its label to $(Th, n + 1)$, where $n = max\{Ord(v) | v \in N(u)$ and $\lambda'(v) = (X, Ord(v)), X \in \{T, Th, D\}\}$.*
4. *Each node u supposed in the "CS" (labeled $St(u) = D$) which finds one of its neighbor v also supposed in the "CS", sets its label to $(Th, n + 1)$, where $n = max\{Ord(v) | v \in N(u)$ and $\lambda'(v) = (X, Ord(v)), X \in \{T, Th, D\}\}$.*

---

[1] $\#M$ denotes the cardinality of the set M.

Concerning the concurrency property, in [13], it is proved that if a vertex v is locally elected in a ball $B(v)$ then there is no other vertices locally elected in $B(v)$, meaning that v can perform its action (v can apply a rule of the GRS) without influence in the behavior of the vertices at a distance greater than 1 from v. That is :

Lemma 3. *The concurrency property is verified by the algorithm.*

To capture the worst cases of the abnormal configurations, we will use the following definition:

Definition 2 *Let $a > 0$ be an integer and let $\mathcal{AB}_a$ denote the set of abnormal vertices, including f_1, f_2 vertices, in G at the ath step of the algorithm.*

1. *Let $P_k = (v_0, v_1, \cdots, v_{k-1})$ be a path in $\mathcal{AB}_a$. Each of the nodes v_0 and v_{k-1} is called the end-path node. Then,*
   (a) *P_k is a D-abnormal path if $\forall j \in [0..k-1]$, $\lambda(v_j) = (D, -1)$.*
   (b) *P_k is a Th-abnormal path if $\forall j \in [0..k-2]$, if $\lambda(v_j) = (Th, i)$ then $\lambda(v_{j+1}) = (Th, i)$.*
2. *Let $\mathcal{P}$ be a set of paths P_k that covers $\mathcal{AB}_a$. In other words, $\mathcal{P}$ covers the set of all abnormal vertices.*

To prove the mutual exclusion we will be interested in studying the possible construction and destruction of the *D-abnormal* configurations and for the liveness property we consider the *Th-abnormal* configurations. To eliminate a *D*-abnormal node v_0 it suffices that v_0 changes its label to (Th, i). That is, we focused on the expected time necessary for v_0 to change its label to (Th, i). When a vertex v is labeled $(D, -1)$, it needs to become locally elected to change its label to (Th, i). Recall that from [13], the expected waiting time for a vertex v to become elected in a ball $B(v)$ is

$$\mu(v) = d(v) + 1. \tag{1}$$

Now, we deal with abnormal paths as Definition 2. Let $\mathcal{P}_D$ be a cover composed of *D*-abnormal paths, so $\mathcal{P}_D = \{P_i^1, P_j^2, \cdots P_k^l\}$. The worst case corresponds to the elimination of the abnormal paths one by one. We will study the longest *D*-abnormal paths because they influence the needed time to achieve the mutual exclusion property. Indeed, if $P_k = (v_0, v_1, \cdots, v_{k-1})$ is such a path, it is clear that the application of the rule $RC5$ by v_0 or v_{k-1} decreases the length of the path by 1. However, the application of the same rule by other nodes decreases the length by 2. The worst case corresponds to the application of the $RC5$ by the end-path nodes. We consider the case when $v_0, v_1, \cdots, v_{k-1}$ are elected locally in that order.

Lemma 4. *Let $P_k = (v_0, v_1, \cdots, v_{k-1})$ be a D-abnormal path of length k. If τ_{Dk} denotes the time to obtain local elections on $v_0, v_1, \cdots$ and v_{k-1} in that order then*

$$E(\tau_{Dk}) = \sum_{i=0}^{k-1} (d(v_i) + 1). \tag{2}$$

Now, we show that after a finite time the computation will reach a configuration deprived of f_2 nodes. So we are interested to study the worst case or the upper bounds. From the previous, the expected waiting time to destroy a D-abnormal path corresponds to the expected waiting time for each of its vertices to become locally elected. Thus, the expected waiting time to destroy a $\mathcal{P}_D$ cover corresponds to the expected waiting time to destroy each of its D-abnormal paths. According to (2), if τ_D denotes the time to obtain local elections on v_0, and then local elections in its D-abnormal paths $P_0, \cdots$ and $P_{d(v_0)-1}$ in that order then we deduce:

$$E(\tau_D) \leq \#\mathcal{P}_D \sum_{i=0}^{dia(G)-1} (d(v_i) + 1). \tag{3}$$

Where $\#\mathcal{P}_D$ denotes the number of D-abnormal paths in $\mathcal{P}_D$.

From the previous and Properties 1, the following holds:

Theorem 2. *Eventually, after a finite time the mutual exclusion property is verified by the algorithm and then both ordering and mutual exclusion properties are verified for ever.*

To prove that the computation reaches a configuration satisfying the liveness property, we use the same reasoning as the previous. First, to eliminate all the Th-abnormal nodes it suffices that each of them changes its label to (Th, max_ord) such that its order becomes the most of its neighbors. In fact, we study the expected time necessary for a vertex v_0 to change its label to (Th, max_ord). To do this, v_0 must become locally elected to change its label (1). For the liveness property we consider Th-abnormal paths in the sense of Definition 2. In fact, if $P_k = (v_0, v_1, \cdots, v_{k-1})$ is such a path, it is clear that the application of the rule $RC4$ by v_0 or v_{k-1} decreases the length of the path by 1. However, the application of the same rule by other nodes decreases the length by 2. The worst case corresponds to the application of the $RC4$ successively by the end-path nodes. We consider the case when $v_0, v_1, \cdots, v_{k-1}$ apply the rule $RC4$ in that order.

Lemma 5. *Let $P_k = (v_0, v_1, \cdots, v_{k-1})$ be a Th-abnormal path of length k. If τ_{Thk} denotes the time to obtain local elections on $v_0, v_1, \cdots$ and v_{k-1} in that order, then*

$$E(\tau_{Thk}) = \sum_{i=0}^{k-1} (d(v_i) + 1). \tag{4}$$

Now, we show that after a finite time the computation will reach a configuration deprived of f_1 nodes. We consider the case of Th-abnormal cover denoted by $\mathcal{P}_{Th}$. The expected waiting time to destroy all the Th-abnormal nodes corresponds to the expected waiting time to destroy each of the Th-abnormal paths in $\mathcal{P}_{Th}$. According to (1) and (4), we deduce:

$$E(\tau_{Th}) \leq \#\mathcal{P}_{Th} \sum_{i=0}^{dia(G)-1} (d(v_i) + 1). \tag{5}$$

After the reach of a configuration satisfying the liveness property, it remains to prove that the liveness property is satisfied for ever. We need to start by proving some lemmas verified by the GRS which encodes the algorithm.

Definition 3 *Let $a > 0$ be an integer and let $\mathcal{H}_a$ denote the set of vertices labeled (Th, i) in G at the ath step of the algorithm. Let $P_k = (v_0, v_1, \cdots, v_{k-1})$ be a path in $\mathcal{H}_a$. Then P_k is a consecutive path if $\forall j \in [0..k-2]$, if $\lambda(v_j) = (Th, i)$ and $\lambda(v_{j+1}) = (Th, l)$ then $l = i + 1$.*

We are interested on the investigations of the consecutive paths because they are the worst cases for our algorithm. Indeed, if $P_k = (v_0, v_1, \cdots, v_{k-1})$ is such a path, it is clear that v_{k-1} will not be allowed to change its label to $(D, -1)$ until all the vertices $v_0, v_1, \cdots$, and v_{k-2} have changed their labels in that order.

A vertex which is labeled $(T, -1)$ must become locally elected a first time to change its label to (Th, i), and a second time to verify that it has the lowest rank of its neighbors and then changes its label to $(D, -1)$. So, if at a step s, a consecutive path $P_k = (v_0, v_1, \cdots, v_{k-1})$ is formed, this means that v_0, v_1, $\cdots$, and v_{k-1} were elected in that order. Using the same reasoning, it is easy to see that the vertex v_{k-1} will change its label to $(D, -1)$ if the vertices v_0, v_1, $\cdots$, and then v_{k-1} were elected again in that order. That is, in this section, we focus on the expected time necessary for $v = v_{k-1}$ to change its label to $(D, -1)$ since it is equal to the expected time for P_k to be constructed.

When a vertex is labeled (Th, i), it needs to become locally elected to change its label to $(D, -1)$. But, the time it will take for the vertex v_{k-1} to become able to change its label to $(D, -1)$ will be the time for the vertices $v_0, v_1, \cdots, v_{k-2}$ to become locally elected respectively.

Lemma 6. *Let $P_k = (v_0, v_1, \cdots, v_{k-1})$ be a consecutive path of length k. If τ_k denotes the time to obtain local elections on v_0, $v_1, \ldots$ and v_{k-1} in that order, then*

$$E(\tau_k) = \sum_{i=0}^{k-1} (d(v_i) + 1). \tag{6}$$

The aim of the next lemmas is to show that the use of a randomized procedure ensures that the probability of obtaining a *long* consecutive path is small.

Lemma 7. *Let v be a vertex and let $P = (v_0 = v, v_1, \cdots, v_{k-1})$ be a consecutive path of length k. Then with probability $1 - \frac{d(v)+1}{\alpha}$, we have $k \leq \alpha$, $\forall \alpha > d(G)+1$.*

That is we have the expected theorem:

Theorem 3. *Eventually, after a finite time the liveness property is verified by the algorithm and then it remains verified for ever.*

5   Discussion and Future Works

There are similarities between our algorithm and the algorithm of [6], except that our algorithm deals with conflict problem in asynchronous and anonymous systems while the algorithm of [6] deals with the mutual exclusion problem in a complete graph. Moreover, the work presented in[6] uses a validating and checking approach whereas our approach is an approach of algorithm analysis. We plan to design a self-stabilizing algorithm with validating and checking aspects. Another extension of our work is the use of weaker model such as message passing model to encode our solution. Recall that the space requirements of our solution is optimal since our algorithm is encoded without adding auxiliary variables. For the time complexity, we are interested in reducing the stabilization times taking into account the message complexity.

References

1. Chandy, K.M., Misra, J.: The drinking philosophers problem. ACM Transactions on Programming Languages and Systems 6(4), 632–646 (1984)
2. Dijkstra, E.W.: Hierarchical ordering of sequential processes. In: Hoare, C.A.R., Perrott, R.H. (eds.) Operating Systems Techniques, Academic Press, London (1972)
3. Welch, J.L., Lynch, N.A.: A modular drinking philosophers algorithm. Distributed Computing 6(4), 233–244 (1993)
4. Barbosa, V.C., Benevides, M.R.F., Oliveira Filho, A.L.: A priority dynamics for generalized drinking philosophers. Information Processing Letters 79, 189–195 (2001)
5. Dijkstra, E.W.: Self stabilizing systems in spite of distributed control. Communications of the ACM 17(11), 643–644 (1974)
6. Lamport, L.: A new solution of dijkstra's concurrent programming problem. Commun. ACM 17(8), 453–455 (1974)
7. Lehmann, D.J., Rabin, M.O.: On the advantages of free choice: A symmetric and fully distributed solution to the dining philosophers problem. In: ACM SIGPLAN-SIGACT Symposium on Principles of Programming Languages, pp. 133–138 (1981)
8. Duflot, M., Fribourg, L., Picaronny, C.: Randomized dining philosophers without fairness assumption. In: International Conference on Theoretical Computer Science TCS, pp. 169–180 (2002)
9. Herman, T.: Probabilistic self-stabilization. Inf. Process. Lett. 35(2), 63–67 (1990)
10. Mosbah, M., Sellami, A., Zemmari, A.: Using graph relabeling systems for resolving conflicts. In: New Technologies for Distributed Systems (NOTERE'2006), Lavoisier, Hermes, pp. 283–294 (June 2006)
11. Hamid, B., Mosbah, M.: An automatic approach to self-stabilization. In: 6th ACIS International Conference on Software Engineering, Artificial Intelligence, Networking, and Parallel/Distributed Computing (SNPD 2005), Baltimore, USA, pp. 123–128. IEEE Computer Society, Los Alamitos (2005)
12. Litovsky, I., Métivier, Y., Sopena, E.: Graph relabeling systems and distributed algorithms. In: Ehrig, H., Kreowski, H.J., Montanari, U., Rozenberg, G. (eds.) Handbook of graph grammars and computing by graph transformation, vol. III, pp. 1–56. World Scientific Publishing, Singapore (1999)
13. Métivier, Y., Saheb, N., Zemmari, A.: Randomized local elections. Inf. Process. Lett. 82(6), 313–320 (2002)

A Metaprotocol Outline for Database Replication Adaptability*

M.I. Ruiz-Fuertes, R. de Juan-Marín, J. Pla-Civera, F. Castro-Company,
and F.D. Muñoz-Escoí

Instituto Tecnológico de Informática
Universidad Politécnica de Valencia
Camino de Vera, s/n
46022 Valencia (Spain)
{miruifue,rjuan,jpla,fmunyoz}@iti.upv.es, fracasco@doctor.upv.es

Abstract. Database replication tasks are accomplished with the aid of consistency protocols. Commonly, proposed solutions use a single replication protocol providing just one isolation level. The main drawback of this approach is its lack of flexibility for changing scenarios –i.e. workloads, access patterns...– or heterogeneous client application requirements. This work proposes a metaprotocol for supporting several replication protocols which use different replication techniques or provide different isolation levels. With this metaprotocol, replication protocols can either work concurrently with the same data or be sequenced for adapting to changing environments. In this line, the use of a load monitor would enable the best choice for each transaction, selecting the most appropriate protocol according to the current system characteristics. This paper is focused on outlining this metaprotocol design, establishing the metadata set needed and the required interaction between the main database replication protocol families.

1   Introduction

Different replication protocols provide different features –consistency guarantees, scalability...– and obtain different performance results. It can be depicted a scenario (workload, access patterns, isolation requirements, etc.) where each concrete protocol can be chosen as the most suitable. So, instead of proposing a consistency protocol with general purposes, we have designed a metaprotocol that allows a set of consistency protocols to work concurrently. With this metaprotocol, applications will be able to take advantage of the protocol that better suits their needs. Not only the best protocol for the general case can be selected for an application but also it can be changed for another one if the application access pattern changes drastically or if the overall system performance changes due to specific load variations or network or infrastructure migrations. These selections and changes can be forced by an administrator following theoretical and empirical studies or they can be automatically triggered according to background performance analysis (response times, abort rates...) executed by a monitoring system.

---

* This work has been partially supported by FEDER and the Spanish MEC under grants TIN2006-14738-C02-01 and BES-2007-17362.

R. Meersman, Z. Tari, P. Herrero et al. (Eds.): OTM 2007 Ws, Part II, LNCS 4806, pp. 1052–1061, 2007.

Also it is possible that a given application knows for sure which protocol is more suitable for its needs or that a given transaction access pattern is best served with a certain protocol.

There exist different architectures and many replication protocols in the literature. As each architecture displays advantages and disadvantages working in different scenarios, recent studies such as [1] and [2] compare protocols performance. Jiménez et al. [1] present the ROWAA approach as the most suitable for the general case and Wiesmann's paper [2] is based on total order broadcast techniques. Our paper follows the classification given in the latter.

The rest of this work has been structured as follows. Section 2 describes the assumed system model for our middleware metaprotocol support. Section 3 explains the metaprotocol and the interaction between data replication protocols belonging to different families in the classification proposed by [2]. Later, Section 4 discusses some related work and finally, Section 5 presents the paper conclusions.

2 System Model

We assume a partially synchronous distributed system –where clocks are not synchronized but the message transmission time is bounded– composed by N nodes where each one holds a replica of a given database; i.e., the database is fully replicated in all system nodes. Each system node has a local DBMS that is used for locally managing transactions. On top of the DBMS a middleware –called MADIS– is deployed in order to provide support for replication. More information about our MADIS middleware can be found in [3,4]. This middleware also has access to a group communication service (abbreviated as GCS, on the sequel). A GCS provides a communication and a membership service supporting virtual synchrony [5]. The communication service features a total order multicast for message exchange among nodes through reliable channels. The GCS groups messages delivered in views [5]. The uniform reliable multicast facility [6] ensures that if a multicast message is delivered by a node (correct or not) then it will be delivered to all available nodes in that view. In this work, we use Spread [7] as our GCS.

3 Metaprotocol

As it has been said, the main goal of our whole system is to use the replication protocol that best behaves in regard to the current requirements and environment of the replicated system. This implies that it must change the working protocol when it is detected that another one would fit better in the replicated system –possibly common in changing environments. Moreover, our proposal is that this protocol exchange is performed without stopping working. Then, when a protocol exchange must be done, transactions already started will end their execution using the protocol they started with, while new ones will use the new protocol.

In this section, we will present a basic element of this system: a metaprotocol which supports real concurrency among several replication protocols. Concurrency can lead to inefficient systems due to the natural differences in the behavior of the protocols, reason for what it should only be exploited to support the protocol exchange phase previously commented. Nevertheless, depending on the system requirements, it could be acceptable to use this concurrency without restrictions.

For designing and developing this system, different aspects must be considered. As several protocols must work concurrently, it will be necessary to know the metadata needed by each supported protocol to work; on the other hand, it is mandatory to study and analyse important issues related to their interaction.

To do so, we will use the protocol families presented in [2]. As a first try, we will only consider those families based on total order broadcast: active, certification-based and weak voting replication. All these families are update-everywhere [8], so they are decentralised replication protocols.

3.1   Targeted Replication Protocol Families

In *active* replication, a client request arrives to one server –the delegate–, which sends the transaction to all replica servers using a total order broadcast. Later, server replicas –including the delegate one– execute and commit the transaction in the order it has been delivered. This way, the total order broadcast ensures that all replicas commit transactions in the same order. This protocol ensures 1-copy-serializability isolation level.

In the *certification-based* replication, the delegate server locally executes the transaction before broadcasting it. Then, when the local transaction arrives to its commit phase, the delegate broadcasts the transaction information to all server replicas. Once it is delivered, a certification phase starts in all replicas to determine if such transaction can be committed or not. The total order established by the broadcast determines the certification result: in case of a pair of conflicting transactions, the newest inside the total order is the one that is aborted. The transactions that commit, do it in the order established by the broadcast. The transaction information that must be broadcast will depend on the guaranteed isolation level. The writeset (set of written objects) is enough for achieving snapshot isolation; for 1-copy-serializability, also the readset (set of read objects) must be broadcast. Some additional metadata is also needed in order to complete the certification; e.g., the transaction start logical timestamp in case of using the *snapshot isolation* [9] level.

The *weak voting* technique is very similar to the *certification-based* replication in their first steps. A client request arrives to one server –the delegate– who starts to process it locally. When the local transaction arrives to its commit phase, the delegate broadcasts the transaction information to all replica servers. Upon delivering this message, the delegate can determine if conflicting transactions have been committed before (i.e., their writesets have been delivered before its own one). If so happens, the transaction being analysed should be aborted. Based on this information, the delegate server does a new broadcast reporting the outcome of the transaction to the replicas. It must be noticed that with this technique, broadcasting just the transaction writeset already allows 1-copy-serializability isolation level, since delegate replicas are able to check for conflicts between their own transactions readsets and the remote transactions' writesets.

3.2 Metadata Set

Before considering the interaction among these protocols, it is important to determine the metadata needed by each supported protocol because when two or more protocols work concurrently the metaprotocol will need to generate, for each executed transaction in the system, the metadata needed by each executing protocol. Therefore, if $P1$ and $P2$ are the two protocols working concurrently, and $MP1$ and $MP2$ are their respective metadata when working alone, the metaprotocol must generate for transactions executed by $P1$ and $P2$ the $MP1 \cup MP2$ metadata.

As active replication applies transactions sequentially in the order obtained by the total order delivery, it does not need any metadata. Certification-based replication uses in an explicit way the *begin of transaction* and writeset as metadata, thus providing snapshot isolation [9]. The readset may be added to provide 1-copy-serializability [9]. Weak voting technique simply uses the writeset as metadata. It must be noticed that active and weak voting techniques do not need additional metadata because the total order broadcast provides them enough information to work.

Therefore, the transaction metadata needed when all protocols are working is compound by: begin of transaction, writeset and readset. However, readsets are only necessary when the certification-based technique is used for providing 1-copy-serializability. Therefore, due to the high cost associated to manage readsets, it would be preferable to use the certification-based technique only for providing snapshot isolation. This way, when talking about certification-based transactions, only the writesets will be considered from now on.

3.3 Metaprotocol Outline

Once the protocols have been described in behavior and metadata, the interaction between them can be detailed. The meeting point where these replication techniques work concurrently is a pair of shared lists: the *log* list with the history of all the system transactions (only needed by the *certification* technique, but including information from all the protocols), and the *tocommit* list, with the transactions pending to commit in the underlying database. These lists are maintained by the metaprotocol in each replica. The outline of this metaprotocol is depicted in Figure 1. A transaction is included in both lists when it is delivered by the GCS and, depending on the protocol class, it has not been rejected during validation phase. Transactions are processed following the list order. Some dependencies may arise due to the behavior of each replication technique. For determining these dependencies, it will be necessary to know how each protocol behaves:

- When a transaction managed by active replication reaches the first position in the *tocommit* list, it is applied in the database, obtaining its associated writeset –and readset when needed.
- Transactions using certification-based replication should be certified before being inserted in the lists. If they pass their certification without being rejected, they are appended to both lists but must wait to arrive to the first position of the *tocommit* list for being applied and committed in the database. In case of an unsuccessful

Initialization:
1. ∀ available protocol P_q
   a. P_q.enabled := suitable(P_q)
   (suitable(P_q) is true if P_q is currently suitable.
   P_q.enabled can be also set later.
   When set to true, P_q is loaded.
   When set to false, P_q is unloaded if there are no
   local transactions already started with it.)
2. tocommit := ∅; log := ∅
3. L-TOI := 0; N-TOI := 1
(TOI = Total Order Index
N-TOI = TOI for the next transaction to be delivered.
L-TOI = TOI of the last committed transaction.)

I. New local request for a transaction in protocol P_q:
1. if P_q.enabled
   a. P_q.locals++
   b. grant request
2. else, deny request

II. A local P_q transaction terminates:
1. P_q.locals--
2. if P_q.enabled == false and P_q.locals == 0
   a. unload P_q

III. P_q asks to send message $M_n = \{t,p,r,c\}$ related to T_i
where t is message type, p is protocol id,
r is replica id and c is message content:
(see message types in table 2)
1. if t == WV-1 or C
   a. T_i.bot := L-TOI   [1]
2. piggyback garbage collection info (L-TOI)   [2]
3. broadcast M_n

IV. Upon delivery of M_n related to transaction T_i:
1. if M_n.type == C
   a. call T_i.protocol for validate(T_i)
   (check conflicts with concurrent transactions)
   b. if validation == negative
   (if T_i conflicts with a resolved)
      i. if T_i.replica == R_k
         (transaction is local)
         call T_i.protocol for rollback(T_i)
      ii. else
         discard T_i
   c. else
      i. if validation == positive
      (if T_i has no conflicts and ∄ w-pending)
         T_i.log_entry_type := resolved
         T_i.committable := true
      ii. if validation == pending
      (if T_i conflicts with a c-pending or ∃ w-pending)
         T_i.committable := false
         establish_dependencies(T_i)
      iii. T_i.toi := N-TOI++
      iv. append to log and tocommit
2. if M_n.type == WV-1
   a. T_i.toi := N-TOI++
   b. T_i.log_entry_type := c-pending
   c. if M_n.replica == R_k
      i. T_i.committable := true
   d. else
      i. T_i.committable := false
   e. append to log and tocommit

3. if M_n.type == A
   a. T_i.toi := N-TOI++
   b. T_i.bot := T_i.toi−1   [1]
   c. T_i.log_entry_type := w-pending
   d. T_i.committable := true
   e. append to log and tocommit
4. if M_n.type == WV-2
   a. if M_n.vote == commit
      i. T_i.committable := true
   b. if M_n.vote == abort
      i. delete T_i from log and tocommit
   c. resolve_c-dependencies(T_i, M_n.vote)
5. do garbage collection   [2]

V. Garbage collection:
1. delete oldest log entry if all nodes applied it

VI. Committing thread:
1. T_i := head(tocommit)
2. if T_i.committable == true
   a. if T_i.replica ≠ R_k
      i. call T_i.protocol for apply(T_i)   [3]
   b. call T_i.protocol for commit(T_i)   [4]
   c. L-TOI := T_i.toi
   d. if T_i.log_entry_type == w-pending
      i. resolve_w-dependencies(T_i)   [5]
   e. delete T_i from tocommit

VII. Metaprotocol procedures:
1. establish_dependencies(T_i):
   T_i.log_entry_type := c-pending
   ∀ concurrent w- or c-pending T_j in conflict with T_i
      add T_i in the dependent list of T_j
      T_i.dependencies++

2. resolve_w-dependencies(T_j):
   ∀ transaction T_i in T_j.dependent
      remove T_i from T_j.dependent
      if no conflicts between T_i and T_j
         T_i.dependencies--
         if T_i.dependencies == 0
            T_i.committable := true
            T_i.log_entry_type := resolved
            resolve_c-dependencies(T_i, commit)
      else
         delete T_i from log and tocommit
         resolve_c-dependencies(T_i, abort)

3. resolve_c-dependencies(T_j, termination):
   ∀ transaction T_i in T_j.dependent
      remove T_i from T_j.dependent
      if termination == commit
         delete T_i from log and tocommit
         resolve_c-dependencies(T_i, abort)
      else
         T_i.dependencies--
         if T_i.dependencies == 0
            T_i.committable := true
            T_i.log_entry_type := resolved
            resolve_c-dependencies(T_i, commit)

Fig. 1. Metaprotocol algorithm at replica R_k (See explanations for right-enumerated lines in Section 3.6)

TYPE	DESCRIPTION
WV-1	weak voting transaction writeset
C	the writeset (and readset if needed) of a transaction executed in a certification-based protocol
A	the whole transaction of an active protocol
WV-2	the voting message of a weak voting protocol

Fig. 2. Message types

TYPE	DESCRIPTION
resolved	a committable (or already committed) transaction with writeset (and readset if needed) info available
c-pending	a transaction with writeset (and readset if needed) info available but not yet committable (e.g. a weak voting transaction waiting for its voting message)
w-pending	a transaction with no writeset info available (e.g. an active transaction not yet committed)

Fig. 3. Log entry types

certification, such transactions are discarded in their remote replicas and aborted in their delegate one.

– A transaction T_i replicated with weak voting has to overcome multiple validations against all remote writesets previously delivered. To this end, each time one of such remote writesets is applied, the delegate replica can check if it has conflicts with such local transaction T_i. If so arises, the local transaction is tagged as *to-be-aborted*, and a second broadcast (reliable, but without total order) is sent to all replicas telling them that the transaction should be aborted. Otherwise, the transaction writeset will finally arrive to the head of the *tocommit* queue. When this arises, the second broadcast is also sent, but notifying in this case that the transaction should be committed. In all other replicas, the transaction writeset is applied as soon as it arrives at the head of the *tocommit* queue, but its terminating action (either commit or abort) is delayed until the second broadcast is delivered. In case of abortion, such transaction data should be removed from the *log* list.

3.4 Dependencies

Several dependencies may arise in the validation phase –in case of a certification-based replication technique– due to these transaction behaviors. A *certification-based* transaction, in order to be certified, must know the writesets of all concurrent and previously delivered transactions that will eventually commit. But this can not be immediately known, as there is some pending information in the entries of the log list (see table 3). First, the writesets of *active* transactions are not known until their commit time (we will identify these *active* transactions as *w-pending* transactions). Second, *weak voting* transactions final termination (commit/abort) is not known until it is delivered in the voting phase (until then, these transactions are identified as *c-pending* transactions). Third, *certification-based* transactions whose certification phase is waiting for the resolution of any of the previous cases, are also treated as *c-pending* transactions.

Note that a *certification-based* transaction T_k may be waiting for a *c-pending* transaction T_j, being T_j also a *certification-based* one waiting for the resolution of, for example, a *weak voting* transaction T_i. Note that T_k and T_i may have no conflicts between them, but they are related through a dependency chain. The metaprotocol must handle properly these dependency chains. It also has to be noticed that a transaction may have several dependencies and it only will be resolved when all its dependencies are resolved.

Once a pending transaction is resolved, this resolution is propagated along the depedecy chain, as detailed in Figure 1 in procedures VII.2 and VII.3.

As it can be seen, these dependencies imply some delays in regard to the protocol original behavior when working alone. So, these delays would imply an extra cost that usually could not be tolerated during long periods. Therefore, our metaprotocol must be designed in order to not introduce any overhead when a protocol works alone, and allowing to work concurrently several replication protocols for short time periods when a protocol change must be done without stopping working. Future work will provide measures of metaprotocol overhead.

Regarding to the protocol exchange possibilities, our metaprotocol offers the required functionality for another system component (such as a load monitor, an administrator, etc.) to enable and disable protocols in order to best fit the current system requirements.

3.5   An Easy Example

In order to make easy the understanding of the metaprotocol pseudocode, let us consider an example situation. Assume that each protocol is currently enabled and working, so transactions from all of them are being delivered at a replica R_k. Suppose also that, at a given moment, the *tocommit* list is empty. Then, four new transactions are delivered to the replica. In our example, these four transactions will be enqueued in the *tocommit* list before the first one commits –in order to show how the dependencies among them are established–. The first delivered transaction, A, is an active one. Due to this, A is directly appended to the *tocommit* list, and marked as committable and w-pending (we haven't got its writeset yet). The commitment process of this transaction can start at any moment from now on, but, as said before, it will not end until the four sample transactions are enqueued. The second delivered one, W, is a weak voting transaction. This way, W is marked as c-pending and appended to the *tocommit* list. Suppose that R_k is not the delegate replica of W. This means that W will remain as c-pending and non-committable until its voting message is delivered to R_k. The next two delivered transactions are certification-based. The first one, C1, ends its validation phase with a *pending* result as there exist w-pending transactions in the *tocommit* list and, let us suppose, it presents conflicts with W. Therefore, C1 is marked as c-pending and non-committable, and we say that C1 is dependent on both A and W. Later, another certification-based transaction C2 is delivered. C2 is also marked as c-pending and non-committable because of a conflict with C1 and the existence of A. In this scenario, as C2 depends on C1, and C1 depends on W, a dependency chain appears between C2 and W –note however that these two transactions do not conflict in our example–. At this moment, A finally commits and its writeset is collected. Now it is time to resolve the w-dependencies with C1 and C2. Assume that A does not conflict with any other transaction, so all the w-dependecies are just removed. Now, the voting message for W is delivered with a commit vote. So W is now committable and its c-dependencies can be resolved. As W presented conflicts with C1 and W is going to commit, C1 must abort. Due to the termination of C1, its c-dependencies are resolved on cascade, thus removing all the dependencies presented by C2, that becomes committable. Any committable transaction is eventually committed when it arrives to the head position of the *tocommit* list.

3.6   Further Details

For completeness purposes, we should describe some details from the metaprotocol outline (set as [number] in Figure 1).

1. Initialization of the *begin of transaction* logical timestamp metadata. This metadata is only needed for certification-based protocols. In weak voting and certification-based replication techniques, the value is set at broadcast time. With a conflict detection mechanism [4], we are able to ensure that this logical timestamp is valid for representing the transaction start. In active replication, this value is set at commit time since it actually starts at that time.
2. Garbage collection. Once a writeset has been applied in all replicas, all the local conflicting transactions have been eliminated, so such writeset can be removed from the log list as it will not be necessary any more [2]. This garbage collection should include a heartbeat mechanism in order to guarantee its liveness. But for simplicity, this heartbeat is not included in the outline.
3. Possible conflicts with local transactions. At this point, is when our conflict detection mechanism eliminates local conflicting transactions allowing the correct remote writeset application.
4. Commitment of a transaction. At this time, a weak voting protocol will broadcast the outcome of the transaction. This outcome can be either *commit* or *abort*, being an *abort* when another conflicting transaction has committed causing the rollback of this weak voting transaction.
5. After commitment, active protocol transactions will be required to collect their writeset in order to resolve possible dependencies.

Finally, if we want to add support in our metaprotocol for the *primary copy* technique replication [2], it is needed to know how it can interact with other protocols. When working alone in our metaprotocol, it can work in its original way, but when it must work concurrently with one of the other protocols some changes must be applied. The main problem relates to the fact that in the primary copy only one node serves client requests, and therefore it is the only one which must perform conflict detections. Therefore, when a primary copy technique works concurrently with another one where all replicas can serve client requests, the conflict detection performed by the primary one would not consider possible conflicts with concurrent transactions that are in their local phase in other replicas. Then, it is needed that all replicas participate in the conflict checking process. To do so, a possible approach would consist in modifying the primary copy technique in order to behave like a certification-based protocol when working concurrently with other protocol families. In this case, it would be a certification-based technique where only one replica processes all client requests. Notice that this artificial behavior would not be necessary if all concurrent protocols are primary copy approaches sharing the same primary copy server.

4 Related Work

This paper is a continuation of the work started by our group with MADIS [3], a platform designed to give support to a wide range of replication protocols. MADIS was thought to support different kinds of pluggable protocols, whose paradigms range from eager to lazy update propagation, from optimistic concurrency control to pessimistic, etc. Consequently, MADIS was though to maintain a wide range of metadata in order to cover the most common database replication protocols requirements. In particular, it was thought for switching from one consistency protocol to another, as needed, without the need of recalculating metadata for the newly plugged-in protocol. The resulting MADIS architecture allowed the administrator to change the used protocol, stopping first the old one and starting later the new one, thus not concurrent execution was supported.

An exchanging algorithm for database replication protocols was presented in [10]. In this work, the authors designed a protocol for supporting a closed set of database replication protocols.

Communication protocols are another distributed system research branch that also has studied techniques for dynamically changing protocols. In this case, the idea is to select each time the group communication system which best fits with the network layer and changing load profile of the replicated system. [11] and [12] are different approaches to provide this support. Another proposal in order to switch between total order broadcast communication protocols is presented in [13]. In this case the switching system does not buffer messages neither blocks their propagation, instead of that the system spreads each message using both protocols during the switching phase. It must be noticed that the third switching mechanism is less aggressive than the other two, but presents some overload in the network, whilst [11] and [12] can reach in a so easy way the inactivity period because they do not care about the existence of transactions.

In regard to supporting concurrently multiple isolation levels in replicated databases –a future work line of this paper–, [14] presents a different approach. [14] authors propose a general scheme for designing a middleware database replication protocol supporting multiple isolation levels. The authors based it on progressive simplifications of the validation rules used in the strictest isolation level being supported, and on local (to each replica) support for each isolation level in the underlying DBMS.

5 Conclusions

The aim of this paper is to design a metaprotocol that will be the key piece of an adaptable system allowing to dynamically change the replication protocol used in a replicated database. The idea is to select each time the replication technique that best fits with the changing requirements and dynamic environment characteristics, trying to provide always the best achievable performance.

As first approach, the goal of the metaprotocol designed in this paper is to provide support for the concurrent execution of the most relevant database replication protocol families based on total order broadcast [2]: active, certification-based and weak voting replication.

We present a basic metaprotocol algorithm for concurrently supporting the targeted database replication protocols, based on the required protocols interaction.

In following works, we will provide a correctness proof and new metaprotocol revisions for supporting other database replication techniques. Another future work will consist in implementing this metaprotocol and studying possible optimizations in order to increase the performance when several protocols work concurrently. Among these optimizations, the holes technique presented in [15] can be generalized and included in our metaprotocol algorithm.

References

1. Jiménez-Peris, R., Patiño-Martínez, M., Kemme, B., Alonso, G.: How to select a replication protocol according to scalability, availability, and communication overhead. In: SRDS, pp. 24–33. IEEE Computer Society, Los Alamitos (2001)
2. Wiesmann, M., Schiper, A.: Comparison of database replication techniques based on total order broadcast. IEEE Transactions on Knowledge and Data Engineering 17, 551–566 (2005)
3. Irún, L., Decker, H., de Juan, R., Castro, F., Armendáriz, J.E.: MADIS: a slim middleware for database replication. In: 11th Intnl. Euro-Par Conf. Monte de Caparica (Lisbon), Portugal, pp. 349–359 (2005)
4. Muñoz-Escoí, F.D., Pla-Civera, J., Ruiz-Fuertes, M.I., Irún-Briz, L., Decker, H., Armendáriz-Iñigo, J.E., de Mendivil, J.R.G.: Managing transaction conflicts in middleware-based database replication architectures. In: SRDS, pp. 401–410. IEEE Computer Society, Los Alamitos (2006)
5. Chockler, G.V., Keidar, I., Vitenberg, R.: Group communication specifications: A comprehensive study. ACM Computing Surveys 4, 1–43 (2001)
6. Hadzilacos, V., Toueg, S.: Fault-tolerant broadcasts and related problems. In: Mullender, S. (ed.) Distributed Systems, pp. 97–145. ACM Press, New York (1993)
7. Spread: The Spread communication toolkit (2007), Accessible in URL http://www.spread.org
8. Wiesmann, M., Schiper, A., Pedone, F., Kemme, B., Alonso, G.: Database replication techniques: A three parameter classification. In: SRDS, pp. 206–215 (2000)
9. Berenson, H., Bernstein, P.A., Gray, J., Melton, J., O'Neil, E.J., O'Neil, P.E.: A critique of ANSI SQL isolation levels. In: SIGMOD Conf. pp. 1–10. ACM Press, New York (1995)
10. Castro-Company, F.: Muñoz-Escoí, F.D.: An exchanging algorithm for database replication protocols. Technical report, Instituto Tecnológico de Informática, Valencia, Spain (2007)
11. Miedes de Elías, E., Bañuls Polo, M.C., Galdámez Saiz, P.: Group communication protocol replacement for high availability and adaptiveness. Thomson Paraninfo, 271–276 (2005)
12. Liu, X., van Renesse, R., Bickford, M., Kreitz, C., Constable, R.: Protocol switching: Exploiting meta-properties. In: 21st International Conference on Distributed Computing Systems, p. 37. IEEE Computer Society, Washington, DC, USA (2001)
13. Mocito, J., Rodrigues, L.: Run-time switching between total order algorithms. In: Nagel, W.E., Walter, W.V., Lehner, W. (eds.) Euro-Par 2006. LNCS, vol. 4128, pp. 582–591. Springer, Heidelberg (2006)
14. Bernabé-Gisbert, J.M., Salinas-Monteagudo, R., Irún-Briz, L., Muñoz-Escoí, F.D.: Managing multiple isolation levels in middleware database replication protocols. In: Guo, M., Yang, L.T., Di Martino, B., Zima, H.P., Dongarra, J., Tang, F. (eds.) ISPA 2006. LNCS, vol. 4330, pp. 511–523. Springer, Heidelberg (2006)
15. Lin, Y., Kemme, B., Patiño-Martínez, M., Jiménez-Peris, R.: Middleware based data replication providing snapshot isolation. In: Ozcan, F. (ed.) SIGMOD Conf. pp. 419–430. ACM, New York (2005)

A Reliable Context-Aware Intrusion Tolerant System

Ayda Saidane

University of Trento, Italy
name.surname@unitn.it

Abstract. The next generation of computing applications will be facing new challenges due the changing environments in which they run. Ambient intelligent and pervasive environments combine heterogeneity, mobility and dynamism and building secure systems in these contexts becomes harder. Moreover the effective security requirements of the system would be identified at runtime and the system should support runtime reconfiguration of its defense capabilities. We propose in this paper an architecting framework for adaptive context-aware intrusion tolerant systems suitable for such environments. The adaptability concerns the runtime re-configuration of the deployed reaction policy together with the active monitors according to the context change.

1 Introduction

Traditional approaches for building secure systems typically consist of measures in three main areas: prevention, detection, and response. Unfortunately, most organizations concentrate their efforts on prevention and neglect the other two areas. During the last years, many case studies demonstrate the limitations of these approaches that are based on unreasonable assumptions: vulnerabilities can and should be avoided or eliminated, and attacks should be prevented. In fact, preventive approaches are becoming insufficient when used on open networks like the Internet or when considering mobile computing, ad-hoc networks or pervasive computing environments also called ambient intelligent environments.

The next generation of computing applications will be facing new challenges due the changing environments in which they run. These new environments raise new challenges compared to traditional IT systems. The context-awareness is a fundamental principle for these new systems that have to adapt their behavior to the context change. The combination of heterogeneity, mobility and dynamism of these new systems imposes new requirements on the intrusion detection systems monitoring them. In particular, these new IDSs should fulfill the requirement on flexibility, adaptability and context-awareness. Since in different contexts, the target system may not be vulnerable to the same threats so it would be useless to have all detection capabilities running the whole system life.

We are considering here systems with limited resources and where the solutions should commensurate with the real threats existing in the environment in order to avoid making the system stronger at the current instant and more

R. Meersman, Z. Tari, P. Herrero et al. (Eds.): OTM 2007 Ws, Part II, LNCS 4806, pp. 1062–1070, 2007.

vulnerable later on. The resources are not available all the time and we should use them only if necessary. We propose in this paper an architecting framework for adaptive context-aware intrusion tolerant systems suitable for ambient intelligent environments. The intrusion tolerance consists of a combination of intrusion detection and reaction steps. The adaptability concerns the runtime re-configuration of the deployed reaction policy together with the active monitors according to the context change.

The rest of the paper is organized as follows. In section 2, we present an overview on the existing intrusion detection techniques. This section ends up with identifying the limitations of these approaches when used in the context of ambient intelligent environments. In section 3, we introduce our novel architecting framework for context-aware intrusion detection and reaction systems and detail the different components. Finally, our conclusions and future work are presented in section 4.

2 Background

The problem of intrusion detection has been studied for several years [1], [2]. An intrusion could be defined as a successful attempt to compromise the integrity, confidentiality or availability of some protected resources. Thus, an intrusion detection system aims to detect and in some cases react to intrusions, whether on one system, a group of systems, or a computer network. Two approaches have been used: signature-based detection and anomaly detection.

Signature-based detection relies on the monitoring of system activity and the identification of behaviors similar to pattern signatures of known attacks or intrusions stored in a signature database. This category of IDS detects accurately known attacks, the signatures are often generalized in order to detect the many variations of a given known attack. But this generalization leads to the increase of false positives (i.e., false alarms). The main limitation of such IDSs concerns their incapability to detect new and/or unknown intrusions that are not already present in the signature database.

Anomaly detection systems detect intrusions by observing deviations from a pre-established normal system or user behavior. This approach allows detecting new or unknown attacks, if these attacks imply an abnormal use of the system. The main difficulty when implementing reliable anomaly detection systems concerns the creation of the normal behavior model. Since it is difficult for complex systems to define correctly this model, we would obtain incomplete or incorrect models, which leads to false negatives or to false positives. During the last years, the emergency of a new generation of computing systems and networks, like P2P systems, ad-hoc networks, pervasive and ubiquitous applications, has changed the way of building secure systems. These environments differ from traditional systems by the unpredictability and frequency of attacks and attackers. The threat model required when designing a secure system is difficult to define in these environments. It is evident that if you deploy all your detection capabilities at once, you will have better detection efficiency but such approach would lead to considerable degradation of quality of service or resource exhaustion in resource-limited systems. It would be wiser to deploy only the detection capability required by a given environment.

Many works have been done on intrusion detection techniques for mobile ad-hoc networks and pervasive networks. The proposed approaches are oriented to both distributed and cooperative architectures [3], [4]. But these propositions do not take into account the requirements on cost-effective solutions and adaptability to the environment where the system is evolving. In fact, we cannot have a unique model for normal behavior but we will have different normal behaviors for different contexts. Considering a global model integrating all the normal behaviors in all contexts would generate a lot of false negatives and considering the minimal normal model would generate a lot of false positives. Concerning the signature-based approaches, we will be facing resource problems in implementing traditional signature-based IDs in pervasive and ubiquitous environments.

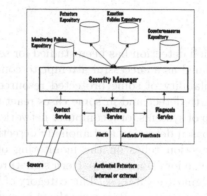

Fig. 1. Architecting Framework for Context-aware intrusion detection system

3   Towards a Reliable Context-Aware Intrusion Tolerant System

The main idea of this paper is to propose a generic architecting framework for adaptive context-aware intrusion detection and reaction systems. The framework defines the basic bricks required for such IDRS. The implementation of real IDRS corresponds to the customization and instantiation of the different components according to the domain specific characteristics. The figure 1 shows the architecting framework. In this section, we describe the architecture as set of black boxes and defining the necessary requirements that the real components implementing the IDRS should meet. The proposed framework can be decomposed into 3 parts: 1) set of diversified detectors, 2) monitoring services and 3) reaction to attacks.

3.1   Security Manager

The fundamental role of the security manager is to synchronize the activities of the different services and according to the monitoring and reaction policies adapt the configuration of the IDRS to the context change.

- According to the current monitoring policy, it sends orders to the monitoring service to activate or deactivate one or more detectors or to reconfigure activated ones.
- It receives information from the context service about the context change, such information are processed as triggers for changing monitoring policy.
- It receives information from the diagnosis service about the detected intrusions and this information is used as triggers to activate the appropriate countermeasures according to the reaction policy.

3.2   Detectors Repository

The reliability of an intrusion detection system depends strongly on the efficiency of the used detectors. This component define the detection capabilities of the Ids. One of the characteristics of ambient intelligent system is related to the continual environment change so it is important that the IDRS adapts its detection capabilities to the current context where the target system is running. The needed detection capabilities will be selected from the detectors repository according to the monitoring policy. The repository includes signature-based or anomaly based IDS, commercial IDS or domain-specific detectors like watchdogs or challenge-response protocols.

In order to integrate these heterogeneous components into a common framework it is important to embed them with a common interface. This interface will allow the monitoring subsystem to invoke them, change their configuration or repair them. In fact, we consider that according to the context a detector may be useful or not.

Example 1. In a trusted and friendly environment we may do not need to deploy the same kind and amount of the detectors as in an untrusted and hostile environment. There is behind this design decision is to make a good compromise between security, performance and cost.

The heterogeneity of these detectors is related to their nature, reliability, functional requirements, cost, performance impact and output type. The detectors should be chosen for their complementarity in such away the ones cover the vulnerabilities of others. In fact, as shown in this simple example, using diversified detectors make the system more resilient to attacks (Figure2) but of course not unvulnerable. In fact, it will be more difficult for an attacker to circumvent the different detectors.

The monitored system could be a stand-alone AmI [1] system or a set of distributed AmI systems. In both case, the detectors could be running on the target

---

[1] Ambient Intelligent.

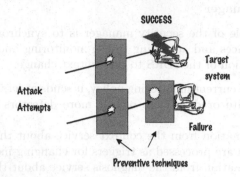

Fig. 2. Diversifying preventive techniques for more resiliency to attacks

system like host-based detectors or outside the target system like network-based detectors.

3.3   Context Service

The adaptability of the intrusion detection system to the environment changes relies on this component. Its main function consists in identifying the context change. It has two source of knowledge: 1) sensors and 2) alerts raised by the detectors.

- The relevant information about the context is highly domain and application dependent. We consider that the sensors send to the context service whatever information since we consider that the context includes context any information that can be used to characterize the situation of the system. Among useful information, we can identify location for mobile devices or connected peers in P2P systems or client profile in service-oriented architectures. Mainly these information may lead to at least two categories of environments known and unknown.
- The alerts raised by the detectors can be used by the context service to classify the environment as friendly or hostile.

Example 2. Let's consider an intrusion tolerant web server publishing security alerts. The Internet is an open environment characterized by its unpredictibility where the risk level is chaging quickly and permanently from very low to vey high. It can be evaluated by counting the attacks attempts and their frequency/severity or by receiving alerts from trusted peers about new attacks on the internet. This component is application-dependent.

3.4   Monitoring Service

This is an important component for the intrusion detection system. In fact, it is responsible for managing the detectors, unifying their output and filtering out

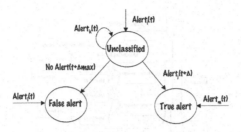

Fig. 3. Alerts' classification

false alerts. The Figure4 represents the architecture of monitoring service. It is composed of three parts: 1) detectors manager, 2) alerts manager and 3) set of wrappers interacting with real detectors.

Detectors manager —it is responsible for executing security managers orders concerning the activation and reconfiguration of the detectors. It is also responsible for maintaining a consistent overview on the detectors status in particular are they active? and are they functioning correctly?

Alerts manager —it is responsible for distinguishing between false and true alerts. As shown by the figure above, the alerts manager has a pre-established classification of the alerts into three categories (Figure3): false alerts, true alerts and unclassified alerts. The classification should take into account the source of the alert and the alerts content. When it receives an un classified alerts there are two different issues:

- This alert may become false alert, if no others alerts have been received within a timeout Dmax or if the threshold, corresponding to the number of alerts from the same category, hasnt been reached within the timeout. The alerts are processed using clustering techniques [5] in order to identify the correlated alerts. In fact, an attack does not happen generally alone, we will observe a set of attack attempts occurring quasi-simultaneously. The parameters of the alert processing algorithm are context dependent and they should defined by the reaction policy for each context.
- This alert becomes true alert if it is confirmed by other sources within the timeout or if we receive an alert classified as true alert within the timeout.

The knowledge used to classify the alerts is initiated by some expert and can be enriched online by some learning techniques. This issue will not be addressed in this paper.

Monitors —the monitors are wrappers interacting with the detectors. They mediate the commands from the detectors manager and unify the output of the different detectors in order to have the same data format processed by the alerts manager.

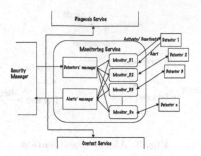

Fig. 4. Monitoring services architecture

3.5   Diagnosis Service

The diagnosis service receives the information about the alerts from the monitoring service including the source of the correlated alerts, their sequence and their content. Combining the alerts information with the context information, it should be able the select the appropriate reactive countermeasures to be executed. Once identified, the diagnosis service extracts the countermeasures from the countermeasures repository and applies them to the system.

3.6   Countermeasures Repository

The countermeasures repository consists of a library of executable patches. Each of them is characterized by the list of corresponding alerts. The countermeasures could be from different kinds:

- System reconfiguration like deactivating wireless connection
- Monitoring policy change
- Reset some subsystems
- Raise alert to administrator
- etc.

Example 3. Let's consider a fault tolerant application service based on replication. When the monitoring service detects a corrupted replica, it should isolate it and restart it. The countermeasure, in this case, would be stored in the repository as a pointer to executable file responsible for requesting a reconfiguration of the system (isolation) and restarting the corrupted entity.

3.7   Monitoring Policies Repository

The monitoring policies repository consists of a database storing different kind of monitoring policies suitable for different contexts. Each monitoring policy should specify for a given context, which detectors should be activated or deactivated or which modifications should be applied to the already activated ones. For example:

- We may need to increase the timeout of a watchdog component because of the change in the context of the application due to the growth of workload. In fact, in such conditions the requests processing will obviously take more time due to growing workload and the IDS should not raise a false DoS alert.
- When some countermeasures actions require the deactivation of some components all the corresponding monitors should be deactivated as well.
- When the system is declared to be in a hostile environment, more efficient monitors should be activated under the assumption that the security of the system is more prior than performance.
- Different sensors should be used when connected to an Ethernet network or Wireless network.
- etc.

3.8 Reaction Policies Repository

The reaction policies repository consists of a database storing the reaction policies corresponding the possible alerts. Each reaction policy should specify the list of alerts (source and content) and the context attributes where it should be applied.

4 Conclusion

Our motivation is to provide an adaptive, reliable and context-aware intrusion tolerant framework in order to deal with the difficult problems that the next generation of computing systems will be facing while evolving in mobile heterogeneous environments. The proposed framework can be decomposed into 3 parts: 1) set of diversified detectors, 2) monitoring and 3) reaction to attacks. All of them are intended to be re-configurable at runtime according to context change.

We are implementing a prototype based on eHealth case study where the patient need to be monitored in real time. Such system is expected to be resilient to both accidental and malicious faults. Two fundamental tests would be interesting to perform concerning: 1) the efficiency of the detection subsystem (false positives/ false negatives) and 2) the feasibility and effectiveness of the reaction subsystem.

References

1. Debar, H., Dacier, M., Wespi, A.: Towards a Taxonomy of Intrusion-Detection Systems. Computer Networks pages 2 (1999)
2. Jones, A.K., Sielken, R.S.: Computer System Intrusion Detection: A Survey. University of Virginia, Computer Science Technical Report pages 2 (2000)
3. Zhang, Y., Lee, W., Huang, Y.: Intrusion detection techniques for mobile wireless networks. A CM MONET Journal pages 3 (2003)

4. Zhou, B., Shi, Q., Merabti, M.: Intrusion Detection in Pervasive Networks Based on a Chi-Square Statistic Test. In: Computer Software and Applications Conference, 2006. COMPSAC 2006. 30th Annual International pages 3 (2006)
5. Zhong, S., Khoshgoftaar, T.M., Nath, S.: A Clustering Approach to Wireless Network Intrusion Detection. In: Proceedings of the 17th IEEE international Conference on Tools with Artificial intelligence pages 6 (2005)

Exploiting Commutativity for Efficient Replication in Partitionable Distributed Systems*

Stefan Beyer, M. I. Ruiz-Fuertes, Pablo Galdámez, and Francesc D. Muñoz-Escoí

Instituto Tecnológico de Informática, Universidad Politécnica de Valencia,
46022 Valencia, Spain
{stefan,miruifue,pgaldamez,fmunyoz}@iti.upv.es

Abstract. In decentralised systems, replication is commonly used to provide a certain degree of fault tolerance. Whereas many systems only consider the failure of individual system nodes, partitionable systems also consider network link faults that can cause the system to be divided into isolated parts. Replication in the presence of network partitioning is problematic, as updates to replicas in different partitions can lead to data inconsistencies that are not detected until the partitioning is repaired. The degree to which temporary or permanent inconsistencies can be tolerated depends heavily on the application.

We exploit commutativity to define a group of replication protocols that improve the performance for operations without order constraints on replicated objects. The protocols provide a way to trade consistency for improved availability in the presence of partitions and also simplify the reconciliation process, when two or more partitions are merged. The protocols have been implemented in the DeDiSys add-on for the CORBA middleware and some performance results are provided.

1 Introduction

Replication is commonly used as the primary technique to provide fault tolerance in distributed systems. As distributed systems are employed more and more in large scale environments and dynamic scenarios such as mobile ad-hoc networks, it is important that the replication protocols employed can provide fault tolerance, not only in the case of node failures, but also in the case of network link failures that lead to partitioning. Replication in the presence of network partitioning is problematic, as updates to replicas in different partitions can lead to data inconsistencies that are not detected until the partitioning is repaired. The degree to which temporary or permanent inconsistencies can be tolerated depends heavily on the application. If strong consistency is required, the solution usually employed is the primary partition model [1]. This model only allows updates to a partition which contains the majority of the system nodes; that is, more than half of the total number of nodes in the system. If inconsistencies are allowed to be introduced during partitioning, conflicts usually need to be removed in the

---

* This work has been partially funded by the European Community under the FP6 IST project DeDiSys (Dependable Distributed Systems, contract number 004152) and by FEDER and the Spanish MEC under grants TIN2006-14738-C02-01 and BES-2007-17362.

R. Meersman, Z. Tari, P. Herrero et al. (Eds.): OTM 2007 Ws, Part II, LNCS 4806, pp. 1071–1080, 2007.

reconciliation phase, when the partitioning is repaired. The restoration of consistency in this phase is heavily application dependent and can be very expensive.

In this paper, we apply the idea of exploiting commutativity to replication. We reduce the cost of inconsistencies that are introduced during network partitioning. This allows for a simplified reconciliation process. Furthermore, we use this idea to improve the performance of a replication protocol by reducing the time it takes to synchronise object state between replicas and we reduce the overhead of messages that are interchanged between nodes during an execution. We believe that this reduction in communication overhead makes our approach more suitable for large or dynamic decentralised systems. We have designed and implemented two replication protocols that make use of our key idea and have used one of them to obtain measurements of the performance improvements provided. Finally, with the second protocol we demonstrate that our approach can be used to trade consistency for increased availability.

A replicated entity can be anything from a single abstract data type to a full database. It has a state and a number of operations, which may or may not alter the state of the replicated entity. For the purpose of this paper we use the term object when referring to a replicated entity. Informally, two operations of an object are commutative if they can be executed in any order with final object state being the same. In Section 2 we introduce a more formal definition of commutativity. We define an object *order-free* if all its operations are commutative. Any object that is invoked in a nested invocation also has to be order-free.

Consider a simple object holding an integer value. If the object has only increment and decrement operations, these can be safely executed in any order and the object is therefore considered an order-free object. If the object provides a set operation it is not order-free, as an increment before or after the setting of a value can lead to very different results. However, we relax the definition of order-free and state that an object is order-free, if it is acceptable for the application to execute them in any order. Essentially we are relaxing consistency, in that we give the application programmer the freedom to allow certain inconsistencies. For some applications, the inconsistency might be acceptable. A real-life example of where such an inconsistency is usually accepted is the way the banks calculate interest rates. The bank calculates the interest rate at a certain fixed time. However, a cheque payment or transfer from a different bank might have occurred previously, but has not yet affected the account, meaning that the interest rate is calculated on a stale value. The operations are essentially executed in the wrong order. This sometimes works in favour of the bank's client, sometimes it works in favour of the bank. Nevertheless, the inconsistency is generally accepted, since it introduces minimal variations on the final object state.

Applying this principle to replicated objects has several advantages for the design of replication protocols. As the order of the invocations on order-free objects does not matter, we can reduce the guarantees usually required on the order of delivery in the underlying communication protocols. Furthermore, we can rely on local concurrency control instead of distributed transactions. This significantly improves invocation performance, as we demonstrate in this paper. Moreover, the reconciliation process after network partitioning is simplified, as operations missed by a replica can just be re-applied, without having to enforce the global order in which they were originally invoked. In addition, by delegating the decision of marking an object as order-free to the application programmer, we essentially define a way to allow to reduce consistency temporarily, in order to increase availability during partitioning. If strict global order is required certain operations have to be disallowed during partitioning, as read values might be stale as the state might have been modified in a different partition.

We provide the results of performance experiments that show a significant increase in performance, compared to a conventional replication protocol.

2   Commutativity

The mathematical definition of commutativity is as follows:

"A binary function $f : A \times A \to B$ is said to be commutative, if $f(x,y) = f(y,x) \; \forall x, y \in A$"

Applying this to operations on objects, we can define the sequence of execution of two operations as a function, where the execution of each operation is one of the operands. The execution is commutative, if the order of the operands, i.e. the execution of each operation, can be reversed. To apply this principle to non-binary functions, i.e. to an execution sequence with more than two operations, we have to exploit the associativity property:

"A binary operation $*$ on a set S is called associative, if $(x*y)*z = x*(y*z)$ $\forall x, y, z \in S$"

This essentially provides us with a property for a sequence of any number of operations, in which the order is irrelevant to the final result. We call an object for which all possible execution orders of its operations fulfil this property *order-free*. Informally, any order of execution of an order-free object's operations leads to the same final object state.

We exploit this property to allow order-free objects to be replicated more efficiently. We allow the application programmer to declare objects order-free. As has been illustrated with the interest rate example in the introduction of this paper, the commutativity property does not have to actually hold for an object to be declared order-free, if small inconsistencies are acceptable for a specific application.

Objects with order-free operations or objects with operations were the inconsistencies introduced by out of order execution are acceptable have the following advantages:

1. *Reduction of Required Message Delivery Guarantees.* Replication requires messages to be multicast with certain reliability and order guarantees. To this end a group communication service is usually employed. The group communication service guarantees that messages are delivered reliably and in the correct order. Different replication protocols require different delivery order guarantees. The two main replication paradigms are active replication [2], in which invocations are directed to all replicas, and passive replication [3] [4], in which a primary replica is invoked and the state changes are propagated to secondary copies. Both paradigms require different order guarantees on the message

delivery of the underlying multicast communication protocols. Active replication requires atomic multicast (i.e., a reliable multicast with total order delivery), whereas for passive protocols reliable FIFO multicast is usually sufficient [5]. If operations are commutative, the delivery guarantees can be reduced to merely reliable multicast, without the additional order requirements. This weaker delivery guaranty should in theory be more efficient than atomic or reliable FIFO multicast.

2. *No Need for Distributed Transactions.* Distributed transactions are used by many middleware systems. Transactions guarantee four properties: atomicity, consistency, isolation and durability (ACID). However, if the execution order of invocations does not matter due to commutativity, consistency and isolation can be guaranteed using more efficient local concurrency control. Furthermore, the atomicity and durability properties can also be achieved easily without distributed transactions.

3. *Increased Availability.* If replicas of the same object are invoked in different partitions it may be impossible to reproduce the correct ordering of operations during reconciliation as they happened in time. To avoid this, the common solution is to disallow operations in minority partitions; that is, in partitions that contain half or less than half of the total number of nodes in the system. Consistency of such strong levels is usually only required for a certain proportion of an application. Order-free objects provide a way to specify objects that can be safely modified in any partition, hence the restriction on invocability in minority partitions can be removed for these objects. This leads to a higher overall availability of the system.

4. *Simplified Reconciliation.* If operations are allowed to continue in all partitions, the state of replicas of the same object in different partitions may diverge. Resolving this inconsistency at reconciliation time can be very expensive and in some cases impossible. If the order of execution does not matter, it is possible to simply re-apply operations that have been executed in a non-local partition at reconciliation time. The reconciliation process is therefore greatly simplified, without compromising consistency.

Order-free objects can improve the performance and availability of some applications. However, not all objects can be declared order-free. The interest rate example presented in the introduction of this paper is a good example of the type of application that can be implemented using order-free objects. Using this application as a starting point we have defined a list of properties an application should have to benefit from our approach. The example works well, as the operations are expressed in terms of increasing a value. Generalising this we can say that the state is manipulated by *increments* and *decrements*. If a numerical value is increased or decreased several times it generally does not matter in which order the increments and decrements are performed. As long as an increment or decrement does not depend on a read value the final result will be the same. For instance, in the interest rate example, the interest calculation can be computed as a multiplication (e.g., if the balance is 2000 monetary units and the rate is 1% the interest amount is $2000 * 0.01 = 20$ monetary units), but should be transformed into an addition (i.e., to add 20 units in our example) in order to ensure consistency when some disconnected replica will apply such operation in its reconciliation process (since the balance in the recovering replica might not be 2000 monetary units when it tries to apply the interest rate), since the common operations on bank accounts are additions and subtractions. Note that the same principle applies if the operation being used is any more complex mathematical expression: the resulting effect should be transformed into the common commutative updating operations on the variable. Perhaps in other examples such operations could be multiplication and division, but this leads to avoid additions and subtractions for such examples since multiplication and addition do not commute. If an increment is calculated using a value of the object that is read before performing the update, as it is the case in the interest rate example, the object might still be declared order free. As long as the designer is aware of the *read dependencies* he might decide that the potential inconsistency is acceptable.

Order-free objects should have no *absolute set operation*. An operation that allows the object state to be set directly, can introduce very different results if operations are invoked in different orders. For the previous properties to hold, it is important that the state can be modified in terms of increment and decrement operations. This works well for objects with a *numerical state* or those that represent *collections of data*, for example sets or hash tables.

It should be mentioned, that if order-free objects are to be used to improve availability in case of partitioning, the application should have *tradable consistency*. For some applications small inconsistencies are acceptable. For these applications it may make sense to declare objects order-free, even though a perfectly consistent result cannot be guaranteed. However, there are also applications that do require strong consistency and in which a slightly inconsistent value is of no use. These application, clearly do not benefit from our approach.

3   Replication Protocols Optimised for Commutativity

We have designed two passive replication protocols making use of order-free objects. The protocols are adaptions of our previous primary-per-partition protocol (P4) [6] and of the primary partition approach. We have implemented our protocols in the DeDiSys middleware add-on for CORBA [7]. We have chosen this middleware, because we have access to it and, as its authors, know it very well. However the protocols can easily be integrated into other middleware, as has been demonstrated with .NET [8] implementation of the original P4 protocol.

The two protocols are very similar. The reason to have two slightly different protocols is to demonstrate how commutativity can be exploited in application scenario with different consistency requirements. Whereas the first protocol provides weaker consistency, it shows how commutativity can be used to improve the performance of a replication protocol. In the second protocol we restrict operations on non-order-free objects in some situations to provide strong consistency. In this example we demonstrate how order-free objects can be used to relax consistency on certain data, and hence be used to configure the trade-off between consistency and availability.

Our protocols use the passive replication model. Invocations are made to primary replicas. The state changes are propagated synchronously. That is, the changes are sent to secondary copies before the result of the call is returned to the client. To provide transparency in the passive replication model, we differentiate between *logical object references* and *replica references*. A logical object reference addresses a group of replicas of the same logical object, irrespective of how many replicas there are. A replica reference identifies a concrete replica of an object. The client application is only ever exposed to logical object references. We relax the passive model for read operations which can be served by any secondary replica, preferably a local copy.

We distinguish between two system modes: *normal mode*, in which all nodes are reachable, and *degraded mode*, in which one or more nodes are unreachable due to node failure or partitioning. To return to normal mode from degraded mode, when failure has been repaired, a reconciliation phase is required. We do not describe a specific reconciliation policy in this paper, but any of the techniques we describe in [9] could be used.

The protocols allow objects to be marked order-free. Invocations to normal objects are wrapped into distributed nested transactions, but order-free objects do not have to be executed within the context of a distributed transaction. A further difference between the two types of objects lies in the way updates are propagated. Changes to the primary copy of a normal object are multicast to the secondary copies using reliable multicast with FIFO delivery guarantees. This is not necessary for order-free objects, which propagate their updates using reliable multicast without order guarantee.

The two protocols differ during degraded mode. In the first protocol, if a primary replica is not reachable, because it is located on a failed or unreachable node, a secondary replica is promoted to a temporary primary and all subsequent invocations in the local partition are directed to this replica. In contrary, the second protocol only allows operations on normal objects in a majority partition. That is, invocations are only allowed if the local partition contains at least one more than half the total number of nodes in the system. However, operations on order-free objects are allowed to be executed in minority partitions as well. At reconciliation time, operations done in different partitions are just re-applied to the replicas in the local partition.

Figure 1 shows a description of the steps involved in an invocation in the first protocol. For the second protocol the steps are the same, apart from the fact that operations are not executed on normal objects in minority partitions.

One detail, which is not shown in the figure is that during degraded mode, the protocols need to keep certain information to allow reconciliation. In order to re-apply operations an operation log is needed. For the purpose of the prototype implementation used in the experiments, we have not implemented a reconciliation technique. However, in [9] and in [6] we describe how we have approached reconciliation in the original P4 protocol.

Our protocols have been implemented as follows: In the DeDiSys architecture calls are intercepted at four different points during an invocation, as described in [7]. Calls are intercepted before and after and invocation on both the client and the sever side. Replication protocols may choose to make use of an interception point or not.

Figure 2 shows the main component of the DeDiSys replication support, which we call replication manager (RM). The main task of the RM is to keep track of objects and their replicas. Logical object references are mapped to replica references. The RM consists of various sub-components of which only the *Replication Object Adaptor (ROA)* is visible to the server application. The ROA is a CORBA object adaptor.

The RM is a distributed component with an instance on each system node. The individual instances use an underlying group membership and communication service to

1. Client invokes object using logical ref.
2. Invocation intercepted on client side.
3. If object non-order-free, start transaction.
4. If primary replica not available,
   elect new primary replica.
5. Register primary replica in transaction.
6. Call forwarded to primary replica.
7. Call intercepted on server side.
8. Updates propagated to secondary replicas.
   If non-order free, add FIFO delivery guarantee
   and register secondary replicas in transaction.
9. If transactional, execute commit protocol.
10. Return control to client.

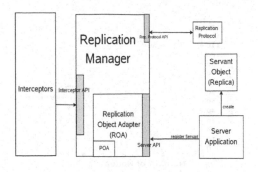

Fig. 1. Replication Protocol

Fig. 2. Replication Support Overview

maintain, on each node, a consistent view of the current partition membership, and for reliable communication between the nodes. DeDiSys uses the Spread group communication service [10]. The RM also interacts with a *replication protocol* component (RP), in which replication protocol details, such as update transfer policies, are implemented. By encapsulating such policy in a separate component with a defined interface, the replication protocol can be changed easily. We have made use of this facility to easily implement our new protocols.

4 Experiments

We have performed a set of experiments to evaluate our protocols. We have measured the average time it takes to invoke order-free objects and have compared the results with the invocation time on normal objects. We performed the experiments in a cluster of nodes. Each node contained a Pentium 4 2.8 GHz CPU and 1GByte of RAM. The nodes were interconnected through a 1 GBit/s Ethernet network. DeDiSys was executed in Sun's Java Virtual Machine version 1.5 on Linux with kernel version 2.4.22. A deliberately simple object was implemented. The objects state consisted of an integer and the only operations were a read and a write operation. Having such a simple object has the advantage that the actual execution time of the object is minimal, and more importantly, constant and does therefore not interfere with the times measured. The object was replicated on all servers. The object was invoked 2000 times, first marked as a normal object and then marked as an order free-object. The experiments were repeated with different number of nodes. All experiments have been performed on the first version of the protocol, which does not disallow operations in minority partitions. However, since we just measured the performance of executed operations, the results should be the same using the second protocol.

We have also performed a simulation of the protocols in a larger setup with up to 40 nodes. The simulation results can be consulted in [11].

Figure 3 shows the average invocation of normal objects and order-free objects, as the number of nodes is increased. The included table shows the actual invocation times

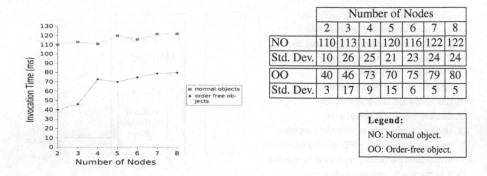

	Number of Nodes						
	2	3	4	5	6	7	8
NO	110	113	111	120	116	122	122
Std. Dev.	10	26	25	21	23	24	24
OO	40	46	73	70	75	79	80
Std. Dev.	3	17	9	15	6	5	5

Legend:
NO: Normal object.
OO: Order-free object.

Fig. 3. Protocol Performance - Invocation Time (ms)

measured, together with the standard deviation. As can be seen, invocations of order-free objects are as much 60 % faster than the invocation of normal objects, with the relative performance improvement showing a decreasing tendency with increasing number of nodes. With eight nodes, the invocation of order free objects was measured 35 % faster than the invocation of normal objects. There is an unexpected spike in the curve at four nodes, but as the standard deviation was relatively high throughout the experiments, the specific numbers cannot be taken as very acurate. Nevertheless, the results still give an indication of the average performance improvement and the improvements measured were well above the standard deviation. The performance inprovement should theoretically be due to a combination of two factors. Firstly, during update propagation involving normal objects, the underlying group membership service is set up to provide FIFO multicast delivery guarantees. This stricter guarantee should in theory be more expensive. Secondly, order-free objects are not invoked in the context of a transaction, which saves several rounds of messages. To investigate this issue further, we have repeated the experiments with a slight variation of the protocol in which order-free objects were also forced to execute in a transaction. The invocation times measured for order-free and normal objects were the same, which means that the above performance improvement was solely due to the fact that no transactions were needed. The lack of a performance difference with lower delivery guarantees can be explained with the implementation of the Spread group communication service. Due the logical token ring nature of the implementation, lesser guarantees than FIFO order are actually not measurably faster. Using a different optimised group communication toolkit would probably further improve the performance of our protocols.

5   Related Work

The idea of exploiting commutativity was first introduced by Gray et al. in the context of mobile databases [12]. The authors state that in certain cases transactions can be designed to be commutative, in order to provide convergence of state, when local updates on a mobile node are applied to the database at reconnection time. As far as we know this concept has not been further explored outside the context of mobile databases.

However, there are different approaches to deal with controlling inconsistencies in replicated systems. In particular, the idea of trading consistency for availability has been explored. Systems that consider this trade-off generally require the application programmer to specify the required consistency or the required availability. The authors of [13] use consistency units (conits) to specify the bounds on allowed inconsistency. A conit is a set of three values representing "numerical error", "order error" and "staleness". The system does not support partitioning. In CoRe [14] the principle of specifying consistency is extended to allow the programmer to define consistency using a larger set of parameters. The system only focuses on data objects; that is, objects that do not cause invocations to other objects. AQua [15] approaches the trade-off from the other direction by allowing availability requirements to be specified. In AQua "quality objects" are used to specify quality of service requirements. AQua considers crash failures, value faults and time faults, but does not consider partitioning. In our previous work on the DeDiSys [16] system, we have based our notion of consistency on integrity constraints. During failure, the system allows certain constraints to be temporarily relaxed, in that they can be evaluated on possibly stale values. We have implemented our protocols on top of the CORBA implementation of DeDiSys [7].

Finally, our P4 protocol taken as the basis for the performance evaluation of the commutativity approach is not the best possible example of a decentralised algorithm (the primary replica coordinates the algorithm and, as a consequence, the result is rather centralised). Indeed, we have used it because it was already implemented and debugged and its deployment did not require any effort. Note however that there are other replication models that can equally benefit from the commutativity principle. For instance, a voting scheme might relax its concurrency control, being possible to obtain the votes of a given node without blocking, since commutative operations must be applied sequentially, but their order does not matter and this removes the need of blocking potentially conflicting requests. An additional example is active replication; if the application can relax the replica consistency when a sequence of operations is being applied, requiring only that their final value in all replicas converges, total order delivery will not be required in active replication. All these issues provide similar performance boosts in their respective replication models to those described here for the P4 protocol. Moreover, the reconciliation simplification obtained with commutative operations is also applicable to all of them.

6   Conclusion and Future Work

In this paper we have shown the advantages commutativity and order free objects can have for some applications. We have described the design of two replication protocols and have provided performance results that suggest a considerable performance improvement in the case of a small number of replicas and a smaller performance improvement in systems with a large number of replicas.

We have also discussed which type of application can benefit most from our approach and given some guidelines for application programmers.

In future work, we plan to perform further experiments with a larger number of real nodes, in order to validate the scalability of the protocols. Furthermore, we would like to test our protocols with a different underlying group communication service, which optimises multicasts that do no require any FIFO order guarantees. Finally, we are planning to perform experiments, in order to evaluate the reconciliation process, as this is were we expect to obtain a big advantage with order free objects.

References

1. Ricciardi, A., Schiper, A., Birman, K.: Understanding partitions and the "non partition" assumption. In: Workshop on Future Trends of Distributed Systems (1993)
2. Schneider, F.B.: Replication management using the state-machine approach. In: Mullender, S.J. (ed.) Distributed Systems, 2nd edn. pp. 17–26. ACM Press, Addison-Wesley (1993)
3. Budhiraja, N., Marzullo, K., Schneider, F.B., Toueg, S.: The primary-backup approach, pp. 199–216. ACM Press, Addison-Wesley (1993)
4. Guerraoui, R., Schiper, A.: Software-based replication for fault tolerance. Computer 30(4), 68–74 (1997)
5. Hadzilacos, V., Toueg, S.: Fault-tolerant broadcasts and related problems. In: Distributed systems, 2nd edn. pp. 97–145. ACM Press, Addison-Wesley (1993)
6. Beyer, S., Bañuls, M., Galdámez, P., Muñoz-Escoí, F.D.: Increasing availability in a replicated partionable distributed object system. In: Guo, M., Yang, L.T., Di Martino, B., Zima, H.P., Dongarra, J., Tang, F. (eds.) ISPA 2006. LNCS, vol. 4330, pp. 682–695. Springer, Heidelberg (2006)
7. Beyer, S., Muñoz-Escoí, F.D., Galdámez, P.: Implementing Network Partition-Aware Fault-Tolerant CORBA Systems. In: Proceedings of the 2nd Int. Conf. on Availability, Reliability, and Security, pp. 69–76 (April 2007)
8. Osrael, J., Froihofer, L., Stoifl, G., Weigl, L., Zagar, K., Habjan, I., Goeschka, K.M.: Using replication to build highly available .NET applications. In: DEXA Workshops, pp. 385–389. IEEE-CS, Los Alamitos (2006)
9. Asplund, M., Nadjm-Tehrani, S., Beyer, S., Galdámez, P.: Measuring Availability in Optimistic Partition-tolerant Systems with Data Constraints. In: 37th Intnl. Conf. on Dependable Systems and Networks (June 2007)
10. Amir, Y., Danilov, C., Stanton, J.R.: A low latency, loss tolerant architecture and protocol for wide area group communication. In: International Conference on Dependable Systems and Networks, pp. 327–336 (2000)
11. Beyer, S., Muñoz-Escoí, F.D.: Exploiting commutativity for efficient replication in partitionable distributed systems. Technical Report ITI-ITE-07/13, Instituto Tecnológico de Informática (2007)
12. Gray, J., Helland, P., O'Neil, P., Shasha, D.: The dangers of replication and a solution. In: Proceedings of the 1996 ACM SIGMOD International Conference on Management of Data, pp. 173–182 (1996)
13. Yu, H., Vahdat, A.: Design and evaluation of a conit-based continuous consistency model for replicated services. ACM Trans. Comput. Syst. 20(3), 239–282 (2002)
14. Ferdean, C., Makpangou, M.: A generic and flexible model for replica consistency management. In: Ghosh, R.K., Mohanty, H. (eds.) ICDCIT 2004. LNCS, vol. 3347, pp. 204–209. Springer, Heidelberg (2004)
15. Cukier, M., et al.: Aqua: An adaptive architecture that provides dependable distributed objects. In: SRDS 1998: Proceedings of the The 17th IEEE Symposium on Reliable Distributed Systems, p. 245 (1998)
16. Osrael, J., Froihofer, L., Goeschka, K.M., Beyer, S.: A system architecture for enhanced availability of tightly coupled distributed systems. In: Int. Conference on Availability, Reliability and Security, pp. 400–407 (2006)

Scheduling in Time-Triggered Networks

Sebastian Voss

EADS Deutschland GmbH,
EADS Innovation Works,
81663 Munich, Germany
Sebastian.Voss@eads.net
http://www.eads.net

Abstract. Modern avionics systems consist of distributed components that communicate over shared networks. Such networks offer — in contrast to point-to-point connections — more system flexibility and cost reduction due to reduced wiring. A key component for increasing the level of performance, and attaining a reliable and predictable communication in distributed avionics systems is an adequate scheduling strategy.

When considering avionics architectures with distributed nodes communicating over a time-triggered shared medium, the problem is to generate a schedule that preserves hard-real time properties. For this, the general task scheduling policies have to be extended by also considering message scheduling to address the whole complexity of such systems.

This paper presents a novel off-line scheduling approach for (hard) real-time systems based on a time-triggered communication network that integrates task scheduling at system level with message scheduling at communication level. The approach augments conventional scheduling rules with algorithms addressing the specific problems of time triggered system design, allowing therefore for the generation of a feasible communication schedule. This leads to an improvement in reliability and predictability of communication.

1 Introduction

The trend for more flexible communication architectures, in particular for safety critical aeronautic applications, respects the growing need for reliable communication in an optimized system configuration. Commercial reasons entail that communication architecture is increasingly realized as shared communication busses instead of several discrete point-to-point connections. This replacement exerts an additional burden of complexity on system designers. Bus arbitration, mapping signals to frames, scheduling, and electrical configuration of a bus with many subscribers get more and more complex. Especially for (hard) real-time systems, additionally design constraints are highly significant for leading to optimal system configuration.

The challenge is to find an optimal system configuration with respect to bandwidth and latency that satisfy the given hard real-time requirements. Such a configuration will guarantee a reliable and predictable communication.

R. Meersman, Z. Tari, P. Herrero et al. (Eds.): OTM 2007 Ws, Part II, LNCS 4806, pp. 1081–1090, 2007.
© Springer-Verlag Berlin Heidelberg 2007

This paper presents a novel off-line scheduling approach for (hard) real-time systems based on a time-triggered communication network that integrates task scheduling at system level with message scheduling at communication level. The approach augments conventional scheduling rules with algorithms addressing the specific problems of time triggered system design, allowing therefore for the generation of a feasible communication schedule. This leads to an improvement in reliability and predictability of communication.

[1] describes a two step approach to generate the message and task schedules. Based on the LET concept [2] first a message schedule and then a task schedule with deadline constraints from the bus schedule is generated. This is a contrast to our approach, where a combined task- and message scheduling approach is presented.

The remainder of this paper is structured as follows. Section 2 provides background information about the applicability of time-triggered protocols for distributed systems. Section 3 describes the basic principles of scheduling in hard real-time systems and the general character of the scheduling problem. Section 4 analyzes the requirements a scheduler for dependable distributed systems has to deal with in the aeronautic field of application. The need of distinction between task- and message scheduling is pointed out in this section. Section 5 describes the problem of scheduling in the development of aeronautic systems and presents and approach for scheduling in time-triggered applications within this field. We conclude in Section 6.

2   Time-Triggered Protocols for Distributed Systems

In distributed system architectures a proper intra-system communication is crucial for reliable operation. From the communication point of view, the subsystems are called nodes. The interconnections between the nodes can be realized as a shared bus. Since more than one node is generally able to transmit on the shared bus, the available bandwidth of the communication channel has to be shared among the nodes. Furthermore, it is conceivable that multiple nodes might attempt to transmit simultaneously. It is therefore necessary to use dedicated protocols to organize bandwidth allocation and arbitration of the shared communication medium. The communication can either be organized according to an event-triggered or a time-triggered principle - both having specific benefits and drawbacks.

In this paper we preset a scheduling approach based on a time-triggered protocol (e.g. FlexRay [3]). The major advantage of a time-triggered protocol is the determinism and therefore a feasible candidate for the use in (hard) real-time systems.

The FlexRay protocol is based on a static communication cycle whose length has to be defined a priori. In order to avoid collisions, two different medium access mechanisms are implemented in FlexRay, providing support for a deterministic time-triggered, as well as a dynamic event-triggered, operation:

- TDMA: Time Division Multiple Access [4]
- FTDMA: Flexible Time Division Multiple Access [3]

These two mechanisms are combined in the communication cycle, where they are implemented in a static segment and a dynamic segment, respectively. So, each communication cycle on each channel consists of a static and a dynamic segment. The static segment is subdivided into slots of equal length. The communication in FlexRay is frame based, i.e. each application message is packed in a FlexRay frame together with other protocol data. This is done before being transmitted on a communication channel. Each FlexRay slot is used to transmit such a FlexRay frame. Upon reception, the frame is first unpacked, before the message is stored in a buffer, where it can be read by the application running on the host controller of the node. The packing and unpacking of messages is carried out by the communication controller, that executes the communication protocol. The message transmission times are specified a-priori. Due to this static configuration a static schedule, computed offline during system design, assigns each node one or more slots on the shared medium, where this node is allowed to transmit its messages, if the clocks are synchronized. Thus, the access to the shared communication medium is collision free. The sending time of each node is pre-specified in a *communication schedule*.

3   Scheduling in (Hard) Real-Time Systems

A real time system may consists of several components for providing various functions. These functionalities can be described as computational activities and are called tasks. In literature the term *task* and *process* are used interchangeable [5]:

A process is a computation that is executed by the CPU in a sequential fashion.

Tasks may be classified as periodic or aperiodic. Periodic tasks are triggered regularly, aperiodic tasks are triggered by events.

A real-time task T_i is characterized by the following parameters [5]:

- Arrival time r_i (request time): the time at which a task becomes ready for execution; it is also referred to request time (release time);
- Computation time C_i: time necessary to the processor for executing the task without interruption;
- Absolute deadline d_i: time before which a task should be completed to avoid damage / degradation;
- Relative Deadline D_i: difference between absolute deadline and the request time: $D_i = d_i - r_i$;
- Start Time s_i: time at which a task starts its execution;
- Finishing Time f_i: time at which a task finishes its execution;
- Response Time R_i: difference between the finishing time and the request time: $R_i = f_i - r_i$

Figure 1 illustrates some of the parameters defined above.

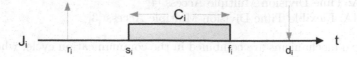

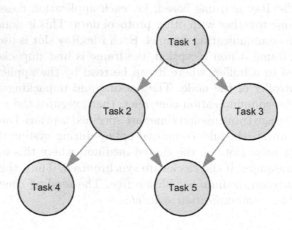

Fig. 1. Typical task parameter

Fig. 2. Precedence Graph

Furthermore the relations between the different tasks have to be defined. Depending on different system requirements tasks often cannot be executed in arbitrary order but have to respect some precedence relations. These relations can be described through a directed acyclic graph, called *precedence graph G*.

Figure 2 illustrates such a precedence graph. This graph can be seen as an abstract model for functional system representation. Each node in this graph represents a task. The arrowed edge showes the direct precedence between the different tasks. Implicitly, this graphs highlights the input and output relations between the different tasks. *Task 2*, for instance needs an input information from the source *Task 1*. After receiving this information the task can be computed on the allocated CPU and sends its output to the *Task 4*, as stated in figure 2.

Schedulers coordinate the execution of system tasks in order to meet the requirements for their functional and temporal behavior. Since a shared system is a multi-processor system, these tasks can overlap in time and may be characterized by different priority and computation duration. The CPUs are assigned to various task according to a scheduling policy. A task is a computation that is executed on a CPU. If a single processor has to execute a set of concurrent tasks, there is the need of a criterion which assigns this CPU to the various tasks. The set of rules that, at any time, determines the order in which tasks are executed on a processor is called a *scheduling algorithm*.

To handle various system requirements, e.g. fault tolerance, determinism, scalability, reliability, validation, and verification, a feasible schedule is needed.

Summarizing the different system requirements, there are many approaches and algorithms (e.g. Earliest Deadline First, Jacksons algorithm, etc.) for optimized task scheduling. Under given system requirements, e.g. periodic / aperiodic functionalities, uni- or multiprocessor systems, a feasible task scheduling algorithm can be found. Buttazzo [5] gives an overview of the predictable scheduling algorithms used for (hard) real-time systems under given system requirements.

However, when considering time-triggered architecture with distributed nodes, task-based scheduling policies have to be extended by taking into account message transmission.

4   Task- and Message Scheduling

In addition to task scheduling at system level, a time-triggered system must also involve communication of messages at the protocol level. This means that the traditional scheduling approach has to be extended by considering message scheduling to address the whole complexity in designing heterogeneous systems based on a time-triggered bus. Hence, a distinction between task and message scheduling is inevitable for correct and proper system design for heterogeneous systems based on a time-triggered bus.

Especially in the aeronautic industry the functional and architectural hard real-time demands placed on aeronautic systems necessitate bounded and acceptable message transmission delays and jitters from the underlying communication. In this context, the relation between task and message schedule is significant.

Before developing a scheduling approach that takes into account both tasks and messages, we proceed first by formulating he scheduling problem in time-triggered networks.

5   Scheduling Strategy

A scheduling strategy is a key component for obtaining high level of performance with time-triggered systems. In order to fulfill system requirements, time-triggered systems must react within precise timing constraints. The correct behavior of these systems depends on the value of the computation and on the time at which the results are produced and transmitted. In the following a schedule is said to be *feasible* for hard real-time systems, if it allows for a task execution that guarantees the hard real time requirements of the systems under consideration.

5.1   Problem Formulation

Given is a set of nodes $N = \{N_1, N_2, ..., N_m\}$ and a set of tasks $T = \{t_1, t_2, ..., t_n\}$ representing the functionality of the system. The mapping from nodes N and tasks T can be described by a function: $\eta : N \to T$. For most aeronautic systems the tasks cannot be executed in a arbitrary order. Precedence relations between the tasks are defined to realize the correct behavior of a system. The set of tasks T

is extended by a precedence graph G, which illustrates the dependencies between the tasks. A set of messages $M = \{m_1, m_2, ..., m_o\}$ is given to describe the input and output data of a task t_i. Therefore, a set of tasks T has to be mapped to the set of messages M through the function $\tau : T \rightarrow M$. This mapping takes into consideration not only system specific parameters but also bus parameters (such as cycle length, number of static slots, slot length, etc.). A scheduler calculates for each task t_i the starting time s_i. In addition to this task schedule the bus schedule has to be calculated. A bus schedule can be described by the allocation of massages to slots: $\sigma : M \rightarrow Slots$.

In the following we illustrate the scheduling problem through an example:

Example 1. We consider a set of given tasks: $T = \{t_1, t_2, t_3\}$ and a set of messages $M = \{m_1, m_2\}$. Furthermore we assume a mapping $\tau : T \rightarrow M$, defined as $\tau(T_1) = \{m_1\}$, $\tau(T_2) = \{m_2\}$, $\tau(T_3) = \{\}$. Each task from T is characterized by a computation duration of 200 μs. As stated in section 3, precedence relations between tasks can be described by an acyclic graph.

Fig. 3. Example: Precedence Graph

For simplicity reasons we consider an easy precedence graph with 3 tasks connected to each other in a consecutive way to describe the functionality of this use case (compare Figure 3). Task t_1 receives a value x, for instance by a special measurement sensor. The output of task t_1 is directly used as input for task t_2. A simple addition is performed by t_2: $f(x) = x + 1$. The result again is used as input for task t_3. Task t_3 receives the value and performs another function: $g(x) = x + 2$. Each task is performed on a different node: $\eta(N_1) = \{T_1\}$, $\eta(N_2) = \{T_2\}$, $\eta(N_3) = \{T_3\}$.

In defining a feasible schedule for this simple example we have to keep in mind the need for diversification between task and message scheduling as stated in Section 4. Thus, on the one hand, the order of tasks, defined by the precedence graph, have to be used to define a proper task schedule (sequence and starting time of the different tasks). On the other hand, a message schedule has to be defined, which assign the messages to specific time slots that are used to transmit the information over the time-triggered bus.

The length of the communication cycle is often bounded in aeronautic applications. (Hard) real-time systems in the aeronautic sector, such as flight control system, are characterized by short repetition times, respectively communication cycles. In addition, when considering a time-triggered protocol, parameters such as cycle length, slot length, etc. have to be taken into account. If the application is synchronized with the bus, communication and application have to be well coordinated. Aeronautic system designers are forced to deal with complex communication aspects, which may lead to a suboptimal configuration of the underlying communication infrastructure.

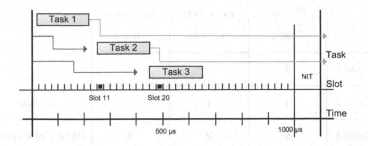

Fig. 4. Problematic configuration between Task and Message Schedule

Commercial scheduling tools for FlexRay request the user (e.g. system designer) manually to specify the task level parameters (e.g. the starting points of each task) as well as the communication level parameters. This implies a complete knowledge of the whole system behavior as well as an higher offline configuration effort, since communication usually depends on schedules that must be calculated a priori. Accordingly, manual configuration becomes difficult or even infeasible very fast, so that an adequate scheduling algorithm is important.

A problem that arises with this approach is described in the following: Assume the system designer specifies the starting points for each task (as illustrated in Figure 4). Furthermore, the system designer states that the result of the task t_1 transmitted with message m_1 in slot 11. In a system, where the application and communication is synchronized, task t_2 is not able to receive the current value of t_1 in the same communication cycle, since this value has not been transmitted at the starting point of t_2. Thus, the current value of task t_1 is logically forwarded to the next communication cycle, where it is available as an input to task t_2. The message m_2 of task t_2 is transmitted in slot 20 and also logically forwarded to the next communication cycle from the same reason as above. The causal end of one consecutive task loop execution has a time delay of 2 communication cycles, as illustrated in figure 5. Task t_1 receives different values by a sensor in the different cycles: $t_1 = 0, 1, 5, 3, 8$. As described above the current value produced by t_1 cannot be used by task t_2 in the same cycle. Therefore, in cycle i t_2 perform its calculation using an old value of t_1 (in this example from cycle $i - 1$). Similarly, t_3 uses in cycle i the pre updated value of t_2 from cycle $i - 1$. This situation is not acceptable in safety-critical aeronautic systems.

The used precedence graphs and the number of tasks used in this simple example are quite small. The shown effect of time delay gets even worse by increasing the number of used tasks and precedence relations between the different tasks. Moreover, when considering parallel task execution the complexity of precedence graphs is further increased.

In the next subsection a novel scheduling approach is presented to handle the problem of time delay and thus to guarantee that task are executed in a timely manner to satisfy hard system requirements.

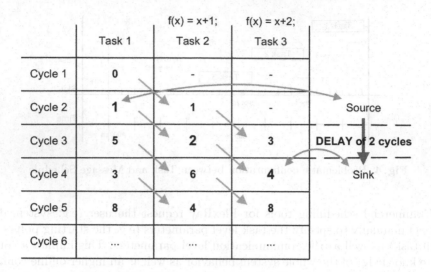

Fig. 5. Communication delay

5.2   Scheduling Approach

In order to define a scheduling approach, additional assumptions/constraints have to be made to bound the complexity.

First of all, the number of processors, the tasks are allocated to, are essential. In general it is distinguished between uniprocessor and multi-processor systems. In an uniprocessor system the architecture would contain out of one "master" node. All tasks are allocated to that CPU. This master node fulfills all function of the network (e.g. calculation). Afterwards the information can be distributed over the network. In multi-processor systems the problem is more complex. Tasks T can be distributed over several nodes, each using a single processor. The communication is performed over a time-triggered shared medium.

We propose a scheduling approach in a multi-processor system with the additional constraint that just one task is allocated to one node. The tasks T are non-preemptive. As an underlying communication protocol we use FlexRay.

Furthermore the precedence graph G is limited in a sequential manner, which is related to the multi-processor constraint above. If just one task t_i is allocated to a specific node N_i and the tasks on the nodes are executed timely sequential, there is no possibility of a parallel task execution.

We assume that the application cycle consists of just one round. Transmission speed is 10MBit/s. Each message m_i is transmitted through a FlexRay frame. In this example a frame is characterized by 20 bytes length, consisting of 12 bytes of payload and the standard 8 byte overhead (header and trailer segment). Thus, the transmission of such a slot would last 1,25 μs. Furthermore we assume 40 static slots. The dynamic segment of FlexRay is not used here. A FlexRay Macrotick (MT) is defined as 1 μs. For the Network Idle Time (NIT) we assume 100MT. Thus, the cycle length is 1100 μs.

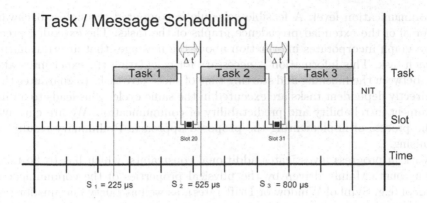

Fig. 6. Combined Task- and Message Schedule with calculated intervals

The proposed scheduling approach is an integrated approach, that schedules the messages M and the tasks T together. The main goal is to calculate intervals $\triangle t$ between the tasks t_1, t_2 and t_3. The size of an interval $\triangle t$ is calculated by the amount of data which is transmitted from one task to the next, according to the precedence graph G. Messages from M describing the amount of data.

The scheduling approach starts from the end of the communication cycle. In the given example in section 3 we start with task t_3, because it is the last task in the precedence graph G. Since t_3 has no output data, it is executed at the end of the communication cycle. In this example, task t_3 has a start time s_3 of: 1000 μs - 200 μs = 800 μs. In this example the amount of data transmitted from parent task t_2 to child task t_3 consist of the single message m_2. The used interval $\triangle t$ can easy be calculated as a single slot, in which the message is transfered. Concerning an additional time margin (e.g. 25 μs) for reading the buffer, the transmission slot of message m_2 is calculated as 31. With the help of the precedence graph the direct parent nodes of t_3 are defined and can be computed. The parent task, according to the precedence graph G, is task t_2. Using the same time margin for writing the information into the buffer the finishing time of t_2 can be calculated and thus, the starting time s_2 of task t_2 is known at 525 μs. Going further backwards this leads to a starting time s_1 of task t_1 of 225 μs and a transmission slot 20 for the message m_1.

The obtained configuration is schematically shown in Figure 6.

The scheduling approach is designed to start from the end of the cycle and going backwards, because there might be a precedence graph where a node has more than one parent node. Thus, the cumulative amount of data, respectively messages, is greater. The proposed scheduling approach is able to handle that and calculate an according interval $\triangle t$.

6   Conclusion

We have presented an off-line scheduling approach for time-triggered avionics applications that integrates task scheduling at system level with message scheduling at communication level. A feasible schedule is computed through a backward traversal of the extended precedence graphs of the tasks. The extended precedence graph incorporates information about the messages that are transmitted by each task. This information is necessary for calculating the exact interval of time between the finishing and starting time of successive tasks to guarantee that all directly dependent tasks are executed in the same cycle. This leads to an improvement in reliability and predictability of communication. We are currently in the process of implementing the proposed approach. Preliminary results are promising.

Work in progress investigates additional constraints, which has to be taken into account. Mainly driven by the physical properties of the communication protocol (e.g. Symbol Window or Drift rates), as well as safety margins for task computation times, which highly depend on the physical architecture.

References

1. Farcas, E.: Transparent distribution of real-time components based on logical execution time (2005)
2. Henzinger, T., Horowitz, B.: Giotto: A time-triggered language for embedded programming. Springer, Heidelberg (2001)
3. FlexRay Consortium: FlexRay communications system protocol specification, version 2.1, revision A. last visited September 2006 (2005),
   URL http://www.flexray.com
4. Kopetz, H.: Real-Time Systems: Design Principles for Distributed Embedded Applications. Kluwer Academic Publishers, Dordrecht (1997)
5. Buttazzo, G.C.: Hard Real-Time Computing Systems. Springer, Heidelberg (2005)

Overview of the Reliability Aspects in the Publish/Subscribe Middleware

Bogumil Zieba

Thales Nederland B.V., AWS Strategy, Technology & Business Development
Haaksbergerstraat 49, Hengelo (O), 7550 GD, The Netherlands
bogumil.zieba@nl.thalesgroup.com
http://www.thales-nederland.nl

Abstract. A publish/subscribe (pub/sub) communication model has recently received significant attention. It has been applied in still more complex distributed systems, wherein high level of a reliability is imposed on the communication infrastructure e.g., Naval Combat Management System (NCMS). Thus far the provisioning of the reliability QoS (Quality-of-Service) in the pub/sub middleware has not been extensively explored. In this paper we make an attempt to identify, comprehensively classify and clearly present all the issues related to the reliability QoS in pub/sub middleware. We classify them into three, proposed by us, reliability aspects: pub/sub clients, events' flow and pub/sub dispatcher(s). The presented reliability issues were identified by means of the following research methodology: case-studies analysis of pub/sub system application in different scenarios e.g., NCMS, the OMG (Object Management Group) industrial standard of pub/sub system, and review of the state-of-art research efforts.

1 Introduction

Some distributed applications communicate in an asynchronous and indirect manner, decoupled by an intermediate software layer, so called **middleware**. Those applications produce or query for large amounts of data without specifying an actual reference or address of other party (**indirect data-centric communication**). This method of communication is different from the "request-reply" method - direct and explicit interaction between applications.

A **publish/subscribe (pub/sub) communication model** has recently received significant attention because it fits well when dealing with distributed data-centric applications. In this model publishers - applications publish events to the distributed systems. Subscribers - applications subscribe to events of interest within the system. An event is a message, containing information, explicitly not addressed to any subscriber. In a case of positive match, between events and subscription, the pub/sub middleware deliveries (dispatch) events from publishers to subscribers and notifies the subscribers about events [1].

The pub/sub distributed system may be seen as **the application-level overlay** network consisting of publishers, subscribers (**pub/sub clients**) and application(s) that facilitate delivery of events (**events dispatcher(s)**). They create a virtual topology

R. Meersman, Z. Tari, P. Herrero et al. (Eds.): OTM 2007 Ws, Part II, LNCS 4806, pp. 1091–1100, 2007.
© Springer-Verlag Berlin Heidelberg 2007

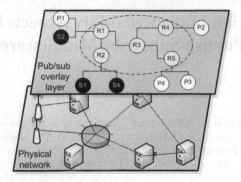

Fig. 1. Mapping an overlay network onto a physical network

above the basic transport protocol level. On figure 1 such overlay is presented, where S, P, R consequently denote subscribers, publishers and events dispatchers.

Many different topology and organizations of pub/sub overlay network have been proposed in literature. Few of them assure **scalability** and **low network load** when applied to the large-scale networks [2][3][4][5][6]. Figure 1 presents the **routers overlay,** in which network of pub/sub **routers** is responsible for scalable events dispatching. Routers selectively dispatch events only to those links, from which they received subscription, preventing from broadcast traffic of events.

In radical mobile scenarios no assumption is made about the topology of the system and the networking infrastructure. The intermediate nodes (routers), in charge of events dispatching, are dynamic and mobile. Such systems blur the distinctions between end nodes (clients) and intermediate ones (dispatchers) assuming that all the network nodes may participate in events dispatching [7]. The **peer-to-peer (p2p)** overlay is symmetric (no distinction between intermediate, and end nodes), decentralized, scalable, and self-organized network [8][9]. For example, Scribe[6], Hermes[1], and Bayeux[5] are pub/sub implementations built on top of two overlay network infrastructures (Pastry and Tapestry). The routing in such network exploits basic lookup service of the p-2-p overlay network, based on independent from network address, application-level addressing, which in very scalable and efficient manner route information from data sources to sinks.

The publish/subscribe (pub/sub) communication model has been applied in still more complex distributed systems, wherein high level of the **reliability** is imposed to the pub/sub middleware communication infrastructure e.g., Naval Combat Management System (NCMS) [10]. Thus far the provisioning of the reliability **QoS (Quality-of-Service)** in pub/sub middleware has not been extensively explored. Nevertheless few publications report provisioning of the reliability QoS in pub/sub middleware in terms of reconfiguration of an application-level pub/sub overlay network [7][11][12][13]. The objective of the reconfiguration is to retain connectivity in case of network topology changes. Other approaches rely on underlying reliable network protocol e.g., TCP or on specific technological solutions e.g., the DDS OMG (Object Management Group) pub/sub specification [14]. DDS (Data Distribution

Service for Real-Time Systems Specification) is the OMG specification for interoperable pub/sub middleware. The purpose of this specification is to offer standardised interfaces and behaviour of pub/sub systems.

In this paper we make an attempt to identify, comprehensively classify and clearly present all the issues related to the reliability in the pub/sub middleware. We classify those issues into three, proposed by us, reliability aspects: **pub/sub clients**, **events' flow** and **pub/sub dispatcher(s)** (see figure 2).

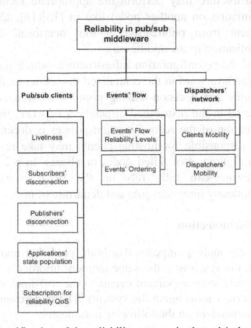

Fig. 2. Classification of the reliability aspects in the pub/sub middleware

The presented classification and reliability issues were identified by means of the following research methodology: case-studies analysis of pub/sub system application in different scenarios e.g., NCMS, the OMG (Object Management Group) industrial standard of pub/sub system, and review of the state-of-art research efforts.

2   Reliability Related with Clients

The role of the pub/sub middleware exceeds beyond the delivery of events. In the modern approach the responsibility for management of a distributed system is shifted to the middleware. Those management activities include monitoring the clients, re-instantiation of clients in case of crash, replication of them (for example for load-balancing purposes) etc. Further in this section we review the different managements techniques related to the pub/sub middleware.

2.1 Liveliness

The pub/sub middleware monitors clients' availability and performs correction actions in case of their failure. Such solution is reported in the DDS specification as the **Liveliness QoS.** The middleware infrastructure, at a client site, periodically publishes the 'heartbeat' event. The frequency of publication is a QoS parameter. The lack of the 'heartbeat' event in a certain deadline denotes the failure of client. The reconfiguration infrastructure may perform the appropriate reconfiguration actions (e.g., client re-instantiation on another node) like in [15][16]. Use of the Liveness QoS releases a client from publication of the 'heartbeat' event shifting the responsibility for publication to the middleware.

Implementation of the reconfiguration infrastructure, which makes benefit of the Liveliness QoS is straightforward in the centralized dispatcher architecture, where the reconfiguration policies and decision-making point is centralized. The problem is going to be complicated in the distributed dispatchers network, where monitoring of clients and a coordination of the reconfiguration polices is decentralized. In routers overlay each router responsible for hosting clients may take responsibility for the monitoring and reconfiguration of failed nodes or clients. In p-2-p overlay network the pub/sub middleware can fully rely on the p-2-p overlay network level management that seamlessly integrates join and departure of nodes or applications.

2.2 Subscribers' Disconnection

The feature of time decoupling imposes that both, publisher and subscriber do not have to be present in the systems at the same moment in order to communicate. The pub/sub middleware may store appointed events to the 'off-line' subscriber for 'late-delivery' (when subscriber joins again the system). The middleware stores the events in its internal buffer dependent on the following parameters:

- **Expire time of events.** The subscriber specifies expire time of event. If an event is not delivered within a specified deadline then the middleware discards such events removing it from the internal buffer [17].
- **Event queue.** The subscriber specifies the length of queue of last published events, which the middleware shall store in case of subscribers' unavailability. The optional value of such parameter is 'all events', what is indication that the middleware shall store all events in account of subscriber[14].
- **Resource limit.** The maximal value of a certain hardware resource e.g., memory allocated for storing events [14].

In the centralized architecture of the event dispatcher, the problem of events storage is obvious. The centralized dispatcher stores all events, which cannot be currently delivered. In the distributed dispatchers' architecture the middleware stores events items in a fully distributed manner. Work [21] reports the service capable of storing events in wireless mesh network. Events are propagated through routers infrastructure, where each router has a partial knowledge about global set of events. Routers update its knowledge about events using gossip algorithms.

2.3 Publishers' Disconnection

In many mission-critical systems (e.g., Naval Command and Control Systems) an additional and redundant publisher is introduced in order to provide continuous publication of events during reconfiguration of the publisher. It takes over the role of the 'main publisher' for the time of its unavailability. When time is a crucial factor the 'hot standby' method is applied, when two publishers publish events simultaneously. The very important issue is to dispatch events from one publisher and discard from the redundant one. The DDS specification introduces QoS **Exclusive Ownership** that determines ownership of events [14]. When two or more publishers publish events that have exclusive ownership, events from one of them (with higher ownership value) are delivered and those from others are discarded.

2.4 Applications State Population - 'Global Data Space'

Clients periodically or state-change-driven publish their internal state as a collection of events. The state of the whole system consists of a collection of events codified as the most current values representing the state of each client in the system. Events representing the state of the system are populated across the dispatchers network creating the illusion of a shared "**global data space**". The idea of 'global data space' is straightforward in centralized dispatcher (like in the DDS [14]). In a fully distributed system, routers store the partial state of the system. They use the epidemic (gossip) algorithms [18][21] in order to propagate the state between each other and detecting inconsistency on a view on systems state. In case of client reconfiguration (re-instantiation, migration to another node), client performs bootstrapping procedure consisting of subscribing to events messages containing its internal state, waiting for delivery, after receiving loading its state and starting the application.

2.5 Subscription for Reliability Values of Publisher

Work [22] advocates that QoS parameters should be embedded on the type or content of the events. The key idea is to subscribe for events not only by subject/content but also according to a certain QoS parameters. In this approach subscribers could subscribe to events published by publishers with a certain level/value of reliability e.g., source redundancy, Mean Time to Failure (MTTF), Mean Time to Repair (MTTR), Mean Time between Failures (MTBF), percentage of time available etc. The example of such subscription in SQL-based language is shown in the table 1.

Table 1. Example of the subscription extended by subscription for reliability of parameters

```
select * from measurements where temperature >20 and
substance='air' and MTTF>10.
```

3 Reliability Related to Events' Flow

The pub/sub model features an indirect communication. Both publishers and subscribers are completely decoupled from each other. They communicate by publishing/subscribing to sequence of events, so called **events' flow**. In this chapter we introduce the concept of assigning the reliability QoS to those flows.

3.1 Events' Flow Reliability Levels

Subscribers require some events to be delivered in a sequence, reliable and precisely-once e.g., **command events' flow** [23]. The pub/sub middleware shall enforce the reliable events' flow and order of delivery. Both, publishers and subscribers may define the required level of reliability giving an explicit directive to the dispatcher about manner of delivery:

- **Unreliable delivery.** The middleware does not enforce the reliable events delivery. This level is applicable for the events' flows generated by real-time sensors that have the following properties: values may change continuously, have short persistence, is time-critical (updates are useless when they are old), idempotent (repeated updates are acceptable), last-is-best (new information is more important than a missed sample) [23]. Two levels of unreliable events delivery are possible: **At most once** – unreliable delivery the events without repetitions. **Best Effort -** unreliable events delivery with possible repetitions.
- **Reliable level.** The middleware enforces the reliable events delivery. This level is essential for e.g., **command events' flow** that requires instructions to be delivered in a sequence, reliable and precisely-once. The possible levels are: **At least once** - reliable delivery, where events updates of the same events are possible. **Exactly once** - reliable delivery. Repeated updates of the same events are not possible.

In [24] we propose the extension to pub/sub syntax appending the QoS to subscription. This makes possible for a subscriber to subscribe to events featured by a certain level of reliability. On the publisher side, appending QoS to advertisement implicitly attach the QoS to every event. We assumed an advertisement-based pub/sub system, in which every event is advertised ahead of its publication. In order to find out the reliability value appended to an event it is necessary to look upon an advertisement that advertised that event before.

The levels of reliability of events delivery are ordered in the sequence: Unreliable delivery< Reliable level. Thus before a dispatcher begins the process of events delivery it needs to check the compatibility of levels required by clients. They must match before the communication occurs.

3.2 Events Ordering

Ordering of events is very important in the command events' flow. The intuitive ordering is by **reception time** [14], where events are presented to the subscriber in order, which were delivered from dispatcher(s). Another possibility assumes events ordering according to **publishing time**. This criterion of ordering is important in case of more then one publisher of the same events' flow.

4 Reliability Related to Distributed Event Dispatcher

The reliability addressed to the distributed events' dispatchers deals with reconfiguration of pub/sub overlay network in a case of changes in network topology. The topology can be changed because of: changes in links connectivity, leaving/ joining distributed dispatchers and clients. We separate our consideration to two cases: mobile clients, where the dispatching infrastructure is assumed to be static and radical mobile scenario, which does not make any assumption about network topology.

4.1 Clients Mobility

The problem of mobile clients (nomadic scenario) concerns how to provide such level of connectivity to clients, that a client can disconnect from the current dispatcher and lately join the network, connecting another dispatcher. Dispatching infrastructure, which remains static, shall follow the location of client and accordingly to it re-address the events that are published or subscribed by/to this mobile client. The reconfiguration of distributed dispatchers is mainly related to the following concerns [7]:

1) Maintain the connectivity in pub/sub overlay in case of clients' disconnection.
   The scope of the problem contains detection of a broken link and determination of a new possible route. For example, routing protocols for mobile and ad hoc network, as: DSR [25] or AODV [26] propose a solution of dealing with detection of broken links. A complete work about the techniques of maintaining the connectivity can be found in [13]. For example, the Bayeux [5] and Gryphon [27] use the mesh network, which redundancy in a network connection guarantees the reliability in case of broken link.
2) Updating the routing information stored on each dispatcher and consistency preservation. Several routing protocols have been proposed in order to maintain the routing information, in consistent way, in case of high-rate of reconfiguration of the overlay [11][12].
3) Detection of lost events and recovering them. Epidemic (gossip) algorithms have been applied for the detection of lost events in distributed dispatcher network [18]. The key idea of this manner of detection is following. Dispatchers communicate with each other (in random manner), exchanging the information about set of events appeared in the system. When any dispatcher reports that its local knowledge about published events in the system differs with knowledge of other dispatcher, then lost events are recovered by sending from other dispatcher, that potentially keeps the history of last events.

4.2 Dispatchers Mobility

MANET (mobile ad hoc network) are scenarios that do not make any assumption about network topology, dispatching infrastructure, and type of mobility. The complicated problem of mobile clients becomes even more challenging by extending it by a mobility of dispatchers. In radical scenarios p-2-p overlay networks seem to provide sufficient solution. They provide the connectivity substrate for pub/sub middleware, which is fully distributed and self-organized (seamlessly integrate leaving and joining nodes/application with high chunk rate). For example, Scribe [6],

Hermes [1] and Bayeux [5] are pub/sub systems built on top of two overlay network infrastructures (Pastry and Tapestry). They do not deal with reconfiguration as it is completely managed by the overlay network level. The routing in such network exploits basic lookup service of the p-2-p overlay network. The p-2-p lookup service is based on independent from network address, application-level addressing, which in very scalable and efficient manner route information from data sources to sinks. P-2-P network fits well dealing with topic-based publish/subscribe, where each topic is associated with multicast tree. Much less intuitive is employing p-2-p overlay network in content-based pub/sub, where each event shall be associated with different multicast tree. However, some attempts to build the content-based pub/sub on p-2-p network have been reported [27][29][30][31].

5   Conclusions and Future Work

The issues related to the reliability QoS in pub/sub systems are tackled in many researches and technologies. We have noticed that they deal with different concerns and views on the reliability. We have not came across the self-contained research work that would give a broad overview on the reliability issues in pub/sub systems. Therefore, we have made an attempt to identify, comprehensively classify and clearly present all the issues related to the reliability in the pub/sub middleware. In order to separate the concerns we introduced the categorization of those issues into three categories: pub/sub clients, events' flow and pub/sub dispatcher(s).

This research is a part of a broader research effort which goal is to develop the reflective QoS provisioning techniques for large-scale and dynamic pub/sub system. This middleware shall be able to adjust itself during the course of execution in order to provide such level of communication between applications that their QoS demands are accomplished.

Acknowledgements

This work was partly supported by Casimir project funded by Netherlands Organisation for Scientific Research (NWO).

I would like to thank Wojciech Mlynarczyk - colleague at Thales Nederland for the valuable comments and his advices regarding this research.

References

[1] Eugster, P.T., Felber, P.A., Guerraoui, R., Kermarrec, A.-M.: The Many Faces of Publish/Subscribe. Technical report, Swiss Federal Institute of Technology in Lausanne (EPFL) (2001)

[2] Pietzuch, P.R., Hermes, A.: Scalable Event-Based Middleware. PhD thesis, Queens' College University Cambridge (February 2004)

[3] Muhl, G.: Large-Scale Content-Based Publish/Subscribe Systems. Dissertation Vom Fachbereich Informatik der Technischen Universitat Darmstadt (2002)

[4] Carzaniga, A.: Architectures for an Event Notication Service Scalable to Wide-area Networks. PhD thesis, Politecnico di Milano, Milano, Italy, (December 1998)

[5] Zhuang, S.Q., Zhao, B.Y., Joseph, A.D., Katz, R., Bayeux, K.J.: An Architecture for Scalable and Fault tolerant Wide-area Data Dissemination. In: 11th Int. Workshop on Network and Operating Systems Support for Digital Audio and Video, 2001. Segall B Arnold D (2001)

[6] Rowston, A., Kermarrec, A., Castro, M., Druschel, P., SCRIBE,: The Design of a Large-Scale Notification Infrastructure. In: Crowcroft, J., Hofmann, M. (eds.) NGC 2001. LNCS, vol. 2233, Springer, Heidelberg (2001)

[7] Cugola G., Murphy A.L., Picco G.P.: Content-based Publish-Subscribe in a Mobile Environment. http://www.elet.polimi.it/upload/picco/papers/mobMwBook.pdf

[8] Rowstron, A., Druschel, P.: Pastry: Scalable, Decentralized Object Location and Routing for Large-Scale Peer-to-Peer Systems. In: Guerraoui, R. (ed.) Middleware 2001. LNCS, vol. 2218, Springer, Heidelberg (2001)

[9] Zhuang, S.Q., Zhao, B.Y, Joseph, A.D, Katz, R., Kubiatowicz, J.: Tapestry: An Infrastructure for Fault-Tolerant Wide-Area Location and Routing. Technical Report UCB/CSD-01-1141, University of California at Berkeley, Computer Science Division (April 2001)

[10] Skowronek, J., van't Hag J.H.: Evolutionary software development. NATO ESD conference (2003)

[11] Cugola, G., Frey, D., Murphy, A.L., Picco, G.P.: Minimizing the Reconfiguration Overhead in Content-Based Publish-Subscribe. In: Proc. of the 19th ACM Symposium on Applied Computing (SAC 2004), pp. 1134–1140. ACM Press, New York (2004)

[12] Picco, G.P., Cugola, G., Murphy, A.L.: Efficient Content-Based Event Dispatching in Presence of Topological Reconfiguration. In: Proc. of the 23rd Int. Conf. On Distributed Computing Systems (ICDCS 2003, pp. 234–243. ACM Press, New York (2003)

[13] Frey, D., Murphy, A.L.: Maintaining publish-subscribe overlay tree in large scale dynamic networks. Technical report, Politecnico di Milano, Submitted for publication (2005), www.elet.polimi.it/upload/frey

[14] Data Distribution Service for Real-Time Systems Specification, ptc/03-07-07

[15] Zieba, B., Sinderen, van M., Wegdam, M.: Technical Report: Reconfiguration Service for Publish/Subscribe Middleware, http://www.bzieba.info

[16] Castaldi, M., Carzaniga, A., Inverardi, P., Wolf, A.A.L.: Lightweight Infrastructure for Reconfiguring Applications - B. In: Westfechtel, B., van der Hoek, A. (eds.) SCM 2001 and SCM 2003. LNCS, vol. 2649, pp. 231–244. Springer, Heidelberg (2003)

[17] OMG. CORBA 3.0 Notification Service. Object Management Group (August 2002)

[18] Costa, P., Migliavacca, M., Picco, G.P., Cugola, G.: Introducing Reliability in Content-Based Publish-Subscribe through Epidemic Algorithms. In: Proceedings of the 2nd International Workshop on Distributed Event-Based Systems (DEBS 2003),

[19] Carzaniga A., Wolf A.L.: Fast Forwarding for Content-Based Networking – University of Colorado, Technical Report CU-CS-922-01 (November 2001)

[20] Stoica, I., Morris, R., Karger, D., Kaashoek, M.F., Chord, B.H.: A Scalable Peer-to-Peer Lookup Service for Internet Applications. In: Proceedings of ACM SIGCOMM (2001)

[21] Gavidia, D., Voulgaris, S., van Steen, M.: A Gossip-based Distributed News Service for Wireless Mesh Networks. In: Proceedings of the Third Annual Conference on Wireless On demand Network Systems and Services, Les Ménuires, France (2006)

[22] Araujo, F., Rodrigues, L.: Quality of Service in Indirect Communication Systems. In: Fourth European Research Seminar on Advances in Distributed Systems (ERSADS 2001) (2001)

[23] Pardo-Castellote, G., Schneider, S., Hamilton, M.-N.: The Real-Time Publish-Subscribe Middleware - White paper - Real-Time Innovations, Inc. http://www.rti.com

[24] Zieba, B., Sinderen, M.: Reflective Approach to QoS Integration into Publish/Subscribe Routing. Submitted for publication

[25] Broch, J., Johnson, D.B., Maltz, D.A.: The Dynamic Source Routing Protocol for Mobile Ad Hoc Networks. Internet Draft (October 1999)

[26] Perkins, C.E., Royer, E.M., Das, S.R.: Ad Hoc On Demand Distance Vector (AODV) Routing. Internet Draft (October 1999)

[27] Bhola, S., Strom, R., Bagchi, S., Zhao, Y.: Exactly-once Delivery in a Content-based Publish-Subscribe System. Dependable Systems and Networks (2002)

[28] Baldoni, R., Marchetti, C., Virgillito, R.A.: Content-Based Publish-Subscribe over Structured Overlay Networks. In: Proceedings of the 25th IEEE International Conference on Distributed Computing Systems (ICDCS 2005) (2005)

[29] Triantafillou, P., Aekaterinidis, I.: Content-based Publish-Subscribe Over Structured P2P Networks. In: Proceedings of International Workshop on Distributed Event-Based Systems (DEBS 2004) (2004)

[30] Terpstra, W., Behnel, S., Fiege, L., Zeidler, A., Buchmann, A.P.: A Peer-to-Peer Approach to Content-Based Publish/Subscribe. In: 2nd International Workshop on Distributed Event-Based Systems (DEBS 2003) (2003)

[31] Muthusamy, V., Jacobsen, H.A.: Small-Scale Peer-to-Peer Publish/Subscribe. In: P2P Knowledge Management Workshop at MobiQuitous (2005)

Workshop on Scalable Semantic Web Knowledge Base Systems (SSWS)

SSWS 2007 PC Co-chairs' Message

SSWS 2007 was the third sequence of the successful Scalable Semantic Web Knowledge Base Systems workshops. This workshop provides a forum for discussing scalability issues for the Semantic Web, with the focus on the development and deployment of knowledge base systems for processing Semantic Web data. We expect that scalability issues are going to challenge the Semantic Web for a long time and significant effort is needed in order to tackle them. This workshop brings together researchers and practitioners to share their recent ideas and advances towards building scalable knowledge base systems for the Semantic Web.

This year we received 15 submissions in total. Each paper was carefully evaluated by three workshop Program Committee members. Based on the reviews, we accepted nine papers. The topics of the selected papers span the areas of large-scale data stores, data integration, semantic mapping, reasoning techniques, and query languages. We sincerely thank the authors for all the submissions and are grateful for the excellent work by the Program Committee members.

August 2007

Achille Fokoue
Yuanbo Guo
Thorsten Liebig
Bijan Parsia

An Evaluation of Triple-Store Technologies for Large Data Stores

Kurt Rohloff, Mike Dean, Ian Emmons, Dorene Ryder, and John Sumner

BBN Technologies
10 Moulton St.,
Cambridge, MA 02138, USA
{krohloff,mdean,iemmons,dryder,jsumner}@bbn.com

Abstract. This paper presents a comparison of performance of various triple-store technologies currently in either production release or beta test. Our comparison of triple-store technologies is biased toward a deployment scenario where the triple-store needs to load data and respond to queries over a very large knowledge base (on the order of hundreds of millions of triples.) The comparisons in this paper are based on the Lehigh University Benchmark (LUBM) software tools. We used the LUBM university ontology, datasets, and standard queries to perform our comparisons. We find that over our test regimen, the triple-stores based on the DAML DB and BigOWLIM technologies exhibit the best performance among the triple-stores tested.

1 Introduction

There has been an explosion recently in the development of new technologies for various areas of the Semantic Web. Some of these manifold new developments include query processing, protocols and triple-store technologies among many others. In this paper we evaluate the most promising new Semantic Web triple-store technologies.

This evaluation was motivated by a client with an existing Semantic Web application who wanted to evaluate other storage alternatives. Therefore this study focused on the particular concerns of the client's application, such as the sheer volume of data, fundamental performance measures (load time and query speed), and very basic inference.

This study leverages the Lehigh University Benchmark (LUBM) [1,2], a widely accepted methodology for evaluating triple-stores that is relevant to the context of this evaluation. The metrics in the LUBM evaluation methodology include load time, repository size, and the response time of 14 different queries. The 14 queries are also designed to assess the soundness and completeness of the reasoning performed.

To date there have only been a handful of attempts at formal evaluations, studies or benchmarking efforts that characterize key aspects of triple-stores. While these studies provide valuable insight into many of the challenges associated with benchmarking triple-stores, most of these studies address specific ranges of performance metrics measured under targeted test conditions. We focus on evaluating what we have identified as the most promising currently available and beta-test triple-store technologies that can handle very large data-sets (on the order of 1 billion triples.)

The most relevant studies to this evaluation include the MIT Scalability Report, the RDF Scalable Storage report, and the 2003 Semantic Web Tools Assessment. The MIT

R. Meersman, Z. Tari, P. Herrero et al. (Eds.): OTM 2007 Ws, Part II, LNCS 4806, pp. 1105–1114, 2007.
© Springer-Verlag Berlin Heidelberg 2007

report [3] was primarily focused on the storage and query response capabilities of triple-stores, and queries with large expected results in particular. The RDF Scalable Storage Report [4] provides an overview of a number of open source/free triple-store implementations that are available and does not include any benchmarks or performance measurements. The Semantic Web Tools Assessment[1], a part of the DARPA Agent Markup Language (DAML) program, documented a qualitative survey of available OWL tools.

The triple-stores we evaluated in this study (coupled with their appropriate query framework) are:

- Sesame 1.2.6 + MySQL 5.0
- Sesame 1.2.6 + DAML DB 2.2.1.2
- Sesame 1.2.6 + SwiftOWLIM 2.8.3
- Sesame 1.2.6 + BigOWLIM 0.9.2
- Jena 2.5.2 + MySQL 5.0
- Jena 2.5.2 + DAML DB 2.2.1.2
- AllegroGraph 1 (Free Java Edition)
- AllegroGraph 2.0.1 (Free Java Edition)

The remainder of the paper is organized as follows: Section 2 presents the methodology and metrics we use to evaluate the triple-store technologies. Section 3 presents the study results, and Section 4 analyzes the results and discusses areas for future work.

2 Evaluation Methodology and Metrics

2.1 Framework

We performed our evaluation in a Java environment using a modified version of the LUBM test harness [1,2]. The modified test harness incrementally loads test data in ten blocks and a full set of queries are run after each incremental load. This allowed us to collect data on how the data load times and query response times vary with the number of triples loaded into the triple-store. The triple stores tested here supported different query languages, and so the LUBM queries were ported into their appropriate query languages (SeRQL and SPARQL.)

Data for our evaluation was generated by the UBA (Univ-Bench Artificial data generator) tool developed by the LUBM authors. The LUBM test data are extensible data created over the Univ-Bench ontology. LUBM datasets represent multiple universities, each with multiple professors, students, graduate students, courses, and departments. The courses, students, graduate students, and professors belong to departments, and students, graduate students, and professors can be assigned to courses.

Most of the triple-stores (MySQL[2], DAML DB[3], SwiftOWLIM, and BigOWLIM[4]) require a query framework. We used Sesame[5] and Jena[6] as query frameworks due to

---

[1] http://www.daml.org/2003/10/tool-assessment/Repository.html
[2] http://www.mysql.com/
[3] http://www.daml.org/2001/09/damldb/
[4] http://www.ontotext.com/owlim/
[5] http://openrdf.org/
[6] http://www.hpl.hp.com/semweb/

their wide acceptance and maturity. Both versions of AllegroGraph[7] came with their own query framework. A brief overview of the features and the capabilities of the paired triple-store technologies and query frameworks is presented in Table 1. Note that an enhanced version of DAML DB called "Parliament", has been released as a part of BBN's Asio tool suite for the Semantic Web[8].

Table 1. Triple-Store Feature Matrix

	Sesame + MySQL	Sesame + DAML DB	Sesame + SwiftOWLIM	Jena + MySQL	Jena + DAML DB	AllegroGraph 1	Sesame + BigOWLIM	AllegroGraph 2.0.1
Query language	SeRQL	SeRQL	SeRQL	SPARQL	SPARQL	SPARQL	SeRQL	SPARQL
Rule support	RDFS	Limited RDFS	RDFS + most of OWL Lite	RDFS + most of OWL Lite	Limited RDFS	RDFS + a bit of OWL Lite	RDFS + most of OWL Lite	RDFS + a bit of OWL Lite
owl:sameAs	No	Using Layered SAIL	Yes	Yes	No	Yes	Yes	Yes
owl:Inverse Functional Property	No	Yes	Yes	Yes	Yes	No	Yes	No
API Extensibility	Yes, SAIL	Yes, SAIL	Yes, SAIL	Yes, Jena	Yes, Jena	No	Yes, SAIL	No
Maturity	Mature	Mature	Mature	Mature	Mature	Mature	Beta	Beta
Price	Free	Free	Free	Free	Free	Not free	Unknown	Not free

We initially generated a data set consisting of 1500 universities written to independent OWL data files, which corresponds to roughly 200 million triples. We found that a number of the triple-store technologies (Sesame + BigOWLIM, Sesame + DAML DB and Jena + DAML DB) performed very well over this range of data, so we generated a second, larger data set to further test the performance of these triple-stores. The data-set used in this second round of testing consisted of 8000 universities written to independent OWL data files, which corresponds to roughly 1 billion triples. Due to time constraints, we were not able to evaluate the triple-stores over the entire larger data-set, but we were able to evaluate the selected triple-stores over most of this data. The developers of AllegroGraph 2.0.1 were unable to provide us a beta-test version of their technology during the first round of testing, but we included this triple-store in our second round of testing with the 8000-university dataset.

---

[7] http://www.franz.com/products/allegrograph/

[8] http://asio.bbn.com/parliament.html

2.2 Test Environment and Hardware

Our evaluation focused on centralized data-stores with a single user making queries. For this reason, we assumed that the test dataset, triple-store, and client software all reside on the same host and that queries are processed serially.

The test hardware used in the study was a Dell PowerEdge 2950 with two (2) Quad Core Intel Xeon E5345 Processors running at 2.33Ghz, a 1333MHz FSB, 16GB of RAM running at 533MHz and 6TB of storage disk storage. The test harness was written in Java and the tests were run using the Sun Java virtual machine, version 1.6.

We found that all of the triple-stores except Sesame + DAML DB required additional heap space to operate on datasets with more than a million triples. We tested multiple settings of the Java VM -Xmx and -Xms options to allocate additional heap space. Setting the virtual machine options of "-Xms 13000m -Xmx13000m", which allocates 13GB of heap to the virtual machine, provided all around good performance for the triple-stores. This large allocation of heap space could potentially be a hindrance to DAML DB which accesses RAM outside of the Java heap.

During initial testing we observed the Java virtual machine halting test execution due to the garbage collection operation exceeding its overhead limit. Through both testing rounds we turned off the garbage collection overhead limit using the "-XX:-UseGCOverheadLimit" JVM option. All triple-stores were then able to run without halting due to memory errors except for Sesame + SwiftOWLIM and AllegroGraphv1, which were able to run at least as long before halting due to an out-of-memory error.

2.3 Metrics

The individual metrics initially used by the LUBM were used as a starting point for the data collection in this evaluation study. Specifically, we collected data on:

- Cumulative Load Time: the time, measured in hours, to load the OWL files describing university departments into the triple-store for a given number of triples. This includes any time spent processing the ontology and source files.
- Query Response Time: query response time is calculated as the mean of the execution time for each of four identical queries.
- Query Completeness and Soundness: a triple-store is complete if it returns all of the correct responses to a query, while a triple-store is sound if it only returns correct responses to a query.
- Disk-Space Requirements: the amount of disk-space required to load the evaluation data. This data was collected only during the second round of testing for the triple-stores that were able to load and perform query operations on the large dataset.

The LUBM queries we used to measure the query response time were a low-volume, low-complexity query (LUBM Query 1), a high-volume, low-complexity query (LUBM Query 2), and a high complexity query (LUBM Query 9). Our descriptions of volume and complexity with respect to query types is taken from the LUBM documentation. A low-volume query is one where the number of query results is very small relative to the number of triples in the triple-store. Conversely, a high-volume query is one that returns a large portion of the stored triples in response to a query. A low-complexity query is one that requires very little processing power to complete, while a high-complexity query is one that requires substantial computing power to complete.

LUBM Query 1 asks for the number of GraduateStudents at a particular university that take a particular course. The correct response for this query over our test data is always 4. LUBM Query 14 requests the number of UndergraduateStudents in the knowledge base. A correct response for this query is a large fraction of the number of triples stored in the triple-store (on the order of 15%). LUBM Query 9 requests all students who take a course taught by their adviser. This query contains 5 atomic clauses that need to be addressed by the triple-stores' query engines. The correct response to this query is on the order of several thousand. Performance issues associated with low- and high-volume queries did not affect Query 9.

During initial testing, we noticed that the ordering of query clauses has a potentially large impact on the performance of the triple-stores. By ordering the queries such that fewer items in a data-set satisfy the initial clauses the query response time of queries can be improved upon otherwise identical queries with query clause reordered so that more elements of a data-set satisfy the reordered initial query clauses.

We attempted to optimize our query orderings as much as possible to minimize query response times during the test process while using the same query orderings for all triple-store evaluations. This was not necessary when using DAML DB because it re-orders queries automatically, but for the sake of consistency we used the manually re-ordered queries for all triple stores.

Queries commonly contain explicit restrictions that can be inferred from the ontol-ogy. In such cases, a clever query optimizer may be able to eliminate the redundant statement from the query, thereby reducing the level of processing required to answer it. We suspect that none of the triple-stores in this study perform this query optimiza-tion. We did not attempt to remove these statements manually, but this might be an avenue for future study.

3 Study Results

During our operations of the triple-stores, we found that not all of the triple-stores we tested were complete. AllegroGraph 2.0.1 was able to correctly respond to all queries. Sesame + SwiftOWLIM, Sesame + BigOWLIM and all combinations of MySQL and DAML DB with Sesame and Jena were able to respond correctly to all queries except for Queries 10-13. The triple-stores were unable to infer the consequences of inverseOf and sameAs relationships. Two of the triple-stores (AllegroGraph 1 and Sesame + BigOWLIM) returned errors with Queries 14 and 13, respectively. Rather than demon-strating incompleteness, this most likely demonstrates that these triple-stores have bugs in their implementations that should be addressed.

During the first round of testing, Sesame + BigOWLIM appeared to be unsound for a large portion of the LUBM queries. During the second round of testing, we found that we could correct these errors by exclusively using SELECT DISTINCT clauses in the queries and not using WHERE clauses in the queries. Interestingly, SwiftOWLIM did not have these difficulties, so they were most likely due to BigOWLIM's beta-test status. These alterations in the queries used for Sesame + BigOWLIM were the

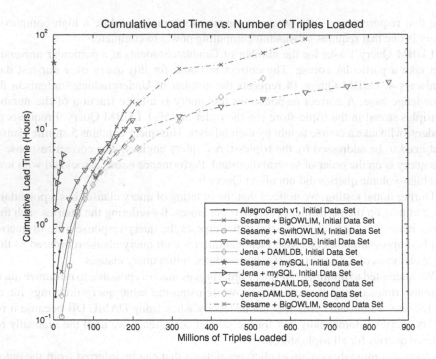

Fig. 1. Cumulative Load Time for Various Triple-Stores for Two Data Sets

only differences between queries in the triple-stores evaluations (except for translations between the SeRQL and SPARQL languages.)

During the first round of testing, Jena + DAML DB was incorrectly configured so that the full inferencing power of DAML DB was unused. This error was corrected during the second round of testing.

A graph of the cumulative load times vs. the number of triples loaded for the various triple-stores is shown in Figure 1. In the graph in Figure 1 and all other graphs in later figures, the results from operations with the initial 1500 university dataset during the first test round is plotted using solid lines and results from operations with the larger 8000 university dataset during the second test round is plotted using dashed lines.

Ideally, the cumulative load time of a triple-sore for this performance metric is linear in the number of triples loaded. (Note that a linear relationship appears to be sub-linear on a semi-log plot, which is what we used in all figures.) All of the triple-stores we tested had cumulative load times that grew linearly with the number of triples loaded.

As can be seen from the graph Jena + DAML DB and Sesame + DAML DB had the best cumulative load time performance from the triple-stores that we evaluated using the larger dataset. Over the initial dataset, Jena + MySQL and Sesame + MySQL exhibited substantially worse performance than the other triple-stores. Due to time constraints associated with their load times, we terminated the test runs of the Jena + MySQL and Sesame + MySQL triple-stores before completion. When provided 13 GB of heap space with 64-bit addressing, AllegroGraph 1 and Sesame + SwiftOWLIM were unable to load all of the triples used in the test run due to insufficient memory errors. AllegroGraph 1

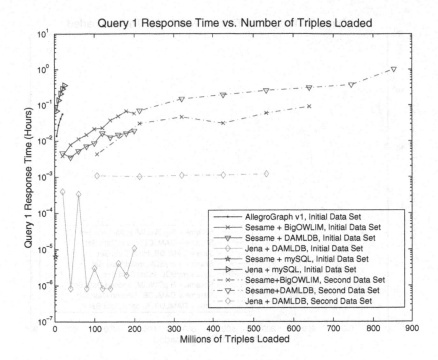

Fig. 2. Query 1 Response Times for Various Triple-Stores for Two Data Sets

was able to load 25 million triples, while Sesame + SwiftOWLIM was able to load just over 80 million triples. In order to achieve reasonable query performance in Allegro-Graph 2.0.1, an indexAll operation needs to be performed after large sets of data are loaded but before query operations are issued. We found that after loading a large dataset, the indexAll operation was unable to complete (most likely due to a bug), so we were unable to obtain any meaningful performance results for AllegroGraph 2.0.1.

We also measured the amount of disk space the various triple stores required during the second testing round (see Table 2). In our evaluation, Sesame + BigOWLIM made more efficient use of disk space than Sesame + DAML DB and Jena + DAML DB.

Table 2. Disk Space Required per Million Triples Loaded

Triple-Store	Disk Space Required per Million Triples Loaded
Sesame + DAML DB	175 MB
Jena + DAML DB	167 MB
Sesame + BigOWLIM	53 MB

Figure 2 shows the Query 1 response performance of the triple-stores. Jena + DAML DB had very good performance and was able to return responses in a nearly trivial amount of time (in less than a second.) The accuracy of our measurements of query

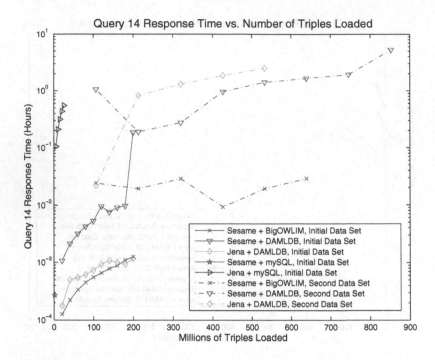

Fig. 3. Query 14 Response Times for Various Triple-Stores for Two Data Sets

response times was limited to 15ms due to the limitations of the system clock. This caused the query response times of Jena + DAML DB to appear to fluctuate wildly on the semi-log plot for the data collected during from the initial dataset. Out of the other triple-stores, Sesame + DAML DB and Sesame + BigOWLIM exhibited performance worse than Jena + DAML DB but comparable to one another.

Ideally, a triple-store should have a response time that does not increase, or increases very little as the number of triples loaded increases. We found that only Jena + DAML DB, Sesame + DAML DB and Sesame + BigOWLIM exhibit this behavior. The jump in Query 1 response time for Jena + DAML DB between rounds is most likely due to the increase in inferencing power for Jena + DAML DB between rounds.

Figure 3 shows the Query 14 response performance of the triple-stores. Sesame + BigOWLIM exhibited the best performers for this metric over both rounds of testing. The performance degradation of Jena + DAML DB between rounds is most likely due to the increase in inferencing power for Jena + DAML DB between rounds. Increased query response times for Jena + DAML DB and Sesame + DAML DB is most likely due to the limits of caching in DAML DB.

Figure 4 shows the Query 9 response performance of the triple-stores. Sesame + BigOWLIM had the best performance, and Sesame + DAML DB and Jena + DAML DB performed roughly an order of magnitude worse than Sesame + BigOWLIM. All of the triple-stores in this study had performance that degraded super-linearly with respect to the data repository size.

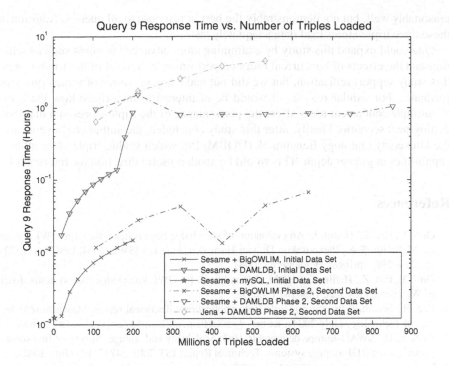

Fig. 4. Query 9 Response Times for Various Triple-Stores for Two Data Sets

4 Analysis and Areas for Future Work

Due to the observed limitations in the ability of Sesame + SwiftOWLIM to load very large datasets, we feel that this triple-store would not be a good performer as compared to the other triple-store technologies when evaluated over large datasets. Admittedly, this is a design limitation of SwiftOWLIM as it was designed to operate over smaller datasets. Due to the observed very long load times, we feel that AllegroGraph 1, Sesame + MySQL, and Jena + MySQL would not be appropriate for use with large datasets, even if load time were a secondary consideration for evaluation. If the indexAll operation in AllegroGraph 2.0.1 could be fixed, this technology could exhibit very good performance, but we were unable to adequately evaluate the query response times of this triple-store due to the non-termination of the indexAll operation.

We feel that the triple-stores that offered the best all-around performance for operations with a large dataset were Sesame + DAML DB, Jena + DAML DB, and Sesame + BigOWLIM. Each of these triple-stores has their own relative merits. Most importantly, all three of them provide adequate query response time performance for various queries, but no one triple-store is clearly better than the other triple-stores in all cases under the conditions evaluated in this study. For instance, Sesame + BigOWLIM provides better response time than the other triple-stores when responding to complex queries. Additionally, Jena + DAML DB provides the best performance for low-complexity queries, at least on the low-complexity queries that we tested. Sesame + DAML DB performs reasonably well, but not demonstrably the best in response to all queries. Additionally, these three triple-stores load data adequately fast.

One could expand this study by examining more advanced features such as reification and the effects of concurrent users on performance. Several of the triple-stores in this study support reification, but we did not make any measures of reified query performance. For similar reasons, it would be of interest to investigate how the actions of multiple concurrent users affect the performance of the triple-stores in a distributed deployment scenario. Finally, after this study concluded, the authors became aware of the University Ontology Benchmark (UOBM) [5], which studies triple-store inference capabilities in greater depth. This would be another useful direction for future work.

References

1. Guo, Y., Pan, Z., Heflin, J.: An evaluation of knowledge base systems for large OWL datasets. In: McIlraith, S.A., Plexousakis, D., van Harmelen, F. (eds.) ISWC 2004. LNCS, vol. 3298, pp. 274–288. Springer, Heidelberg (2004)
2. Guo, Y., Pan, Z., Heflin, J.: LUBM: A benchmark for OWL knowledge base systems. Journal of Web Semantics 3(2), 158–182 (2005)
3. Lee, R.: Scalability report on triple store applications. Technical report, Massachusetts Institute of Technology (July 2004), http://simile.mit.edu/reports/stores/
4. Beckett, D.: SWAD-Europe deliverable 10.1: Scalability and storage: Survey of free software / open source RDF storage systems. Technical Report IST-2001-34732, EU (July 2002), http://www.w3.org/2001/sw/Europe/reports/rdf_scalable_storage_report
5. Ma, L., Yang, Y., Qiu, Z., Xie, G., Pan, Y., Liu, S.: Towards a complete OWL ontology benchmark. In: Sure, Y., Domingue, J. (eds.) ESWC 2006. LNCS, vol. 4011, pp. 125–139. Springer, Heidelberg (2006)

Hawkeye: A Practical Large Scale Demonstration of Semantic Web Integration

Zhengxiang Pan, Abir Qasem, Sudhan Kanitkar,
Fabiana Prabhakar, and Jeff Heflin

Department of Computer Science and Engineering, Lehigh University
19 Memorial Dr. West, Bethlehem, PA 18015, U.S.A.
{zhp2,abq2,sgk205,ffp206,heflin}@cse.lehigh.edu

Abstract. We discuss our DLDB knowledge base system and evaluate its capability in processing a very large set of real-world Semantic Web data. Using DLDB, we have constructed the Hawkeye knowledge base, in which we have loaded more than 166 million facts from a diverse set of real-world data sources. We use this knowledge base to demonstrate realistic integration queries in e-government and academic scenarios. In order to support Hawkeye, we extended DLDB with additional reasoning capabilities. At present, the Semantic Web consists of numerous independent ontologies. We demonstrate that OWL can be used to integrate these ontologies and thereby integrate the data sources that commit to them. In terms of performance, we show that the load time of our system is linear on the number of triples loaded. Furthermore, we show that many complex queries have response times under one minute, and that simple queries can be answered in seconds.

1 Introduction

The 2005 index of Swoogle [2] contained 850,000 of Semantic Web documents, in 2006 this index had 1.5 million SW documents and at the time of writing this paper it boasts a staggering 2.1 million SW Documents. The Semantic Web is growing and clearly scalability is an important requirement for Semantic Web systems. Furthermore, the Semantic Web is an open and decentralized system where different parties can and will, in general, adopt different ontologies. Thus, merely using ontologies, does not reduce heterogeneity: it just raises heterogeneity problems to a different level. Without some form of alignment, the data that is described in terms of one ontology will be inaccessible to users that ask questions in terms of another ontology. In this paper we present a scalable system (166 million triples) and a knowledge base that has been integrated using only OWL axioms (as opposed to special purpose mapping languages).

We put forward that, in addition to providing semantics to the data, OWL can also be used to establish alignments between these heterogeneous web sources. Using map ontologies, ones that contain OWL axioms that align the concepts of two ontologies, we have integrated many autonomous data sources and successfully demonstrated useful queries. For example, consider a researcher looking for colleagues to collaborate with her in a paper. One heuristic she may apply in her search is to look for people who

R. Meersman, Z. Tari, P. Herrero et al. (Eds.): OTM 2007 Ws, Part II, LNCS 4806, pp. 1115–1124, 2007.
© Springer-Verlag Berlin Heidelberg 2007

have cited a paper that has been cited by her in her other publications. Obviously, this can be done using Google, but it will require several intermediate steps to meet her specific information need. Our system can get her the answer from two different sources (Citeseer and DBLP) in just a few seconds. We discuss this query and others that we have tested in Section 3.2

Before we present our work, we would like to briefly review the state of the public Semantic Web. We note that there are several traits that we have observed in the existing Semantic Web data (indexed by Swoogle) that influenced our design choice.

First, we observe that if we account for minor syntactic errors (e.g. missing a type declaration) most of the ontologies in the current Semantic Web have an expressivity equivalent or less than OWL DL. As these syntactic issues can be programmatically resolved, most of the OWL Full ontologies can be easily converted to OWL DL, which is most likely what the developer had intended [1]. In a recent survey of ontologies, Wang et al. [13] report similar syntactic errors leading to OWL Full ontologies. Therefore, our system's overall focus is to support OWL DL as opposed to OWL Full.

Second, we have observed that the ontologies and data from the social network domain are currently dominating the Semantic Web landscape. The most frequently used ontology in the Semantic Web is the Friend of A Friend (FOAF) ontology. It is interesting to note that although FOAF was originally designed for individuals to make their profiles available to public, the prevalence of FOAF data is due to Blog sites and social network sites (LiveJournal, etc.) which generate FOAF data from users' public profile. Each site generates its own URI for an individual and therefore we have several different URIs pointing to the same object. This is essentially an entity resolution problem. In order for us to have a plausible integration of the Semantic Web, we needed to resolve these duplicate entities, establish alignments and add instance equality reasoning to DLDB system. The *owl:InverseFunctionalProperty* has helped us in this task. Basically if a property, p, is annotated as InverseFunctionalProperty, then $\forall$ x, y, z p(y,x) $\wedge$ p(z,x) $\rightarrow$ y = z. With the FOAF data we have used InverseFunctionalProperty to state for example if two individuals (two distinct URIs) have the same email address then they essentially are the same individual.

Third, we have observed that it is important to support the *TransitiveProperty* attribute of OWL properties. There are several ontologies in the Semantic Web that describe properties in terms of this characteristic. For example, many ontologies have made use of transitive properties such as *hasPart* and *subLocationOf*. SKOS, the World Wide Web Consortium's recent effort in describing a controlled vocabulary for thesauri, classification schemes, subject heading systems and taxonomies within the framework of the Semantic Web, makes extensive use of transitive properties

In what follows we first describe our enhanced DLDB system. We present its architecture, design and implementation with a focus on the additional reasoning and optimizations that we have added to the system based upon the characteristics of the Semantic Web. After presenting the system we then describe our Hawkeye knowledge base and at the end present related work and conclude. Note: in this paper we build on our initial work [11] in this area and now present a more comprehensive demonstration on a larger set of Semantic Web data.

2   DLDB: A Semantic Web Query Answering System

The initial architecture of DLDB is presented in [10]. It is a knowledge base system that extends a relational database management system with additional capabilities for partial OWL reasoning. The DLDB core consists of a Load API and a Query API implemented in Java. Any DL Implementation Group (DIG) compliant DL reasoner and any SQL compliant RDBMS with a JDBC driver can be plugged into DLDB. This flexible architecture maximizes its customizability and allows reasoners and RDBMSs to run as services or even clustered on multiple machines.

It is known that the complexity of complete OWL DL reasoning is NEXPTime-complete. Our pragmatic approach is to trade some completeness for performance. The overall strategy of DLDB is to find the good balance of precomputation of inference and run-time query execution via standard database operations. Whenever DLDB loads an RDF/OWL file it determines if it is an ontology or instance data document. Ontologies are first processed by the DL reasoner in order to compute implicit subsumptions and the results are used to create tables and views in the database. Instance data are directly loaded into the database tables. The consideration behind this approach is that DL reasoners are optimized for reasoning over ontologies, as opposed to instance data.

DLDB facilitates integration by supporting the perspectives presented in Heflin and Pan [5]. Ontology perspectives allow the same set of data sources to be viewed from different contexts, using different assumptions and background information. Each perspective will be based on an ontology. Thus, data sources that commit to the same ontology have implicitly agreed to share a context. When it makes sense, we also want to maximize integration by including data sources that commit to different ontologies. We require that each Semantic Web query to be associated with a particular perspective, so that the answers to a query depend on the entailment of the perspective. For example, when the perspective is based on different mapping ontologies, a query would get different answers resulting from different mapping axioms.

In DLDB, creating tables corresponds to the definition of classes or properties in ontology. Each class and property has a table named using its URI. This means new tables are created as new ontologies are discovered. In DLDB, class hierarchy information is stored through database views. The view of a class is defined recursively. It is the union of its table and all of its direct subclasses' views. Hence, a class's view contains the instances that are explicitly typed, as well as those that can be inferred.

DLDB currently supports conjunctive queries. In terms of query language, DLDB supports a subset of SPARQL, namely the SELECT query form, combined with the triple pattern and filter that allows numeric types. We think this subset covers most of the frequently posed extensional queries. We plan to support some useful modifiers such as "ORDER BY" and filters on date.

During query execution, predicates and variables in the query are substituted by table names and field names through translation. Depending on the perspective being selected, the table names are further substituted by corresponding database view names. Finally, a standard SQL query sentence is formed and sent to the database via JDBC. Then the RDBMS processes the SQL query and returns appropriate results.

Instance Equalities. OWL does not make the unique names assumption, which means that different names do not necessarily imply different objects. Given that many individuals contribute to the Web, it is highly likely that different IDs will be used to refer to same object.

In DLDB, each unique URI is assigned a unique integer id. Our approach to equality is to designate one id as the canonical id and globally substitute the other id with this canonical id in the knowledge base. The advantage of this approach is that there is effectively only one system identifier for the (known) individual, nevertheless that identifier could be translated into multiple URIs. Since reasoning in DLDB is based on these identifiers instead of URIs, the existing inference and query algorithms do not need to be changed to support equalities.

However, in many cases, the equality information is found much later than the data that it "merges". Thus, each URI is likely to have been already used in multiple assertions. Finding those assertions is especially difficult given the table design of DLDB, where assertions are scattered into a number of tables. It is extremely expensive to scan all the tables in the knowledge base to find all the rows that use a particular id. We devised auxiliary tables to keep track of the tables that each id is appeared in. An `Individual_Occurrence` table is used to record the the occurrences of each id. To substitute an id with another id, the procedure queries those auxiliary tables to find the set of tables (and columns) upon which an update is issued to perform the substitution.

Often times, the knowledge on equality is not given explicitly. Equality could result from inferences across documents: *owl:FunctionalProperty* , *owl:maxCardinality* and *owl:InverseFunctionalProperty* can all be used to infer equalities. DLDB is able to discover equality on individuals using a simple approach. If two URIs have the same value for an *owl:InverseFunctionalProperty*, they are regarded as representing the same individual. A naive approach is to check it every time a value is being inserted into an inverse functional property table. However, this requires a large number of queries and potentially a large number of update operations. In order to improve the throughput of loading, we developed a more sophisticated approach which queries the inverse functional property table periodically during the load. The specific interval is specified by users based upon their application requirements and hardware configurations (we used 1.5 million in our experiment). This approach not only reduces the number of database operations, but also speeds up the executions by bundling a number of database operations as a stored procedure.

Transitive Closure and Its Interaction with Other Reasoning. One of the ABox reasoning tasks is to infer implicit property assertions through the transitive property. This task can be regarded as computing a transitive closure over a directed acyclic graph. In DLDB, we must also address how these algorithms interact with other reasoning. One of the existing algorithms is to maintain the transitive closure from scratch, i.e. starting from when the table is empty [3]. Each time a new pair is added, the maintenance algorithm will compute new relations and add them to the table so that the table corresponds to its transitive closure. However, this algorithm would not work under some circumstances. For example, the property *isIn* is transitive, whereas its two subproperties: *isInState* and *isInRegion*, are not (and should not be) transitive. If the data only contains instances of *isInState* and *isInRegion*, no transitive closure algorithm could be invoked. When *isIn* is queried, its transitive closure has not been computed. Note, it is difficult to continuously maintain the transitive closure of a view because the insertions go through its underlying tables.

Our adapted algorithm, which runs periodically, joins the view iteratively until a fixed point is reached. This algorithm uses a *temp* table to record the results of joining the view of a property with itself and a *delta* table to record the results that are new to the property table. Then the iterations only join the property view with the *delta* table, which is usually much shorter than the property table. This algorithm also takes care of the perspectives, which allows different ontologies to independently describe a property as transitive or not. Details about the algorithms mentioned in this section can be found in our technical report [9].

3 The Hawkeye Knowledge Base

This section has two main objectives. First, we want to evaluate the new reasoning capabilities of DLDB. Second, we want to demonstrate the ability to answer realistic queries in the Semantic Web from distributed and heterogeneous data sources. Unfortunately, existing data on the Semantic Web tends to be unrelated. In order to make the queries interesting we needed to augment existing Semantic Web data with some new data sources. We focus on two scenarios which involve data associated with academic publications and government activities.

3.1 Data Sources and Maps

Table 1 describes the data sources used in our scenarios. The first column lists the shorthand prefixes we assigned to each data source.

The e-government scenario uses **g** (daml.org), **c** (house.gov), **b** (govtrack.us) and **n** (govtrack.us) data sources. The academic publication scenario uses **d** (dblp.uni-trier.de), **s** (citeseer.ist.psu.edu), **a** (aigp.csres.utexas.edu), **w** (nsf.gov) and **f** (found from Swoogle's crawl) data sources.

To augment our Semantic Web data we transformed the data from these sources to RDF which commits to valid ontologies. For sources originally in XML format, we developed an ontology for each of them based on their XML schema and developed domain specific scripts to translate the XML to conforming RDF. For each of those sources originally in HTML pages, we developed a crawler to collect the pages and a scraper which extracts the desired information from these pages. We then developed ontologies for each of them and generated conforming RDF for the scraped data.

The purpose of using data from multiple sources is that a single source doesn't hold all the information for a certain individual. When different URIs are used to refer the same individual, we must explicitly annotate that they are equivalent using the owl:sameAs property. There are a number of techniques with varying accuracy for automatic co-reference resolution. We try to match the names of individuals (e.g. Authors) from each of the sources using simple string matching techniques. In the case of FOAF, our system relies on inverse functional properties to infer equivalence of individuals. Totally, we created 109,790 sameAs statements between 4 pairs of data sources.

Table 1. Data sources summary

Pre	Data Source	Original Format	Classes	Properties	Triples
a	AIGP	No Ontology Set of HTML Pages	AIResearcher	hasAdvisor influencedBy hasInfluenced	5973
b	Bill Data	No Ontology RDF	Politician Bill	name sponsoredBy	75711
c	107th Congress	No Ontology XML Data	Member USCD	party, isIn fromUSCD	2628
d	DBLP	No Ontology XML Data	Article foaf:Person	author coauthor	15523209
f	FOAF	RDF Schema Ontology	Person	knows	11601453
g	Geographic Data	DAML Ontology DAML Data	USRegion USState	memberstate region	578
n	Census Data	No Ontology N3 File	State	population landarea	314
s	Citeseer	No Ontology XML Data	Article foaf:Person	author, coauthor references	7630021
w	NSF Awards Data	No Ontology Set of HTML Pages	NSFAward	principalInvestigator state	462102

To use the concepts and properties of different ontologies, we must explicitly specify the relationships between them. For this we need to use OWL axioms in a separate ontology which we call the map ontology. For example, for the academic domain the property *dc:creator* used in the Citeseer ontology is a super property of the property *author* used in the DBLP ontology.

3.2 Performance

We have used Swoogle's 2006 index as our dataset along with the data we prepared that described above. The DLDB main program runs on a workstation featuring dual 64-bit CPUs and 10GB main memory. The RDBMS is MySQL 5.0 and the DL reasoner is RacerPro. It took 650 hours to process the 1.7 million urls from Swoogle. Many of these had not been successfully downloaded due to various network issues, such as HTTP404 and connection timed out. We successfully retrieved 759,834 SW documents; the time to load and process just these documents is about 350 hours. In total 16,280 among them are identified as ontologies by Hawkeye. It takes approximately 18 gigabytes disk space for the RDBMS to store the 166M resulting triples. To the best of our knowledge this is the largest load of diverse, real-world Semantic Web data. This once again validates that the DLDB approach scales fairly well.

Load Performance. The chart in Figure 1 shows the cumulative load time after each million triples loaded into the system. In general we see that the load time increases gradually and our system scales well. Note the "local" time is defined as the total time minus the time spent in transferring the documents from a remote host. The "limited reasoning" time is the local time minus the time spent in batch processing of ABox

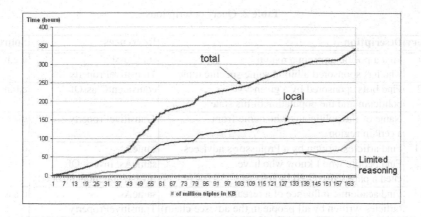

Fig. 1. Hawkeye Cumulative Load time

reasoning (including *InverseFunctionalProperty* and transitive closure inference). The steep slope from 43 to 49 million triples corresponds to the identification of a bug in our code and its subsequent correction. The steep slope at the end of the curves is contributed by a large number of explicit sameAs statements in the DBLP and Citeseer data. These statements were loaded after the data they mapped, thereby requiring a lot of substitutions. In a typical data load, the sameAs statements would be interspersed with the data and processing would be faster.

Query Performance. In order to evaluate the query performance of Hawkeye, we used the six query templates described in Table 2. For each query template, we listed the inferences and the data sources. Note the mappings between different sources are indispensable too. Table 3 shows the query performance of our system. For each query template, we issued a number of variations by changing the constants in the query. We then calculated the average and standard deviation of the response time. Most of the queries finished very quickly (in less than a minute). We also calculated the percentage of the non-zero answers and the average number of results for each query template. The high percentages of the non-zero answers demonstrates a significant degree of integration of the sources, particularly with respect to entity resolution. If our mapping ontologies or individual maps were insufficient, we would get many (perhaps even all) queries with 0 answers.

4   Lessons Learned

In working with real Semantic Web data, we have encountered many interesting challenges. In this section, we summarize those challenges and our experiences.

In looking at the real data and our experiment, we found out that for a practical Semantic Web knowledge base system, the key inference capabilities are individual equivalence (both explicit and inferred), subclass / subproperty inference, inverse property, transitive property. These were sufficient to do integration and testing of the many

Table 2. Query descriptions

Query	Description	Inferences	Sources
Pol1	Find a politician from a region who has sponsored a bill in some specific topic	subClassOf TransitiveProperty	g,c,b
Pol2	Find bills sponsored by a given politician and the population of his state	equivalentClassOf	c,b,n
Pol3	Name of the politicians who come from a certain region	TransitiveProperty	g,c
Aca1	Find articles written by a Professor's advisees	sameAs	d,s,a
Aca2	Find people who I know who have cited a paper also cited by me	sameAs, inverseOf InverseFunctionalProperty	d,s,f
Aca3	Find academic influence of a researcher (articles written by all people in the advisee chain)	sameAs TransitiveProperty	d,s,a
Aca4	Find publications of the AIResearchers from a certain state who have been awarded NSF grants	sameAs subClassOf	d,s,a,w

Table 3. Query performance

		Response time (ms)		No. of results	
Query template	No. of variations	Avg.	Stdev.	%of non-zeros	Avg
Pol1	300	871	635	90	63
Pol2	300	1018	897	100	115
Pol3	300	53	22	95	89
Aca1	200	24151	632	95	326
Aca2	20	345906	32149	87	1626
Aca3	200	24659	654	95	3549
Aca4	51	25318	2063	90	10245

web sources that we looked at. Note, arithmetic would probably be critical too (for unit conversion), but OWL doesn't support it.

Our DLDB design occasionally forms queries that the underlying database management system finds difficult to optimize. For example, one of our queries (Aca 2 in Table 2) took about 6 minutes to complete. This is due to the fact that it uses the foaf:knows property which has 16 million instances in our database. We performed some tests where we manually re-wrote the view for the foaf:knows property. We noticed that the query over this view had a constant expression in the where clause that would reduce the view scope. By moving the constant expression from the where clause of the query to the where clause of the view we were able to reduce the query execution to 3 minutes. We believe that future work can be done in order to create an algorithm to automatically optimize the queries which will increase the performance considerably.

During the implementation of DLDB, we observed that if precomputation of ABox reasoning is needed, it is faster to do it in batch, rather than doing it per assertion. For example, when discovering individual equalities, our batch approach not only reduces the number of database operations, but also speeds up the executions by bundling a number of database operations as a stored procedure.

5 Related Work

We claim that this is the first attempt to load the real Semantic Web data into a single knowledge base system. We should however note that there are several projects that process the Semantic Web in various other ways. For example, Swoogle [2] is the largest index of Semantic Web documents. However, Swoogle's query and retrieval mechanism is basically an information retrieval system. This does not exploit the reasoning that can be done over Semantic Web data.

There are some ongoing efforts to bootstrap the Semantic Web by providing ontologies and reusable knowledge bases, such as TAP [4]and "CS AKTive Space" [12]. Although they have large amounts of data, they each assume a common ontology.

In the past few years there has been a growing interest in the development of systems that will store and process large amount of Semantic Web data. The general design approach of these systems is similar to ours, in the sense that they all use some database systems to gain scalability while supporting as much inference as possible by processing and storing entailments. However, most of these systems emphasize RDF and RDF(S) data at the expense of OWL reasoning. Some systems resemble the capabilities of DLDB, such as KAON2 [6], which uses a novel algorithm to reduce OWL DL into disjunctive datalog programs. OWLIM [7] uses a rule engine to support a limited OWL-Lite reasoning. Minerva [14] uses DL reasoner to do TBox reasoning and a rule engine to do ABox reasoning. It is claimed to be sound and complete on DHL (a subset of OWL-DL) ontologies and reportedly less scalable than DLDB in terms of load time [14]. Both OWLIM and Minerva chose the "vertical" table design. To the best of our knowledge, none of the systems above have been used with a real world Semantic Web data at this scale (166M triples) , though BigOWLIM [8] has been claimed to support 1 billion triples of artificially generated data.

6 Conclusion and Future Work

In this paper we present an enhanced version of our DLDB system.We have extended our previous work by identifying and implementing critical inference capabilities and optimizing the system so that it can now handle at least 166 million facts as opposed to the 45 million of the previous version. The performance on query response time remains highly scalable, most of the queries in our experiment can be finished in less than one minute. We use ontology alignments expressed in OWL to provide a uniform view of the Semantic Web to the user We defer integration until query time and thus provide a framework where the user is not bound by a predetermined schema. Our ontology perspective mechanism gives the user the flexibility to choose the type of integration (s)he desires. Our maps do not need any additional language primitives beyond OWL. Therefore the maps as created and published become part of the Semantic Web. We put forward that scalability is more critical in processing the data sources as opposed to ontologies, because data sources will substantially outnumber the ontologies in the Semantic Web.

Although we believe our work is a first step in the right direction, we have discovered many issues that remain unsolved. First, although our system scales well to the current size of the Semantic Web, it is still unknown if such techniques will continue to scale well as the Semantic Web grows. Second, we will investigate query optimization techniques that can improve the query response time.

Acknowledgment

This material is based upon work supported by the National Science Foundation (NSF) under Grant No. IIS-0346963. We sincerely thank Tim Finnin of UMBC for providing us access to Swoogle's index of URLs.

References

1. Bechhofer, S., Volz, R.: Patching syntax in OWL ontologies. In: Proceedings of the Third International Semantic Web Conference (2004)
2. Ding, L., Finin, T., Joshi, A., Peng, Y., Pan, R., Reddivari, P.: Search on the semantic web. IEEE Computer 10(38), 62–69 (2005)
3. Dong, G., Libkin, L., Su, J., Wong, L.: Maintaining transitive closure of graphs in SQL. Int. Journal of Information Technology (1999)
4. Guha, R.: Tap: Towards the semantic web. Demo on World Wide Web 2002 Conference At: http://tap.stanford.edu/www2002.ppt
5. Heflin, J., Pan, Z.: A model theoretic semantics for ontology versioning. In: Proc. of the 3rd International Semantic Web Conference, pp. 62–76 (2004)
6. Hustadt, U., Motik, B., Sattler, U.: Reducing shiq description logic to disjunctive datalog programs. In: Proc. of the 9th International Conference on Knowledge Representation and Reasoning, pp. 152–162 (2004)
7. Kiryakov, A.: Owlim: balancing between scalable repository and light-weight reasoner. In: Developer's Track of WWW2006 (2006)
8. Ognyanoff, D., Kiryakov, A., Velkov, R., Yankova, M.: A scalable repository for massive semantic annotation. Technical Report D2.6.3, SEKT project (2007)
9. Pan, Z., et al. Hawkeye: A practical large scale demonstration of semantic web integration. Technical Report LU-CSE-07-006, Lehigh University (2007)
10. Pan, Z., Heflin, J.: DLDB: Extending relational databases to support semantic web queries. In: Fensel, D., Sycara, K.P., Mylopoulos, J. (eds.) ISWC 2003. LNCS, vol. 2870, pp. 109–113. Springer, Heidelberg (2003)
11. Pan, Z., Qasem, A., Heflin, J.: An investigation into the feasibility of the semantic web. In: Proc. of Twenty First National Conference on Artificial Intelligence (AAAI 2006) (2006)
12. Schraefel, M.C., Shadbolt, N.R., Gibbins, N., Harris, S., Glaser, H.: CS AKTive space: representing computer science in the semantic web. In: WWW 2004: Proceedings of the 13th international conference on World Wide Web (2004)
13. Wang, T.D., Parsia, B., Hendler, J.: A survey of the web ontology landscape. In: Cruz, I., Decker, S., Allemang, D., Preist, C., Schwabe, D., Mika, P., Uschold, M., Aroyo, L. (eds.) ISWC 2006. LNCS, vol. 4273, Springer, Heidelberg (2006)
14. Zhou, J., Ma, L., Liu, Q., Zhang, L., Yu, Y., Pan, Y.: Minerva: A scalable owl ontology storage and inference system. In: Mizoguchi, R., Shi, Z., Giunchiglia, F. (eds.) ASWC 2006. LNCS, vol. 4185, pp. 429–443. Springer, Heidelberg (2006)

URI Identity Management for Semantic Web Data Integration and Linkage

Afraz Jaffri, Hugh Glaser, and Ian Millard

Dependable Systems and Software Engineering Group
School of Electronics and Computer Science
University of Southampton
{a.o.jaffri,hg,icm}@ecs.soton.ac.uk

Abstract. The Semantic Web vision involves the production and use of large amounts of RDF data. There have been recent initiatives amongst the Semantic Web community, in particular the Linking Open Data activity and our own ReSIST project, to publish large amounts of RDF that are both interlinked and dereferenceable. The proliferation of such data gives rise to millions of URIs for non-information resources such as people, places and abstract things. Frequently, different data providers will mint different URIs for the same resource, giving rise to the problem of coreference. This paper describes the phenomenon of coreference, where it occurs in other disciplines and how it is relevant to the Semantic Web. We propose a 'Consistent Reference Service' for URI identity management and describe how this is being used in the infrastructure of a scalable Semantic Web system.

1 Introduction

The Semantic Web is growing and evolving. The increased adoption of RDF and OWL as knowledge representation formats is enabling the production of Semantic Web systems that can manage, manipulate and display data in novel ways [1, 2]. However, there are also those who believe that the Semantic Web is merely a dream that will never be fulfilled [3]. In order to encourage the adoption of Semantic Web technologies there has been an increased amount of activity in providing a linked data backbone that can be used to bootstrap the Semantic Web [4]. The Linking Open Data project [5] has been a catalyst for such activity, producing data sources that expose their knowledge as RDF and assert links between datasets. Current information sources include Geonames, DBLP, MusicBrainz, The CIA Factbook and US Census data.

1.1 Linking Open Data

The production of the first tutorial on how to link Open Data [6] means that many more information providers are likely to make their knowledge available. Such activity will allow a formidable mass of knowledge to be used by Semantic Web applications. The linked data methodology has also introduced the use of additional

R. Meersman, Z. Tari, P. Herrero et al. (Eds.): OTM 2007 Ws, Part II, LNCS 4806, pp. 1125–1134, 2007.
© Springer-Verlag Berlin Heidelberg 2007

techniques to publish Semantic Web data, such as using HTTP 303 redirects to dereference URIs about non-information resources, which have already allowed a new breed of Web browser to be built that can analyse and explore linked data [7].

The first set of data that is being used as a base for all subsequent data linkage is the DBpedia [8] dataset. The DBpedia dataset reportedly contains over 91 million RDF triples and has knowledge covering over one million concepts. The knowledge has been extracted from Wikipedia info boxes that appear on Wikipedia pages. Consequently there have been over one million URIs created corresponding to each Wikipedia page that contains an info box. DBpedia URIs take the form http://dbpedia.org/ resource/*resourceName* where *resourceName* is the name of a Wikipedia article. DBpedia has a lightweight ontology that has predicates derived from infobox data such as *name, placeofbirth, placeofdeath* and *capital*. There are also predicates used from other ontologies that link into the dataset including *foaf:page, rdfs:label* and *geonames:featureCode*.

1.2 The Problem of Coreference

The explosion in the number of information sources being exposed as RDF has also led to an explosion in the number of URIs used to identify different entities. It is often the case that data in different repositories will hold information regarding identical entities. The multiplicity of URIs leads to the problem of *coreference*, where different URIs are used to describe the same entity. On an open Semantic Web this presents a problem when there is a need to link together knowledge from disparate information providers. The present approach, used by the Linking Open Data community, is to use various equivalence mining techniques in order to assert *owl:sameAs* relations between entities that are considered to be the same [9]. DBpedia has, for example, made an assertion that: *<http://dbpedia.org/resource/Berlin>* is *<owl:sameAs>* *<http://sws.geonames.org/2950159/>*.

In this paper we will argue why this is not the best approach for dealing with coreference, and propose a system for dealing with consistent reference across multiple knowledge bases. Section 2 describes the problem of coreference and how it fits into the current Web architecture, Section 3 gives the implementation details of our solution, Section 4 presents an application that is using the system and Section 5 gives a summary and presents areas for future work.

2   Coreference and URI Identity

The term 'coreference' is used in the field of linguistics to define the situation where different terms are used to describe the same referent. This is often done using words such as 'he', 'she', 'we', 'them' or 'it'. On the Semantic Web we use the term coreference to define the situation where different URIs are used to describe the same non information resource. This section gives a brief description of coreference in information science and databases and then goes on to discuss the meaning of a URI and the importance of giving coreference due importance in Web architecture.

2.1 Coreference in Information Science

The problem of coreference within the field of information science has existed for many years and the solution to the problem is based around the use of controlled vocabularies. Such a solution is possible because of the closed world nature and human processing characteristics of a library system. The most popular closed vocabulary is the Library of Congress Subject Headings (LCSH). This vocabulary gives a defined and precise meaning to each subject in the vocabulary which contains over 280 000 terms. Thus it is not possible for people to make their own subject headings, keyword descriptions or tags as is prevalent on the Web today.

A more relevant case of coreference occurs in digital libraries when the author of a publication has to be disambiguated. There are many authors who share the same name and matters are made more complex by the use of initials, different naming formats and spelling errors. For example, the author 'Hugh Glaser' could be represented with his full name or by using 'H. Glaser', or 'Glaser, H.'. The task of author disambiguation is an active area of research in information science and many solutions have been proposed. Some solutions use Web based searches in order to determine if one author is the same as another [10, 11]. Such techniques will be needed on the Semantic Web if a consistent web of data is to be created.

2.2 Coreference in Databases

Within the database community the problem of coreference is referred to as record linkage. The need for record linkage arises when records or files from different databases need to be joined or merged. Each database could have duplicate records of the same person or thing which, when amalgamated, would make the data inconsistent or 'dirty' [12].

Record linkage has a well defined mathematical theory as proposed by Fellegi and Sunter [13]. The theory is based on records referring to the same entity having a number of characteristics in common. If a and b are elements from populations A and B and some elements are common to A and B then two disjoint sets can be created. The first set, M, is the set of elements that represent identical entities and the second set, U, is the set of elements representing different entities. Now if $\alpha(a)$ and $\beta(b)$ refer to records from databases A and B respectively and each record has k characteristics then a comparison vector, γ is defined that contains the coded agreements and disagreements on each characteristic:

$$\gamma[\alpha(a), \beta(b)] = \{\gamma^1[\alpha(a), \beta(b)], \dots, \gamma^k[\alpha(a), \beta(b)]\} \tag{1}$$

The theory then gives the probability of observing a specific vector given $(a,b) \in M$ as:

$$M(\gamma) = \sum_{(a,b) \in M} P\{\gamma[\alpha(a), \beta(b)]\} \cdot P[(a,b) \mid M] \tag{2}$$

This theory can also be used for identifying the probability of having identical resources on the Semantic Web. If A and B are two RDF graphs which have URIs a and b identifying the same resource such that $a \in$ A and $b \in$ B, then if the graphs of each URI are made such that ?ap and ?ao correspond to the properties and objects of URI a, and ?bp and ?bo correspond to the properties and objects of URI b, α(a) can be substituted by the result ?ao and β(b) can be substituted by the result ?bo. The characteristics k^i can then be substituted by the union of the result of ?ap and ?bp. The equations (1) and (2) will then hold with these substituted values. The result only shows the probability of two URIs referring to the same entity and can only be used as a basis for coreference resolution as it has been used in the database community for the same purpose.

The limitation with this theory is that it is assumed that there is at least one characteristic that is in common across the different graphs. In the Semantic Web this is not always the case as different graphs will have different predicates for a particular resource and there may be little or no overlap between predicates. For example the URIs http://sws.geonames.org/2950159 and http://dbpedia.org/resource/Berlin are URIs for Berlin that each have over 100 predicates, yet only one of them is in common. There is also the problem that predicates in the Semantic Web are not like database field names that have common names. Therefore, ontology matching techniques need to be used in order to match predicates between different resources, so that characteristics (instances) can then be matched.

2.3 Coreference in the Semantic Web

The subject of coreference on the Semantic Web has been raised previously [14], but it was not pressing or therefore studied for many years because of the lack of real scalable RDF data that was freely available. However, the Linking Open Data project and our own ReSIST project [15] are highlighting the need to have some form of URI management system. For example, the following are all URIs for Spain:

```
http://dbpedia.org/resource/Spain
http://www4.wiwiss.fu-berlin.de/factbook/resource/Spain
http://sws.geonames.org/2510769/
http://www.daml.org/2001/09/countries/fips#SP
http://www4.wiwiss.fu-berlin.de/eurostat/resource/countries/Espa%C3%B1a
```

These URIs come from 5 different sources. There are also at least 9 URIs for Hugh Glaser that originate from 6 different sources:

```
http://acm.rkbexplorer.com/rdf/resource-P112732
http://citeseer.rkbexplorer.com/rdf/resource-CSP109020
http://citeseer.rkbexplorer.com/rdf/resource-CSP109013
http://citeseer.rkbexplorer.com/rdf/resource-CSP109011
http://citeseer.rkbexplorer.com/rdf/resource-CSP109002
http://dblp.rkbexplorer.com/rdf/resource-27de9959
http://europa.eu/People/#person-0ff816fa
http://resist.ecs.soton.ac.uk/wiki/User:hugh_glaser
http://www.ecs.soton.ac.uk/info/#person-00021
```

We have grouped these URIs together because we believe they all refer to the same non-information resource. However, the standard way of dealing with such a plethora of URIs is to use *owl:sameAs* to link between them. The semantics of *owl:sameAs* mean that all the URIs linked with this predicate have the same identity [16], this means that the subject and object must be the same resource. The major disadvantage with this approach is that the two URIs become indistinguishable even though they may refer to different entities according to the context in which they are used. For example, consider the case where a person has a URI at one institution and then moves to another institution that provides another URI. If the person makes an *owl:sameAs* link between them then it will not be possible to differentiate between the person as they were at the first institution and the person as they are at the second institution. The knowledge about the person at institution 1 and institution 2 effectively become merged so, for example, the addresses would not be able to be separated.

Even worse, an incorrect equivalence can cause other incorrect equivalences to be inferred. For example, we found that one of the project investigators (Tom Anderson) had extra information which appeared plausible, but was not correct. We finally tracked this information down to DBLP, where the two Tom Andersons had been conflated.

We subscribe to the belief that the meaning of a URI may change according to the context in which it is used [17]. For example the URIs that refer to Spain given above could refer to 'Spain the political entity', or 'Spain the geographic location', or 'Spain the football team'. Some people would be happy to use each URI interchangeably because they do not care about the precise definition, whereas others will want a URI that specifically matches their intended meaning. There is a requirement to have some form of a system that deals with URIs about the same resource that are not exactly identical. The semantics of *owl:sameAs* are too strong and other alternatives like rdfs:seeAlso do not fit the intended purpose. Such a requirement is vital if data is to be cleanly linked together in a consistent fashion. The next section details our initial attempt to handle URI management called the Consistent Reference Service (CRS).

3   The Consistent Reference Service

The Consistent Reference Service (CRS) has been created in order to manage coreference between the millions of URIs that are accumulating on the Semantic Web. This section will describe the concept of a *bundle* that groups together URIs referring to the same resource, and also describe the implementation and architecture of the CRS.

3.1   URIs and Bundles

The CRS service has been implemented as both an RDF knowledge base and a relational database with RDF export. The CRS sits in the Semantic Web as any other knowledge base or database would. Each data provider maintains one or more CRSs for their own knowledge. In the ReSIST project there are over 15 repositories each with their own CRS.

The CRS introduces the concept of a *bundle* to group together resources that have been deemed to refer to the same concept within a given context. Different bundles may be used to group together URIs of the same resource in different contexts. For example, there may be a bundle containing all of the URIs about a person in the context of institution 1; and another bundle containing all of the URIs about the same person in the context of institution 2. Each CRS can use different algorithms to identify equivalent resources. For example, the algorithms to detect equivalence amongst authors are different from the algorithms used to detect equivalence between countries. To begin with, each URI in a repository has its own bundle in the CRS. When an equivalence is detected the bundles containing the URIs are merged together to create a new bundle. In this way successive iterations group together larger bundles, with each bundle having an anonymous URI.

The concept of a bundle is defined as a class in a coreference ontology used by the CRS. There is also a database schema that maps onto the ontology. Every resource that is defined as being of *rdf:type coref:Bundle* can have the following properties:

coref:hasCanonicalReference – One URI in a bundle can be made to be the canonical representation i.e. the preferred URI that one should use.

coref:hasEquivalentReference – The URIs in a bundle are grouped together using this predicate.

coref:updatedOn – The date of the last update to the bundle.

To illustrate let us take the example of the URIs referring to Hugh Glaser in the previous section. If we assume that we want to group together all the URIs that Citeseer has referring to Hugh then the triples asserted in RDF/XML format would look like:

```
<rdf:RDF xmlns:coref=http://www.resist.ecs.soton.ac.uk/ontology/coref#
         xmlns:rdf="http://www.w3.org/1999/02/22-rdf-syntax-ns#">
    <coref:Bundle
    rdf:about="http://www.rkbexplorer.com/crs/coref#bundle1">
        <coref:hasEquivalentReference rdf:resource=
        "http://citeseer.rkbexplorer.com/rdf/resource-CSP109020"/>
        <coref:hasEquivalentReference rdf:resource=
        "http://citeseer.rkbexplorer.com/rdf/resource-CSP109013"/>
        <coref:hasEquivalentReference rdf:resource=
        "http://citeseer.rkbexplorer.com/rdf/resource-CSP109011"/>
        <coref:hasEquivalentReference rdf:resource=
        "http://citeseer.rkbexplorer.com/rdf/resource-CSP109002"/>
        <coref:hasCanonicalReference rdf:resource=
        "http://citeseer.rkbexplorer.com/rdf/resource-CSP109002"/>
    </coref:Bundle>
</rdf:RDF>
```

The bundle mechanism provides an easy method to manage URI identities without having to incorporate expensive inference mechanisms. When dereferencing a resolvable URI the RDF document returned contains additional predicates identifying CRS services that may provide further information regarding the resource. If the user wishes, then they can assert explicitly *owl:sameAs* or *rdfs:seeAlso* links between the equivalent URIs. The next section will look at how the CRS is used in conjunction with multiple knowledge bases and how bundles can be linked to other open data.

3.2 The CRS and Web Architecture

There have been many discussions in the Semantic Web community regarding the actual meaning of a URI. Does it refer to a sequence of bits? A Web page? A concept? These questions and others arising from the URI identity crisis [18] are outside the scope of this paper. We will use the definitions as given by the W3C Technical Architecture Group (TAG) to show how a coreference mechanism can be included in the current Semantic Web infrastructure.

The CRS can be treated as any other knowledge base, in that it contains knowledge about a particular URI. Our infrastructure implements the current best practice on how to serve linked data [6] and uses cool URIs [19]. As an example we will use the URI   http://southampton.rkbexplorer.com/id/person-21   to   represent   the ´ non-information resource, 'Hugh Glaser'. When a request is given to the server for a description of the URI, an HTTP 303 redirect is issued to one of two locations, depending on the accept headers sent by the client. If the requested content is application/rdf+xml then the server will generate an RDF description detailing the properties of the requested URI by issuing SPARQL CONSTRUCT queries to the appropriate knowledge base. The resulting description is cached and the 303 is issued to http://southampton.rkbexplorer.com/description/person-21. However, if the accept header is set to text/html then a 303 See Other is returned identifying an html description of Hugh Glaser at http://southampton.rkbexplorer.com/browse/peson-21. The server architecture conforms to the latest httpRange-14 [20] recommendation of the TAG that involves HTTP 303 Redirects from the URI of a non-information source to an RDF or HTML information resource. To find all possible equivalences for a URI the following algorithm can be performed:

```
findEquivalence(URI u) {
    Dereference u;
    while (u coref:hasEquivalentReference a) {
        add a to equivalences;
    findEquivalence(a);
    }
}
```

Finding all equivalences is entirely at the discretion of the application wishing to process the results of the search. If only one CRS is required, then only one iteration is necessary. Computing equivalences in this manner gives a considerable amount of flexibility in choosing duplicate URIs for a resource. Taking the URI management into a separate layer without fixing *owl:sameAs* links is an efficient and controllable way to manage coreference between URIs. The next section will describe an application that has been built using the CRS infrastructure.

4   A CRS Application: The Resilience Knowledge Base Explorer

Resilience Knowledge Base (RKB) Explorer is a Semantic Web application that is able to present unified views of a significant number of heterogeneous data sources regarding a given domain. We have developed an underlying information infrastructure that utilises the CRS architecture outlined in Section 3. Our current dataset totals many tens of millions of triples, and is publicly available through both SPARQL endpoints and resolvable URIs. To realise the synergy of disparate information sources we are using the CRS system and have devised an architecture to allow the information to be represented and used.

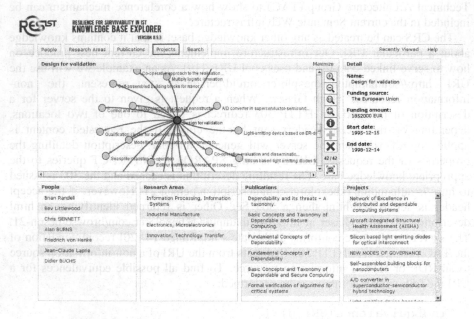

Fig. 1. The figure above shows the single window interface of the faceted browser available at http://www.rkbexplorer.com/explore/

Figure 1 shows the user interface for the RKB Explorer. The main pane shows a chosen concept and related concepts of the same type that the system has identified as being related. In this figure, the ReSIST Project itself is under consideration, with its details on the right, and related projects are shown around it. These are chosen according to the relative weight given to ontological relationships, and the number of those relationships to each concept. The weight of the lines gives a visual ranking. They represent a project `Community of Practice' (CoP) for the project. Clicking on a resource will show the detail for it, while double-clicking will add the CoP for the new resource to the pane. This will then allow a user to see how different projects are related, and see the projects that provide linkage between them.

The panes in the lower half of the display show the related people, research areas, publications and projects, identified by similar ontologically informed algorithms, and are ranked by decreasing relevance. Thus the lower right-hand pane gives a list of the related projects found in the main pane, while the lower left-hand pane shows those people involved in the currently selected project.

The RKB Explorer is based on the implementation described in Section 3. The CRS system manages the URIs for each knowledge base. There are many URIs from each knowledge base that refer to the same resource, for example there are hundreds of the same authors and papers in different knowledge bases, such as the ACM, IEEE and DBLP. Managing these millions of URIs has led to increased scalability and performance benefits as compared with taking an *owl:sameAs* approach. The RKB Explorer is being expanded and integrated with existing linked data and it is envisioned that the CRS system behind the explorer will also follow the same route.

5 Summary and Future Work

This paper has shown that managing URI identities at Semantic Web scale is a serious issue and problem. The main aim is to stimulate debate and discussion about how URI identities need to be managed and dealt with. The issue is becoming all the more relevant with the linked data that is now available online as RDF. This has caused a real debate on how to manage coreference in the Semantic Web [21]. Our proposed solution is meant to highlight the issues involved in building a URI management system and how such a system can be used to handle millions of triples of RDF data.

Our future plan is to distribute all our knowledge bases such that the ownership is delegated to a third party, together with an accompanying CRS or several CRSs. The next stage of research will then highlight exactly what issues need to be resolved when linking and reasoning about such a highly distributed system. The idea is to be able to mimic the stability and robustness of the document Web to the data Web in order to make the Semantic Web a truly scalable system.

References

1. Shadbolt, N.R., Gibbins, N., Glaser, H., Harris, S., Schraefel, M.C.: CS AKTive Space or how we stopped worrying and learned to love the Semantic Web. IEEE Intelligent Systems, 41–47 (2004)
2. Lei, Y., Uren, V.S., Motta, E.: SemSearch: a search engine for the Semantic Web. In: Proceedings of 15th International Conference on Knowledge Engineering and Knowledge Management, Podebrady, Czech Republic, pp. 238–245 (2006)
3. Shirky, C.: 2001. The Semantic Web, Syllogism, and Worldview [15 February 15, 2007] [online], http://www.shirky.com/writings/semantic_syllogism.html
4. Suchanek, F.M., Kasneci, G., Weikum, G.: YAGO:A Core of Semantic knowledge. In: Proceedings International WWW Conference 2007, Banff, Alberta, Canada, pp. 697–706. ACM Press, New York (2007)
5. Linking Open Data Project, http://www.linkeddata.org/
6. Bizer, C., Cyganiak, R., Heath, T.: How to Publish Linked Data on the Web [July 20, 2007] [online], http://sites.wiwiss.fu-berlin.de/suhl/bizer/pub/LinkedDataTutorial/
7. Berners-Lee, T., Chen, Y., Chilton, L., Connoly, D., Dhanara, R., Hollenbach, J., Lerer, A., Sheets, D.: Tabulator:Exploring and Analyzing Linked Data on the Web. In: Proceedings 3rd International Semantic Web User Interaction Workshop, Athens, Georgia (2006)
8. DBpedia [July 1, 2007] [online], http://dbpedia.org/docs
9. Equivalence Mining and Matching Frameworks [July 1, 2007] [online], http://esw.w3.org/topic/TaskForces/CommunityProjects/LinkingOpenData/EquivalenceMining

10. Yang, K., Jiang, J., Lee, H., Ho, J.: Extracting Citation Relationships from Web Documents for Author Disambiguation, Technical Report No.TR-IIS-06-017,Institute of Information Science, Academia Sinica, Taipei, Taiwan (December 2006)
11. Tan, Y.F., Kan, M.-Y., Lee, D.: Search Engine Driven Author Disambiguation. In: Proceedings 6th ACM/IEEE-CS Joint Conference on Digital Libraries, pp. 314–315. ACM Press, New York
12. Hernandez, M., Stolfo, S.: Real-world Data is Dirty: Data Cleansing and the Merge/Purge Problem. Data Mining and Knowledge Discovery 2(1), 9–37
13. Fellegi, I.P., Sunter, A.B.: A Theory for Record Linkage. Journal of the American Statistical Association 64(328), 1183–1210 (1969)
14. Alani, H., Dasmahapatra, S., Gibbins, N., Glaser, H., Harris, S., Kalfoglou, Y., O'Hara, K., Shadbolt, N.: Managing Reference: Ensuring Referential Integrity of Ontologies for the Semantic Web. In: Proceedings of 13th International Conference on Knowledge Engineering and Knowledge Management, Sigenza, Spain, pp. 317–334 (2002)
15. Resilience for Survivability in IST (ReSIST) Network of Excellence, http://resist-noe.eu
16. Bechofer, S., Van Harmelen, F., Hendler, J., Horrocks, I., Mcguiness, D.L., Schneider, P.F., Stein, L.A.: OWL Web Ontology Language Reference, Technical Report, W3C [online], http://www.w3.org/TR/owl-ref/
17. Booth, D.: URIs and the Myth of Resource Identity. In: Proceedings of the Workshop on Identity. Meaning and the Web (IMW 2006) at International World Wide Web Conference 2006, Edinburgh, Scotland (2006)
18. Halpin, H.: Identity, Reference and Meaning on the Web. In: Proceedings of the Workshop on Identity, Meaning and the Web (IMW 2006) at International World Wide Web Conference 2006, Edinburgh, Scotland (2006)
19. Berners-Lee, T.: Cool URIs Don't Change [online], http://www.w3.org/Provider/Style/URI
20. Fielding, R.: W3C Technical Architecture Group mailing list (June 18, 2005) [online] http://lists.w3.org/Archives/Public/www-tag/2005Jun/0039
21. W3C Mailing List Discussion Thread, Terminology Question Concerning Web Architecture and Linked Data, http://lists.w3.org/Archives/Public/semantic-web/2007Jul/0049.html

Parallelizing Tableaux-Based Description Logic Reasoning

Thorsten Liebig and Felix Müller

Inst. of AI, University of Ulm, D-89069 Ulm, Germany
thorsten.liebig@uni-ulm.de,
felix.mueller@uni-ulm.de

Abstract. Practical scalability of Description Logic (DL) reasoning is an important premise for the adoption of OWL in a real-world setting. Many highly efficient optimizations for the DL tableau calculus have been invented over the last decades. None of them aimed at parallelizing the tableau algorithm itself. This paper describes our approach for concurrent computation of the nondeterministic choices inherent to the standard tableau procedure. We discuss how this interrelates with the well-known optimization techniques and present first promising performance results when benchmarking our prototypical reasoner *UUPR* (*Ulm University Parallel Reasoner*) with a selection of established DL systems.

1 Motivation

Tableaux-based algorithms have shown to be an adequate method in order to implement Description Logic (DL) reasoning services for many practical use-cases of moderate size. However, scalability of OWL reasoning is still an actual challenge of DL research [6]. Recent optimizations have shown significant increase in speed for answering queries with respect to large volumes of individual data under specific conditions. Unfortunately, almost all optimizations typically do come with some restriction in expressivity and end-users have to take care which approach to choose for a particular language fragment.

On the other hand, current processor families typically pool more than one processing unit on a single chip. Recent consumer desktops even have two quad-core processors on board. Today's reasoning engines unfortunately do not distribute their work load in such a setting. This is an unnecessary waste of computing power. Clearly, parallel computation can only reduce processing time by a factor which is limited by the available processing units but has the potential of being applicable without any restriction especially to the most "costly" cases.

This paper describes how to parallelize the well-known tableau algorithm as utilized sequentially within reasoning systems such as RacerPro, FaCT++, or Pellet. Our approach aims at parallelizing the tableau procedure itself rather than executing various instances of this procedure in parallel. The latter is a naive kind of parallelization whose synchronization may create some problems. This is because an optimized computation of the concept hierarchy does not consist of independent tasks as it will exploit previous subsumption results.

R. Meersman, Z. Tari, P. Herrero et al. (Eds.): OTM 2007 Ws, Part II, LNCS 4806, pp. 1135–1144, 2007.
© Springer-Verlag Berlin Heidelberg 2007

In contrast, parallelizing the nondeterministic choices within the standard DL tableau procedure has several advantages. First of all, nondeterminism is inherent to the tableau algorithm due to logical operators such as disjunction, at-most, or qualified cardinality restrictions. The generated alternatives from these expressions are completely independent of each other and can be computed concurrently. In case of a positive result the other sibling threads can be aborted. The parallel computation of nondeterministic alternatives also makes the algorithm less dependent on heuristics which otherwise have to choose the next alternative to process. For instance, a bad guess within a sequential algorithm inevitably will lead to a performance penalty. A parallel approach has the advantage of having better odds with respect to at least one good guess.

2   An Approach to Distributed DL Tableaux Proofs

Our approach aims at parallelizing the sequential algorithm proposed in [4] for $\mathcal{ALCNH}_{R+}$ (also referred to as $\mathcal{SHN}$) ABoxes with GCIs.

Every standard reasoning task can be reduced to a corresponding ABox unsatisfiability problem. A tableau prover will then try to create a model for this ABox. This is done by building up a tree (the tableau) of generic individuals a_i (the nodes of the tableau) by applying tableaux expansion rules [1]. Tableaux expansion rules either decompose concept expressions, add new individuals or merge existing individuals.

2.1   Parallel Processing

The most obvious starting point for parallel evaluation are nondeterministic tableaux rules. Nondeterministic branching yields multiple alternatives, which can be seen as different possible ABoxes to continue reasoning with. In our setting, the following nondeterministic rules are covered:

The disjunction rule. If for an individual a the assertion $a : C \sqcup D$ is in the ABox $\mathcal{A}$, then there are two possible ABoxes to continue with, $\mathcal{A}' = \mathcal{A} \cup \{a : C\}$ or $\mathcal{A}'' = \mathcal{A} \cup \{a : D\}$.

The number restriction merge rule. If at an any point in the tableau there are m r successors of a in $\mathcal{A}$, $a : (\leq n\ r)$ is an assertion in $\mathcal{A}$ and $m > n$, the existing successors need to be merged to fulfill the restriction. An ABox $\mathcal{A}^i$ results for every possible combination.

As there are no dependencies between the alternatives, they can be evaluated within parallel threads.

To realize parallelism without recursively creating an overwhelming number of threads, we decided to implement a *work pool* design: A fixed number of threads is generated at the start of the tableau proof. This number typically will be equal to or less than the number of available processing units. These threads have synchronized read and write access to a common pool of jobs (i.e. the ABoxes to evaluate). In an initial step the tableaux root node (the original

ABox) is added to the pool. The executor starts the workers and one of them will fetch this job. In case of a nondeterministic rule application a worker will generate the necessary alternative ABoxes by creating copies of the preceeding ABox which are then submitted to the pool. These jobs will be processed by the next available workers. Figure 1 illustrates the resulting components within UUPR. The process is stopped when either

i) an ABox that represents a complete tableau is found, or
ii) no satisfiable alternative was found and there are no alternatives left to process.

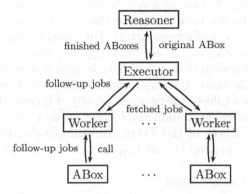

Fig. 1. Component interaction within work pool design of UUPR

An important decision in this design is the choice of the underlying pool data structure. The commonly used queue is unsuitable in this setting as it promotes a breadth-first style evaluation order. Thus, ABoxes which were created earlier (generated by fewer applications of nondeterministic rules) are preferred, and the discovery of complete ABoxes is delayed. The usage of a stack would not reliably lead to a depth-first oriented processing order either, because several threads can access the pool to put jobs into it.

We therefore chosed to use a priority queue in order to be able to explicitly influence the processing order. A simple heuristic to control the processing order:

– The priority of the original ABox is set to 0.
– ABoxes generated from an ABox with priority n are given the priority $n + 1$.

This allows for a controlled depth-first oriented processing order. More sophisticated heuristics or even some kind of A*-algorithm would also be possible. For example, FaCT++ also utilizes a priority queue for its ToDo list [9], weighting tableaux rules with different priorities. The difference is that FaCT++'s ToDo list contains all tableaux rules, while ours is restricted to nondeterministic rules (the other rules have a fixed order).

2.2 Data Representation

We also tried to design our internal data representation to be as efficient as possible. The main idea is to use integers to represent concepts and expressions (FaCT++ seems to use a similar encoding). Logical negation is realized through integer negation. The TBox is an array of concepts, with the most general concept $\top$ having index 1. When parsing a concept, numbers are recursively assigned to all subconcepts and subexpressions. Each indexed expression is represented as an integer array, where constructor, cardinalities and roles are all encoded into the first integer of this array. More precisely, the 32 bit of the first integer are split into three parts (aabbbccc in hexadecimal encoding). The first 8 bit encode the logical constructor. In case of a role constructor the next two chunks of 12 bit encode the cardinality value as well as the referenced role. A TBox, finally, is represented as an array of integer arrays. For example, Figure 2(a) shows the TBox containing the definition $C \equiv \neg \forall r.A \sqcap (\geq 2\ s)$ (note that the first integer for each indexed expression is shown in hexadecimal representation).

Assertions are collections of arrays for individuals and their role connections. I. e. for an individual a role-specific connection object is created in case of one (or more) fillers. In addition, each individual stores an associated concept assertion set which refers to indices of the TBox. An example assertion containing the assertions $(a_1, a_2) : r$, $(a_1, a_3) : r$ and $(a_3, a_4) : s$ is shown in Figure 2(b).

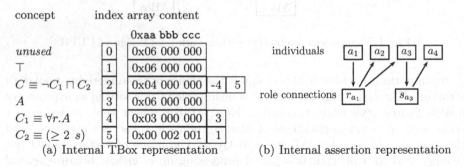

concept	index	array content		
		0xaa bbb ccc		
unused	0	0x06 000 000		
$\top$	1	0x06 000 000		
$C \equiv \neg C_1 \sqcap C_2$	2	0x04 000 000	-4	5
A	3	0x06 000 000		
$C_1 \equiv \forall r.A$	4	0x03 000 000	3	
$C_2 \equiv (\geq 2\ s)$	5	0x00 002 001	1	

(a) Internal TBox representation   (b) Internal assertion representation

Fig. 2. Example of UUPR's KB data structures

This compact representation guarantees low memory consumption and high processing speed. For instance, detecting a syntactic clash between two indexed expressions is reduced to a simple integer addition operation.

2.3 Optimizations

Today's state of the art reasoners achieve performance mainly through many highly efficient optimizations. Therefore, it is necessary to explore whether existing optimizations can be applied to our parallel architecture.

According to [9], DL tableau optimizations can be classified as follows:

- *Preprocessing and simplification*: As these optimizations are applied before the actual reasoning process is started, they are easy to combine with our approach. The most prominent optimizations of this kind is *GCI absorption*.

- *Optimizations in classification*: Current reasoners offer services to compute a taxonomy for a given ontology. Here, the number of subsumption tests can be reduced by exploiting implicitly computed subsumption relations between classes or by cheap syntactical (but incomplete) tests, such as pseudo model merging. For hierarchy computation, the applied subsumption algorithm is irrelevant, and thus a parallel subsumption algorithm can be used.
- *Optimizations in core satisfiability testing*: Optimizations that work directly in the reasoner core obviously are those which may interact with a parallel reasoner architecture. However, many of them are very important even in prototypical reasoners, since a naive implementation will often lead to practical nontermination even for small knowledge bases.

As a proof of concept we added the following known (mostly core reasoner) optimizations to our UUPR implementation:

Naming and Lazy Unfolding. These two techniques are fundamental to DL reasoning engines and integrated deeply in our reasoner core. Naming is done be recursively assigning names to all occurring subconcepts and is reflected by the internal data representation described in Figure 2(a). Lazy unfolding means that these names are only expanded when needed.

Lexical Normalization. This is a preprocessing optimization and aims to normalize input data, such that inconsistencies are detected as early as possible. Lexical normalization includes a number of syntactical simplification rules.

Semantic Branching. Semantic Branching is a technique similar to the DPLL procedure used in propositional satisfiability testing. It influences the way in which alternatives are generated during reasoning. The main idea is to avoid having to solve the same sub-problems in multiple alternatives by explicitly making them distinct: For instance, for $A \sqcup B$, A and $\neg A$ (respectively) are added to the alternatives.

Simplification. Simplification tries to reduce the amount of nondeterminism by avoiding unnecessary branching. It is a technique similar to boolean constraint propagation (BCP). For example, when $\neg A \sqcap (A \sqcup B)$ is contained in a tableau node, no branching is necessary and B can be added to the node.

Caching. This is applied to all calculated results of subproblems encountered during the reasoning process.

From the above optimizations, only caching leads to additional synchronization overhead, as cache accesses are mutually exclusive. All other optimizations were integrated without any special adaption into our reasoner core.

2.4   Implementation

Our parallel ABox reasoner is implemented in C++ as a shared memory program using the boost.Threads library[1]. More precisely, UUPR is developed for the SMP (symmetric multi processor) architecture, where all processors have access to one main memory. Unlike in Java or Lisp, programmers using C++ can

---

[1] http://www.boost.org/

influence the way memory allocation is managed. The standard template library (STL) container classes normally use the std::allocator. However, using the latter, no parallel speed-up could be achieved. In fact, using more processors only resulted in decreasing program performance as shown in Figure 3.

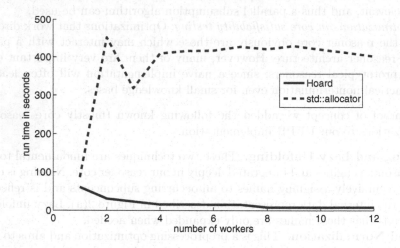

Fig. 3. Impact of memory manager

Fortunately, the STL classes can be parameterized to use a different memory allocator. It turned out that the superior heap organization utilized in memory allocators such as Hoard[2] [2] is an essential premise for any performance gain in a parallel shared memory environment. Consequently, UUPR can be compiled with one of two memory managers specifically developed for use in parallel programs, the Hoard library as well as Intel's thread building blocks (TBB)[3]. The number of parallel workers can be specified as a parameter at run time.

3   Experimental Results

Our performance tests were run on the following platforms:

- A Sun compute server with 12 UltraSPARC IV+ dual core processors, running at 1.8 GHz each, and 96 GB of main memory. The processor load was about 50%, so effectively, at best we had about 12 processors during testing.
- An ubuntu Linux system with two AMD Opteron dual core processors, which run at 2.2 GHz and 16 GB of main memory.
- As a standard desktop computers we used a 2.4 GHz dual core AMD desktop computer with 1 GB main memory running Suse Linux.

---

[2] http://www.hoard.org/

[3] http://www.intel.com/cd/software/products/asmo-na/eng/294797.htm

- For comparison, a 3 GHz single core computer with 1 GB of main memory is also included. We also made spot tests on a MacBook Core 2 Duo, which yielded results similar to the dual core desktop machine.

Three test cases were selected for evaluation:

Filler merging. Test case 1 is taken from [7] (2b). Checking satisfiability of $X2$ will create a lot of role successors, which then must be merged due to a maximum cardinality restriction: $X2 \equiv \exists r.C1 \sqcap \ldots \sqcap \exists r.C15 \sqcap (\leq 2\ r)$. Three of the Ci are mutually disjoint, so that all possible combinations need to be examined. This leads to a lot of nondeterminism and many small ABoxes have to be checked, making the synchronized pool access a possible performance limitation. Results are shown in Figure 4.

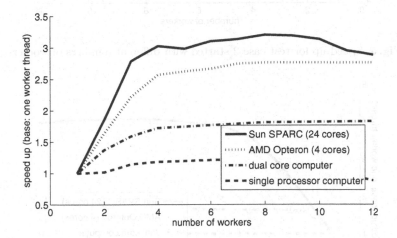

Fig. 4. Speed-up for test case 1 started with different numbers of workers

Disjunction. Test case 2 is an extended version of case 28 from [7]. It is designed to be a costly satisfiability test of a concept A without any nondeterminism. Here we check for satisfiability of a concept C, defined as a disjunction of eight concepts similar to A: $C \equiv A_1 \sqcup \ldots \sqcup A_8$. Since semantic branching was disabled here, the result is an equal distribution of 8 costly tasks on workers with low synchronization overhead. Therefore, Figure 5 shows the effect of a step-wise speed-up whenever the number of workers is a divider of 8.

Realistic ontology. To determine performance on a more realistic ontology and to demonstrate the applicability of our approach to all of $\mathcal{SHN}$, we took the example ontology given in [4]. It models a family ontology, using TBox and ABox knowledge as well as GCIs. We added a subtle contradiction and performed an ABox satisfiability check. Results are given in Figure 6.

A considerable speed-up can be observed in all cases. Except for the Sun the performance increase is almost linear up to number of available processing units. The performance decline for the Sun platform presumably has two

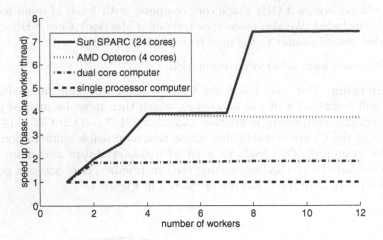

Fig. 5. Speed-up for test case 2 started with different numbers of workers

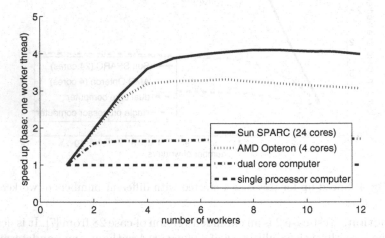

Fig. 6. Speed-up for family ontology started with different numbers of workers

reasons. First, the Sun was heavily used by many other users during testing. Second, a lot of nondeterminism occurs in the cases 1 and 3. This results in many memory allocation operations (when ABoxes are created) and requires a lot of synchronization (for work pool access). This extra effort is proportional to the number of concurrently executed worker threads exclusively running on their own processing core. Whether memory bandwidth or thread synchronization is the limiting factor here is subject to further investigations.

Table 1 shows a comparison of UUPR's respective best performance with a selection of other reasoners. For test case 3 we reproducibly got a segmentation fault for UUPR when deactivating the optimizations of sec. 2.3 (except naming and lazy unfolding) on one of our test environments (Sun), while there (again

Table 1. Comparison with other systems on Sun

System	test case 1	test case 2	test case 3
KAON2	*memout*	*timeout*	2.490s
Pellet 1.4	*timeout*	144.921s	2.375s
UUPR with optimizations	56.834s	8.152s	13.122s
UUPR without optimizations	60.401s	8.794s	*failed*

reproducibly) was a time out (10 min) on the other platforms. A plausible reason for this could not be determined. The most noticable difference was the different memory manager (Hoard on Sun, Intel TBB for the others).

4 Related Work

As far as we know there is no approach aiming at parallelizing the tableau calculus particularly for DL reasoning. However, there are a couple of tableau-based theorem provers capable of parallelizing obviously independent parts of the search tree, such as or-parallelism (e. g. Meteor, PartabX, Parthenon, SETHEO) [8]. Current work seems to be only in the niche of temporal reasoning and is at best expected to become an integral part of provers in a couple of years [10].

Older DL-related work deals with parallel processing of the structural algorithms utilized by the FLEX system. According to [3], the most promising processing phases for parallelization is the execution of propagation rules for ABox realization as opposed to the structural algorithm for TBox reasoning such as normalization or comparison. They argue that within the setting of a MIMD (Multiple Instruction, Multiple Data) system the basic operations during structural classification are too fine-grained for efficient parallelization.

5 Outlook

Our parallel reasoner has shown encouraging first results. The approach can be combined with well-known optimizations and bears no restriction which would prevent it from being extended to more expressive language fragments (even $\mathcal{SROIQ}$). At the same time there are many extensive optimizations conceivable.

An obvious way to further increase the performance of our approach is to employ techniques to reduce the amount of synchronized work pool access to a ratio compatible with our design. This can, for example, be done by cutting off parallel computation above a given problem size or branching factor and by implementing known optimizations such as GCI absorption. Another optimization is also missing: dependency directed backtracking. The latter is more difficult to implement because it requires to keep track of references to previous nodes of the tableau, which are currently missing within our design.

A particular parallel optimization refers to cascading work pools and/or caches, so that only a small number of threads share one work pool or cache. This would drastically reduce synchronization efforts, especially with larger numbers of worker threads. In addition, it would potentially allow to execute the threads of one work pool on a different machine. Although not an issue within our evaluation, memory could be saved by replacing ABox cloning by ABox structure sharing across threads in case of nondeterministic alternatives.

Extensions to the parallel evaluation itself are another option. For instance, qualified cardinality restrictions, as offered by OWL 1.1, add nondeterminism due to their choose-rule. Another idea is the parallel evaluation of conjunctions. In principle, the conjuncts C and D of a conjunction $C \sqcap D$ can be evaluated in parallel. The problem here are mutual dependencies between the conjuncts. For example, in the conjunction $\forall r.A \sqcap \exists r.\neg A$ the clash would not be detected if the conjuncts were processed in parallel. Therfore, a dependency test is needed to determine whether parallel evaluation is possible, that is, to check whether two conjuncts do interact in some way. A technique similar to pseudo-model merging [5] of sub-expressions (not only root nodes) could be used to achieve this.

References

1. Baader, F., Calvanese, D., McGuinness, D.L., Nardi, D., Patel-Schneider, P.F.: The description logic handbook: theory, implementation, and applications. Cambridge University Press, New York (2003)
2. Berger, E.D., McKinley, K.S., Blumofe, R.D., Wilson, P.R.: Hoard: A scalable memory allocator for multithreaded applications. In: International Conference on Architectural Support for Programming Languages and Operating Systems (ASPLOS-IX), Cambridge, MA, pp. 117–128 (November 2000)
3. Bergmann, F.W., Quantz, J.J.: Parallelizing Description Logics. In: Wachsmuth, I., Brauer, W., Rollinger, C.-R. (eds.) KI-95: Advances in Artificial Intelligence. LNCS, vol. 981, pp. 137–148. Springer, Heidelberg (1995)
4. Haarslev, V., Möller, R.: Expressive ABox Reasoning with Number Restrictions, Role Hierarchies, and Transitively Closed Roles. In: Int. Conf. on Principles of Knowledge Representation and Reasoning (KR2000), pp. 273–284 (2000)
5. Haarslev, V., Möller, R., Turhan, A.Y.: Exploiting Pseudo Models for TBox and ABox Reasoning in Expressive Description Logics. In: Goré, R.P., Leitsch, A., Nipkow, T. (eds.) IJCAR 2001. LNCS (LNAI), vol. 2083, pp. 29–44. Springer, Heidelberg (2001)
6. Horrocks, I.: Applications of description logics: State of the art and research challenges. In: Dau, F., Mugnier, M.-L., Stumme, G. (eds.) ICCS 2005. LNCS (LNAI), vol. 3596, pp. 78–90. Springer, Heidelberg (2005)
7. Liebig, T.: Reasoning with OWL – system support and insights –. Technical Report TR-2006-04, Ulm University, Ulm, Germany (September 2006)
8. Schumann, J.: Tableau-Based Theorem Provers: Systems and Implementations. Journal of Automated Reasoning 13, 409–421 (1994)
9. Tsarkov, D., Horrocks, I., Patel-Schneider, P.F.: Optimising Terminological Reasoning for Expressive Description Logics. Journal of Automated Reasoning 39(3), 277–316 (2007)
10. Voronkov, A.: Automated Reasoning: Past Story and New Trends. In: Proc. of the Int. Joint Conf. on AI (IJCAI-2003, pp. 1607–1612 (2003)

Scalability of OWL Reasoning:
Role condensates

Sebastian Wandelt[1] and Ralf Möller[1]

Institute for Software Systems,
TU Hamburg-Harburg
wandelt@tuhh.de, r.f.moeller@tuhh.de

Abstract. In the last years, there has been an increasing interest in the performance of reasoning on the Semantic Web in presence of large ABoxes. Traditional reasoners make heavily use of in-memory structures and are therefore not suitable to deal with large ABoxes directly.

We propose an approach that, informally speaking, works on a composite representation of the role-part of a SHIQ knowledge base. With respect to conjunctive queries, this helps us to provide a kind of proxy that restricts the set of possible bindings for a variable in advance. Furthermore we can use this proxy to reject several queries with no answer substitution immediately. Most notably our approach is query independent.

1 Introduction

As the Semantic Web evolves, scalability of inference techniques becomes increasingly important. Even for basic inference techniques, e.g. concept satisfiability, it is only recently understood on how to perform reasoning on huge input in an efficient way. This is not yet the case for problems that are too large to fit into main memory. For more complex reasoning problems like conjunctive query answering, this problem becomes even harder.

It has been shown recently [GHLS07] that answering conjunctive queries over a SHIQ knowledge base is decidable. The data complexity (complexity w.r.t. the size of the ABox) was shown to be co-NP complete. Given that result, dealing with large ABoxes (10^6 and more assertions) is likely to be unfeasible in practice for the general/worst case.

In this paper we present a complete, but unsound approach to answer conjunctive queries in a proxy-like manner. For our approach we have a look at the role part of a knowledge base. For a given SHIQ knowledge base, we create a condensed role graph. Usually, this graph is several orders of magnitudes smaller than the ABox. Furthermore, we convert a conjunctive query into a number of forest-like structures. Then we show that the conjunctive query is only entailed by the knowledge base if one query forest is isomorphic to a subgraph of the condensed graph.

One might claim that the subgraph isomorphism problem is NP-complete and that we have not much gain for the price of unsoundness. Yet, we advocate, that our data set is smaller. Thus, the impact of the high worst-case complexity is not

R. Meersman, Z. Tari, P. Herrero et al. (Eds.): OTM 2007 Ws, Part II, LNCS 4806, pp. 1145–1154, 2007.

so dramatic, and the structures fit into main memory. Furthermore, subgraph isomorphism checking is a very fundamental problem of computer science and well-known heuristics can be applied to speed up the algorithm in practice.

This paper is structured as follows. Section 2 presents some background on query answering over SHIQ ontologies. In Section 3 we introduce the notions of a role condensate and in Section 4 we show an appropriate role-encoding of conjunctive queries. Section 5 provides our main reasoning result. We conclude the paper in Section 6 and also point at some ideas for further work.

2   Basics

First, we briefly introduce the syntax and semantics of the description logic SHIQ. We assume a collection of disjoint sets: a set of *concept names* N_C , a set of *role names* N_{RN}, with a subset $N_{RN+} \subseteq N_{RN}$ of transitive role names, and a set of *individual names* N_I. The *set of roles* N_R is $N_{RN} \cup \{R^-|R \in N_{RN}\}$. We also define the following functions for inverse and transitive roles:

$$Inv(R) = R^- \iff R \in N_{RN}$$
$$Inv(R) = S \iff R = S^-$$
$$Trans(R) \iff R \in N_{RN+} \lor Inv(R) \in N_{RN+}$$

SHIQ-concepts are built inductively by using the grammar:

$$C ::= \top|\bot|D|\neg C|C_1 \sqcap C_2|C_1 \sqcup C_2|\forall R.C|\exists R.C| \leq_n S.C| \geq_n S.C,$$

where $D \in N_C, R \in N_R, S \in N_R$
N_{RN+} and $n \in \mathbb{N}$.

We define the semantics by using a standard Tarski-style semantics with an interpretation $\mathcal{I} = (\Delta^{\mathcal{I}}, \bullet^{\mathcal{I}})$, consisting of a non-empty set (*domain*) $\Delta^{\mathcal{I}}$ and a function (*valuation*) $\bullet^{\mathcal{I}}$ which maps every concept name D to a subset $D^{\mathcal{I}} \subseteq \Delta^{\mathcal{I}}$, every role name $R \in N_R$ to a binary relation $R^{\mathcal{I}} \subseteq \Delta^{\mathcal{I}} \times \Delta^{\mathcal{I}}$, while satisfying the following equations:

$$(C \sqcap D)^{\mathcal{I}} = C^{\mathcal{I}} \cap D^{\mathcal{I}}, (C \sqcup D)^{\mathcal{I}} = C^{\mathcal{I}} \cup D^{\mathcal{I}}, (\neg C)^{\mathcal{I}} = \Delta^{\mathcal{I}} \setminus C^{\mathcal{I}},$$
$$(\exists R.C)^{\mathcal{I}} = \{x|\exists y.(x,y) \in R^{\mathcal{I}} \land y \in C^{\mathcal{I}}\},$$
$$(\forall R.C)^{\mathcal{I}} = \{x|\forall y.(x,y) \in R^{\mathcal{I}} \implies y \in C^{\mathcal{I}}\}$$
$$(\geq_n R.C)^{\mathcal{I}} = \{x| |\{y|(x,y) \in R^{\mathcal{I}} \land y \in C^{\mathcal{I}}\}| \geq n\},$$
$$(\leq_n R.C)^{\mathcal{I}} = \{x| |\{y|(x,y) \in R^{\mathcal{I}} \land y \in C^{\mathcal{I}}\}| \leq n\},$$

A *general concept inclusion* is an expression $C \sqsubseteq D$, where C and D are concepts. A finite set of general concept inclusions is called *TBox*. A *role inclusion* is of the form $R \sqsubseteq S$ with R and S roles. A *role hierarchy* is a finite set of role inclusions. With $\sqsubseteq^*$ we denote the transitive closure of $\sqsubseteq$. An *assertion* is of the form $C(a), \neg C(a), R(a,b), \neg R(a,b)$ or $a \neq b$, where C is a concept, R is a role and $a, b \in N_I$. A finite set of assertions is called *ABox*. With $Ind(\mathcal{A})$ we denote the set of individuals occurring in $\mathcal{A}$. A *knowledge base* KB is a triple $\langle \mathcal{T}, \mathcal{R}, \mathcal{A} \rangle$ that consists of a TBox $\mathcal{T}$, a role hierarchy $\mathcal{R}$ and an ABox $\mathcal{A}$.

An interpretation $\mathcal{I}$ is a *model* of a concept C, if $C^{\mathcal{I}} \neq \emptyset$. An interpretation $\mathcal{I}$ is a model of a general concept inclusion $C \sqsubseteq D$, if $C^{\mathcal{I}} \subseteq D^{\mathcal{I}}$. An interpretation is a model of a TBox T if it satisfies all axioms in T. Being model of an RBox and ABox is defined as usual. An interpretation is a model of a knowledge base, if it is a model of $\mathcal{T}$, $\mathcal{R}$ and $\mathcal{A}$.

In the following we use a common definition of conjunctive queries, as e.g. given in [GHLS07]. Let N_V be a countably infinite set of variables. A *concept atom* is an expression $C(v_1)$ and a *role atom* is an expression $R(v_1, v_2)$, where C is a concept name, R is a role, and $v_1, v_2 \in N_V$. A *conjunctive query* Q is a non-empty set of atoms. Let $Var(Q)$ denote the set of variables in a query Q.

We write for an interpretation $\mathcal{I}$, a conjunctive query Q and a total function $\pi : Var(Q) \to \Delta^{\mathcal{I}}$:

1. $\mathcal{I} \models^{\pi} C(v)$ if $(\pi(v)) \in C^{\mathcal{I}}$
2. $\mathcal{I} \models^{\pi} R(v_1, v_2)$ if $(\pi(v_1), \pi(v_2)) \in R^{\mathcal{I}}$

Furthermore, we write $\mathcal{I} \models^{\pi} Q$ if we have $\mathcal{I} \models^{\pi} A$ for all atoms A in Q. If $\mathcal{I} \models^{\pi} Q$ for all models $\mathcal{I}$ of a knowledge base KB and some π, we write $KB \models Q$.

Given a knowledge base KB and a query Q, the *query entailment problem* is to decide whether $KB \models Q$. It is folklore, that query entailment and query answering problems are equivalent [HT00].

A conjunctive query Q is *connected* if, for all $v, v' \in Var(Q)$, there exists a sequence $v_0, ..., v_n$ such that $v_0 = v$ and $v_n = v'$ and for all $i < n$, there exists a role R, s.t. $R(v_i, v_{i+1}) \in Q$.

A *role hierarchy graph* is a directed graph $\mathcal{G}_{\mathcal{H}} = \langle N, E \rangle$, where roles are the nodes and there is a directed edge from R to S, whenever $R \sqsubseteq S$. We assume that the graph is reflexive, that is, each node has an edge directed to itself. Usually, a role hierarchy graph will contain several unconnected subgraphs. We call two roles S, T *hierarchically connected*, written $S \cong T$, if they belong to the same subgraph.

With $[R]_H$ we denote the equivalence class (the subgraph of $\mathcal{G}_{\mathcal{H}}$) R is contained in. That is, $R \cong S \iff [R]_H = [S]_H$. We extend our definition of $Trans$ in such a way, that $Trans([R]_H) \iff (\exists S \in [R]_H . Trans(S))$. Let $N_{[R]_H} = \{[R]_H | R \in N_R\}$.

3   Determining Role Condensates

Let $KB = \langle \mathcal{T}, \mathcal{R}, \mathcal{A} \rangle$ be a SHIQ knowledge base. We want to build a graph G from KB, that has the following property: Whenever $KB \models R_1(i_1, i_2) \wedge R_2(i_2, i_3) \wedge ... \wedge R_{n-1}(i_{n-1}, i_n)$, for named or fresh (=not occuring in the ABox) individuals $i_1...i_n$, then we have a path $f(i_1) \to_{[R_1]_H} \cdots \to_{[R_n]_H} f(i_n)$ in G, where f is a homomorphism and $\to_X$ stands for a directed X-edge.

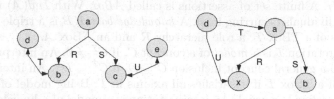

Fig. 1. Possible merging of individuals

The intuition is to find a graph, s.t. anything that can be observed about role-transitions in a knowledge base, can also be observed about role transitions in our graph, but not vice versa. We create this graph G in two separate steps:

1. We create a small graph, that creates a superset of possible roles between named individuals.
2. We extend this graph by applying a kind of worst-case algorithm to derive further information about all possible roles that can be inferred by $\exists R$- and $\geq_n R$-constraints

Let us define the graph structure for the first step:

Definition: A *role condensate* (of a knowledge base KB) is a graph $\mathcal{G_A} = \langle V, E, \phi, \omega \rangle$, where $V \subseteq \mathbb{N}$, $E \subseteq V \times V \times 2^{N_{[R]_H}}$, $\phi : V \to 2^{Ind(\mathcal{A})}$ and $\omega : V \to 2^{clos(KB)}$.

In the rest of this paper we will often use the inverse of ϕ. This is a function as well, since we enforce $\forall x, y \in V.\phi(x) \cap \phi(y) = \emptyset$. Whenever we call a role condensate *grounded*, we assume, that ω is empty.

To build the graph, notice that in a SHIQ-knowledge base one never removes existing roles between two named individuals. The known algorithms only infer additional roles between named individuals based on given constraints. These constrains are transitivity, role hierarchies and maximum cardinality constraints. Let us have a look at these three cases:

- **Transitivity**
  Whenever we have $KB \models R(a,b)$, $KB \models R(b,c)$ and $Trans(R)$, then we can conclude, that $KB \models R(a,c)$.
- **Role hierarchies**
  Whenever we have $KB \models S(a,b)$ and $S \sqsubseteq^* R$, then we can conclude, that $KB \models R(a,b)$.
- **Maximum cardinality constraints**
  In the following we will discuss the case of two mergable nodes. This can be easily extended for the case of n nodes. There are two distinct situations to consider.
  1. We have to merge two named individuals b and c (Figure 1, left). This can only happen, if $a :\leq_n V.X$ and $R \cong V \cong S$. The merged individuals

will have all the neighbors of b plus all the neighbors of c. Thus, if we replace all roles R by $[R]_H$, then we can merge named individuals based on role equality.

2. We have to merge a named individuals b and an unnamed individual x (Figure 1, right). The unnamed individual x can only be connected to another (arbitrary) named individual d, if $T \cong U \cong R$ and $Trans([U]_H)$. Furthermore, x and b will only be merged if $R \cong S$. Thus, if we replace all roles R by $[R]_H$, then we can simulate named-unnamed merging by explicitly closing the graph for transitivity.

The above facts tell us what we have to do to create a superset of possible roles between named individuals for our role condensate $\mathcal{G}_\mathcal{A}$. Yet, we also want to reduce the number of nodes in $\mathcal{G}_\mathcal{A}$. We want to merge similar individuals if they are connected to the same individual via a role $[R]_H$. The key is to have an adequate measure of equality of 2 (or n) named individuals. Let us define an abstract *individual similarity relation* $Sim(a, b) : Ind(\mathcal{A}) \times Ind(\mathcal{A})$. Three possible instances of this relation (among others) are:

1. **Same set of in-/outgoing roles:** We could define $Sim(a, b) \iff (S_{Role}(a) = S_{Role}(b))$, where $S_{Role}(a)$ is the union of incoming and outgoing roles of i.
2. **Same concept sets:** We could define $Sim(a, b) \iff (S_{Conc}(a) = S_{Conc}(b))$, where $S_{Conc}(a)$ is the set of concepts associated to an individual/node a.
3. **All individuals are similar:** We could define $Sim(a, b) = Ind(\mathcal{A}) \times Ind(\mathcal{A})$.

In the following we will use Sim without making a commitment to particular instances. Next, we propose a way to create a role condensate from a given SHIQ KB. The algorithm is shown in figure 2.

Input: SHIQ knowledge base $KB = \langle \mathcal{T}, \mathcal{R}, \mathcal{A} \rangle$
Output: Grounded role condensate $\mathcal{G}_\mathcal{A}$
Algorithm:
    1. **For each** $R(a, b) \in \mathcal{A}$ **do**
        (a) $v_a = getOrCreate(\mathcal{A}, \mathcal{G}_\mathcal{A}, a)$
        (b) $v_b = getOrCreate(\mathcal{A}, \mathcal{G}_\mathcal{A}, b)$
        (c) **Add** $[R]_H(v_a, v_b)$ to $\mathcal{G}_\mathcal{A}$ and add $Inv([R]_H)(v_b, v_a)$ to $\mathcal{G}_\mathcal{A}$
        (d) **While** $\mathcal{G}_\mathcal{A}$ is changed below
            i. **If** $(\exists c, d, e.S(c, d) \in \mathcal{G}_\mathcal{A} \wedge S(c, e) \in \mathcal{G}_\mathcal{A} \wedge d \neq e \wedge (Sim(d, e) \vee \leq_n S.X \in clos(KB))$ then $Merge(\mathcal{G}_\mathcal{A}, d, e)$
            ii. **If** $(\exists c, d, e.S(c, d) \in \mathcal{G}_\mathcal{A} \wedge S(d, e) \in \mathcal{G}_\mathcal{A} \wedge S(c, e) \notin \mathcal{G}_\mathcal{A} \wedge Trans([S]_H))$ **then add** $[S]_H(c, e)$ to $\mathcal{G}_\mathcal{A}$

Fig. 2. Grounded role condensates: algorithm

The algorithm should be self-explaining. We will only have a look at the while-loop of the main algorithm.

The first part of the loop merges individuals if they are similar or might have to be merged due to a maximum cardinality restriction. The latter is determined in a rather coarse-grained approach. We only look at the closure of all concepts

Function *merge*	Function *getOrCreate*
Parameter: $\mathcal{G}_{\mathcal{A}}$, nodes d, e	**Parameter:** $\mathcal{A}$, $\mathcal{G}_{\mathcal{A}}$, individual a
Algorithm:	**Returns:** node n
1. **For each** $R(x, e) \in \mathcal{G}_{\mathcal{A}}$ **do**	**Algorithm:**
(a) **Add** $R(x, d)$ to $\mathcal{G}_{\mathcal{A}}$	1. **If** $a \notin \bigcup_{v \in V} \phi(v)$ **then**
(b) **Remove** $R(x, e)$ from $\mathcal{G}_{\mathcal{A}}$	(a) **Add** new node n to $\mathcal{G}_{\mathcal{A}}$
2. **For each** $R(e, y) \in \mathcal{G}_{\mathcal{A}}$ **do**	(b) **Set** $\omega(n) = \{C \mid a : C \in \mathcal{A}\}$
(a) **Add** $R(d, y)$ to $\mathcal{G}_{\mathcal{A}}$	(c) **Set** $\phi(n) = \{a\}$
(b) **Remove** $R(e, y)$ from $\mathcal{G}_{\mathcal{A}}$	(d) **Return** n
3. **Set** $\phi(d) = \phi(d) \cup \phi(e)$	2. **else**
4. **Set** $\omega(d) = \omega(d) \cup \omega(e)$	(a) **Return** v, s.t. $\phi(v) = a$
5. **Remove** e from $\mathcal{G}_{\mathcal{A}}$	

Fig. 3. Grounded role condensates: helper functions

in the knowledge base. We could do better by applying a similar strategy as in [FKM$^+$06]. They carefully analyze the ABox and determine to which individuals maximum cardinality restrictions can be forwarded.

The second part of the while-loop creates all possible transitive edges between individuals.

Lemma (*homomorphic monotonicity*): Let $\mathcal{G}_{\mathcal{A}}$ be the role condensate of $\mathcal{A}$. Whenever the algorithm adds to $\mathcal{G}_{\mathcal{A}}$ an R-edge from $\phi^-(a)$ to $\phi^-(b)$, then $\mathcal{G}_{\mathcal{A}}$ will also have an R-edge from $\phi^-(a)$ to $\phi^-(b)$ after the algorithm terminates.

Proof: Can be done by case analysis of the algorithm. $\qquad\qquad\square$

Theorem 1. *Let $\mathcal{G}_{\mathcal{A}}$ be the grounded role condensate of $\mathcal{A}$. Whenever we have $\langle \mathcal{T}, \mathcal{R}, \mathcal{A} \rangle \models R(a, b)$, for two named individuals a and b, then we have that $[R]_H(\phi^-(a), \phi^-(b)) \in \mathcal{G}_{\mathcal{A}}$.*

Proof: This can be done by induction. The base case is that the theorem is true for all existing role relationships in the source ABox. The induction steps are based on possible reasons for adding an R-edge between two named individuals and the use of homomorphic monotonicity lemma. $\qquad\qquad\square$

We emphasize that the opposite direction of this theorem is not true. We do have complete, but unsound, reasoning on the role-relationships between named individuals in the ABox. The above result can be easily lifted to the case of paths between named individuals.

To make our approach clear, we will demonstrate it with a simple example. Let us consider the 'messed up' snapshot of the british royal family shown in Figure 4(a). We assume the following role information: $has\_father \sqsubseteq has\_parent$, $has\_mother \sqsubseteq has\_parent$ and $Inv(has\_parent) = has\_child$. The result of the algorithm is shown in 4(b) (inverse roles are not explicitly drawn). Please note that in the grounded role condensate all individuals are grouped w.r.t. their level in the family's genealogical tree.

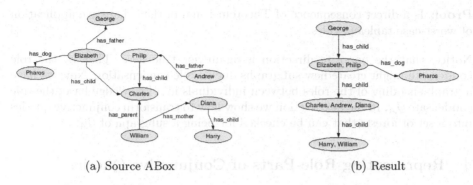

(a) Source ABox   (b) Result

Fig. 4. Example 1: Grounded role condensates

With the grounded role condensates we have a kind of base skeleton for the distribution of roles in a SHIQ knowledge base. Additional roles can only exist between named and unnamed individuals, and also between two unnamed individuals. These role relationships can only be created by concept expressions such as $\exists R.X$ and $\geq_n R.X$. We call these two expressions R-*generators* from now on.

To extend the role condensates to the non-grounded case, we apply a kind of worst-case tableaux algorithm to a grounded role condensate $\mathcal{G}_A$. Note that after running the tableau algorithm, we set blocked nodes equal to their respective blockers, to capture infinite paths in the tree.

First, we initialize ω of $\mathcal{G}_A$ as follows: $\forall v \in V.(\omega(v) = \{C | \exists a \in \phi(v).C(a) \in \mathcal{A}\})$. The tableau-like algorithm is shown in figure 5.

⊓-rule	If $C_1 \sqcap C_2 \in \omega(x)$, x is not indirectly blocked and $\{C_1, C_2\} \not\subseteq \omega(x)$, then $\omega(x) = \omega(x) \cup \{C_1, C_2\}$
⊔-rule	If $C_1 \sqcup C_2 \in \omega(x)$, x is not indirectly blocked and $\{C_1, C_2\} \not\subseteq \omega(x)$, then $\omega(x) = \omega(x) \cup \{C_1, C_2\}$
∃-rule	If $\exists R.C \in \omega(x)$, x is not blocked and x has no $[R]_H$-neighbor y with $C \in \omega(y)$, then create a new node y with $\omega(y) = \{C\}$ and $E = E \cup \{(x, y, [R]_H)\}$
∀-rule	If $\forall R.C \in \omega(x)(x)$, x is not indirectly blocked and there is an $[R]_H$-neighbor of y of x with $C \notin \omega(x)(y)$, then $\omega(y) = \omega(y) \cup \{C\}$
$\geq_n$-rule	If $\geq_n R.C \in \omega(x)(x)$, x is not blocked and x has no $[R]_H$-neighbor y with $C \in \omega(x)(y)$, then create a new node y with $\omega(y) = \{C\}$ and $E = E \cup \{(x, y, [R]_H)\}$
$\forall_+$-rule	If $\forall R.C \in \omega(x)$, x is not indirectly blocked, we have $Trans(R)$ and there is an R-neighbor y of x, with $\forall R.C \notin \omega(y)$, then $\omega(y) = \omega(y) \cup \{\forall R.C\}$

Fig. 5. Worst-case role tableau algorithm

Theorem 2. *Let $KB = \langle \mathcal{T}, \mathcal{R}, \mathcal{A} \rangle$ be a SHIQ knowledge base and let $\mathcal{G}_A$ be the role condensate of $\mathcal{A}$. Whenever $KB \models R_1(i_1, i_2) \wedge R_2(i_2, i_3) \wedge ... \wedge R_{n-1}(i_{n-1}, i_n)$, for named or fresh individuals $i_1...i_n$, then we have a path $\phi^-(i_1) \to_{[R_1]_H} ... \to_{[R_n]_H} \phi^-(i_n)$ in $\mathcal{G}_A$.*

Proof: Is a direct consequence of Theorem 1 and of the exhaustive application of worst-case tableaux rules.   □

Notice, that the opposite direction is again not true here. That is, the role condensate might create new subgraphs due to role condensation. Now we have a graph encoding of the roles between individuals in a knowledge base, the role condensate $\mathcal{G}_\mathcal{A}$. In the next section we show how to transform conjunctive queries into a set of forests that can be checked for being a subgraph of $\mathcal{G}_\mathcal{A}$.

4   Representing Role-Parts of Conjunctive Queries

This section shows how to use role condensates to answer conjunctive queries in an unsound, but complete manner. Without loss of generality we assume that the conjunctive query Q is connected. Otherwise we split Q into n connected queries q_i and answer each of them separately.

Definition: A *query forest match* (of a conjunctive query Q) is a forest $\mathcal{G}_Q = \langle V, E \rangle$, where $V \subseteq Var(Q) \cup \mathbb{N}$ and $E \subseteq V \times V \times N_R$. A query forest match $\mathcal{G}_Q$ is created non-deterministically by the algorithm shown in figure 6.

The idea of a query forest match is to encode one possible forest structure (role-wise) which a conjunctive query can enforce. In contrast to *query graphs* [GHLS07], we do not (in general) only regard role query atoms, but also possible unfoldings of the concept query atoms. This is done by the helper function *conceptPathway*. It creates one tree representing role paths for each possible (non-deterministic) unfolding. The non-determinism is introduced by the disjunction of concepts. The function uses three not further explained helper functions to create trees: *mergeRoots* (merges two sibblings), *leaf* (creates a leaf of a tree) and *newRoot* (creates a new root node for a subtree).

Function: $getQueryForestMatch$
Input: Knowledge base KB, Conjunctive query Q, Max depth d
Output: Query forest match $\mathcal{G}_Q$
Algorithm:
    1. **Let** $\mathcal{G}_Q$ be the graph created by the role query atoms in Q
    2. **For each** $C(X) \in Q$ do
        (a) **Add** one t, s.t. $t \in conceptPathway(KB, C, d)$, to node X in $\mathcal{G}_Q$

Fig. 6. Query forest match algorithm

Notice that we restrict the unfolding to a particular depth d. First, since we might have cyclic GCIs, we guarantee termination that way. Second, when we combine the set of unfoldings for each concept of a named individual, we obtain a possibly exponential amount of forests. This is not desirable for practical reasoning. Still, we want to emphasize that our approach is complete for each $d \in \mathbb{N}$. For $d = 0$ we obtain the standard query graph.

Function: *conceptPathway*
Input: Knowledge base KB, Concept C, Max depth d
Output: Tree T
1. **If** $d = 0$ **then**
   return $T = leaf()$
2. **If** $C = C_1 \sqcap C_2$ **then**
   return $T = mergeRoots(conceptPathway(C_1, d), conceptPathway(C_2, d))$
3. **If** $C = C_1 \sqcup C_2$ **then**
   return $T = conceptPathway(C_n, d - 1)$, for one $n \in \{1, 2\}$
4. **If** $C = \exists R.C_1$ or $C = \geq_n .C_1$ **then**
   return $T = newRoot(R, conceptPathway(C_1))$
5. **If** $(C \sqsubseteq C_1) \in T$ **then**
   return $T = conceptPathway(C_1, d - 1))$
6. **else return** $T = leaf()$

Fig. 7. Query forest match algorithm - *conceptPathway* function

5   Answering Conjunctive Queries

Let Q_F be the set of all possible query forest matches for a query Q, that is,
$Q_F = \{\mathcal{G}_Q | getQueryForestMatch(KB, Q, d)\}$.

Theorem 3. *Let $KB = \langle \mathcal{T}, \mathcal{R}, \mathcal{A} \rangle$ be a SHIQ knowledge base and Q be a conjunctive query. If we have that $KB \models Q$, then we can find a query forest match $\mathcal{G}_Q$, that is isomorphic to a subgraph of $\mathcal{G}_\mathcal{A}$.*

Proof: The proof follows directly from Theorem 2.   □

We applied our approach to the LUBM Ontology [GPH05], an ontology benchmark in the setting of universities. It is accompanied by a set of 14 test queries. The second query of that set is interesting for our approach, since it has no solution for most data generations of the LUBM Generator. This query is shown in Figure 8, using a SPARQL-like syntax.

```
(type GraduateStudent ?X), (type University ?Y), (type Department ?Z),
(memberOf ?X ?Z), (subOrganizationOf ?Z ?Y), (undergraduateDegreeFrom ?X ?Y)
```

Fig. 8. LUBM query 2

First, we have created a role condensate for several LUBM(X,0) (where X is the number of universities). The result is shown in figure 9. For the individual similarity relation Sim, we have used the one with 'same set of incoming/outgoing roles'.

Universities	Individuals	Sim_1:nodes in $\mathcal{G}_\mathcal{A}$
5	102K	25
10	205K	24
20	407K	26

Fig. 9. LUBM statistics

Let Q be the conjunctive query shown in Figure 8. Setting the maximum depth for the concept unfolding $d = 0$, then we obtain a single query forest match $\mathcal{G}_Q$, that corresponds to the query graph of Q. When we check whether $\mathcal{G}_Q$ is isomorphic to a subgraph of $\mathcal{G}_A$, we get a *no* immediately, for all tested number of universities. This small example shows already that our approach *can* dramatically speed up the rejection of queries with no answers.

6   Conclusion and Future Work

The decision procedure presented in this work makes it possible to answer conjunctive queries in a complete, but unsound, way. We obtain a proxy-like decision system which can possibly immediately reject conjunctive queries having no answer replacements. Moreover, it can be used to provide a set of obvious non instances for each variable in the grounded setting (e.g. the query language NRQL [HMW04]). This will be discussed in future work.

Until now, our approach works only for the description logic SHIQ. For OWL DL, that is SHOIQ, we need to further add handling of nominals. This is part of future work. We will also investigate an upper bound for the complexity of role condensate creation. Furthermore, we will apply the approach to additional test ontologies (especially non-LUBM) in order to provide detailed statistics about its usefulness in practice.

References

[FKM+06]   Fokoue, A., Kershenbaum, A., Ma, L., Patel, C., Schonberg, E., Srinivas, K.: Using Abstract Evaluation in ABox Reasoning. In: SSWS 2006, Athens, GA, USA, November 2006, pp. 61–74 (2006)

[GHLS07]   Glimm, B., Horrocks, I., Lutz, C., Sattler, U.: Conjunctive query answering for the description logic shiq. In: IJCAI 2007, AAAI Press (2007)

[GPH05]   Guo, Y., Pan, Z., Heflin, J.: Lubm: A benchmark for owl knowledge base systems. J. Web Sem. 3(2-3), 158–182 (2005)

[HMW04]   Haarslev, V., Möller, R., Wessel, M.: Querying the semantic web with racer + nrql. In: Proceedings of the KI-2004 International Workshop on Applications of Description Logics (ADL 2004), September 24, Ulm, Germany (2004)

[HT00]   Horrocks, I., Tessaris, S.: A conjunctive query language for description logic aboxes. In: AAAI/IAAI, pp. 399–404 (2000)

A Pragmatic Approach for RDFS Reasoning over Large Scale Instance Data

Tuğba Özacar, Övünç Öztürk, and Murat Osman Ünalır

Department of Computer Engineering,
Ege University
Bornova, 35100, Izmir, Turkey
{tugba.ozacar,ovunc.ozturk,murat.osman.unalir}@ege.edu.tr

Abstract. In this paper, we propose a pragmatic approach where time consumption of RDFS reasoning remains fixed with increasing sizes of instance data. The approach infers facts about schema, but prevents producing facts about individuals. At the time of query answering the queries are rewritten. The query rewriting process does not result in complex or disjunctive queries due to the syntactic transformation that we made on the ontology and the rules. The most prominent contribution of this work is reducing reasoning to schema level without increasing query complexity. Thus, query execution performance improves in a considerable manner.

1 Introduction

RDF Schema [1] reasoning is heavily related with computing the implied statements about individuals. Therefore it is important for RDFS reasoners to cope with realistic sets of instance data [2]. Although current reasoners are capable of schema reasoning with real world ontologies, they break down when the number of instances becomes large [3].

Present works focus at query rewriting techniques, but these techniques end in complex and disjunctive queries. In [4], it is proposed to exclude *type* statements from the closure for scaling up RDFS reasoning. The reasoning about *type* statements is performed at the time of query answering by replacing parts of the query that ask for instance level statements by a more complex query expression that involves some schema elements. Wilbur's RDF(S) reasoner [5] is implemented by rewriting access queries. Rewriting is done by recursively substituting all occurrences of those RDF [6] properties that have a semantic theory with a more complex query expression. [5] also has problems in dealing with *domain* and *range* restrictions. A recent solution, namely summary-Abox [7], reduces reasoning to a small subset of the original ABox, using a filtering technique. But the filtering is dynamic and it must be computed for each incoming query.

Similar to related works, we infer facts about schema and we prevent producing facts about individuals. In brief, reasoning is reduced to schema level by preventing instance data to participate in reasoning. It is not important how

R. Meersman, Z. Tari, P. Herrero et al. (Eds.): OTM 2007 Ws, Part II, LNCS 4806, pp. 1155–1164, 2007.

much the instance data of an ontology increases, the reasoning process always finishes in a constant time. But different from related works, the query rewriting process does not result in complex or disjunctive queries due to the syntactic transformation that we made on the ontology and the rules. Consequently, the query execution performance improves in a considerable manner.

We have also used similar constructs for scalability in a previous work [8]. But, this work differs from the former one in more than one way. First of all, this work has a novel contribution. It makes the cost of inference constant with increasing sizes of instance data and it does not increase the query complexity. Secondly, the previous work uses only *class instance grouping constructs*, but this work introduces *property instance grouping constructs* and modifies the *class instance grouping constructs* used in the previous work. Finally, this work supports reasoning with a limited set of rules (see section 3.2) on RDFS documents satisfying conditions[9] for an RDF graph to be a DL ontology. On the contrary, the previous work supports all RDF based languages with no restrictions.

The rest of the paper is organized as follows: Section 2 describes the transformation model using grouping constructs. Section 3 presents the application of the transformations on triples, rules and queries. Section 4 contains a basic analysis on how much space is saved by the proposed approach. Section 5 presents the implementation phase and evaluation of the model with large scale ontologies. Finally, section 6 concludes the paper with some potential future research. Please, refer to http://aegont.bilmuh.ege.edu.tr/ssws07/running_example.pdf for a running example that uses the approach.

2   The Transformation Model Using Grouping Constructs

The model is based on a simple principle, which is about inverting direction of some existing relations or extracting the hidden relations. These relations are the ones that relate an individual to its concept. The set of individuals of a concept is named as *extension* in [9]. These extensions are not defined explicitly in the ontology. Consequently, there isn't any relation type to define the elements of the extension, namely individuals of the concept.

Our main contribution is to define these extensions and its elements explicitly in the ontology. In order to do this, some extra language constructs, which are named as *grouping constructs*, are used. Grouping constructs simply relate an extension to the corresponding concept and its individuals.

2.1   Types of Grouping Constructs

The grouping constructs are classified into four;

Class Instance Grouping Construct, namely *hasClassExtension*, relates every class with class extensions, which hold the individuals of the class. *hasClassExtension* has two subproperties; *hasExplicitClassExtension* and *hasInferredClassExtension*. *hasExplicitClassExtension* relates the class with its explicit class extension, which holds all explicit individuals of the class. *has-InferredClassExtension* relates the class with one of its inferred class extentions, which partially hold the inferred individuals of the class.

Property Subject Grouping Construct, namely *hasSubjectExtension*, relates every property, except *type* property, with subject extensions, which hold the subjects of individuals of the property. *hasSubjectExtension* has two subproperties; *hasExplicitSubjectExtension* and *hasInferredSubjectExtension*. *hasExplicitSubjectExtension* relates the property with its explicit subject extension, which holds subjects of all explicit property individuals. *hasInferredSubjectExtension* relates the property with one of its inferred subject extentions, which partially hold the subjects of inferred individuals of the property.

Property Object Grouping Construct, namely *hasObjectExtension*, relates every property except *type* property, with object extensions, which hold the objects of individuals of the property. *hasObjectExtension* has two subproperties; *hasExplicitObjectExtension* and *hasInferredObjectExtension*. *hasExplicitObjectExtension* relates the property with its explicit object extension, which holds objects of all explicit property individuals. *hasInferredObjectExtension* relates the property with one of its inferred object extentions, which partially hold the objects of inferred individuals of the property.

Property Instance Grouping Construct, namely *hasPropertyExtension*, relates every property except *type* property, with property extensions, which *symbolize* the individuals of the property. *hasPropertyExtension* has two subproperties; *hasExplicitPropertyExtension* and *hasInferredPropertyExtension*. *hasExplicitPropertyExtension* relates the property with its explicit property extension, which symbolizes all explicit individuals of the property. *hasInferredPropertyExtension* relates the property with one of its inferred property extentions, which partially symbolizes the inferred individuals of the property.

The first three grouping constructs connect concepts to extensions having identical structures. The items of these extensions are either class individuals or datatype values (only in *objectExtensions*). An extension is connected to each of its items with a *contains* predicate.

The forth grouping construct differs from the others in that it doesn't have a concrete extension. Its extension is an empty and virtual list and it is not related to any of its items with a *contains* predicate. The reason for not keeping the individuals of this extension is not to increase the triple count too much, after transformation.

2.2   Motivation for the Use of Grouping Constructs

We also need to emphasize why we transform the ontology to a model that uses grouping constructs. It may seem that the whole approach could be rephrased without using any grouping constructs and making any transformations at all. So, the approach would not produce facts about individuals and computes these facts at the time of query answering like in the related works. But, the lack of grouping constructs leads to complex and disjunctive queries. Lower performance is the price that we could pay if we support disjunctive queries [10]. For example, the query rewriting mechanism in [4], replaces triple patterns that contain a type statement by the union of five possible rewritings of the respective pattern. This rewriting mechanism will obviously increase the query complexity. On the other hand, the use of grouping constructs provides a way to write concise queries. As a result, both inference and query answering performances improve.

3   Transforming Triples, Rules and Queries

This section is about the implementation of the transformations. First, we describe the transformation of ontology triples and language axioms. Then, we represent the transformation of rules. Finally, the query rewriting process is described.

3.1   Transforming Triples

The following transformations are applied to all triples in the ontology $\mathcal{O}$. It is worth to note that the same transformations are done whenever a triple is added to or removed from the ontology.

- For every class[1] $\mathcal{C}$, *add* the following triple to the ontology;
  - $hasExplicitClassExtension(\mathcal{C}, \mathcal{C_E\_C})$

- For every property[2] $\mathcal{P}$, except *type*, *add* the following triples[3] to the ontology;
  - $hasExplicitSubjectExtension(\mathcal{P}, \mathcal{S_E\_P})$
  - $hasExplicitObjectExtension(\mathcal{P}, \mathcal{O_E\_P})$
  - $hasExplicitPropertyExtension(\mathcal{P}, \mathcal{P_E\_P})$

- *Substitute* every triple having a *type* predicate (e.g. $type(\mathcal{I}, \mathcal{C})$) with the following triple, where $hasExplicitClassExtension(\mathcal{C}, \mathcal{C_E\_C}) \in \mathcal{O}$;
  - $contains(\mathcal{C_E\_C}, \mathcal{I})$

- For every triple(e.g., $\mathcal{P}(\mathcal{A}, \mathcal{B})$), which has a predicate other than *type*, *add* the following triples to the ontology(without removing $\mathcal{P}(\mathcal{A}, \mathcal{B})$), where

---

[1] The subjects of triples in the form of $type(\mathcal{C}, Class)$ or a class from *rdf* or *rdfs* namespace(e.g.,*rdf:Class, rdf:Property*).

[2] The subjects of a triples in the form of $type(\mathcal{A}, Property)$ or a property from *rdf* or *rdfs* namespace(e.g.,*rdfs:subPropertyOf, rdfs:domain, rdfs:range*).

[3] In fact, there is no need to add a subjectExtension or an objectExtension to *rdfs:subClassOf, rdfs:subPropertyOf, rdfs:domain* or *rdfs:range*, because the approach does not allow to redefine the built-in properties or classes. Thus, we do not define a domain or range for these properties and subject and object extensions of them become useless. But for the sake of simplicity, we do not handle these properties in a different way, and we also add a subject and object extension to them, although these extensions are never used.

$hasExplicitSubjectExtension\,(\mathcal{P}, \mathcal{S_E}\_\mathcal{P}) \in \mathcal{O} \land hasExplicitObjectExtension$
$(\mathcal{P}, \mathcal{O_E}\_\mathcal{P}) \in \mathcal{O};$

- $contains(\mathcal{S_E}\_\mathcal{P}, \mathcal{A})$
- $contains(\mathcal{O_E}\_\mathcal{P}, \mathcal{B})$

3.2   Transforming Rules

This work supports reasoning[4] with a limited set of rules (Table 1) on RDFS documents satisfying conditions [9] for an RDF graph to be a DL ontology. These conditions are the rules of thumb for OWL DL ontologies.

Table 1. Rules before transformation

Rule No	Rule
1	$\dfrac{domain(?p,?c) \land ?p(?a,?b)}{type(?a,?c)}$
2	$\dfrac{range(?p,?c) \land ?p(?a,?b)}{type(?b,?c)}$
3	$\dfrac{subPropertyOf(?p_1 ?p_2) \land subPropertyOf(?p_2,?p_3)}{subPropertyOf(?p_1,?p_3)}$
4	$\dfrac{type(?p, Property)}{subPropertyOf(?p,?p)}$
5	$\dfrac{subPropertyOf(?p_1,?p_2) \land ?p_1(?a,?b)}{?p_2(?a,?b)}$
6	$\dfrac{subClassOf(?c_1,?c_2) \land type(?a,?c_1)}{type(?a,?c_2)}$
7	$\dfrac{type(?c, Class)}{subClassOf(?c,?c)}$
8	$\dfrac{subClassOf(?c_1,?c_2) \land subClassOf(?c_2,?c_3)}{subClassOf(?c_1,?c_3)}$

Rule 3 and *rule 8* are applied without transformation. Other rules are transformed in the following way;

(1) $\dfrac{domain(?p,?c) \land ?p(?a,?b)}{type(?a,?c)} \Rightarrow \dfrac{domain(?p,?c) \land hasSubjectExtension(?p,?e)}{hasInferredClassExtension(?c,?e)}$

(2) $\dfrac{range(?p,?c) \land ?p(?a,?b)}{type(?b,?c)} \Rightarrow \dfrac{range(?p,?c) \land hasObjectExtension(?p,?e)}{hasInferredClassExtension(?c,?e)}$

(4) $\dfrac{type(?p, Property)}{subPropertyOf(?p,?p)} \Rightarrow \dfrac{hasExplicitPropertyExtension(?p,?e)}{subPropertyOf(?p,?p)}$

(5) $\dfrac{subPropertyOf(?p_1,?p_2) \land ?p_1(?a,?b)}{?p_2(?a,?b)} \Rightarrow$

$\dfrac{subPropertyOf(?p_1,?p_2) \land hasSubjectExtension(?p_1,?e)}{hasInferredSubjectExtension(?p_2,?e)}$

$\dfrac{subPropertyOf(?p_1,?p_2) \land hasObjectExtension(?p_1,?e)}{hasInferredObjectExtension(?p_2,?e)}$

$\dfrac{subPropertyOf(?p_1,?p_2) \land hasPropertyExtension(?p_1,?e)}{hasInferredPropertyExtension(?p_2,?e)}$

(6) $\dfrac{subClassOf(?c_1,?c_2) \land type(?a,?c_1)}{type(?a,?c_2)} \Rightarrow \dfrac{subClassOf(?c_1,?c_2) \land hasClassExtension(?c_1,?e)}{hasInferredClassExtension(?c_2,?e)}$

(7) $\dfrac{type(?c, Class)}{subClassOf(?c,?c)} \Rightarrow \dfrac{hasExplicitClassExtension(?c,?e)}{subClassOf(?c,?c)}$

---

[4] This approach does not allow a property to be defined as a *subPropertyOf rdf:type*. Since we aim a pragmatic reasoning approach, and defining a property in this way does not have importance in practical use, we do not consider this situation as an important defect.

We also add the following rules to support the transformation;

$$\frac{hasExplicitClassExtension(?c, ?e) \lor hasInferredClassExtension(?c, ?e)}{hasClassExtension(?c, ?e)}$$

$$\frac{hasExplicitSubjectExtension(?c, ?e) \lor hasInferredSubjectExtension(?c, ?e)}{hasSubjectExtension(?c, ?e)}$$

$$\frac{hasExplicitObjectExtension(?c, ?e) \lor hasInferredObjectExtension(?c, ?e)}{hasObjectExtension(?c, ?e)}$$

$$\frac{hasExplicitPropertyExtension(?c, ?e) \lor hasInferredPropertyExtension(?c, ?e)}{hasPropertyExtension(?c, ?e)}$$

3.3 Query Rewriting

The query rewriting process replaces some query conditions in the following way;

– All query conditions with a *type* predicate (e.g., $type(?i, ?c)$) are replaced
   with the following expression;
   - $hasClassExtension(?c, ?e) \land contains(?e, ?i)$
– All query conditions having a predicate, which is neither a *type* predicate
   nor a property from *rdf* or *rdfs* namespaces, (e.g., $?p_1(?a, ?b)$) are replaced
   with the following expression;
   - $hasPropertyExtension(?p_1, ?e) \land hasExplicitPropertyExtension(?p_2, ?e) \land ?p2(?a, ?b)$

The first replacement aims to find individuals of a class in the transformed version of the ontology. First all *classExtension*s of the class are found, then the individuals in them are returned. These *classExtension*s include the *explicitClassExtension* of the class and all *inferredClassExtension*s that are related to that class after inference (inferred via a subclass, domain or range relation).

The second replacement aims to find individuals of a property in the transformed version of the ontology. First all *propertyExtension*s of the property are found. These *propertyExtension*s include the *explicitPropertyExtension* of the property and all *inferredPropertyExtension*s that are related to that class after inference (via a subProperty relation). In order to obtain the individuals of a *propertyExtension*, we first find the property that the *propertyExtension* is related to with an *hasExplicitPropertyExtension* predicate, because *propertyExtension*s do not have concrete extensions (see Section 2). Then using this property, we obtain the property individuals, in other words triples that use this property as a predicate.

4 Analysis of the Space Complexity

This section provides a basic analysis on how much space is saved by the proposed transformation. This is important to better understand the rationale behind the method and to identify further improvements.

Let $\mathcal{O}$ be an ontology, $n(\mathcal{O})$ be the number of triples, $\mathcal{R}_{\mathcal{O}}^{type}$ be the number of *type* relations, $\mathcal{R}_{\mathcal{O}}^{type^+}$ be the number of *type* relations with an object that is from *rdf* or *rdfs* namespace, $\mathcal{R}_\mathcal{O}^{type^-}$ be the number of *type* relations with an object that is not from *rdf* or *rdfs* namespace, $\mathcal{R}_\mathcal{O}^{\mathcal{P}^+}$ be the number of instances of a property that is not a *type* property but is from *rdf* or *rdfs* namespace, $\mathcal{R}_\mathcal{O}^{\mathcal{P}^-}$ be the number of instances of a property that is not from *rdf* or *rdfs* namespace, $\mathcal{C}_\mathcal{O}$ be the number of classes and $\mathcal{P}_\mathcal{O}$ be the number of properties except *type* property. Then;

$$n(\mathcal{O}) = \mathcal{R}_\mathcal{O}^{type} + \mathcal{R}_\mathcal{O}^{\mathcal{P}^+} + \mathcal{R}_\mathcal{O}^{\mathcal{P}^-}$$

After transformation, every *type* relation between a class and its individual is replaced with a *contains* relation between the *classExtension* of the class and the individual. So the number of *contains* relations in the transformed ontology is equal to $\mathcal{R}_\mathcal{O}^{type}$. But, since a *hasExplicitClassExtension* relation is added to every class, we add $\mathcal{C}_\mathcal{O}$ new triples to the transformed ontology. Then, for every instance of a property other than *type* (e.g. $A(\mathcal{U}, \mathcal{B})$), we add two new relations to the ontology ($contains(\mathcal{S}_\mathcal{E\_A}, \mathcal{U})$, $contains(\mathcal{O}_\mathcal{E\_A}, \mathcal{B})$). So the number of instances of a property other than *type* ($\mathcal{R}_\mathcal{O}^{\mathcal{P}^+} + \mathcal{R}_\mathcal{O}^{\mathcal{P}^-}$), is tripled ($3 \times \mathcal{R}_\mathcal{O}^{\mathcal{P}^+} + 3 \times \mathcal{R}_\mathcal{O}^{\mathcal{P}^-}$) in the transformed ontology. Since three relations are added (*hasExplicitSubjectExtension*, *hasExplicitObjectExtension* and *hasExplicitPropertyExtension*) to every property that is not a *type* property, we add $3 \times \mathcal{P}_\mathcal{O}$ triples to the transformed ontology.

Let $\mathcal{O}^t$ be the transformed version of ontology $\mathcal{O}$ and Δ_1 be the increase in the number of triples after transformation, in other words the difference between $n(\mathcal{O})$ and $n(\mathcal{O}^t)$, then;

$$n(\mathcal{O}^t) = \mathcal{R}_\mathcal{O}^{type} + \mathcal{C}_\mathcal{O} + 3 \times \mathcal{R}_\mathcal{O}^{\mathcal{P}^+} + 3 \times \mathcal{R}_\mathcal{O}^{\mathcal{P}^-} + 3 \times \mathcal{P}_\mathcal{O}$$

$$\Delta_1 = \mathcal{C}_\mathcal{O} + 2 \times \mathcal{R}_\mathcal{O}^{\mathcal{P}^+} + 2 \times \mathcal{R}_\mathcal{O}^{\mathcal{P}^-} + 3 \times \mathcal{P}_\mathcal{O}$$

After transformation, only a small amount of triples participates in reasoning. These triples can be defined as the instances of *rdfs:domain*, *rdfs:range*, *rdfs:subPropertyOf*, *hasExplicitClassExtension*, *hasExplicitPropertyExtension*, *hasExplicitSubjectExtension*, *hasExplicitObjectExtension* properties.

After transformation, we still make reasoning on the schema but we close the individual reasoning. Since the schema level reasoning is not affected by the transformation, the decrease in the number of inferred triples equals to the decrease in the number of inferred triples about individuals. Additional facts about *grouping predicates* are derived, but these derivations are small enough to be negligable.

In a nutshell, the transformation leads to an increase in the triple count. But since, a small amount of these triples participates in reasoning, the memory and time consumptions of reasoning decrease dramatically. Besides, instance data does not participate in reasoning, thus memory and time consumptions of reasoning and the number of inferred triples remain fixed, even if the size of instance data increases.

The following factors increase the utilization of the approach; (1) higher amount of instance data (2) deep class and property hierarchies, especially the ones where individual population intensifies at the bottom of the hierarchy (3) higher amount of domain/range relations.

5   Implementation and Experiments

In the implementation phase, all triples are loaded to a database table called *statementsTable*, after parsing the ontology file(s). Then, the triple transformation is applied on each record of this table. Using triple transformation, the triples participating in reasoning are filtered and loaded to the reasoner (see Section 4). All *contains* relations are placed in a seperate table called *containsTable*, in order to improve the query performance. The database schema resembles the vertical table design described in [11].

In order to evaluate the proposed model with large scale instance data, we conduct tests using Lehigh University Benchmark (LUBM) [12]. We tested the model on a desktop computer with a AMD Athlon 64 500+ processor, 2 GB RAM. We used Mono Runtime Version 1.11 for development and Sqlite Mono Plug-in 2.1 (using in-memory database) for database management. Table 2 shows that inference time and the number of inferred triples are constant even if the size of the instance data increases. This table compares the approach (PCIGC, namely **P**roperty and **C**lass **I**ndividual **G**rouping **C**onstructs) with two other approaches. The first one is NIGC (**N**o **I**ndividual **G**rouping **C**onstruct) where no grouping construct is used. The second one is CIGC (**C**lass **I**ndividual **G**rouping **C**onstructs) where only class instance grouping constructs are used [8].

Table 2. Loading and Inference performances of PCIGC for 50 universities

Metrics \ Data Set	LUBM (1,0)			LUBM (5,0)			LUBM (10,0)	LUBM (20,0)	LUBM (50,0)
	PCIGC	NIGC	CIGC	PCIGC	NIGC	CIGC	PCIGC	PCIGC	PCIGC
Loading	51.47	39.66	42.67	05:36.21	12:20.18	13:21.29	11:47.10	26:33.43	01:22:45.70
Inference	0.20	23.45	8.28	0.19	7:37.86	1:48.17	0.19	0.19	0.20
Inferred Triples	303	26442	4922	303	159901	30504	303	303	303

In Table 3, we evaluated the query performance using test queries of LUBM. As expected, the soundness or completeness of the queries does not change, in other words the answer counts of the queries remain same, after applying transformation. Query 10, 11, 12, 13 are omitted, because our target is RDFS reasoning and these queries are intended to verify OWL [13] reasoning capabilities. We moved a big percentage of triples to database, thus answers of query conditions about schema are searched in the small set of remaining triples. The answers of conditions about instances are searched in database by trivially transforming them to sql queries.

Table 3. Query performances of PCIGC for 50 universities (in milliseconds)

Query \ Data Set	LUBM(1,0)	LUBM(5,0)	LUBM(10,0)	LUBM(20,0)	LUBM(50,0)
1	19	84	147	300	926
2	268	1889	3378	7955	22623
3	62	472	836	1857	5581
4	45	289	516	1189	3308
5	547	3629	6582	15169	43156
6	146	1968	6026	15002	32574
7	150	1040	1897	4313	12253
8	477	1507	2507	5471	14959
9	1662	11495	20442	47899	136343
14	131	1792	5239	10714	34270

6   Conclusion and Future Work

In this work, we proposed a model with grouping constructs, which scales up RDFS reasoning by preventing instance data to participate in reasoning. The model improves loading, reasoning and querying times, especially with large scale instance data. Opening ontologies now largely depends on parsing ontologies and building triples rather than reasoning. Although transforming triples and filtering them bring an extra workload, this becomes much more negligible as the size of the ontology grows.

In the future work, we will improve the query performance of the model by using a more complex database schema and some query optimization techniques. According to [14], our database schema is at the structure level which we can change to semantic level for improving the performance. Another work is to increase the expressivity of the language that the model supports.

References

1. Brickley, D., Guha, R.: RDF vocabulary description language 1.0: RDF Schema. Technical report (2004)
2. Horrocks, I., Li, L., Turi, D., Bechhofer, S.: The instance store: Dl reasoning with large numbers of individuals. In: Description Logics (2004)
3. Haarslev, V., Möller, R.: High performance reasoning with very large knowledge bases: A practical case study. In: IJCAI, pp. 161–168 (2001)
4. Stuckenschmidt, H., Broekstra, J.: Time - space trade-offs in scaling up rdf schema reasoning. In: WISE Workshops, pp. 172–181 (2005)
5. Lassila, O.: Taking the rdf model theory out for a spin. In: Horrocks, I., Hendler, J. (eds.) ISWC 2002. LNCS, vol. 2342, pp. 307–317. Springer, Heidelberg (2002)
6. Hayes, P.: Rdf semantics (2004)

7. Fokoue, A., Kershenbaum, A., Ma, L., Schonberg, E., Srinivas, K.: The summary abox: Cutting ontologies down to size. In: Cruz, I., Decker, S., Allemang, D., Preist, C., Schwabe, D., Mika, P., Uschold, M., Aroyo, L. (eds.) ISWC 2006. LNCS, vol. 4273, pp. 343–356. Springer, Heidelberg (2006)
8. Öztürk, Ö., Özacar, T., Ünalir, M.O.: Reducing the inferred type statements with individual grouping constructs. In: International Semantic Web Conference, pp. 573–582 (2006)
9. Dean, M., Schreiber, G., van Harmelen, F., Hendler, J., Horrocks, I., McGuinness, D., Patel-Schneider, P., Stein, L.(eds.): Owl web ontology language reference. W3C Recommendation, February 10, 2004 (2003)
10. Mena, E., Illarramendi, A., Kashyap, V., Sheth, A.: OBSERVER: An approach for query processing in global information systems based on interoperation across pre-existing ontologies. International journal on Distributed And Parallel Databases (DAPD) 8, 223–272 (2000)
11. Pan, Z., Heflin, J.: Dldb: Extending relational databases to support semantic web queries. In: Workshop on Practical and Scaleable Semantic Web Systms, ISWC 2003, pp. 109–113 (2003)
12. Guo, Y., Pan, Z., Heflin, J.: An evaluation of knowledge base systems for large owl datasets. In: International Semantic Web Conference, pp. 274–288 (2004)
13. Smith, M.K., Welty, C., McGuinness, D.L.: Owl web ontology language guide. Technical report, OWL Web Ontology Language Guide (2004)
14. Broekstra, J., Kampman, A., van Harmelen, F.: Sesame: A generic architecture for storing and querying rdf and rdf schema (2002)

Modular Web Queries—From Rules to Stores

Uwe Aßmann[2], Sacha Berger[1], François Bry[1], Tim Furche[1],
Jakob Henriksson[2], and Jendrik Johannes[2]

[1] Institut für Informatik, Ludwig-Maximilians-Universität München
{sacha.berger,francois.bry,tim.furche}@ifi.lmu.de
[2] Fakultät für Informatik, Technische Universität Dresden
{uwe.assmann,jakob.henriksson,jendrik.johannes}@tu-dresden.de

Abstract. Even with all the progress in Semantic technology, accessing Web data remains a challenging issue with new Web query languages and approaches appearing regularly. Yet most of these languages, including W3C approaches such as XQuery and SPARQL, do little to cope with the explosion of the data size and schemata diversity and richness on the Web. In this paper we propose a straightforward step toward the improvement of this situation that is simple to realize and yet effective: Advanced module systems that make partitioning of (a) the evaluation and (b) the conceptual design of complex Web queries possible. They provide the query programmer with a powerful, but easy to use high-level abstraction for packaging, encapsulating, and reusing conceptually related parts (in our case, rules) of a Web query. The proposed module system combines ease of use thanks to a simple core concept, the partitioning of rules and their consequences in flexible "stores", with ease of deployment thanks to a reduction semantics. We focus on extending the rule-based Semantic Web query language Xcerpt with such a module system though the same approach can be applied to other (rule-based) languages as well.

1 Introduction

As the amount and diversity of data available on the Web is constantly increasing, querying this great abundance of information is becoming more and more important. In fact, it is becoming less important to possess certain knowledge, but more important to know how to acquire it—know how to formulate a precise *query* to find the desired information. Query languages for different purposes are emerging in multitude. [2] surveys some existing query and transformation languages for Web and Semantic Web data, identifying 14 textual XML query languages and 24 for RDF metadata.

Yet, most of these languages provide very little support to the user to cope with the dramatic increase in information size and diversity. Increasing information diversity results in increase of query size and complexity, which can weigh down even experienced query programmers. It must be easy for users to partition (both conceptually and from an evaluation point of view) query programs and to make such partitioning flexible enough to allow for reuse in different contexts. This is not the case unless the query language provides some means to separate large and complex queries into smaller, properly isolated, and reusable fragments—*modules*. Such modules allow to "localize" the effect of the introduction of additional data sources or query tasks in query programs.

R. Meersman, Z. Tari, P. Herrero et al. (Eds.): OTM 2007 Ws, Part II, LNCS 4806, pp. 1165–1175, 2007.

Thus, modules allow a *separation of concern* not just on the basis of single rules but on the basis of larger conceptual units of a query program. For example, one part of a Web application is often concerned with extracting data from a set of sources, such as a set of Web pages. At the next step, the data might have to be syndicated into a common view and format. From this syndicated data, some new implicit data could possibly be derived. Finally, the resulting data set should be displayed in an appropriate human-readable form, for example, by being displayed in a well-structured Web page (see Section 3 for an example). These different steps taken by the application have to do with different concerns of the overall realization, such as data extraction, data management and data display. Furthermore, each of the concerns deals with different schemata, but the knowledge of the schemata can be hidden and encapsulated within each concern – within each module. In contrast, exposing all these concerns in one monolithic query program not only becomes very hard to understand, but is also impossible to manage as a change in some part may affect any other part.

This work is based on ideas from [8,1] where we propose a flexible approach (demonstrated along Datalog examples) for augmenting arbitrary languages with new levels of *abstractions*, and new constructs for authoring reusable entities. The only requirement we put on the newly introduced constructs is that their realization is already expressible in the original language, i.e., that they have a reduction semantics. In doing this we take advantage of existing software composition techniques[1] to realize the added reuse abstractions [8]. However, in this paper we do not focus on the details of composition systems, but show an application of the ideas to a concrete query language, viz. Xcerpt [11]. The semantics is derived from the formal semantics in [1].

For that language, we propose a module system that (a) demonstrates how Web query languages can profit from modules by partitioning the query program as well as its execution; (b) provides an easy, yet powerful module extension for Xcerpt that shows how well-suited rule-based languages are for component-based reuse; (c) is based on a single new concept, viz. "stores"; and (d) uses a reduction semantics exploiting the power of a language with views. This semantics enables the reuse of the existing query engine making the design of the module system easier and its deployment less time consuming.

The rest of this paper is organized around these contributions: Following a brief introduction to Xcerpt we demonstrate the need for modules or similar reuse and partitioning mechanisms by a use case on integrating (Semantic and plain old) music data on the Web. Then we introduce the module extension for Xcerpt by implementing part of the aforementioned use case. We conclude with a discussion of the semantics and realization of the module extension.

2   Introducing Xcerpt

We choose to demonstrate our ideas using the rule-based, Web and Semantic Web query language Xcerpt [11], which has been co-developed by some of the authors and is particularly well-suited for reuse due to its rule-based nature. This chapter is not intended

---

[1] Developed within the Reuseware Composition Framework (http://reuseware.org).

as a full introduction to Xcerpt but merely recalls some of its most relevant features for this article. For a proper introduction please see [11].

An Xcerpt program consists of a finite set of Xcerpt *rules*. The rules of a program are used to derive new, or transform existing, XML data from existing data (i.e. the data being queried). *Construct rules* are used to produce intermediate results while *goal rules* form the output of programs.

While Xcerpt works directly on XML or RDF data, it has its own data format for modeling XML documents or RDF graphs, viz. Xcerpt *data terms*. For example, the XML snippet <book><title>White Mughals</title></book> corresponds to the data term book [title ["White Mughals"]]. The data term syntax makes it easy to reference XML document structures in queries and extends XML slightly, most notably by also allowing unordered data.

For instance, in the following query the construct rule defines data about books and their authors which is then queried by the goal. Intuitively, the rules can be read as deductive rules (like in, say, Datalog): if the body (after **FROM**) holds, then the head (following **CONSTRUCT** or **GOAL**) holds. A rule with an empty body is interpreted as a fact, i.e., the head always holds.

```
1 GOAL
    authors [ var X ]
3 FROM
    book [[ author [ var X ] ]]
5 END

7 CONSTRUCT book [ title [ "White Mughals" ], author [ "William Dalrymple" ] ] END
```

Xcerpt *query terms* are used for querying data terms and intuitively describe patterns of data terms. Query terms are used with a pattern matching technique[2] to match data terms. Query terms can be configured to take partiality and/or ordering of the underlying data terms into account during matching (indicated by different types of brackets).

Query terms may also contain logic variables. If so, successful matching with data terms results in variable bindings used by Xcerpt rules for deriving new data terms. Matching, for instance, the query term book [title [var X]] with the XML snippet above results in the variable binding {X / "White Mughals" }.

Construct terms are essentially data terms with variables. The variable binding produced via query terms in the body of a rule can be applied to the construct term in the head of the rule in order to derive new data terms. For the example above we obtain the data term authors ["William Dalrymple"] as result.

3   Use Case: Music Aggregation with the Web Music Library

The use case illustrated in Figure 1 presents a library (called MusicLibrary) of functionality useful for coping with music and information about music found on the (Semantic) Web. At an (arguably) lower layer, information is extracted from various established

---

[2] Called *simulation unification*. For details of this technique, please refer to [10].

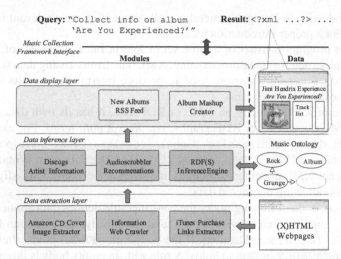

Fig. 1. Many query languages only allow writing monolithic queries, while modular query development greatly increases reuse and ease of programming

Web sites like amazon.com or discogs.org. The extraction has to be handled differently for every web site, but is valuable for many users and applications. For example, many of the currently established desktop music players exploit the album or CD images of Amazon to display cover art while playing back music. Encapsulating reusable queries dealing with a particular information source allow for flexible maintenance and propagation to a larger user base. The legacy information as found on external Web sites is then converted to an internal representation loosely based on the Music Ontology [7]. Music Ontology is an RDFS-based standard, hence knowledge inference and reasoning on—possibly incomplete—Music Ontology data can be achieved using an RDFS reasoner. Since such a reasoner is usable in many different fields of applications, it is implemented and provided as an Xcerpt module and included in the main library, hence allowing for its reuse. Perhaps more interesting to the end user, various modules providing pleasant visualizations of gathered information or predefined query skeletons can be provided in the library. Such modules can also be provided by third parties or, last but not least, as part of an application using the Web Music Library. We show only small extracts of the actual modules for space and presentation reasons.

3.1   Realizing Musical Modules in Xcerpt

How can we today realize this application in Xcerpt? In the absence of modules we have to carefully craft a single query program with a considerable number of rules (well over three dozens if we follow the basic design presented below) at each step taking great care that the rules do not, by chance, interfere with each other. Furthermore, we have to update the whole query program as soon as any information source changes, since this information is hard-coded in the program.

In the presence of a module extension, the task becomes a lot less daunting: Let us start from the top with a user program that gathers information about Jimi Hendrix from all the sources described in Figure 1. For that, it relies on a module called MusicLibrary (discussed above). The library is not a mere database, it is an interface to various ways of reasoning about musical information available on the Web. To the user the complexity remains hidden. The user just poses his query to the module without caring whether the data is extensional or intensional and how it is obtained. The module system ensures that, regardless of the actual rules and their distribution between modules, there is no chance for interference by rules of different sub-modules used within MusicLibrary.

```
1 IMPORT "MusicLibrary"

3 GOAL
    html [ body [
5     h1 [ "Records by Jimi Hendrix" ],
      table [ tr [ td [ "Record" ], td [ "Year" ] ],
7         all tr [ td [ var R ], td [ var Y ] ] ]
      ] ]
9 FROM
    in "MusicLibrary" (
11    desc record { artist [ "Jimi Hendrix" ],
                    title [ var R ], year[ var Y ] } )
13 END
```

The MusicLibrary module itself is integrating data and knowledge of other modules the same way as the user program. It has to provide the information, and only the desired information, to the user of the module. Some rules may be necessary internally in the module to achieve the task, but should not be directly visible to the user of the module. The visible parts of the module are hence *public*, the others (implicitly) *private*.

Apart from using knowledge of other modules, modules may also receive data provided by importing modules. MusicLibrary accesses data extracted by a module gathering MusicBrainz metadata, feeds it to a module for converting that data to Music Ontology knowledge (Musicbrainz2MOFacts), and finally injects that knowledge to an RDFS reasoner (using the MO-Ontology-Reasoner module). It also accesses discogs.org directly and feeds the acquired data into another instance of the MO-Ontology-Reasoner. To distinguish multiple instances of the reasoner, each instance is given an alias (using the @

```
1 MODULE "MusicLibrary"
  IMPORT "MusicBrainz"
3 IMPORT "Musicbrainz2MOFacts"
  IMPORT "MO-Ontology-Reasoner" @ "reasoner-for-musicBrains"
5 IMPORT "MO-Ontology-Reasoner" @ "reasoner-for-discogs"

7 CONSTRUCT public var KNOWLEDGE
  FROM in "reasoner-for-musicBrains" ( var KNOWLEDGE ) END
9
  CONSTRUCT public var KNOWLEDGE
11 FROM in "reasoner-for-discogs" ( var KNOWLEDGE ) END

13 CONSTRUCT to "reasoner-for-musicBrains" ( var FACTS )
  FROM in "Musicbrainz2MOFacts" ( var FACTS ) END
15
  CONSTRUCT to "Musicbrainz2MOFacts" ( var METADATA )
17 FROM in "MusicBrainz"( metadata [[ var METADATA ]] ) END
  ...
19 CONSTRUCT discogs-document-for-crawler[ all HREF ]
  FROM in document(iri="http://www.discogs.org") ( desc a [[ href [ var HREF ] ]] ) END
```

construct), which can be used the same way as the module identifier when querying, or sending data to, a module. In this way, modules also give rise to *scoped* reasoning where consequences only apply in a certain scope (or module), but are not (automatically) propagated outside of that scope. In particular, knowledge in different scopes may, if considered globally, be inconsistent, but within each scope be consistent.

Finally, let us glance at the MO-Ontology-Reasoner module which is one of the modules that not only extracts data but is injected with data to operate on. Hence, one of the queries is adorned with the **public** keyword, indicating that chaining is to be performed against the rules of the importing module that pass input data to the reasoner. Those facts, together with the ontology definition (and any domain dependent reasoning we would like to perform on the music ontology data) are sent to an RDFS reasoner module, whose consequences are then made publicly available. This RDFS reasoner is an example of a highly reusable module that can be shared among many different modules. It implements the RDF semantic in the (graph-based) query language Xcerpt (cf. [5] for details).

```
  MODULE "MO-Ontology-Reasoner"
2 IMPORT "RDFS-Reasoner"

4 CONSTRUCT public var KNOWLEDGE
  FROM in "RDFS-Reasoner" ( var KNOWLEDGE ) END
6
  CONSTRUCT to "RDFS-Reasoner" ( var FACTS )
8 FROM public var FACTS END

10 CONSTRUCT to "RDFS-Reasoner" ( var MO )
   FROM in document(type="xmlrdf" iri="http://purl.org/ontology/mo/") ( var MO ) END
```

4   Modular Xcerpt—Requirements and Constructs

We have seen that modules can greatly ease the development of complex Web queries (as observed increasingly) and how to apply them in examples. Before we discuss the principles of the semantics in Section 5, let us first summarize the core concepts and constructs introduced. We divide the presentation of the concepts in two parts: from the perspective of the module programmer and of the module user.

Module programmers need constructs for defining sets of rules and ways of declaring appropriate access to the module—interfaces for proper encapsulation. To allow module authors to encapsulate modules, *visibility constructs* are employed. For each rule of the module, the construct term and the query term (if present) is associated with a visibility concept: *public* or *private*. Only public visibility is specifically specified, otherwise the default visibility *private* is used to encourage encapsulation.

Module declaration: We can group sets of rules into modules and give such a set an identifier. This module can than be imported into other modules or programs.
$\langle module \rangle ::= $ 'MODULE' $\langle module\text{-}id \rangle \ \langle impart \rangle * \ \langle rules \rangle *$
Module interfaces: We can declare allowed access points to a module to facilitate encapsulation and proper interfaces. Any construct term can be annotated with

public to indicate that it can be queried by importing modules (see below).

⟨*interface-out*⟩ ::= 'public' ⟨*construct-term*⟩

Conversely, importing modules may provision data to an imported module (see 'module provision' below). This data is exclusively queried by query terms marked with **public** in the imported module.

⟨*interface-in*⟩ ::= 'public' ⟨*query-term*⟩

In other words, a module programmer defines the name and the in- and output interfaces of a module. The input of a module is accessed or queried by public query terms, the output of a module is formed by public construct terms. A module should also be complemented by documentation for the user describing its task and interfaces.

Module users need to be able to (a) declare which modules they want to use in a program, to (b) query the public interfaces of such modules, and to (c) provide data to such modules.

Module importation: We can import modules into other modules or programs. The only effect of a module is that the module identifier (or its alias, if an alias is used) becomes available for use in module querying or provision statements. In practice, module identifiers are often rather long and complex URIs which makes the use of (short and easy to read) aliases advisable in most cases.

   ⟨*import*⟩ ::= 'IMPORT' ⟨*module-id*⟩ ('@' ⟨*alias-id*⟩)?

Module querying: We can query the consequences of the public construct terms of a module. The given query term is matched only against the results from *public* rules of the given module but neither against those from that module's *private* rules nor against other rules from the current module.

   ⟨*module-access*⟩ ::= 'in' ⟨*module-id*⟩ ('(' ⟨*query-term*⟩ ')'

Module provision: We can feed or provision data to the public query terms of a module. The result of a rule with such a construct term is only considered for *public* query terms in the given module, not for query terms in the current module or for query terms from the given module that are not marked *public*.

   ⟨*module-provision*⟩ ::= 'to' ⟨*module-id*⟩ ('(' ⟨*construct-term*⟩ ')'

With only these three operations, a module user can flexibly compose modules (even multiple instances of the same module) while all the encapsulation is taken care of by the module system without further user intervention.

So far, all module access is always explicitly scoped with the module identifier. In a language with views such as Xcerpt, this suffices as we always can add a bridging rule (such as the first rule in the MusicLibrary module from Section 3) that makes all data obtained from the public interface of an imported module available to other rules in the importing module (without need for qualification). We provide two additional variants of module import for convenience that cover this case. They only differ in the way they affect module cascading: 'import public' ⟨*module-id*⟩ makes all data provided by the public interface of module *module-id* available to all unqualified rules in the importing module and also adds it to the public interface of that module whereas 'import private' ⟨*module-id*⟩ only makes it available to the unqualified rules.

5   Reducing Xcerpt Modules—Stores

The dual objectives of our approach are to (a) keep the module system simple and easy to use and to (b) allow the reuse of existing language tools and engines without modification. These two objectives actually go hand in hand, as a reduction semantics for modules (i.e., a semantics that is based on the semantics of the module-free language) proves to be elegant and easy to understand and naturally fulfills the second objective.

To allow users to truly think in terms of modules and make use of this abstraction, it is important to ensure proper and valid module interactivity *statically* before applying the module-unaware query engine to the involved rules. Thus, only the intended rule dependencies must be present in the merged rules—we have no way of enforcing rule separations during rule execution.

For the Xcerpt module system we ensure proper rule dependencies using the notion of **stores**. Intuitively, a store is a designated data area where data and queries are appropriately redirected to adhere to the proper access of rules as specified by the module programmer. A store is associated with an identifier and consists of a *private*, *in* and *out* part. Intuitively, the *private* part is intended for data access internal to the module only and the *in* and *out* parts for input and output data of that module. That is, data to be processed by the module will be injected into the *in* part of the store and data constructed by the module—upon request from another module—will reside in the *out* part of the store and can be queried by an importing module.

Stores can already be simulated using the existing Xcerpt mechanisms. Let us first assume that for each module we have one associated store that is identified by the same (unique) identifier. The construct terms and query terms of each rule in an imported module as well as rules using **in** or **to** for module access or provision in an importing module are modified such that the appropriate store is referenced:

```
in <module-id> ( <query> )        ⟶    store [ id [ <module-id> ], access [ "out"], <query> ]

to <module-id> ( <construct> )    ⟶    store [ id [ <module-id> ], access [ "in"], <construct> ]

CONSTRUCT <c> FROM <q> END        ⟶    CONSTRUCT store [ id [<module-id>], access["private"], <c>]
                                        FROM store [ id [<module-id>], access["private"], <q> ] END
```

Some rules in the imported module are exempted from this transformation, viz. construct terms in goals (producing results for the end user), query terms specifically referencing an external resource (such as an XML document or other module) rather than the internal module store. Also, if the query term is a complex query it might be necessary to propagate the store specification inside the query (e.g., over disjunctions, negation, etc.). However, these details are omitted here for space reasons.[3]

5.1   Refining *Stores:* Instance Stores

The store concept described above ensures basic encapsulation capabilities for Xcerpt modules and is attractive for its simplicity. However, there are certain situations where associating one store per module is not sufficient. Consider the situation where two

---

[3] But available with examples at http://www.reuseware.org/modularxcerptexample.

modules (A, B) imports a third one (C) and both A and B injects data into the store associated with C. In such a case, after module C has processed the data, module A *may* receive data initially injected by module B. As such, modules A and B are not kept separate violating one of the core premises of our desire for modules. This is not a limit of the store approach, but due to the assumption of the existence of one store per module.

To address this problem, we associate stores not with a module but with a module *import*. This can be seen as instantiating a store for each module import with the identifier of the importing module. We thus end up with two stores C<A> and C<B>, due to two import operators. A similar case where this is needed is when we use the same module but with different "feeds" using aliases. This is the case in the Music Library module presented in Section 3 where aliases (using @) were used to force such separations.

Implementation. Not only is it an advantage to reuse the query engine in executing the transformed and merged rules, it is also beneficial if existing technology can be used to realize the above-described transformations. To achieve this, we realize the module system via composition in the Reuseware Composition Framework [8]. The composition framework allows for the development of a light-weight composition system responsible for handling the augmented constructs related to modules. The composition framework allows both to extend the Xcerpt language with the additional syntactic constructs and to handle the transformation and merging of the involved rules in the manner described above to enforce encapsulation. The details of this implementation are left out for space reasons, but are available at http://www.reuseware.org/modularxcerptexample.

6 Related Work

Practical Web *query* languages need to provide support for some form of reuse and modules as evidenced by (though somewhat limited) module support in languages such as XSLT and XQuery. *Rule* languages for the Web, on the other hand, show an apparent lack of module support, despite considerable research on module extensions for classical logic programming. One of the reasons that modules are still not in the "standard repertoire" of rule languages may be the complexity of many previous approaches.

Representative and, arguably, the most comprehensive treatment of modules in logic programming is presented in [4]. It is far more expressive than our approach but at the price of a complex semantics and several operations with, in our opinion, little practical use (such as module intersection or renaming). We believe that a single well-designed union operation with clear interfaces together with a strong reliance on views as a core feature of rule languages is not only easier to grasp but also easier to realize.

Though many rule languages for the Web fail to provide modules, this is not true for the two preeminent Web query languages, XSLT and XQuery. XSLT [6] can be considered a rule language, however using precedence rather than union semantics for multiple applicable rules. Rule precedence is also the dominating issue for XSLT's module system which provides intricate mechanisms for determining the precedence of rules from different modules. Nevertheless, the resulting module system is considerably less powerful (no scoped import, limited parameterization: apply-imports) yet needs a more complex semantics than module-free XSLT, quite in contrast to our approach.

It is worth mentioning that XQuery [3] also provides a module system, however without parameterization, but as a function programming language requires explicit flow control in all cases. Thus, issues such as private or public import (or the difference between import and include in XSLT) do not apply for XQuery. SPARQL [9], finally, the recently proposed RDF query language, has no concept of user defined program units (such as rules, functions, procedures, etc.) and thus no use for a module concept in the sense of our approach. However, rule-based extensions for SPARQL (in the spirit of Datalog) could certainly profit from the module system illustrated here using Xcerpt.

7   Conclusions and Outlook

We argue that one ingredient to cope with size and diversity of information on the (Semantic) Web is *modular* query authoring and execution. We show advantages along a concrete use case dealing with music information aggregation on the Web. Furthermore, we demonstrate how it is possible to augment existing query languages—here focused on the language Xcerpt—with new constructs while reusing already developed semantics and query engines thanks to a reduction semantics approach. The proposed module system is simple to use (in contrast to many approaches from logic programming) yet provides better encapsulation and more advanced features (such as scoping and paramterization) than module systems for XSLT or XQuery.

The proposed module system has been formalized [1] and implemented using the Reuseware Composition Framework. Integration with upcoming revisions of Xcerpt is planned. Furthermore, we would like to exploit existing techniques and tools such as Xcerpt's type system [12] for improving module composition. We are also investigating how similar techniques can be applied to add or improve module systems for other (non-rule based) query languages (for example, the module system of XSLT).

Acknowledgement. This research has been co-funded by the European Commission and by the Swiss Federal Office for Education and Science within the 6th Framework Programme project REWERSE number 506779 (cf. http://rewerse.net).

References

1. Aßmann, U., Berger, S., Bry, F., Furche, T., Henriksson, J., Patranjan, P.-L.: A generic module system for web rule languages: Divide and rule. In: Proc. Int'l. RuleML Symp. on Rule Interchange and Applications (2007)
2. Bailey, J., Bry, F., Furche, T., Schaffert, S.: Web and Semantic Web Query Languages: A Survey. In: Eisinger, N., Małuszyński, J. (eds.) Reasoning Web. LNCS, vol. 3564, Springer, Heidelberg (2005)
3. Boag, S., Chamberlin, D., Fernández, M.F., Florescu, D., Robie, J., Siméon, J.: XQuery 1.0: An XML Query Language. Working draft, W3C (2005)
4. Brogi, A., Mancarella, P., Pedreschi, D., Turini, F.: Modular logic programming. ACM Trans. Program. Lang. Syst. 16(4), 1361–1398 (1994)
5. Bry-Haußer, F., Furche, T., Linse, B.: Data Model and Query Constructs for Versatile Web Query Languages: State-of-the-Art and Challenges for Xcerpt. In: Alferes, J.J., Bailey, J., May, W., Schwertel, U. (eds.) PPSWR 2006. LNCS, vol. 4187, pp. 90–104. Springer, Heidelberg (2006)

6. Clark, J.: XSL Transformations, Version 1.0. Recommendation, W3C (1999)
7. Giasson, F., Raimond, Y.: Music ontology specification. Specification, Zitgist LLC (2007)
8. Henriksson, J., Johannes, J., Zschaler, S., Aßmann, U.: Reuseware – adding modularity to your language of choice. In: Proc. of TOOLS EUROPE 2007: J. of Object Technology (2007)
9. Prud'hommeaux, E., Seaborne, A.: SPARQL Query Language for RDF. Candidate recommendation, W3C (2007)
10. Schaffert, S.: Xcerpt: A Rule-Based Query and Transformation Language for the Web. Dissertation/Ph.D. thesis, University of Munich (2004)
11. Schaffert, S., Bry, F.: Querying the Web Reconsidered: A Practical Introduction to Xcerpt. In: Proc. Extreme Markup Languages (Int'l. Conf. on Markup Theory & Practice) (2004)
12. Wilk, A., Drabent, W.: A Prototype of a Descriptive Type System for Xcerpt. In: Alferes, J.J., Bailey, J., May, W., Schwertel, U. (eds.) PPSWR 2006. LNCS, vol. 4187, Springer, Heidelberg (2006)

Leveraging the Expressivity of Grounded Conjunctive Query Languages

Alissa Kaplunova, Ralf Möller, and Michael Wessel

Hamburg University of Technology (TUHH)

Abstract. We present a pragmatic extension of a Semantic Web query language (including so-called grounded conjunctive queries) with a termination safe functional expression language. This addresses problems encountered in daily usage of Semantic Web query languages for which currently no standardized solutions exist, e.g., how to define aggregation operators and used-defined filter predicates. We claim that the solution is very flexible, since users can define and execute *ad hoc extensions* efficiently and safely on the Semantic Web reasoning server without having to devise and compile specialized "built-ins" and "plugins" in advance. We also address the scalability aspect by showing how aggregation operators can be realized efficiently in this framework.

1 Introduction

Nowadays, Description Logics (DLs) provide the basis for Semantic Web technology, and in particular, for the de facto standards for Semantic Web ontology languages such as OWL [1]. DL systems can thus be used as Semantic Web repositories. They offer a set of standard inference services, such as consistency checking, automatic computation of the concept (class) hierarchy (the so-called taxonomy), and the basic retrieval services (e.g., instance retrieval) [2]. An example of a prominent and widely-used DL system is the RACERPRO system [3] which implements the expressive DL $\mathcal{ALCQHI}_{\mathcal{R}+}(\mathcal{D}^-)$, also known as $\mathcal{SHIQ}(\mathcal{D}^-)$. In the context of the Semantic Web, especially expressive semantic query languages (QLs) are of great importance to realize the vision of semantic information retrieval, which is at the heart of the Semantic Web idea. These QLs realize retrieval functionality that goes beyond the retrieval functionality offered by the basic retrieval services (e.g., instance retrieval).

Today, the most prominent Semantic Web QLs are (extended) RQL dialects, SWRL (the Semantic Web Rule Language) [4], SPARQL [5] and OWL-QL. RQL [6] is primarily an RDF QL and thus lacks many important expressive means and inference capabilities needed to query ontologies / documents written in more expressive Semantic Web languages, e.g. OWL. The same holds for SPARQL; many nowadays existing SPARQL implementations do not consider the inferred (axiomatic) triples in an RDFS document at all and from the standard it is not clear whether they should or "how complete" a SPARQL implementation w.r.t. the RDFS semantics should be. Only recently attempts are made to augment and enhance the expressivity and *inference-awareness* of SPARQL in such a way that it will become useful to query, e.g. OWL documents. This extension will be called SPARQL-DL. SWRL was primarily designed as a rule language, but can also be used as a query language. Full SWRL is undecidable, but

R. Meersman, Z. Tari, P. Herrero et al. (Eds.): OTM 2007 Ws, Part II, LNCS 4806, pp. 1176–1186, 2007.
© Springer-Verlag Berlin Heidelberg 2007

restricted subsets (so-called DL-safe SWRL) exists. Considering the nowadays available SWRL implementations, the situation is similar as for SPARQL (e.g., sometimes, simple forward chaining rule engines are used to implement SWRL, so most of the inferences are missed). OWL-QL is very complex and does not seem to get much support from implementors, since its semantics is quite involved.

The native QL of the RACERPRO description logic system is called NRQL (new RACERPRO Query Language, [7]). It was primarily designed as a DL ABox query language and was later extended to also address specific aspects of OWL (despite its name, NRQL is not an extended RQL dialect). Due to its OWL capabilities, NRQL is also a Semantic Web QL. Being primary a QL for a DL system, NRQL was always fully inference-aware and thus provides expressive means not found in the QLs discussed above. Among others, nRQL offers classical and non-monotonic negation (so-called "negation as failure" or NAF-negation), extended concrete domain querying facilities, and a projection operator. Since NAF-negation and projection are available, both closed world and open world (universal and existential guarded) quantifications are available. Classical negation is not only applicable to (possibly complex) classes or concepts, but also to properties or roles. Classical and NAF-negation can even be mixed (e.g., one can retrieve those instances which are not known to be instances of $\neg mother$; these could be potential *mothers*).

Although NRQL is not standardized, it was and still is being used in many Semantic Web research projects and applications, not only because of its expressive means, but also due to its efficient and stable implementation which offers some unique features (queries are maintained as objects, multi-threading, etc.). However, in our own projects we found a need to extend the expressivity and practical relevance of NRQL even further. For example, *aggregation operators* were missing. Certain RQL dialects offer the standard SQL aggregation operators (sum, max, min, count, avg, etc.). Also for SWRL, there is the principle possibility to realize these through so-called "built-ins". For example, one could think of a specialized built-in atom such as

```
sum(?car,?weigth-of-parts,has-part;weight)
```

Given a binding for ?car, the variable ?weigth-of-parts is then bound to the sum of the weight datatype fillers of the has-part object property fillers of that ?car. However, the semantics of such extensions is unclear (questions such as "which further predicates apply to ?weight-of-parts?", "are variables typed?" etc. arise). Obviously, the wish list of such conceivable "extensions" is endless. The list of built-ins is therefore extensible in SWRL – user-defined built-ins can be added. The implementations of these atoms must be provided by users – a plugin architecture of the SWRL engine is thus required. The same idea applies of course to *filter* atoms or predicates (here, filter(?x) is true iff ?x satisfies a certain user-defined filter predicate).

In order to offer a greater deal of flexibility and to allow for *ad hoc extensions,* we designed a more general extension mechanism for NRQL– (almost) arbitrary procedural extensions can be specified as part of a NRQL query. These extensions are written in a functional expression language called MINILISP. Even though MINILISP per se is purely functional we say *procedural* extension (see the Conclusion for further discussions).

The semantics of the MINILISP extension is defined in such a way that it does not interfere with the NRQL semantics. The nRQL query body is thus "kept clean". In the relational database realm, so-called stored procedures are well-known. However, stored procedures can result in *unsafe, non-terminating queries* (an unsafe query may run forever). Since decidability is crucial in the Semantic Web context, MINILISP is not a programming language, but a termination-safe functional expression language. MINILISP "programs" are executed efficiently on the RACERPRO server and thus offer the required efficiency and flexibility to "implement" arbitrary aggregation operators, filter predicates that cannot be formulated solely in the query language, create XML or HTML reports from query results, etc. In order to realize aggregation operators, often sub-queries have to be evaluated from within MINILISP programs. We introduce a new optimization technique (so-called promises) tailored for this purpose.

2   Theoretical Background

The class of conjunctive queries is well-known and established in the literature, e.g. see [9]. A *conjunctive query (CQ)* has the form $ans(X) \leftarrow atom_1, \ldots, atom_n$, where $X = (x_1, \ldots, x_m)$ is a variable vector. The expression $ans(X)$ is called the *head*, and $atom_1, \ldots, atom_n$ is the *body* of the query (interpreted as a conjunction). The $atoms$ in the query body reference variables and/or individuals. All variables listed in X must also be mentioned in the body. The vector of body variables is denoted by Y. Let Z denote those variables in Y which do not appear in X.

Most Semantic Web QLs offer at least concept and role query atoms. If x and y are variables or individuals, then *concept query atoms* are unary atoms (e.g., $C(x)$), whereas *role query atoms* are binary atoms (e.g., $R(x, y)$). Often, also a (binary) *equality atom* is offered (either written as $x = y$, $= (x, y)$ or $same\_as(x, y)$).

A *query answer* is a set of (m-) tuples. Each tuple represents bindings for the head variables X. In general, a head variable (in X) is bound to an RDF node (representing an ABox individual), or an ABox individual. These individuals are denoted with $inds(\mathcal{O})$ in the following. In order to compute the bindings for the variables in X, all possible substitution functions $\alpha : X \rightarrow inds(\mathcal{O})^m$ are considered and applied to the query body. In case the resulting variable substituted query is entailed by the ontology, α denotes a result tuple. In *unrestricted conjunctive queries*, the variables in Z are bound to individuals in the interpretation domain $\Delta^{\mathcal{I}}$ of the logical models of the ontology. These variables are therefore considered as *existentially quantified*. In so-called *grounded (or restricted)* conjunctive queries (GCQs), the following simplification is made: Not only are the X variables bound to $inds(\mathcal{O})$, but also the Y variables (and thus, all variables in that query).

The answer of a CQ can now be specified by the following simple set comprehension; please note that variables not mentioned in X and individuals ($inds(\mathcal{O})$) remain unaltered by α (for such i, $\alpha(i) = i$ holds):

$$\{ (i_1, \ldots, i_m) \mid \exists \alpha : X \mapsto (i_1, \ldots, i_m), (i_1, \ldots, i_m) \in inds(\mathcal{O})^m,$$
$$\mathcal{O} \models \exists Z. \alpha(atom_1) \wedge \cdots \wedge \alpha(atom_n) \}.$$

For grounded conjunctive queries, we simply change the domain of α from X to Y and remove "$\exists Z$." from the first-order body formula. In grounded conjunctive queries, the standard semantics can be obtained for so-called *tree-shaped queries* by using corresponding existential restrictions in query atoms [8] (e.g., if y is not in X, then the atom $R(x, y)$ can be replaced by $\exists R.\top(x)$; $\exists R.\top$ is a complex anonymous concept).

From a theoretical perspective, NRQL goes beyond grounded conjunctive queries. NRQL provides additional expressive means (see [7]), especially, NAF-negation, projection and union operators, concrete domain reasoning facilities (so-called *constraint checking atoms*), as well as a novel lambda-based expression language to be discussed in this paper.

Let us consider the following example query which will also explain some basics of lambda expressions and introduce the constraint query atoms as well. Suppose we want to retrieve the *woman* instances which are at least 40 and which have children whose fathers are at least 8 years older than their mothers. Let us start with the query body

$$(woman \sqcap \geq_{40} age)(x), has\_child(x, y), has\_father(y, f), has\_mother(y, m),$$
$$age(f, age_1), age(m, age_2), (\lambda(v_1, v_2) \bullet (v_2 + 8 \leq v_1))(age_1, age_2)$$

Please note that the atom $(\lambda(v_1, v_2) \bullet (v_2 + 8 \leq v_1))(age_1, age_2)$ specifies an anonymous predicate with formal parameters v_1, v_2 which is applied to the actual arguments age_1, age_2. Unlike in SWRL or SPARQL, NRQL does not allow variables to be bound to anything else than ABox individuals in order to prevent semantic problems. Thus, the age_i cannot be used as variables. We therefore have to rewrite the body using a more complex lambda expression, utilizing $age(v_i)$ *terms* in the comparison predicate instead of simple v_i variables:

$$(woman, \geq_{40} age)(x), has\_child(x, y), has\_father(y, f), has\_mother(y, m),$$
$$(\lambda(v_1, v_2) \bullet (age(v_2) + 8 \leq age(v_1)))(f, m)$$

This translates more or less directly into concrete NRQL syntax:

```
(and (?x (and woman (min age 40))) (?x ?y has-child)
     (?y ?f has-father) (?y ?m has-mother)
     (?f ?m (constraint age age
                        (<= (+ age-2 8) age-1))))
```

Thus, a constraint query atom is very similar to a lambda expression. It is applied to ABox variables ?f, ?m, whose actual values of the attribute age are then bound to the formal arguments in the constraint "lambda body", age-1, age-2. Concrete domain reasoning is used to check whether the concrete domain predicate holds. Note that there may be no concrete known values for the age attributes, or their values need not be unique, if only constraints are specified on them, e.g., only $age(betty) + 10 < age(charles)$ is known. This also explains why NRQL does not offer variables ranging over the concrete domain (their solutions resp. bindings could not be computed in all cases). Finally, a complete NRQL query (including a query head) $ans(x) \leftarrow \ldots$ is written as (retrieve (?x) ...).

3   The Power of λ

Unfortunately, the number of predicates which can be constructed with constraint query atoms is quite limited, either in order to ensure decidability in the concrete domain reasoning engine in RACERPRO, or simply because the required predicate is missing. Unfortunately, RACERPRO does not offer user-defined concrete domains. Thus, it is even impossible to query for persons having a certain firstname (e.g., "Betty"), given that only the concrete domain attribute fullname exists. This is unfortunate, since in many cases full concrete domain reasoning is not required (i.e., if concrete datatype values are specified as "told values" in the ontology). This is where MINILISP comes into play.

The basic idea is simple: A NRQL query head may not only contain variables, but also *lambda applications*. Thus, a head is a vector $X = (h_1, \ldots, h_m)$, where h_i is either a variable or a lambda application. Such a lambda application has the syntax $((\lambda(v_1, \ldots, v_p) \bullet \ldots) y_1, \ldots, y_p)$; the y_i are again variables, which also have to appear in the body of the query: $y_i \in Y$. The answer of a GCQ with body $atom_1, \ldots, atom_n$, head $(h_1, \ldots, h_m)$ and $Y = (x_1, \ldots, x_k)$ is then specified by the following set comprehension:

$$\{ (j_1, \ldots, j_m) \mid \exists \alpha : Y \mapsto (i_1, \ldots, i_k), (i_1, \ldots, i_k) \in \mathsf{inds}(\mathcal{O})^k,$$
$$\mathcal{O} \models \alpha(atom_1), \ldots \mathcal{O} \models \alpha(atom_n),$$
$$\text{such that for all } l \in 1 \ldots m:$$
$$j_l = \alpha(h_l) \quad \text{if } h_l \text{ is a variable,}$$
$$j_l = ((\lambda(v_1, \ldots, v_p) \bullet \ldots) \alpha(y_1), \ldots, \alpha(y_p))$$
$$\text{if } j_l = ((\lambda(v_1, \ldots, v_p) \bullet \ldots) y_1, \ldots, y_p)$$
$$\text{and } j_l \neq \perp \}.$$

So, instead of just returning the tuple $(i_1, \ldots, i_k)$, the h_l "functions" are applied and its results included in the constructed answer tuple at that position. This is very similar to the mapcar operation in COMMON LISP. In case h_k is a variable, its binding $\alpha(h_k)$ is included. Otherwise, h_k denotes a lambda application: $((\lambda(v_1, \ldots, v_p) \bullet \ldots) \alpha(y_1), \ldots, \alpha(y_p))$. Its result is included in the answer tuple at that position in case the lambda did not return $\perp$. Whenever $\perp$ is returned by the lambda, the whole answer tuple is rejected instead. Thus, lambdas can be used to specify arbitrary filter predicates. Answer sets consisting of unary tuples can also be considered as flat sets, and thus, the structure of the elements in the answer set can be defined completely by means of lambda expressions. Moreover, by posing sub-queries in lambda bodies, we can easily implement arbitrary aggregation operators, as demonstrated in Section 5.

Obviously, the expressivity of that extension depends on the admissible lambda bodies. It is well-known that an unrestricted use of lambda results in undecidability (e.g., consider the classical textbook example $((\lambda(x) \bullet (x\ x))(\lambda(x) \bullet (x\ x))$ which specifies an endless loop). Lambda applications are specified in MINILISP. In order to grant termination, lambdas itself are not first class citizens (which is the case in languages such as Scheme). We will now describe the flexibility and added value of MINILISP in concrete syntax by means of examples.

4   MINILISP by Example

Consider an ABox representing objects in a geographic information system having width and length, and we want to compute and return the area of these objects with a query. The individual box1 has a width of 10 and a length of 20:

```
(define-concrete-domain-attribute width :type integer)
(define-concrete-domain-attribute length :type integer)
(instance box1 (and (equal width 10) (equal length 20)))
```

We can then query for the areas of the objects in this ABox as follows:

```
(retrieve (?x ((lambda (box)
                 (let ((w (told-value-if-exists (width box)))
                       (l (told-value-if-exists (length box))))
                    (if (and w l) (* w l) :reject)))
               ?x))
          (?x (and (a width) (a length))))
```

The answer thus is: `((( ?x box1) 200))`. The query body `(?x (and (a width)` `(a length)))` selects all ABox individuals which have – possibly only implicit – fillers of the concrete domain attributes `width` and `length`. However, only in case these fillers are "told" in the ontology (= syntactically explicit available) it is possible to also retrieve them. Their retrieval is then supported by means of functional expressions such as `(told-value-if-exists (width box))`. In case these attribute values are told, w and l are bound and `(* w l)` is computed and returned; otherwise, the `:reject` symbol is returned, so the result tuple is rejected (see $j_l \neq \bot$ in the set comprehension in Section 3). Of course, we could easily reject certain boxes (whose size is too big or too small). Thus, almost arbitrary ad hoc filter predicates can be specified.

We claim that MINILISP is easy to understand and use for readers which have some COMMON LISP experience. In a nutshell, MINILISP offers the following data types: numbers, symbols, strings and lists (and thus also trees). It supports conditional execution (`if`, `cond`, `when`, `unless`, `case`), structure mapping functions such as `maplist` (like `mapcar` in COMMON LISP, e.g., `(maplist (lambda (x) (1+ x))` `'(1 2 3))` returns `(2 3 4)`), as well as standard functions borrowed from the host language COMMON LISP (arithmetic functions, list function, string processing functions, comparison and sorting functions, etc.). In order to grant termination, lambdas (as required for higher-order functions such as `maplist`) are not first class citizens (not data objects). Thus, `((lambda (x) (x x)) (lambda (x) (x x)))` simply gives a syntax error (x cannot be bound to the function object `(lambda (x) (x x))`). No unbounded loops can be specified (only mappings over finite structures). MINILISP is purely functional, although there is a notion of a state exploited in counting variables (there is no variable assignment, but `(incf count)` and `(decf count)`).

In principle, all RACERPRO API functions can be called from within a lambda body. This also applies to `retrieve` itself. Thus, nested queries can be posed. We will illustrate how aggregation operators can be implemented using nested queries.

5   Aggregation Operators in MINILISP

Consider the following example KB in which the compositional structure of a car is modeled, see Figure 1. A car has certain parts, and each part has a certain weight:

```
(define-primitive-role has-part :transitive t)
(define-concrete-domain-attribute weight :type real)
```

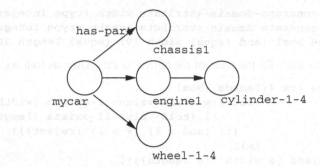

Fig. 1. Compositional structure of mycar

```
(instance mycar car)
(related mycar engine1 has-part)
(related engine1 cylinder-1-4 has-part)
(related mycar wheel-1-4 has-part)
(related mycar chassis1 has-part)
(instance engine1 (= weight 200.0))
(instance chassis1 (= weight 400.0))
(instance wheel-1-4 (= weight 30.0))
```

Suppose we want to compute the overall weight of the car as well as the number of its components. Thus, using MINILISP, for each ?car found we are going to construct two sub-queries. For a given ?car, the first sub-query retrieves the components of that ?car as well as their told weights, and the second sub-query simply counts the number of components. The two sub-queries are constructed and executed from within a MINILISP lambda body. Their results are then appropriately processed and returned.

Note that the bodies of the two sub-queries are almost identical for every ?car, but the considered ?car obviously changes. Thus, the bodies of the sub-queries are not fix. Sub-queries thus have to be constructed based on variable bindings which are established by outer (sub)queries. For this purpose, query templates can be constructed using the "backquote (`) and comma (,) mechanism" from COMMON LISP. For example, if the variable car is bound to mycar, then the expression `(,car ?part hast-part) evaluates to (mycar ?part has-part). The query

```
(retrieve
  (((lambda (car)
    (let ((car-weight
            (reduce '+ (flatten
              (retrieve `(((lambda (car-weight) car-weight)
                            (told-value-if-exists
                             (weight ?part))))
                        `(and (,car car) (,car ?part has-part)
                             (?part (a weight)))))))
          (car-parts (length
            (retrieve `(?part)
                      `(,car ?part has-part)))))
```

```
      `((?car ,car) (?no-of-parts ,car-parts)
         (?total-weight ,car-weight))))
      ?car))
   (?car car))
```

then returns

```
(((((?car mycar) (?no-of-parts 4) (?total-weight 630.0))))).
```

The query works as follows. The body of the outer query consists of the single concept query atom (`?car car`). The `lambda` expression is then applied to the current binding of `?car`. So, within the `lambda` body, car is bound to the binding of `?car`. First, the total weight is computed. For this purpose, a sub-query is constructed. If `?car` = mycar, then the query body `(and (mycar car) (mycar ?part has-part) (?part (a weight)))` is constructed and posed, asking for the parts of mycar. The head of the sub-query consists of yet another `lambda`, which simply applies the `told-value-if-exists` head projection operator to retrieve the told values of the weight attribute of `?parts`. The sub-query result is returned as a nested list; the list is flattened and its items are simply summed using (`reduce '+ ...`). We have computed the overall weight; this result is bound to the local variable car-weight. Similarly, the number of car-parts is computed (by posing yet another sub-query). Finally, the result of the `lambda` expression is constructed and returned. The constructed and returned value will become the result tuple. So, if car is mycar, and no-of-parts is 4, and car-weight is 630.0, then the template `((?car ,car) (?no-of-parts ,car-parts) (?total-weight ,car-weight))` constructs the result tuple (((?car mycar) (?no-of-parts 4) (?total-weight 630.0))). We can easily construct and return a string instead of a tuple by using (`format nil "Car ~A has ~A parts and weights ~A kg." car no-of-pars car-weight`). Thus, the answer is "mycar has 4 parts and weights 630.0 kg".

6   Efficient Aggregation Operators Using Promises

Although the previous query demonstrated the power and utility of MINILISP, the aggregation was not computed efficiently, since for each binding of `?car`, two new sub-queries were constructed. Thus, if 10 cars are present, 20 sub-queries had to be parsed, optimized, compiled and finally executed in order to compute the aggregation. NRQL supports the pre-compilation of queries; in fact, queries are maintained as first order objects which have a complex life cycle [7]. However, the two sub-queries cannot be simply precompiled since their bodies are not fixed. The bodies contain a variable part, `?car`, whose binding can only be established at execution time by the outer query. From the perspective of the inner sub-queries, `?car` is in fact an individual. Unfortunately, `?car` cannot be treated as an individual at pre-compilation time, since the query compiler would then produce special code which will treat `?car` as an individual, but after all, there is not individual `?car` in the KB. This obviously prevents naive pre-compilation. Note that this situation does not arise in SQL engines.

A new optimization technique can help here which is a general technique that does not only apply to the NRQL engine. To the best of our knowledge, the technique has not been proposed or implemented before. A *promise* declares that certain variables have to be treated as individuals during query (pre-)compilation time. Thus, ?car can be treated as an individual by the optimizer and compiler. At the same time, it is *promised to* NRQL that this query will only be executed if a binding for ?car is established in advance, in this case, by the outer query. Thus, although the optimizer and compiler see and treat ?car as an individual, during execution time that individual can change (and so remains a variable). Thus, the query bodies are constant again, and only 2 bodies are needed instead of 20. Using promises, the example aggregation query looks as follows:

```
(with-future-bindings (?car)
    (prepare-abox-query (?part)
        (and (?car car) (?car ?part has-part))
        :id :parts-of-car-query)
    (prepare-abox-query
        (((lambda (weight) weight)
                (told-value-if-exists (weight ?part))))
        (and (?car car) (?car ?part has-part) (?part (a weight)))
        :id :weights-of-parts-of-car-query))
```

This prepares two queries named :parts-of-car-query and :weights-of-parts -of-car-query. The queries are compiled and optimized, but not executed yet. Since NRQL supports full life cycle management for queries, these queries are from now on available as query objects, ready for execution. Due to the surrounding lexical promise with-future-bindings, the query optimizer has treated the ?car variable as an individual, although it must be handled as a variable at execution time. Thus, we have "promised" NRQL that we will only execute these queries if we supply a binding for ?car in advance. We can establish such a binding during query execution using with-nrql-settings as follows:

```
(retrieve
    (((lambda (car)
        (with-nrql-settings (:bindings `((?car ,car)))
            (let ((car-weight
                    (reduce '+
                        (flatten
                            (execute-or-reexecute-query
                                :weights-of-parts-of-car-query)))))
                (car-parts
                    (length
                        (execute-or-reexecute-query
                            :parts-of-car-query))))
            `((?car ,car) (?no-of-parts ,car-parts)
                (?total-weight ,car-weight)))))
        ?car)))
    (?car car))
```

This results in a very efficient query execution, since (re)execution of a prepared query is immediate (only a function call to the compiled query evaluation function is required). No query parsing, optimization and compilation time is needed during query execution, and much less memory is used as well.

7 Conclusion

We have presented a pragmatic extension of a Semantic Web QL by lambda expressions. The termination safe functional expression language MINILISP offers solutions to problems encountered in daily usage of Semantic Web QLs for which currently no standardized solutions exist. The proposed solution is technically sound, since the query body is kept clean from user defined predicates or procedural extension which might result in unsafe queries and semantical problems. The solution is flexible, since users can define and execute ad hoc extensions efficiently on the server without having to compile specialized "plugins" in advance. We have also addressed the scalability aspects by showing how aggregation operators can be realized efficiently in this framework by exploiting the novel notion of promises. Standard aggregation operators could also be made accessible as macros in this framework. We believe that the flexibility offered by MINILISP enhances the applicability of NRQL to real world problems. For example, the generation of HTML is directly supported in MiniLisp with "syntactic sugar".

Is MINILISP a *declarative* extension? The answer is *yes and no*. In case the specified lambda bodies are purely functional, the answer is truly yes. There are not states in MINILISP itself; for example, there is no variable assignment operator. Thus, typical MINILISP use cases will be fully declarative, e.g., if MINILISP is used for the realization of aggregation operators or filter predicates. However, for pragmatic reasons, MINILISP offers full access to all RACERPRO API functions. As such, it is of course possible to alter the state of a knowledge base while the query is still running. Since there is a notion of state involved, such queries can no longer be called fully declarative. One could argue that this kind of extension could also be executed on the client side. However, this will result in a bad performance and prevent ad hoc extensions. Moreover, optimization techniques such as promises cannot be used then.

Acknowledgments

We like to thank Atila Kaya for very valuable and thoughtful comments. This work was partially supported by the EU funded projects TONES (Thinking ONtologiES FET-FP6-6703) and BOEMIE (Bootstrapping Ontology Evolution with Multimedia Information Extraction, IST-FP6-027538).

References

1. van Harmelen, F., Hendler, J., Horrocks, I., McGuinness, D.L., Patel-Schneider, P.F., Stein, L.A.: OWL Web Ontology Language Reference (2003),
   http://www.w3.org/tr/owl-guide/
2. Baader, F., Calvanese, D., McGuinness, D., Nardi, D., Patel-Schneider, P.F. (eds.): The Description Logic Handbook: Theory, Implementation, and Applications. Cambridge University Press, Cambridge (2003)
3. Haarslev, V., Möller, R.: RACER System Description. In: Goré, R.P., Leitsch, A., Nipkow, T. (eds.) IJCAR 2001. LNCS (LNAI), vol. 2083, Springer, Heidelberg (2001)

4. Horrocks, I., Patel-Schneider, P.F., Boley, H., Tabet, S., Grosof, B., Dean, M.: SWRL: A Semantic Web Rule Language Combining OWL and RuleML. Technical Report, World Wide Web Consortium (2004)
5. Prud'hommeaux, E., Seaborne, A.: SPARQL Query Language for RDF. Technical Report, World Wide Web Consortium (2006)
6. Karvounarakis, G., Alexaki, S., Christophides, V., Plexousakis, D., Scholl, M.: RQL: A Declarative Query Language for RDF. In: The Eleventh International World Wide Web Conference (WWW 2002) (2002)
7. Wessel, M., Möller, R.: A High Performance Semantic Web Query Answering Engine. In: Proc. of the 2005 Description Logic Workshop (DL 2005) (2005)
8. Horrocks, I., Tessaris, S.: Querying the Semantic Web: a Formal Approach. In: Horrocks, I., Hendler, J. (eds.) ISWC 2002. LNCS, vol. 2342, Springer, Heidelberg (2002)
9. Calvanese, D., De Giacomo, G., Lenzerini, M.: On the Decidability of Query Containment under Constraints. In: Proc. of the 17th ACM SIGACT SIGMOD SIGART Symp. on Principles of Database Systems (PODS 1998) (1998)

Adaptive Semantic Interoperability Strategies for Knowledge Based Networking

Song Guo, John Keeney, Declan O'Sullivan, and David Lewis

Knowledge & Data Engineering Group (**KDEG**)
Centre for Telecommunications Value Chain Research (**CTVR**)
School of Computer Science & Statistics, Trinity College, Dublin, Dublin, Ireland
{gsong,John.Keeney,Declan.OSullivan,Dave.Lewis}@cs.tcd.ie

Abstract. A Knowledge Based Network is a type of ontological content based network. As Knowledge Based Networks scale, semantic interoperability becomes an important issue since larger populations of users result in more heterogeneity in the content of messages. This paper examines the content heterogeneity problem in a KBN and proposes a mapping service scheme for distributed and heterogeneous knowledge-based applications. It compares a number of strategies that use pre-existing semantic mapping information stored in KBN routers. Evaluation results show that this scheme can effectively solve the heterogeneity problem.

1 Introduction

Given the rapid evolution and dynamism of networking, there is increasingly a desire to allow applications which were designed independently and using different information structures to communicate that information without the necessity of custom building gateways. Publish-Subscribe (Pub / Sub) systems [1][2] provides decoupling of identify between producers and consumers of transmitted information, but requires messages to be categorised into predefined types. In response, Content-Based Network (CBN) have been developed [3][4][5]. These match messages to consuming client interests by specifying a filter on the messages' attribute values. The limitation of current CBNs is that they only support a very limited range of datatypes and operators for use in matching consumer subscriptions to message attributes, typically: Strings, Integers, Booleans, and associated equality, greater than, less than, and regular expression matches on strings. For a CBN to work on a large scale it needs to support a richer expressiveness that can cope with the widely heterogeneous and frequently changing range of message content and consumer subscriptions.

In previous papers [6][7][8] we have described a semantic-based CBN called the Knowledge Based Network (KBN). Producers of knowledge express the semantics of their available information based on an ontological representation of that information. Consumers express subscriptions upon that information as simple semantic queries. An implementation of such a KBN, based on the Siena CBN [3], is available, and enables the efficient distributed routing of distributed heterogeneous knowledge to, and only to, nodes that have expressed a specific interest in that knowledge. We have

R. Meersman, Z. Tari, P. Herrero et al. (Eds.): OTM 2007 Ws, Part II, LNCS 4806, pp. 1187–1199, 2007.
© Springer-Verlag Berlin Heidelberg 2007

been investigating the applicability of such Knowledge-Based Networking in the areas of Network & Telecoms Service Management, Autonomic Systems and Pervasive Services, Context & Management. In particular we have focused on semantic inter-operability [9], self-managing networks [8][9], autonomic communications [10], context distribution [8][9], distributed service discovery [6][11], and the management of efficient knowledge routing mechanisms [12].

The majority of these investigations to date have assumed that the publisher and subscriber applications share a single common ontology. In this paper we go further and report upon how we have initially extended and tested the KBN so that it can cope with the situation where applications may be using multiple diverse ontologies. We have previously demonstrated [9] how the use of ontology and ontology mapping techniques enabled applications built according to different standards could interchange fault alarms over a CBN (Elvin [4]) using an ontology based approach. In that work ontology mappings were used in a generic gateway which was external to the network, co-located with the application. The work described in this paper is similar in intent but differs significantly in design and implementation, in that the ontology mappings are used directly in support of information routing within the network. Thus in this work the mappings are used to help route information within the KBN as well as supporting the model translation for applications at the edges of the network.

The incorporation of semantic interoperability strategies within the KBN routers means that applications that subscribe to information according to one ontology can expect to receive information published according to a different ontology, if there exists a mapping between the ontologies. Although this feature lowers the barrier for participation by applications in any particular KBN, it will potentially increase the workload of an individual KBN router. Our hypothesis reported here is that the impact of the extra processing is far outweighed by the benefits from enabling semantic interoperability between applications.

The enhanced KBN described in this paper supports the following requirements: (i) dynamic networks: where new applications can join in and leave dynamically and frequently; (ii) application autonomy: each application is responsible for its own information specification and representation; (iii) absence of a need for a-priori agreement between applications about information specification and representation.

Section 2 discusses semantic mapping and outlines the scenario used in the paper relating to the exchange of policies related to Dynamic Spectrum Access in a self-managing network. In section 3 the KBN router is described and several strategies for coping with mappings in the KBN router are identified. Section 4 presents some performance comparisons of the mapping strategies and discusses the factors that impact strategy selection and performance. Finally, conclusions and future work are outlined in section 5.

2 Semantic Mappings and Scenario

Semantic Mapping is defined as the establishment of correspondences between a set of source ontologies. In our work we assume that the ontologies are expressed in the web ontology language (OWL) [13]. OWL is also used to describe the mappings of the ontologies. In particular, we use *equivalence*, *subsumes* and *subsumed by* relationships to express the mappings[1]. In order to generate mappings, various matching techniques can be applied to the ontologies. However, the fully automatic generation of mappings from different ontology information is generally considered impractical [14]. This is because there is a degree of uncertainty in any automatic approach to matching two ontologies, with this uncertainty caused by the different syntactic representation of the ontologies, the combination of the similarity measures produced by different matchers, and the heuristic approaches inherent in some matchers [15]. For now, semi-automatic techniques for creating mappings from ontology matching information will continue to dominate [16] and in our work we have used the OISIN tool [17] to support the generation of the mappings.

As an example scenario, we have chosen the exchange of policies between service providers in a Dynamic Spectrum Access (DSA) environment. The XG policy language is developed by DARPA XG group for DSA management [18]. This is a rich scenario as the policy approach advocated by DARPA XG is ontology based, but it is unlikely that all service providers will develop policies according to one single ontology. Thus in order to support policy exchange between the service providers, an information delivery mechanism capable of supporting heterogeneous ontologies is required. Service providers may be arbitrarily located in different regional areas, so they need to subscribe their interests for policies relating to specific geographical locations.

In order to illustrate this approach, we give a concrete mapping example within our scenario, however, the approaches discussed in this paper are independent of the actual contents of the ontologies. Fig. 1 describes a hierarchical structure of mapping

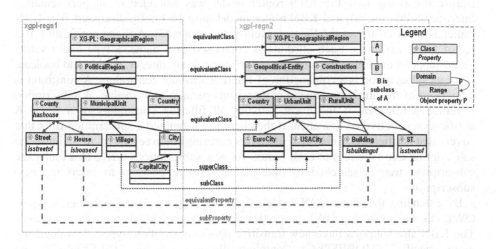

Fig. 1. Mappings between XG region ontologies

---

[1] The *subsumes* relationship describes the super-class and super-property relationships. *Subsumed by* captures the sub-class and sub-property relationships. Equivalence can be used with classes, properties and individuals.

relations between classes and properties from two ontologies. The ontologies *xgpl-regn1* and *xgpl-regn2* are region description ontologies, made by the authors, for delivering XG policies between service providers (acting as KBN clients) in a KBN. To make a mapping between these two ontologies we first need to import both so that the rest of ontology description will be able to refer to the existing elements that are previously defined in an involved ontology. Second, we establish mappings between elements of the involved ontologies. For instance, one class of an ontology may be considered as a *subclass* of another class of another ontology (*xgpl-regn1:Village* is subclass of *xgpl-regn2:RuralUnit* in Fig. 1). Finally, two relations (*subsumption* and *equivalence*) between properties from the involved ontologies can be determined by comparing their members (*xgpl-regn1:ishouseof* is a subProperty of *xgpl-regn2:isbuildingof* in Fig. 1).

Now, let us assume that *xgpl-regn1* is the main application ontology distributed among some KBN routers. If there is a service provider interested in polices within a city range, it subscribes a query expressed by concept *xgpl-regn1:City* to its closest KBN router. If this router receives a notification that has the same queried attribute name but the concept is *xgpl-regn2:EuroCity* ("*EuroCity*" is not defined in *xgpl-regn1* but rather in *xgpl-regn2*), this KBN router needs to explore the mappings to find mapping relations containing "*EuroCity*" and "*City*" to resolve this unknown concept. In this case, the region mapping ontology is explored, where the concept "*City*" is identified as superclass of the concept "*EuroCity*".

3   KBN Router Model and Semantic Mapping Strategies

Before discussing how the KBN router model was extended to support semantic interoperability, the original KBN router model must be briefly discussed. The KBN router is an extension of the Siena content-based router [3]. A Siena notification is a set of typed attributes. Each attribute is comprised of a name, a type and a value. Siena supports the following attribute types: string, long, integer, double and boolean. A Siena subscription is a conjunction of filtering attribute constraints. A constraint is comprised of the attribute name, a comparison operator, and a value. A subscription covers a notification if the event satisfies all filtering constraints of a filter. A notification is delivered to a client if the client has submitted a subscription filter that covers that notification. Siena also discovers coverings between filters to optimise the subscription tree (subtree) at each router. As new subscriptions arrive at a router the subscription tree is searched to find the appropriate position to insert the new subscription.

By extending the Siena CBN model, the KBN supports 3 new ontological types: OWL classes (concepts); OWL properties; and OWL individuals (concept instances). The KBN also supports three new transitive operators for these types: "ontologically more specific" (MORESPEC); "ontologically less specific" (LESSSPEC); and "ontologically equivalent" (EQU). To achieve this, each KBN router holds a copy of an ontology, within which each ontological class, property and individual is described. A more detailed discussion of the KBN router model, and how it is extended from the Siena CBN router, is presented in [7].

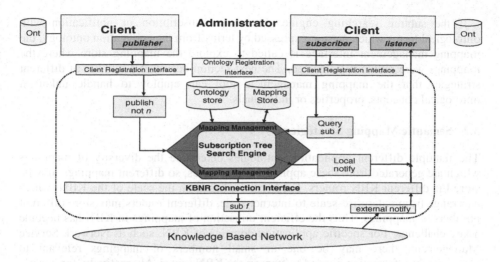

Fig. 2. Enhanced Knowledge Based Network (KBN) router architecture

3.1   Extended KBN Router Model

This section describes how the KBN router was further extended to support heterogeneous ontologies in the network. Each extended KBN router (fig. 2) is implemented with two ontology repositories: the main application ontology store provides the ontology for the KBN operation, whereas the mappings in the mapping ontology store are used for helping the KBN router achieve semantic interoperability. In the original KBN router, every router had a copy of the same main application ontology, however in this extension each router can have a different local main application ontology, and a different set of mapping ontologies to support interoperability between application ontologies. All ontologies are provided by the administrators of the network. The ontology registration interface allows administrators to register both application ontologies and mappings with KBN routers. Both publishers and subscribers register with KBN router via a client registration interface, and they need to provide their own ontology that defines the knowledge bases used by the clients in their subscriptions and notifications.

Subscriptions can arrive at the KBN router either directly from a client or from another node in the KBN network. The query subscription, using terms from the subscriber's local ontology, is passed to the subscription tree (subtree) searching engine, which searches the subtree and inserts the subscription in the appropriate position. However, the subscription may use ontological terms that are not contained in the router's local ontology, and so the position to insert the subscription into the subtree cannot be immediately resolved.

Similarly, when a publication arrives at a KBN router, either directly from a client or from another KBN node, the subtree searching engine walks the subtree to find appropriate matching subscriptions to find the set of subscribers (clients and other KBN nodes) that should be notified with the publication. Again, the publication may use ontological terms that are not contained in the router's local ontology, so the set of matching subscriptions cannot be immediately resolved.

If the subtree searching engine receives a subscription or notification with ontological terms which are not expressed by terms from the application ontology, the mapping management interface is called to explore the mapping store where the mappings were previously injected. The next section discusses a number of different strategies that the mapping management tool can employ to handle unknown ontological concepts, properties or individuals.

3.2 Semantic Mapping Strategies

The multiple different application ontologies determine the diversity of mappings which are generated from these application ontologies, so different mappings may be stored in different KBN routers. Furthermore, based on the scale of the KBN, which can range from enterprise-scale to internet-scale, different routers may store different numbers of mappings. Thus the efficient utilisation of mappings in KBN has become a key challenge. For specific application domains of KBN, such as Network Service Management, there may be one or small numbers of mappings relevant to management information needed to store in the KBN router. Alternatively, for generic applications in a large-scale deployment, there may exist many mappings. Our previous work has indicated that the loading of new ontologies into the reasoner embedded in a KBN node is computationally expensive; the ontological reasoning is memory intensive; and memory usage is proportional to the number of concepts and properties loaded into reasoner [12]. Thus different strategies are required to cope with different situations in the KBN, and it must be possible to alter mapping strategies to cater differing and changing scenarios.

Several strategies have been developed and implemented in KBN routers, aiming at efficiently exploiting mappings to achieve semantic interoperability. In Fig. 3, we

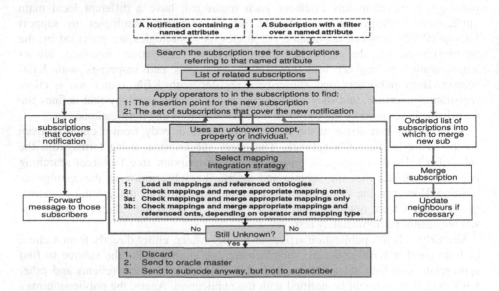

Fig. 3. Workflow of Mapping Strategies

outline the workflow of mapping strategies that are implemented in the KBN router. These strategies are focussed selecting available mapping ontologies (and ontologies referenced in the mappings), which can then be loaded to handle unknown concepts (or properties, or individuals). Strategy selection should be based on a set of rules, for instance, the rules might be defined according to characteristics of application ontologies or the rate of subscriptions and publications. This aspect of the work is not further discussed in this paper, but is a focus on ongoing work.

Strategy1: When a KBN router encounters an unknown concept (or property, or individual), it loads all known mapping files and their imported ontologies at once. This strategy maximises the exploration of mappings to tackle the heterogeneity problem; especially when there is a small number of mapping files stored in KBN router. This strategy would also reduce the probability that further unknown concepts will be encountered at a later stage, at the expense of a high once-off cost. However, this strategy may be wasteful if there are a lot of mappings in different files, or if the occurrences of unknown concepts are rare.

Strategy2: This strategy searches the mapping files stored in the KBN router in order to select mapping files which contain at least one concept used by the conflicting subscription or notification. The router then loads those selected mapping ontology files without necessarily loading, reasoning and merging all of the available mapping files. Furthermore, when compared with strategy 1, the KBN does not load the ontologies imported by the mapping ontologies. Because many of these imported ontologies may not be relevant and to load all imports could introduce considerable overheads. This strategy should be beneficial where KBN router stores a large number of specific and fine grained mapping ontologies.

Strategy3a: This strategy also searches the set of available mappings to try to determine which mappings to load into the KBN router's application ontology. Here only the individual mapping is incorporated into the application ontology, rather than the whole mapping file it was contained in. Ontologies referred to or imported by the mapping ontology are again not loaded. This strategy only succeeds if a direct mapping exists between the unknown concept and a known concept. For example, based on the ontologies introduced in section 3 above, where the *xgpl-regn1* ontology is already loaded, but the *xgpl-regn2* ontology is not loaded. If the unknown concept *xgpl-regn2:EuroCity* from the *xgpl-regn2* ontology is encountered, the mappings are searched. A mapping is found linking the unknown *xgpl-regn2:EuroCity* concept to the previously known *xgpl-regn1:City* concept. In this scenario, only the mapping is loaded, but neither the *xgpl-regn2* ontology, nor referenced ontologies, are loaded.

Strategy3b: This strategy is very similar to strategy 3a above. However, unloaded ontologies referred to in the mapping may also be loaded, depending on the combination of mapping relation found, and the operator to be applied to the unknown concept (for subscription sorting or subscription matching). This strategy could be used where there may not exist a direct mapping relationship between the unknown concept and any known concept, yet the unknown concept may be present in the referenced ontology. Fig. 4 describes when the router needs to load the second referenced ontology. In Fig. 4, **r** is the unknown concept, and a mapping exists

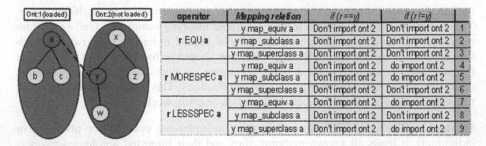

operator	Mapping relation	if (r==y)	if (r!=y)	
r EQU a	y map_equiv a	Don't import ont 2	Don't import ont 2	1
	y map_subclass a	Don't import ont 2	Don't import ont 2	2
	y map_superclass a	Don't import ont 2	Don't import ont 2	3
r MORESPEC a	y map_equiv a	Don't import ont 2	do import ont 2	4
	y map_subclass a	Don't import ont 2	do import ont 2	5
	y map_superclass a	Don't import ont 2	Don't import ont 2	6
r LESSSPEC a	y map_equiv a	Don't import ont 2	do import ont 2	7
	y map_subclass a	Don't import ont 2	Don't import ont 2	8
	y map_superclass a	Don't import ont 2	do import ont 2	9

Fig. 4. Conditions for loading a referenced / imported ontology

between **a**, a known concept in the loaded ontology, and **y**, a concept in an ontology referenced by the mapping. We have considered the equivalence, subclass and superclass relationships and the three new KBN operators, EQU, MORESPEC and LESSSPEC. Lines 4, 5, 7 and 9 shows the cases when the router needs to load second ontology to deal with the unknown concept **r**.

For example, the KBN router gets a subscription containing a named attribute filter using the LESSSPEC operator over a known concept *xgpl-regn1:City* (**a**), and a notification arrives containing an attribute with the same name with its value an unknown concept (**r**) *xgpl-regn2:RuralUnit*. After checking the mapping files, the router finds that *xgpl-regn1:City* (**a**) has a subclass *xgpl-regn2:EuroCity* (**y**). At this stage, since the mapping has not yet been incorporated, it is unknown if *xgpl-regn2:RuralUnit* is related to *xgpl-regn2:EuroCity* (i.e. **r** = **w**, or **r** = **x**, or **r** = **z**). Thus the *xgpl-regn2* ontology (**ont2**) referenced in the mapping must be loaded. Afterwards, operation (*xgpl-regn2:RuralUnit* LESSSPEC *xgpl-regn1:City*) can complete, returning false, and the KBN router's operation can proceed as normal.

If none of the strategies above resolve the unknown concept, property or individual, the KBN is left with a number of possible alternatives for dealing with the unknown ontological data. These include: discard the message with a warning; forward the message to a predetermined "oracle" node that would be responsible for such unrecognised knowledge; or forward the message (publication, subscription, unsubscription etc) to its neighbouring nodes (but not directly to subscribers), hoping that they can handle it. This aspect of KBN routers' operation is not further discussed in this paper, but is a focus on ongoing work.

4   Performance Comparisons of Strategies

In order to determine which of the implemented strategies suit different applications scenarios, it was decided to undertake an experiment to compare performances of each of the strategies. Several metrics are measured in our experiments. Load-time overhead is given as the times taken for the reasoner to load, parse and consistency check the ontology, perform TBox classification, perform ABox realisation and an initial query of all concepts [12]. We also measured the time taken to check mapping

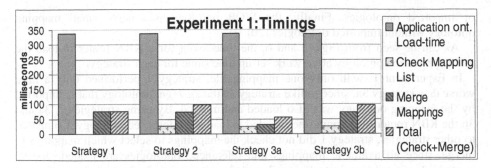

Fig. 5. Experiment 1: Two *xg-regn* ontologies with one mapping ontology

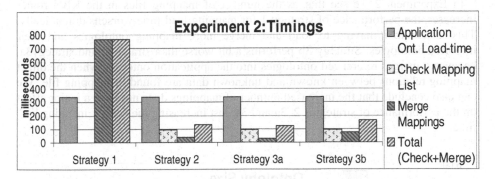

Fig. 6. Experiment 2: Three *xg-regn* ontologies with multiple mapping ontologies

files in the mapping store in order to find correct mappings ("check mapping list") and the time taken to merge these mappings. We also measured the size of the application ontology (number of statements in the ontology after it was reasoned over) after performing each strategy to measure the relative memory usage by different strategies[2].

The first experiment compared the strategies by using the two *xg-regn* ontologies and the mapping ontology described in section 3. The *xg-regn1* ontology is stored in the application ontology store in a KBN router. The mapping store of the KBN router contains the region mapping ontology, whereas the *xg-regn2* ontology is used as a source of unknown concepts for the experiment.

In the second experiment, the *xg-regn1* ontology is again used as the application ontology and region mapping ontology is stored in mapping store, another three mapping ontologies were chosen for testing. Firstly, we created another region mapping ontology, which was generated from the *xg-regn2* ontology and a third *xg-regn3* ontology. The second mapping data set is the owl-s mapping ontology [19]. It is used for policy management in semantic web services, and contains a large number

---

[2] All tests were taken on a Dell Inspiron 9300 laptop with 1.73 GHz Intel processor, 1GB of RAM, running Windows XP SP2. For java-based tools, Sun's JSDK 1.6.1 was used. Jena 2.3, with Pellet 1.3, was used for ontology manipulation. Tests were repeated at least 10 times. Reported timings are averages.

of imported ontologies. Finally, the WOB ontology is relative small mapping ontology without imported ontologies [20].

As can be seen from Figs. 5 and 6, the time taken for a KBN router to load the main application ontology at network set-up time same for each strategy.

In Experiment 1, with only one mapping file, strategy 1 performed only a little worse than strategy 3a, since unlike strategy 3a the *xg-regn2* ontology that is imported by the mapping ontology was also loaded and merged. By using strategies 2, 3a and 3b the KBN router only sometimes needed to load the referenced ontology. Compared to other strategies, strategy 1 did not check the mappings to select which mappings to merge, since all available mappings and referenced ontologies are loaded, and so strategy 1 outperformed strategies 2 and 3a. Strategies 2, 3a and 3b performed similarly for the "check mapping list" time.

In Experiment 2 we see that as the number of mapping files in the KBN router increases, the performance of strategy 1 for mapping load time worsens dramatically. This is due to it having to load and merge more mapping ontologies and their imported ontologies. Strategy 3b performs a bit worse than strategy2 and strategy3a, because it merges referenced ontologies into the application ontology when the direct mapping relations between known and unknown data are found in mapping file. We can also conclude that the increased number of mappings does not have a major effect on the performance of strategies 2, 3a and 3b, just increasing the "check mapping list" time.

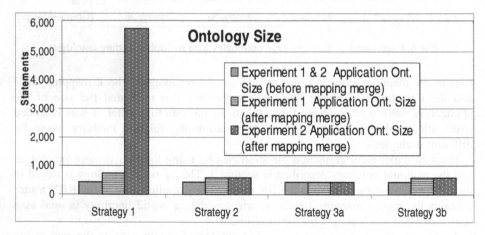

Fig. 7. Experiment 1 & 2: Ontology size

From Fig. 7 we can see the dramatic effect on resources when strategy 1 is used. For this reason it is obvious that strategy 1 should not be used where there is a large number of mapping ontologies, with a large number of imported ontologies, especially where the ontologies are particularly large or resources are scarce. With respect to size, strategies 2, 3a and 3b are much more efficient as only the correct mapping ontology is loaded. Strategy 3a performs best since the mapping ontology is not merged, but just the individual mapping relationship. Strategy 3b performs very similar to strategy 2, except depending on the mapping relationship and operator used an imported ontology may also need to be loaded.

Overall, across all of the experiments, strategy1 seems to have performed worst, with each of the other strategies performing similar to each other. However, it must be remembered that for strategy 1, despite the once off cost of loading all of the mappings and referenced ontology, the KBN is much less likely to encounter unknown concepts in the future, thereby reducing the need to incorporate mappings at a later time. Similarly, although strategy 3b performs slightly worse than strategy 2 or strategy 3a, strategy 3b is more likely to load a complete referenced ontology. Overall, strategy 3a slightly out-performs strategy 2 and strategy 3b since it performs the least amount of ontology loading, querying and merging. This strategy is preferable where the occurrence of unknown concepts is rare and localised. Another point to remember is that the strategies that do not load the imported or referenced ontologies can result in the case where the unknown concept remains unknown after checking the mappings, and so the publication or subscription cannot be handled locally in a satisfactory manner.

5 Conclusions and Further Work

In this paper, a semantic mapping service scheme has been proposed to solve the heterogeneity problem in knowledge based networks by appropriately and efficiently selecting a mapping strategy efficiency of knowledge delivery in KBN. Several strategies are identified to explore how semantic mapping information can incorporated into the router's knowledge base. Each of the strategies were discussed and compared, with scenarios identified to suggest when different strategies should be selected. The comparative evaluation of each strategy indicates that different mapping strategies should be configured in different KBN routers depending on the particular characteristics of the routers' ontologies, the characteristics and number of available mappings, and the type of application operating over the KBN. In a small scale or enterprise scale scenario it may be possible to examine the application running over the KBN to statically determine which strategy is most appropriate. However, in a large scale deployment, or where the characteristics of applications using the KBN may change, then it is necessary to dynamically manage and adapt which strategy is most appropriate.

Work is ongoing to build on these initial evaluations to design and implement a flexible mapping strategy management framework. Work is also ongoing to investigate how mappings can be dynamically distributed around the network as the knowledge bases of clients joining and leaving the network affect the spread of knowledge across the network. It foreseen that the KBN itself would be ideal for such a distribution mechanism.

This work also builds on and validates work previous work by the authors [12] which discusses the impact of changing a KBN router's application ontology at runtime. From that work, and the findings presented in this paper, it is clear that where possible, the introduction of new knowledge into the KBN routers ontology should be minimised where possible, or performed at load time. However, this paper focuses on scenarios where this is not possible, either due to the dynamic nature of the applications, publishers and subscribers that use the KBN, or indeed where it is impractical or impossible (perhaps due to resource constraints) to have the entire knowledge base of the network statically replicated on every KBN router.

Ongoing work is also focussing on how the semantics of the knowledge in the network can be exploited to cluster nodes that focus on semantically similar knowledge [12]. In this way the interoperability of clusters' knowledge bases can be localised to edge of clusters, thereby localising the overhead required for semantic interoperability and mapping.

Acknowledgement. This work is funded by Science Foundation Ireland under Grant No. 03/CE3/I405: Centre for Telecommunications Value-Chain Research (CTVR) and Grant No 05/RFP/CMS014.

References

1. Meier, R., Cahill, V.: Taxonomy of Distributed Event-Based Programming Systems. The Computer Journal 48(5), 602–626 (2005)
2. Eugster, P.T., Felber, P.A., Guerraoui, R., Kermarre, A.-M.: The many faces of publish/subscribe. ACM Computing Surveys 35(2), 114–131 (2003)
3. Carzaniga, A., Rosenblum, D.S., Wolf, A.L.: The Design and Evaluation of a Wide-Area Event Notification Service. ACM Transactions on Computer Systems 19(3) (2001)
4. Segall, B., et al.: Content-Based Routing in Elvin4. In: proc AUUG2K, Canberra (2000)
5. Pietzuch, P., Bacon, J.: Peer-to-Peer Overlay Broker Networks in an Event-Based Middleware. In: DEBS 2003 at ACM SIGMOD/PODS Conference, California (June 2003)
6. Lynch, D., Keeney, J., Lewis, D., O'Sullivan, D.: A Proactive approach to Semantically Oriented Service Discovery. In: Workshop on Innovations in Web Infrastructure (IWI 2006), at Int'l World-Wide Web Conference, Edinburgh, Scotland (May 2006)
7. Keeney, J., Lynch, D., Lewis, D., O'Sullivan, D.: On the Role of Ontological Semantics in Routing Contextual Knowledge in Highly Distributed Autonomic Systems. Tech. Report TCD-CS-2006-15, Dept of Computer Science, Trinity College Dublin (2006)
8. Keeney, J., Lewis, D., O'Sullivan, D.: Ontological Semantics for Distributing Contextual Knowledge in Highly Distributed Autonomic Systems. Journal of Network and System Management 15(1) (2007)
9. Keeney, J., Lewis, D., O'Sullivan, D., Roelens, A., Boran, A., Richardson, R.: Runtime Semantic Interoperability for Gathering Ontology-based Network Context. In: Network Operations and Management Symp (NOMS 2006), Vancouver, Canada (April 2006)
10. Lewis, D., O'Sullivan, D., Feeney, K., Keeney, J., Power, R.: Ontology-based Engineering for Self-Managing Communications. In: Workshop on Modelling Autonomic Communications Environments (MACE 2006), at Manweek 2006, Dublin, Ireland (October 2006)
11. Roblek, D.: Decentralized Discovery and Execution for Composite Semantic Web Services, M.Sc. Thesis, Trinity College Dublin, Tech. Report TCD-CS-2006-66 (December 2006)
12. Lewis, D., Keeney, J., O'Sullivan, D., Guo, S.: Towards a Managed Extensible Control Plane for Knowledge-Based Networking. In: International Workshop on Distributed Systems: Operations and Management (DSOM 2006), at Manweek 2006, Dublin, Ireland (October 2006)
13. W3C (2003) Ontology Web Language, Visited June 2007, http://www.w3.org/2001/sw/

14. Noy, N.F., Musen, M.A.: Algorithm and tool for automated ontology merging and alignment. In: 17th National Conference on Artificial Intelligence (AAAI-2000) (2000)
15. Cross, V.: Uncertainty in the automation of ontology matching. In: 4th Int'l Symp. on Uncertainty Modeling and Analysis, 2003. ISUMA 2003, pp. 135–140 (September 2003)
16. Uschold, M., Grüninger, M.: Ontologies and Semantics for Seamless Connectivity. SIGMOD Record 33(4), 58–64 (2004)
17. O'Sullivan, D., Wade, V., Lewis, D.: Understanding as We Roam. IEEE Internet Computing 11(2), 26–33 (2007)
18. DARPA XG Working Group "DARPA XG Policy Language Framework, RFC", version 1.0, prepared by BBN Technologies, Cambridge MA, USA (April 16, 2004)
19. Owl-s map ontology, http://ontology.ihmc.us/SemanticServices/KAoS-OWL-S-Mapping.owl
20. Ding, L.: Web of Belief Ontology http://daml.umbc.edu/ontologies/webofbelief/0.8/wob.owl

IFIP WG 2.12 & WG 12.4 International Workshop on Semantic Web and Web Semantics (SWWS)

SWWS 2007 PC Co-chairs' Message

The Web has now been in existence for more than a decade and it is influencing in profound ways all aspects of society and commerce. Equally importantly for researchers, the Web has caused a major shift in our thinking about the nature and scope of information processing. However, by its very technological nature and the supporting theoretical foundations, the Web has remained relatively rudimentary, being currently largely suitable for coarse-grain information dissemination. Thanks to Tim Berner-Lee's vision of the Semantic Web, Web technologies are rapidly moving toward a new theoretical foundation founded on semantics and geared toward application processing of Web data and knowledge deployment. Not surprisingly, there has been a tremendous upsurge of research activity addressing problems associated with semantics for the Web. The International IFIP Workshop On the Semantic Web & Web Semantics (SWWS 2007) is in its third year, and is intended to provide a forum for presenting original, unpublished research results, and innovative ideas related to the Semantic Web.

This year, a total of 44 papers were submitted to SWWS. A very large number of these submissions were of high quality. Each of them was rigorously reviewed by at least two experts. For each review, papers were judged according to their originality, significance to theory and practice, readability and relevance to workshop topics. The reviewing process resulted in the selection of 12 regular papers (for a 27% acceptance rate) and 5 short papers for presentation at the workshop. Accepted papers for the workshop touch mainly on the areas of ontology development, ontology management, ontology evolution, process semantics, semantic interoperability, ontology matching and biomedical ontologies. We feel that SWWS 2007 papers will inspire further research on topics that relate to the Semantic Web and its applications.

We would like to express our deepest appreciation to authors who submitted their papers, and also wish to thank all workshop attendees. Last but certainly not least, we are grateful to the Program Committee members and external reviewers for their time and efforts in reviewing, with an eye toward maintaining high quality for accepted papers and turning SWWS 2007 into a success. Thank you all.

August 2007   Elizabeth Chang
                                                                        John Mylopoulos

From Database to Semantic Web Ontology: An Overview

Shuxin Zhao and Elizabeth Chang

Digital Ecosystems & Business Intelligence Institute, Curtin University of Technology
GPO Box U1987
Perth WA 6845, Australia
{s.zhao,e.chang}@curtin.edu.au

Abstract. This paper intends to provide an overview of automated knowledge extraction and transformation from relational databases and their related sources into Semantic Web ontologies. Issues and challenges in this area are addressed. Knowledge embedded in each part of a relational database is analysed and defined. Corresponding techniques for extracting, acquiring and transforming the knowledge are highlighted. In this paper, we classify previous approaches on this work into two types. A comparison table of the first type of approach is also given.

Keywords: database to ontology, domain ontology development, knowledge acquisition from database, knowledge discovery from application code.

1 Introduction and Motivation

The Semantic Web [1] provides explicit meaning for information and data on the web in the future. One of the goals it pursues is to allow intelligent search agents to process and integrate data from heterogeneous resources at the conceptual level. Ontologies capturing and representing various domain knowledge and information, as one of the key factors to the success of the Semantic Web, need to be well developed. However, the current state of domain ontology development for the Semantic Web is still in its infancy in terms of both quantity and quality. One of the main issues is the high development cost associated with manual knowledge acquisition for ontology construction. Acquiring domain knowledge requires many resources and is time-consuming. The knowledge acquisition process for most existing ontology construction, such as Cyc [2] and SENSUS [3], is mainly conducted on a manual basis. This has become one of the bottlenecks of the ontology development. For this reason, how to effectively acquire knowledge from available resources for ontology construction in order to reduce the knowledge acquisition effort has become a hot topic in the ontology research community [4].

While the Semantic Web is waiting for a large amount of knowledge and information represented with explicit and shared semantics, a vast quantity of relational databases contain valuable business information and implicitly embed knowledge cannot participate in the Semantic Web directly. This type of knowledge resource is too valuable to be neglected for ontology development for three reasons: *firstly,* compared to other knowledge sources such as documents, databases are carefully designed and developed data repositories which support business process

R. Meersman, Z. Tari, P. Herrero et al. (Eds.): OTM 2007 Ws, Part II, LNCS 4806, pp. 1205–1214, 2007.

applications. The conceptual model of a relational database is analogous to an ontology model such as EER and UML. Some researchers refer to UML as a lightweight ontology [5]. Some components of the database model could be directly mapped into ontology constructs; *secondly*, these databases are well-maintained. They contain up-to-date data instances and reflect timely business information; and *thirdly*, these databases are usually deployed within a network or on the internet, which are already physically available to the Semantic Web.

On the other hand, there are increasing demands for databases to become sharable and searchable across organisational and application boundaries. This kind of request is driven by constantly increasing collaboration among organisations and by the business need for their service and product information to be searched, integrated and processed without prior negotiation. For instance, patient data from individual health care systems needs to be integrated in the health care domain in order to provide better quality health care services; and travel information needs to be shared amongst travel service providers and agents in order to provide integrated travel services regarding transportation, accommodation and so forth. Ontologies as a means for knowledge sharing provide a solution to this problem. *"Ontologies give a concise, uniform, and declarative description of semantic information, independent of the underlying syntactic representation or conceptual models of information bases"* [6]. What we need to do is to transfer implicit knowledge from databases into Semantic Web ontologies, intelligent search agents can then integrate them like human agents.

Realising the needs of both the Semantic Web and vast databases, this paper intends to provide an overview and insight into the work of transforming the knowledge from databases into the Semantic Web ontologies. The rest of this paper is organised as follows: Section 2 analyses and defines the knowledge embedded in a relational database and its related sources; then brief descriptions of corresponding techniques that are used to extract the knowledge are followed in Section 3; in Section 4, the challenges of automated knowledge extraction from the resources are discussed; then a classification of previous approaches in this area with a comparison table is given in Section 5; Section 6 concludes this paper and indicates future work.

2   Implicit Knowledge Embedded in Relational Databases

This section firstly provides a description of the basic ontology constructs specified in OWL [7] and then identifies the knowledge embedded in a relational database and its related sources that can be used for ontology development.

2.1   The Semantic Web Ontology

Ontology is defined as *"a formal, explicit specification of a shared conceptualisation"* [8]. Ontologies define basic concepts in a domain, assert properties of concepts and represent relationships among concepts in a computer-usable way such as in logic-based language so that *"detailed, accurate, consistent, sound, and meaningful distinctions can be made among the classes, properties, and relations"* [9] by machines. OWL [7], as the W3C recommendation for the Semantic Web ontology language, has three basic concepts: class, properties of class and relationships among classes which are the target of the knowledge acquisition process from databases. The class hierarchy of an ontology is formed by the construct: '*rdfs:subClassOf*' in OWL which models the '*is-a*' type of relationships between classes. Other associations between classes are realized by defining and restricting on properties of classes such as using '*objectProperty*' or by defining class axioms.

In a relational database and its related sources, the knowledge of classes, their properties and relationships in a domain can be obtained from three parts which are the database schema, the data instances, and the applications that are built upon the underlying database.

2.2 Relational Database Schema

Relational database schema specifies the structure of the data held in a database and many business constraints. The conceptual model such as an ER model or an UML class diagram that a database schema implements describes a collection of domain concepts and their relationships which are analogous to the classes and their relationships of an ontology. For this reason, database schema has become the dominant source in previous approaches for acquiring knowledge for ontology development. Domain concepts and their properties and relationships are implied in the forms of relations, attributes, attribute data types, primary keys and foreign keys, and referential integrity constraints etc in the database schema. According to [6, 10-13] a relational schema consists of the following constructs that may be used for ontology development :

- *relations*
- *attributes of relations*
- *atomic data type of attributes*
- *constraints of attributes such as unique, not null*
- *primary keys/foreign keys*

As mentioned in [14], these represent a model of the real world appropriate to the database. One important limitation of database schema is that it restricts its attention only to these parts of the real world of direct relevance to the stored information and it is a static model of the real world. The input form of database schema is usually the logical model which can be easily generated from most of the DBMSs. Through analysis of these constructs and their correlations, a major concept frame in the domain can be obtained.

2.3 Database Instance

Database instances contain concrete and timely data and business information, and implicitly embed up-to-date domain knowledge. It is the input for creating ontology instances of the resulting ontology generated from database schema. Besides, database instances can aid in clarifying the semantics of poorly named attributes through the analysis of data value and data correlations. Furthermore by applying data mining techniques, database instances can also be used to reveal patterns, association rules in the information. This type of previous unknown knowledge is not specified in the relational database schema and may be used to generate axioms of the ontology.

One issue is worth to mention regarding to the frequent changing nature of database instances, therefore, the ontology instances need to reflect the changes in time. A process for generating ontology instances from database instances dynamically on the fly is thus desirable.

2.4 Application Source Code

Applications built on an underlying database are another important means of verifying and identifying domain knowledge. In general, a database supported application that was developed to facilitate a certain business process consists of three components: *database, data manipulation code* and *user interface*. The database provides a persistent data repository needed by the business processes at one end and the user interface displays the *intended meaning* of the data to the user on the other end. Data manipulation code sits in the middle, links user interface and database, carries out data manipulation of the backend database via SQL queries and pre-defined stored procedures upon user requests according to predefined business rules, and performs interpretation of data in the database to their intended meanings in user interface [15]. In this architecture, data instance in databases is only a codification of some facts [15]. It alone, without the specification of the database schema and the interpretation of application source code, does not specify any explicit meaning. On the other hand, the user interface provides users with domain agreed and user-friendly terms for the data in the database which is processed and populated by the data manipulation code. This can be used to verify and clarify the semantics of database schema. Besides, the application code itself embeds business rules which are most likely beyond the definition of the underlying database. For example, the knowledge of calculating the total amount of a customer order which is spread over more than one table may be obtained from the application code. This kind of hard coded knowledge can be extracted for axiom construction in an ontology.

The three parts of a database and its related sources are consistent in a database supported system. They represent the domain knowledge from different aspects and complement and verify one another during the process of knowledge acquisition. However, knowledge embedded in a database and its related sources rarely covers a complete range of the knowledge about a domain. Therefore, once a domain ontology is created from one input database and its related sources, it needs to incorporate knowledge obtained from other sources such as other databases or documents.

3 Techniques of Knowledge Acquisition from Relational Database

The knowledge embedded in each part of a relational database has its own structure, syntax and characteristics. Thus, techniques used to extract the knowledge are based on the input form of each part of the relational database. We can only provide a highlight of some commonly used techniques in this area due to the page limit.

3.1 Database Reverse Engineering

Database reverse engineering aims to recover the data model of an existing database in order to apply the data model to a new application setting. An EER model or an

UML model is the commonly used target of the reverse engineering process in previous approaches. Based on [16, 17], we define database reverse engineering as:

"The process of analyzing a specific database implementation and to perform concept abstraction in order to reconstitute the data asset of an existing system and apply it to a different context"

Through database reverse engineering on the logical model of an existing database, relations can be classified as base relation, dependent relation and composite relations [10, 13], which may be indicators of a concept, specialisation of a concept, or a relationship between concepts. There are three types of correlations (key correlation, attribute correlations and data correlation [10]) of an input database that may be used to identify relationships among concepts and to decide on whether a concept or attribute should be derived. Key correlations, i.e. primary keys and foreign key, are the primary means of identifying relationships among concepts [6, 10-13]. Analysing attribute correlation including attributes equity, overlap, inclusion and disjoint and the like, across relations [10] can help identify different types of relationships among concepts including subtype of a concept; and finally, perform data correlation analysis on instant data may also assist to verify uncertain concepts, attributes and relationships.

3.2 Mapping Technique

The mapping technique is usually used in conjunction with reverse engineering in the context of developing an ontology from databases. It maps the data model derived from the reverse engineering process or the logical model to an ontology language such as F-logic [18], RDF [19] or OWL [7] by specifying the corresponding counterparts between two languages. The mapping is usually performed through a set of predefined mapping rules which specifies the semantics of the mapping between the two models. For example, some atomic data types in relational schema may be mapped to XML schema 'datatypes'.

3.3 Data Mining

The emergence of data mining techniques is resulted from the need to transfer huge amounts of available data into useful and meaningful information and knowledge. It is also known as Knowledge Discovery in Databases (KDD) [20]. Data mining is *"to apply data analysis in order to discover previously unknown, useful patterns and relationships in large data sets"* [21]. Data mining can be used for marketing, fraud detecting, and terrorist detecting and for intelligent e-agents gathering and associating information in an information-rich environment such as the Semantic Web. The goal of data mining can be prediction or description [20]. By applying data mining on database instances, one may discover 'human-interpretable patterns describing the data' or may use existing information to make reasonable prediction regarding future activities [20, 21]. Some commonly used data mining methods include classification, association, clustering and sequence and path analysis and so forth. In database to ontology context, axioms of the resulting ontology may be created from the knowledge discovered from data mining on database instances.

3.4 Information and Knowledge Extraction

Information Extraction (IE) and Knowledge Extraction (KE) are emerging research areas over the past decade as a consequence of the dramatically increasing volume of electronic text including plain text and semi-structured text. IE transforms input text into information that is more readily digested and analysed. *"It isolates relevant text fragments, extracts relevant information from the fragments, and then pieces together the targeted information in a coherent framework"* [22]. IE has yielded NLP and machine learning techniques to extract useful information from text. IE and KE techniques may be used to extract useful information and knowledge from application source code. Approaches for extracting knowledge from web pages have been proposed by automatically detecting extraction rules, useful patterns [23, 24].

3.5 Application Reverse Engineering

Application reverse engineering aims to comprehend legacy systems for system maintenance and reconstruction purposes in the instance that the analysis and design documents are not available. Reverse engineering research on application code level analysis has been successful since early 1990s [17]. It has produced the capabilities for decomposing a system into subsystem, concept synthesis, program slicing and dicing, and analysing static and dynamic dependencies and the like[17]. Current trends of application reverse engineering include the object-oriented approach, component-base approach and the incremental approach [25].

As we can see from the above description, the techniques used to extract knowledge from different parts of a database and its related sources are divergent. Each of the techniques has their own focus. It is important to effectively integrate and combine them into a consistent framework and apply it to the context of automated knowledge acquisition from a database and its related sources.

4   Challenges in Automated Database to Ontology

Knowledge acquisition from a relational database and its related sources in an automated fashion presents many challenges. The first challenge comes from the vagueness of the input source of a database. The semantics of data is not explicitly defined in the relational schema. Knowledge of a specific domain is user-oriented and ad hoc. Necessary assumptions on the original design model of the database and its related sources must be made and user intervention or user verification cannot be avoided during or after the knowledge acquisition process. Therefore, fully automated knowledge acquisition is almost impossible. *Secondly*, knowledge embedded in the input database and its related sources is incomplete as databases are usually designed to support a certain business process or to solve a particular problem in a domain. They most likely only cover a certain scope of the domain knowledge. Therefore, knowledge acquired from this input should be enriched by other knowledge sources in order to represent full domain knowledge. The input forms of other knowledge sources can be divergent in many aspects such as their physical presentation, structure, coding languages and so on. To automatically obtain integrated knowledge from these heterogeneous sources requires many joint efforts from each of those areas. *Thirdly*, there exist many issues regarding the implementation of a running database in the real world. This includes database redundancy, poor naming of database relations and attributes, and poor database design which did not follow good design principles and so forth. This kind of problem is hard to foresee for each individual database implementation. *Finally*, knowledge acquisition from a database system is an interdisciplinary research area. The automated knowledge acquisition process from databases involves many interrelated disciplines and requires a synthesis of techniques from different areas in order to discover the knowledge accurately without information loss. One of the key issues to be solved is that of how to integrate the output from different techniques on different inputs of a database system.

5 Classification of Previous Approaches on Database to Ontology

The previous approaches on database to ontologies can be grouped into two types. The first type of approach generates an ontology model from an input database model through reverse engineering, mapping techniques and other techniques, then create ontological instances from database instances based on the previously generated ontology. This group of approaches includes [6, 10, 11, 26-28]. Key correlations are a major means of identifying relationships between concepts. Attributes correlation [10] and the use of application source code such as HTML forms [27] have begun to be examined and combined into the database reverse engineering process while data correlation is rarely examined. In Kashyap [6], more than one database of the same domain is used to extract the domain knowledge. This approach also utilised a domain specific thesaurus which contains standardised vocabularies and used user queries to refine the ontology model generated from database schemas. An overview of these approaches is shown in Table 1. As we can observe from the overview table, there is no approach that has examined the knowledge embedded in a relational database in its full dimensions. In addition, some approaches need to be tested with real world examples.

The second type of approach proposed mapping languages that directly map database into Semantic Web data syntax such as OWL [7] without analysing database schema. This includes approaches such as R2O [29] and D2R MAP [30]. The mapping languages specify the semantics of the mapping between the relational database and the ontology and intend to be highly expressive and fully declarative. The main drawback of this type of approach is two-fold: *firstly,* database instances will keep updating over time, as a result, constant synchronisation between the published ontological instances and the data in the original database is required; *secondly*, the meaning of the original data model such as relationships among concepts are not reflected in the transformed ontological instances. However, these languages may be effectively used to automatically populate ontology instances from databases when an ontology model of that database is available.

Table 1. An overview of approaches for database to ontology

| Approach | Purpose | Database Schema | | Correlation analysis | | | Data instance | | Application source code | Other source |
		Input	Output	Key	Attribute	Data	Ontology instance	Data mining		
Kashyap (1999) [6]	Create an domain ontology	Logical model of more than one DB	Ontology	✓			✓			User query, Data dictionary
Stojanovic et. al. (2002) [11]	Migrate data-intensive website into SW[1]	Logical model	Ontology in Frame logic	✓						
Jarra & Meersman (2002) [26]	Domain ontology development	Conceptual model	lexon part of DOGMA				✓			Yellow pages
Astrova (2004) [10]	Migrate data-intensive website into SW[1]	Logical model	Ontology in Frame logic		✓	✓	✓			
Astrova (2005) [27]	Migrate data-intensive website into SW[1]	Relational schema is treated as only a look up in knowledge extraction					Data instances contained in HTML forms		Extract from HTML forms to a form model schema	Depends on domain expert to extract a HTML form model
Li et. al (2005) [28]	Learning SW[1] ontology from DB	Logical model	Ontology in OWL	✓			✓			

[1] Semantic Web.

5 Conclusion and Future Work

This paper was motivated by the need for transformation of the knowledge embedded in relational databases and their related sources to Semantic Web ontologies. The vast quantity of databases has many advantages for the Semantic Web ontology development and is too valuable to be omitted in the Semantic Web. However, automatic knowledge acquiring from databases and their related sources requires joint efforts and integrated technologies from several areas, each of which presents different challenges. On the other hand, databases are designed only for specific purposes that only represent limited range of the knowledge in a domain. Therefore, it's important to incorporate other domain knowledge sources such as other existing ontologies, documents and other databases. By discussing these issues and challenges and examination of previous work, we expect an effective and integrated framework to accommodate the needs for capturing the knowledge accurately without information loss and for transferring the knowledge from databases to Semantic Web ontologies. Thus allow knowledge to be better retrieved, reused and searched.

References

1. W3C: Semantic Web. Accessed May 30, 2006 (2006), http://www.w3.org/2001/sw/
2. CYC: Cyc. Accessed April 25, 2006 http://www.cyc.com/cyc/technology/whatiscyc
3. Swartout, B., Patil, R., Knight, K., Russ, T.: Toward Distributed Use of Large-Scale Ontologies. In: R, G.B., A, M.M. (eds.) KAW 1996, the 10th Knowledge Acquisition Workshop, SRDG Publications, University of Calgary, Banff, Canada pp. 32.31-32.19 (1996)
4. Gómez-Pérez, A., Manzano-Macho, D., Alfonseca, E., Núñez, R., Blacoe, I., Staab, S., Corcho, O., Ding, Y., Paralic, J., Troncy, R.: A Survey of Ontology Learning Methods and Techniques. In: Gómez-Pérez, A., Manzano-Macho, D. (eds.) OntoWeb Consortium (2003)
5. Gomez-Perez, A., Fernandez-Lopez, M., Corcho, O.: Ontological Engineering: With Examples from the Areas of Knowledge Management, E-Commerce and the Semantic Web. Springer, London (2004)
6. Kashyap, V.: Design and Creation of Ontologies for Environmental Information Retrieval. In: 12th Workshop on Knowledge Acquisition, Modelling and Management (1999)
7. W3C: Ontology Web Language (Accessed June 2006), http://www.w3.org/2004/OWL/
8. Gruber, T.R.: A Translation Approach to Portable Ontology Specifications. Knowledge Acquisition 5, 199–220 (1993)
9. W3C: Web Ontology Language Use Cases and Requirements. In: Heflin, J. (ed.) (2004)
10. Astrova, I.: Reverse Engineering of Relational Database to Ontologies. In: Bussler, C.J., Davies, J., Fensel, D., Studer, R. (eds.) ESWS 2004. LNCS, vol. 3053, pp. 327–341. Springer, Heidelberg (2004)
11. Stojanovic, N., Stojanovic, L., Volz, R.: A Reverse Engineering Approach for Migrating Data-Intensive Web Sites to the Semantic Web. In: Musen, M.A., Neumann, B., Studer, R. (eds.) The IFIP 17th World Computer Congress - TC12 Stream on Intelligent Information Processing, vol. 221, pp. 141–154. Kluwer, B.V. Deventer, The Netherlands, Montréal (2002)

12. Meersman, R.: Ontologies and Databases: More Than a Fleeting Resemblance. In: d'Atri, A., Missikoff, M. (eds.) OES/SEO Workshop, Rome. Luiss, Rome (2001)
13. Chiang, R.H.L., Barron, T.M., Storey, V.C.: Reverse Engineering of Relational Databases: Extraction of an EER Model from a Relational Database. Data & Knowledge Engineering 12, 107–142 (1994)
14. Dillon, T.S., Tan, P.L.: Object-Oriented Conceptual Modeling. Prentice Hall, Englewood Cliffs (1993)
15. Silva, C., Cullell, J.A.: Knowledge Coordination (2003)
16. Davis, K.H., Aiken, P.H.: Data Reverse Engineering: A Historical Survey. In: Seventh Working Conference on Reverse Engineering, pp. 70–78 (2000)
17. Müller, H.A., Jahnke, J.H., Smith, D.B., Storey, M.-A., Tilley, S.R., Wong, K.: Reverse Engineering: A Road Map. In: The Future of Software Engineering, USA, Limerick, Ireland, pp. 47–60. ACM Press, New York (2000)
18. Wikipedia: F-Logic. Accessed August 2006 (2006), http://en.wikipedia.org/wiki/F-logic
19. W3C: RDF. Accessed June 2006 (2006), http://www.w3.org/RDF/
20. Fayyad, U., Piatetsky-Shapiro, G., Smyth, P.: From Data Mining to Knowledge Discovery in Databases. American Association for Artificial Intelligence, 37–54 (1996)
21. Seifert, J.W.: Data Mining: An Overview. Congressional Research Service, The Library of Congress (2004)
22. Cowie, J., Lenhnert, W.: Information Extraction. Communication of the ACM 39 (1996)
23. Chang, C.-H., Lui, S.-C.: IEPAD: Information Extraction Based on Pattern Discovery. In: The 10th Conference of WWW, Hong Kong, China, pp. 681–688 (2001)
24. Arasu, A., Garcia-Molina, H.: Extracting Structured Data from Web Pages. In: The 2003 ACM SIGMOD international conference on Management of data, pp. 337–348. ACM Press, San Diego, California (2003)
25. Garcia, V.C., Lucrédio, D., Prado, A.F.D.: Towards an Effective Approach for Reverse Engineering. In: The 11th Working Conference on Reverse Engineering (WCRE 2004) (2004)
26. Jarrar, M., Meersman, R.: Formal Ontology Engineering in the Dogma Approach. In: Meersman, R., Tari, Z. (eds.) CoopIS 2002, DOA 2002, and ODBASE 2002. LNCS, vol. 2519, pp. 1238–1254. Springer, Heidelberg (2002)
27. Astrova, I., Stantic, B.: An Html Forms Driven Approach to Reverse Engineering of Relational Databases to Ontologies. In: Hamza, M.H. (ed.) the 23rd IASTED International Conference on Databases and Applications (DBA), Innsbruck, Austria, pp. 246–251 (2005)
28. Li, M., Du, X.-Y., Wang, S.: Learning Ontology from Relational Database. In: The 4th International Conference on Machine Learning and Cybernetics, IEEE exlorer, Guangzhou, China, pp. 3410–3415 (2005)
29. Barrasa, J., Corcho, Ó., Gómez-Pérez, A.: R2O, an Extensible and Semantically Based Database-to-Ontology Mapping Language. In: Bussler, C., Tannen, V., Fundulaki, I. (eds.) SWDB 2004. LNCS, vol. 3372, Springer, Heidelberg (2005)
30. Bizer, C.: D2R Map – a Database to RDF Mapping Language. In: The 12th International World Wide Web. ACM, Budapest, Hungary (2003)

An Ontology-Driven Architecture for Re-using Semantic Web Services

Carlos Granell, Dolores María Llidó, Rafael Berlanga, and Michael Gould

Department of Information Systems, Universitat Jaume I, Castellón, Spain
{carlos.granell,doloresmaria.llido,berlanga,gould}@uji.es

Abstract. As more semantic web services become on the Internet, it is feasible that users collaborate among them to save efforts in complex web solutions by sharing and reusing existing semantic web services, rather than building them from scratch. In this paper we focus on the problem of discovering and reusing semantic web services at a high level of abstraction centered on the concept of abstract pattern, that provides a logical view of a service composition and a great power of classification because it uses ontology concepts and relations. To support semantic discovery and reuse, we present an ontology-driven architecture that enables users to specify structured queries against to knowledge base of abstract patterns and services.

Keywords: Ontology-driven architecture, semantic web services, abstract patterns, knowledge reuse.

1 Introduction

The Model Driven Architecture (MDA) [8] has recently become one of the most emerging techniques which are applied to solve the problem of web service (WS) composition. In short, MDA defines three level of abstraction or viewpoints from which a service composition can be represented, namely Computer-Independent Model (CIM), Platform-Independent Model (PIM) and Platform-Specific Model (PSM). However, most MDA-based approaches for WS composition are focused mainly in the PIM and PSM levels [2][4][5], following a top-down approach in which initial PIM models are transformed into PSM models until reaching the required implementation, ignoring then the advantages of reusing CIM models. For example, the same CIM model can be applied to different contexts, though the generated set of PIMs requires different configurations according to each context features.

In this paper we address the lack of reuse of services and knowledge presenting an ongoing ontology-driven architecture that provides a CIM's Knowledge Base capable of reasoning –classifying, composing, and querying– over services in an abstract way. Also we describe a querying-driven process at CIM level that achieves structured semantic answers given user goals or queries, underlying the importance of CIM for improving the reuse of services and knowledge between CIM and PIM levels.

This architecture is being applied to emergency management scenario. In this scenario, a new PIM is usually required when a new emergency arise, and it should be

R. Meersman, Z. Tari, P. Herrero et al. (Eds.): OTM 2007 Ws, Part II, LNCS 4806, pp. 1215–1221, 2007.

designed and deployed as quick as possible. Thus, the reuse of already built models (at different level of abstraction) can be crucial. In the current implementation, final PSM models are WSBPEL processes over existing (composite) WSs.

The paper is structured as follows. Next section describes the proposed ontology-driven architecture. Section 3 and 4 are devoted to the CIM level, describing the knowledge base structure and the query processing. Section 5 concludes the paper.

2 Ontology-Driven Architecture

Figure 1 illustrates the ongoing ontology-driven architecture that enables using Semantic Web mechanisms in conjunction with MDA techniques to facilitate both Semantic Web Service (SWS) composition and reuse. We first propose the use of DL-semantic descriptions to deal with two key aspects of WSs: the classification of services (with respect to their IOPEs) and the classification of abstract patterns that can be applied over them. The former enables reason and compose properly candidate semantic service descriptions (concepts) which may form part of the target solution, whereas the latter allows us to validate possible abstract patterns that satisfy the selected service concepts. In our approach, both elements that are fully expressed as logic-based ontologies (with OWL-DL expressivity [1]) constitute the semantic required to build the CIM's Knowledge Base (see Section 3).

At CIM level, user goals are specified in terms of semantic descriptions and abstract patterns, disregarding all details about concrete services, their specification languages (e.g. OWL-S [6] or WSML [9]) and execution platforms. Thus, the CIM's Knowledge Base (KB) is aimed at representing just the concepts that can be required for specifying goals, pattern structures (i.e. workflow) and the abstract services and operations they involve. Users will be able to define abstract requests in terms of these concepts to discover and re-use existing concrete solutions.

The CIM's KB is populated with existing semantic services and their compositions. In this way, it is necessary to define the transformations from PIM languages (e.g. OWL-S and WSML) to CIM abstract specifications. Basically, these transformations must drop all PIM concepts not regarded at the CIM KB and classify PIM models according to CIM KB concepts.

The CIM also provides a pattern matching tool able to retrieve all the abstract specifications that best fit user requests. Currently, this matching tool is based on approximate XML retrieval techniques [10]. This kind of techniques allows users to define the similarity function they want to apply to retrieve abstract patterns. The similarity measure can rely on a reasoner to check component subsumption with respect to the CIM KB. As this part goes beyond the scope of this work, we will not give further details.

Once users have selected the abstract patterns of their interest, they proceed to build a PIM model, which can include concrete services and adaptors taken from the CIM KB. As shown in Figure 1, we regard three possible paths at the PIM layer. The first one consists of WSMO specifications, which use the representation language WSML. The second one consists of OWL-S specifications. Finally, the third path represents an ad-hoc platform-independent method for building composite services. The two first paths use the specific tools for matching and validating semantic services, which also include specific execution engines at the PSM level. They constitute two big infrastructures that regard all the stages for managing and executing SWS. Instead, the third one is aimed at combining independent components for the composition and validation of SWS that do not completely fit in with these platforms. More specifically, we have adapted our previous research works [3] and [7] to implement this path.

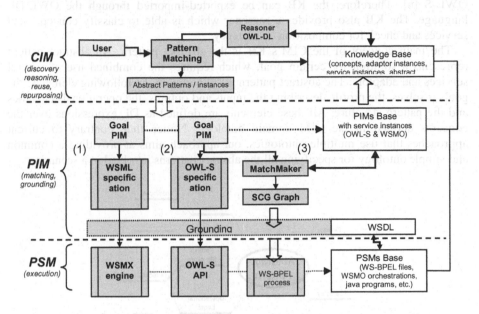

Fig. 1. Ontology-driven architecture for semantic service discovery and reuse

The PIM layer also includes a knowledge base aimed at storing all the details of concrete PIM models that were omitted at the CIM layer. These details are mainly the concrete IOPEs that are specified in the concrete services and their orchestrations. Notice that PIM models can be expressed with different languages. Therefore horizontal transformations between PIM models could be necessary if parts of them are merged to create new ones. Although this can be partially at the CIM layer, concrete IOPEs specified at PIM layer will require such transformations.

Regarding the third path of the architecture, SWS composition is performed through a Semantic Composition Graph (SCG) [7] whose construction is guided by the abstract pattern. The SCG relates CIM concepts with semantic service descriptions stored in the PIM base. From a SCG we can extract several execution plans, which must be validated with respect to the IOPEs of the involved SWS. Finally, the SCG is grounded, and by applying a transformation PIM-PSM an executable WS-BPEL is built [3]. Finally, if the resulting PSM is executed successfully, then the corresponding models at different levels are registered for future use. So, the abstract pattern (knowledge) is included in the CIM's DL-Knowledge Base, whereas the PIM model updates the PIMs Base.

3 CIM's Knowledge Base

The CIM's KB comprises the underlying domain ontology (e.g. a geospatial ontology or a biomedical ontology), the semantic descriptions of the available services and adaptors, and the semantics of the composition abstract patterns. We assume that the KB is expressed in DL [1], which provides the basis for the OWL DL and also for the OWL-S [6]. Therefore, the KB can be exported-imported through the OWL-DL language. The KB also provides a reasoner which is able to classify concepts and services and check for composition consistency.

The main concept of the CIM's KB is the abstract pattern. An abstract pattern represents a solution for certain goal, which requires the combined use of several services and adaptors. The abstract pattern is defined with the following elements: the goals it solves, the profile describing the inputs and outputs, the operational semantics and the pattern structure. All these elements are defined as DL expressions over the concepts defined in the CIM domain ontology. Notice that, contrary to current approaches that use multiple ontologies, our approach aims at providing a common and simple ontology for specifying all the abstract patterns involved in a scenario.

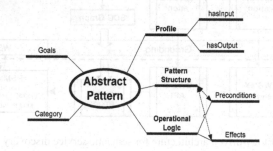

Fig. 2. Structure and concepts for defining abstract patterns

In order to simplify as much as possible the semantic description of abstract patterns, we only introduce a few concepts, namely: Actor, Service, Adaptor, Profile and OpLogic. These concepts have associated the following properties: hasProfile, hasInput, hasOutput, hasOpLogic, hasPreCondition and hasEffect. The former defines the semantics of the services according to their inputs (hasInput) and outputs (hasOutput). The property hasOpLogic expresses the relation between each service and the state concepts it requires (hasPreCondition) and modifies (hasEffect).

As an example, the following axioms in the KB describe a generic Service that provides nearest geospatial objects:

NearestObject ⊑Service ⊓∃hasProfile.PosObject ⊓∃hasOpLogic.PaymentLogic
PosObject ⊑ Profile ⊓∃hasInput.GeoPosition ⊓∃hasOutput.GeoObject
PaymentLogic ⊑ OpLogic ⊓∃hasPrecondition.(User ⊓hasCredit.>5000)

The CIM's KB is completed with the abstract patterns. In general, workflow patterns serve as composition operators to specify how services should be combined. Since most current languages for WS composition like WSBPEL come from the workflow area, we have derived a set of abstract patterns suitable for WSs [3], which provide several advantages with respect to common control patterns available in other languages. First, intentionality, we have maintained the number of redundant workflow patterns to a minimum, in contrast to overlapping and alternative patterns present for example in WSBPEL, leading to simpler but complete set of abstract patterns. In addition, such abstract patterns are quite general to be independent enough of concrete control patterns used in composition languages. For example, the abstract pattern SEQ that represents a sequence of services can be transformed into the corresponding construct either in WSBPEL or in OWL-S.

Although abstract patterns are applicable to concrete services at PSM level, they can also be semantically characterized to be used both within the CIM's KB for defining semantic queries and as part of PIM models. In contrast to other approaches, introducing abstract patterns at CIM level is a key aspect in our approach because it lets users describe more complex and structured queries. Users can not only specify semantic queries in terms of IOPEs but also determine how candidate services should be combined. Table 1 shows the axioms associated to each of these abstract patterns.

Table 1. Semantics for abstract patterns. A and B are concepts denoting services, whereas c is a concept denoting a condition over state variables (operational logic).

Abstract Patterns	Axioms associated to abstract patterns
S=SEQ(A,B)	hasOutput^{-1}.A ⊓ hasInput^{-1}.B is satisfiable
	S ⊑ Service* ⊓∃hasProfile.(Profile ⊓∃hasInput.(∃hasInput^{-1}.A) ⊓ ∃hasOutput.(∃hasOutput^{-1}.B))
S=AND(A, B)	S ⊑ Service* ⊓∃hasProfile.(Profile ⊓∃hasInput.(∃hasInput^{-1}.A ⊓∃hasInput^{-1}.B) ⊓∃hasOutput. (∃hasOutput^{-1}.A ⊔∃hasOutput^{-1}.B))
S=XOR(A,B)	The same as AND (Semantics included at PIM level)
S=AND-DISC(A)	S ⊑ Service* ⊓∃hasProfile.(∃hasProfile^{-1}.A)
	Here A denotes the bag of services involved in the pattern
S=LOOP(c, A)	The same as AND-DISC (Details included at PIM level)
S=IF(c, A, B)	The same as AND (Details included at PIM level)

As an example, consider the following abstract pattern whose goal is to buy stock units of certain market if its value decreases after two observations:

IF(SEQ(AND(A1,A1), ValueDifference)>$Diff, A2)

Here, A1 is the abstract service that provides the stock value of certain market. A2 is the service that permits buying a number of stocks of certain market. These can be defined as follows:

A1 ⊑Service ⊓∃hasProfile.(∃hasInput.Market ⊓∃hasOutput.Number)
A2 ⊑Service ⊓∃hasProfile.(∃hasInput.Market) ⊓∃hasOpLogic.BankClient
BankClient ⊑ OpLogic ⊓∃hasPrecondition.(User ⊓hasCertificate.X112)

ValueDifference is an adaptor which takes two numbers as input and gives its difference. It is worth mentioning that adaptors are treated at the CIM layer in the same way as abstract services. They are distinguished at the PIM layer. Finally, $Diff denotes the input parameter of the abstract pattern that indicates the tolerance for buying stocks. Notice that this abstract pattern can be re-used in any situation where the difference of two observations is used as criterion for a decision. For example, we can use a similar abstract pattern over meteorological services for a fire scenario.

4   Transforming Abstract Patterns

This section describes briefly how the logic structure of the abstract patterns at CIM level can be transformed into specific OWL-S control constructs [6] and WSMO choreographies [9] at PIM. Both approaches define different formal semantics for WS composition. While OWL-S declares a set of controls to define a procedural specification of the services involved in a composition, the representation of WSMO choreographies is based on a state-based machine that consists of states and transition rules to express how the states evolve. Applying transitions rules provokes state changes that directly change the values of concepts and relations between ontologies. Then, OWL-S control constructs are closely related to the semantics of the *structural* part of the abstract patterns (see Table 2), while WSMO choreography is rather concerned with the operational logic of abstract patterns (see Table 3).

In Table 1, the semantic meaning of some abstract patterns englobe multiple OWL-S control constructs [3]. For example, SEQ pattern serves for the sequence, any-order

Table 2. Relationships between abstract patterns and OWL-S

Abstract Patterns	Control Constructs in OWL-S Process
S=SEQ(A, B)	Sequence (A, B); Any-Order (A, B) restricted to a certain order.
S=AND(A, B)	Split-Join (A, B); Split (A, B) with synchronization barrier
S=XOR(A, B)	Choice (A, B)
S=AND-DISC(A)	Split (A, B) without synchronization barrier + Choice (A, B)
S=LOOP(c, A)	Repeat-While (A); Repeat-Until (A); Iterate (A). All control constructs instances iterate until a condition c holds true or false specified with a whileCondition or an untilCondition properties.
S=IF(c, A, B)	If-Then-Else (A, B) where condition c is translated in a ifCondition property

Table 3. Relationships between abstract patterns and WSMO

Abstract Patterns	WSML choreographies
S=SEQ(A, B)	Piped rule + update rule (add, delete, or update facts).
S=AND(A, B)	Choice rule + update rule (add, delete, or update facts).
S=XOR(A, B)	Choice rule + update rule (add, delete, or update facts).
S=AND-DISC(A)	*Not available*
S=LOOP(c, A)	For-all rule with a logical expression condition + update rule (add, delete, or update facts).
S=IF(c, A, B)	If-Then rule with a logical expression condition + update rule (add, delete, or update facts).

and unordered control construst. Also, the conditions expressed in the abstract patterns are used for the control constructs conditions. Finally, abstract patterns can be nested in order to provide structured patterns, which can be viewed as OWL-S composite processes that maintain certain state and includes control constructs.

These transformations can be implemented by means of XSLT. Whereas CIM to OWL-S transformation can be done straightforward as both are based on OWL, CIM to WSMO must be made through WSML templates, which are filled with the proper CIM elements. Notice that new transformations can be performed at the PIM layer to include concrete transitions rules and other details not regarded at the CIM layer.

5 Conclusions and Future Work

The proposed ontology-driven architecture in a MDA-like framework allows us to take advantage of reusing CIM models for composing and reusing SWS. First, we have described a CIM's DL KB capable of reasoning –classifying, composing, and querying– over services in an abstract way, and a querying-driven process at CIM level that achieves structured semantic answers given user goals, underlying the importance of the CIM models for improving the reuse of services and knowledge between CIM and PIM levels. Future work includes rule-based reasoning to simulate the evolution of state variables for checking that all compositions are consistent with respect to the corresponding business logic conditions.

References

1. Baader, F., Calvanese, D., McGuinnes, D., Nardi, D., Patel-Scheneider, P.: The Description Logic Handbook: Theory, Implementation and Applications. CUP, Cambridge (2003)
2. Gannod, G.C., Timm, J.T.E., Brodie, R.J.: Facilitating the Specification of Semantic Web Services Using Model-Driven Development. J. of Web Services Research 3, 61–81 (2006)
3. Granell, C., Gould, M., Grønmo, R., Skogan, D.: Improving Reuse of Web Service Compositions. In: Bauknecht, K., Pröll, B., Werthner, H. (eds.) EC-Web 2005. LNCS, vol. 3590, pp. 358–367. Springer, Heidelberg (2005)
4. Grønmo, R., Jaeger, M.C.: Model-Driven Semantic Web Service Composition. In: Proceedings of the APSEC 2005, IEEE Press, Los Alamitos (2005)
5. Marcos, E., Acuña, C.J., Cuesta, C.E.: Integrating Software Architecture into MDA Framework. In: Gruhn, V., Oquendo, F. (eds.) EWSA 2006. LNCS, vol. 4344, pp. 127–143. Springer, Heidelberg (2006)
6. Martin, D., et al.: Bringing semantics to web services: The owls approach. In: Cardoso, J., Sheth, A.P. (eds.) SWSWPC 2004. LNCS, vol. 3387, pp. 26–42. Springer, Heidelberg (2005)
7. Paraire, J., Berlanga, R., Llidó, D.M.: Resolution of Semantic Queries on a Set of Web Services. In: Andersen, K.V., Debenham, J., Wagner, R. (eds.) DEXA 2005. LNCS, vol. 3588, pp. 385–394. Springer, Heidelberg (2005)
8. Sigel, J.: Developing in OMG's Model-Driven Architecture Object Management Group White Paper (2001), Available at ftp://ftp.omg.org/pub/docs/omg/01-12-01.pdf
9. Roman, D., Scicluna, J., Nitzsche, J., Fensel, D., Polleres, A., de Buijin, J.D.: Ontology-based Choreography. WSMO Final draft (February 2007)
10. Sanz, I., Mesiti, M., Guerrini, G., Berlanga, R.: ArHeX: An Approximate Retrieval System for Highly Heterogeneous XML Document Collections. In: Grust, T., Höpfner, H., Illarramendi, A., Jablonski, S., Mesiti, M., Müller, S., Patranjan, P.-L., Sattler, K.-U., Spiliopoulou, M., Wijsen, J. (eds.) EDBT 2006. LNCS, vol. 4254, Springer, Heidelberg (2006)

OntoGame: Towards Overcoming the Incentive Bottleneck in Ontology Building

Katharina Siorpaes and Martin Hepp

Digital Enterprise Research Institute (DERI), University of Innsbruck, Innsbruck, Austria
katharina.siorpaes@deri.at, mhepp@computer.org

Abstract. Despite significant advancement in ontology learning, building ontologies remains a task that highly depends on human intelligence, both as a source of domain expertise and for producing a consensual conceptualization. This means that individuals need to contribute time, and sometimes other resources, to an ontology project. Now, we can observe a sharp contrast in user interest in two branches of Web activity: While the "Web 2.0" movement lives from an unprecedented amount of contributions from Web users, we witness a substantial lack of user involvement in ontology projects for the Semantic Web. We assume that one cause of the latter is a lack of proper incentive structures of ontology projects, i.e., settings in which the perceived benefits outweigh the efforts for people to contribute.

As a novel solution, we (1) propose to masquerade collaborative ontology engineering behind on-line, multi-player game scenarios, in order to create proper incentives for humans to help building ontologies for the Semantic Web. Doing so, we adopt the findings from the already famous "games with a purpose" by von Ahn, who has shown that pres-enting a useful task, which requires human intelligence, in the form of an on-line game can motivate a large amount of people to work heavily on this task, and this for free. Then, we (2) describe our OntoGame prototype, and (3) prov-ide preliminary evidence that users are willing to invest a lot of time into those games, and, by doing so, unknowingly weave ontologies for the Semantic Web.

1 Introduction

One can observe only limited involvement of users in building ontologies, which is one of the reasons that may explain the shortage of current, high-quality domain ontologies. While the technical aspects of collaborative ontology engineering are already an established research topic, little attention has so far been dedicated to the incentive structures of ontology construction and usage, i.e. research on the motivations for people to contribute to an ontology or to adopt it. However, since building ontologies is a task that depends on human intelligence, both as a source of domain expertise and for producing a consensual conceptualization, it cannot be taken for granted that a sufficient amount of individuals engages in ontology projects just on the basis of altruism. Also, it is important to stress that ontology building is inherently a collaborative task, for two reasons: first, ontologies are supposed to be community

R. Meersman, Z. Tari, P. Herrero et al. (Eds.): OTM 2007 Ws, Part II, LNCS 4806, pp. 1222–1232, 2007.

contracts [1, 2]; second, the combination of required domain expertise and modeling skills is more likely to be found in a group than in a single individual.

In short, producing ontologies consumes resources; thus people with respective expertise must have a sufficient motivation to contribute. Unfortunately, the incentives for ontology building have not been a popular research topic so far. In this context, it can be observed that there is a significant difference in the incentive structures of formal ontologies vs. e.g. collaborative tagging. When tagging a resource, someone adding tags to data achieves two things at the same time: (1) improving the public vocabulary (i.e. the set of tags) and (2) improving his own access to the knowledge assets. Traditional ontology engineering, in contrast, detaches the effort from the benefits: by building an ontology alone, one does not improve one's own access to existing knowledge, while others may enjoy the added value of an ontology without having invested into its construction. Thus, it is not per se granted that those investing resources in the creation or improvement of an ontology will materialize sufficient benefits out of the usage of this ontology.

As a novel solution, we propose to masquerade user contributions to collaborative ontology engineering behind on-line, multi-player game scenarios in order to establish proper incentives for humans to help building ontologies for the Semantic Web. Doing so, we adopt the findings from the already famous "games with a purpose" by von Ahn, who has shown that presenting a useful task, which requires human intelligence, in the form of an on-line game can motivate a large amount of people to work heavily on this task, and this for free. Our goal is that users mastering the intellectual challenges of our games unknowingly weave ontologies for the Semantic Web.

1.1 The Motivational Divide: Web 2.0 Is Fun, Ontology Engineering Is Not

We can observe that "Web 2.0" applications enjoy great popularity and comprise strong user incentives [3]: Tagging, i.e., users describing objects with freely chosen keywords (tags) in order to retrieve content more easily, is immediately rewarding and provides some sort of community spirit. It is immediately rewarding, because each tagging action improves my personal access to the knowledge assets, and it provides community spirit, since seeing others agree with my own favorite tags gives some positive feeling of being in alignment. In the case of Wikipedia, even though arguably not a core "Web 2.0" application, we can see similar patterns. Wikipedia currently contains more than 1.7 million articles[1] with very high quality [4]. Kuznetsov [5] has traced this back to a multiplicity of social motivations for people to contribute to Wikipedia. In parallel, von Ahn's ESP game [6] has demonstrated how the high amount of hours invested in playing games on the Internet everyday can be exploited for useful purposes. The ESP game, which is about finding consensual tags for images, is now extremely popular: Von Ahn observed that some people are playing the game more than 40 hours per week. Within a few months after the initial deployment on October 25, 2003, the game collected more than 10 million consensual image labels, and this without paying a cent to the contributors.

---

[1] http://wikipedia.org/, retrieved on May 11, 2007.

Our idea is to present various tasks of ontology engineering, which require human intelligence, in the form of multi-player game scenarios. Eventually, we have preliminary evidence that the proposed games are (1) sufficiently interesting for users and (2) that the contributions are a valuable input for building ontologies.

1.2 Related Work

Von Ahn and colleagues have already described a series of games with different purposes: The ESP game [7] aims at labeling images on the Web. The idea is that two players, who do not know each other, have to come up with identical tags describing the image. Peekaboom [8] is a related game for locating objects within images. Verbosity [9] is a game for collecting common sense facts. Finally, Phetch [10] is a computer game that collects explanatory descriptions of images in order to improve accessibility of the Web for the visually impaired. Apart from Verbosity, we do not know of any other work on exploiting computer game scenarios for the conceptualization of domain knowledge, in particular none that is using a game for soliciting human intelligence in order to create and maintain ontologies.

1.3 Contribution and Overview

In this paper, we (1) derive from popular ontology engineering methodologies a set of tasks that require a substantial amount of human intelligence and thus user contributions, (2) suggest multiple game scenarios that represent these tasks, (3) describe our OntoGame prototype, and (4) give preliminary evidence that OntoGame is not only fun to play but that players are also able to produce useful and ontologically correct results. Finally, we summarize our work and describe pending and future extensions.

2 Multi-player Games for Ontology Construction

The tasks in ontology construction have been analyzed in depth by work on ontology engineering methodologies, see e.g. [2]. For the purpose of this paper, we focus on a subset of tasks, which are classified according to the lightweight Uschold and King's methodology [11]. In the following, we first outline the most important tasks. Then, we suggest various multi-player game scenarios that represent these tasks.

2.1 Tasks in Ontology Engineering and Ontology Usage

Uschold and King's methodology describes four main activities: (1) *identification* of the purpose, (2) *building* the ontology, (3) *evaluation*, and (4) *documentation*. The second activity is divided into three sub-processes: (a) *ontology capture*, which comprises the identification of key concepts and relationships; (b) *ontology coding* as "committing to basic terms [...] and writing the code"; (c) *integrating existing ontologies*, i.e. the reuse of or the alignment with existing ontologies.

Within those four main activities, the following are steps that we think are particularly suited for representation as game scenarios.

Collecting named entities: Relevant conceptual elements of the domain of discourse must be identified and a unique key assigned.

Typing named entities according to the ontology meta-model: With typing named entities we mean the process of defining the type of conceptual element for each named entity, based on the distinctions from the applicable ontology meta-model. Many popular ontology meta-models support *classes*, *properties*, and *individuals* as core types. Classes "provide an abstraction mechanism for grouping resources with similar characteristics" [12]. Properties are a means for describing individuals of such classes in more detail; a popular distinction is between *object properties* relating an individual to another individual and *data type properties* relating an individual to a data value. Individuals can be viewed as instances of classes, i.e. entities with similar characteristics.

Adding taxonomic and non-taxonomic relations: A flat collection of ontological elements can be enriched by adding taxonomic and non-taxonomic relations. The most prominent form of this task is arranging the concepts into a subsumption hierarchy by introducing subClassOf relations.

Alignment between multiple ontologies: There may be multiple ontologies with a partial overlap in scope or content. In order to increase interoperability between data, such ontologies should be semantically related to each other at a conceptual level, which is known as ontology alignment.

Modularization: Depending on the domain of discourse, it is often useful to define subsets of concepts based on their ontological nature or target applications, since they may be more manageable.

Lexical enrichment: Ontology engineering methodologies tend to focus on formal means for specifying ontologies. However, in order to describe the *intended* semantics of ontology elements, informal means, like natural language labels or synonyms are also needed. However, relating a conceptual element to terms or synonym sets requires careful human judgment, since otherwise, inconsistencies between the informal part and the formal part of the ontology may result.

While not part of the actual ontology development, the **population of an ontology with instance data** (also known as annotation) is also included in here, because it is another task that can be addressed in a gaming scenario. In fact, one may assume that the absolute amount of human intelligence needed for annotating data will be much higher than that for building respective ontologies, and motivating many Web users to contribute to that is thus very promising.

2.2 Suggested Scenarios

As said, our core idea is to develop games that are fun to play and at the same time acquire human judgments relevant for specifying a conceptualization of a domain, so that there is a proper incentive for users to build and maintain ontologies. In Table 1, we present different gaming scenarios which address the tasks from the previous section.

Table 1. Suggested Scenarios

Scenario	Task	Intellectual Challenge
Classifying conceptual entities according to types of an ontology meta-model: Two players are faced with the same Wikipedia article and need to agree on whether what the article describes is a class, an individual, or a relation.	Typing named entities	Assessing whether something is a significant (abstract or tangible) individual, an abstraction over multiple individuals (i.e. a class), or a type of relationship between two individuals.
Finding attributes for a class: Two players are faced with the description of a class and need to agree on an attribute for instances of this class.	Collecting and typing named entities (attributes)	Spotting and naming typical characteristics of instances of a given class.
Determining the range of an attribute: Two players are faced with an attribute and need to agree on the proper range for values, i.e. a class or datatype.	Typing named entities	Spotting the proper range of values, either as a class or a datatype.
Finding a super-class: Two players are faced with a class and need to come up with (and agree upon) a super-class of this class.	Adding taxonomic relations	Finding a proper, consensual abstraction.
Identifying typed relations between two classes: Two players are faced with two different ontology classes and need to enter matching names for typical non-taxonomic relations between those two.	Collecting and typing named entities (relations), adding non-taxonomic relations	Spotting and naming types of relations between instances of two classes.
Identifying the class for an instance: Two players are faced with an instance and need to come up with (and agree upon) the label for a class suitable for that instance.	Ontology population	Spotting a proper class for a given entity.
Annotation of a resource: Two players are faced with a Web resource and have to agree on a proper annotation using a given ontology as quickly as possible.	Ontology population	Grasping the semantic essence of a resource and expressing it using a given vocabulary.
Mapping two ontology elements: Two players are faced with two conceptual elements of the same type from two different ontologies and need to agree on the existence and type of semantic correspondences between the two.	Alignment	Judging upon the existence of a semantic correspondence between two conceptual elements and selecting the type of correspondence (e.g. sameClass, subClassOf,...)
Ontology modularization: Two players are presented with the name and description of a domain ontology and need to agree on which elements from a given set of conceptual elements belong into this module.	Modularization	Assessing the domain relevance of a given conceptual element.

Most scenarios can be made more challenging and at the same time directed towards less obvious solutions by a "taboo word" list, representing solutions that are already known, which ensures that new valuable information is gained. One can also make solving more difficult or more needed problems more rewarding by using an adaptive bonus schema that gives more points for such solutions. Also, for free text entries and for choosing elements from an existing ontology, users should be supported by auto-complete and suggest functions in order to avoid friction (and frustration) caused by lack of consensus between players just due to lexical variants or spelling mistakes.

2.3 Input Data for the Games

There are two different types of input for the game scenarios described in the previous section. We distinguish the following two types of data sources:

Core Data: With that we mean input data that is presented to the players. In most cases it is necessary to have content to start from; otherwise, the gaming fun is limited. We identify several external resources that can be used as such input data: First, Wikipedia, which can be regarded as a huge collection of conceptual entities identified by a URI, for which the conceptual stability has been recently demonstrated [13]. Second, lightweight ontologies extracted from folksonomies [14], and third ontologies and instance data from Semantic Web search engines like Swoogle [15, 16] and Watson[2]. Fourth, we can used popular upper-level ontologies like Proton [17].

Complementing Data: With that we mean additional data that can be helpful for improving the gaming fun or the usefulness of the results. For example, we can use various lexical resources for tolerating lexical variants, foreign language entries, or synonyms as correct answers. This simplifies the process of consensus finding between two users at a conceptual level and likely increases the gaming fun (e.g. that you do not fail a task just because of one user using British English and the other American English). To mitigate such problems, we propose to make use of the several dictionaries that are online accessible such as Leo Dictionary[3] and the Wordnet synsets [18].

3   OntoGame

In this section, we provide a description of our OntoGame prototype, outline the architecture of the implementation, and discuss the handling of cheating and mischievous users.

3.1 Overview

For the first prototype of our game approach, we use Wikipedia articles as conceptual entities, present them to the players, and have the users (1) judge the ontological nature and (2) find common abstractions for a given entry. The game is played by multiple players in parallel by teams of two players. The pairs are defined by random and are anonymous, i.e. players do not know each other and have no way of communicating with each other. Using the Wikipedia "random article" functionality[4], both players will see the initial part of a Wikipedia article (**Fig. 1**), which should be enough to grasp the intention of the article. The goal of a user playing our first OntoGame prototype is then two-fold: The first challenge is to guess whether the co-player judges the Wikipedia article as an **instance** or as a **class**, a distinction which is

---

[2] http://kmi-web05.open.ac.uk/WatsonWUI/
[3] http://dict.leo.org
[4] http://en.wikipedia.org/wiki/Special:Random

common in many popular ontology meta-models. If their answers do not match, both players are taken to the next article and no points are awarded. If they agree it to be an instance, the second challenge is to propose the label for a suitable **class** to which this instance belongs. If both players agree upon it being a class, the second challenge is to propose the label for a **super-class** (Fig. 2). In case they manage to reach consensus, they get points awarded and are taken to another article. The users are given 2 minutes time to agree on as many Wikipedia articles as possible.

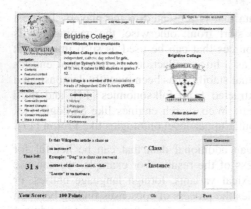

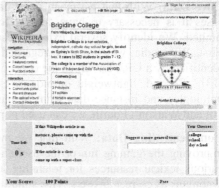

Fig. 1. OntoGame: Phase I   Fig. 2. OntoGame: Phase II

As said, users can always decide to skip one article and proceed to the next one. If they choose to do so in phase II, points earned for mastering the first phase will remain. In other words, players can always choose to pass without losing the points they earned in phase I. In our opinion, this is an important feature, because (1) it is possible that poor or unsuitable articles are presented, (2) some articles might be too specialized to be understood by the current users, and (3) players may simply be unable to reach consensus. Instead of encouraging random guesses, we rather motivate users to proceed to another article.

3.2 Implementation

The architecture of OntoGame is based on Java and includes a game server, a Java applet serving as a client, and several Java servlets enabling communication between the game server and the applet using an object stream over an http tunnel. The game server implements the singleton pattern, which is used to restrict instantiation of a class to one object because in OntoGame exactly one object is needed to coordinate actions across the system including the games, discovering matches, etc. Four different servlets perform the following tasks: login, communication flows for phases I and II of the game, handling user input, matching and passing.

3.3 Cheating and Bad Input

Theoretically it is possible that users try to cheat or undermine the system intentionally. Von Ahn has already described some approaches that aim at minimizing the impact of such behavior. They have shown to be sufficient to make cheating unattractive, and to effectively minimize the impact on the results of the overall task. First, the players are paired anonymously and have no way to communicate directly with each other. Answers entered are visible for the other player only when agreement is reached. Thus, players cannot simply exchange instant messaging contacts or similar as guesses in order to prepare for later cheating. Second, IP address of partners must be different. Third, a massive global agreement (e.g. bots naming all concepts "XY") can be detected in two ways: (1) by checking if the answer time is significantly faster than the average and (2) by letting players play several games with known solutions and compare whether the successful matches are a subset of the known solutions. If not, one can either just ignore the results from the game as not trustworthy, or ban the users temporarily.

4 Evaluation

In this section, we describe our evaluation methodology, the participants and test data, and present preliminary evidence on the relevance of our approach.

4.1 Methodology and Participants

We recruited nine individuals with different backgrounds and asked them to play several rounds of OntoGame. We requested them to do so at a pre-defined point in time in order to ensure that there are enough players to play OntoGame, as there is no single player mode implemented yet. The subjects were instructed to use English words in lower case only. All details of all the games played were recorded. After the experiment, we interviewed the participants about their experiences with the game and analyzed the recorded games. **Participants:** All individuals have experience using the Web, 5 out of 9 hold a degree (bachelor or master) in computer science. Two of them are working in research in the areas of ontologies and the Semantic Web. None had special training in building ontologies. The participants were in different rooms and did not communicate with each other during playing. The game was explained to the participants orally before playing it online.

Table 2. Participants

ID	Age and Gender	Background and Education
1	30, Male	Bachelor in computer science
2	26, Male	Master in computer science
3	24, Female	PhD student in computer science, ontologist
4	26, Male	Bachelor in computer science, ontologist
5	26, Female	Master in computer science
6	20, Female	Student of medicine
7	46, Female	PhD in history
8	17, Male	Student, regular Web user
9	19, Female	Student of economics, regular Web user

4.2 Results

In this section we present the results of our preliminary evaluation regarding the users' experiences as well as the soundness of the output of the games.

Individuals and their experiences: The most important observation is that all of the players liked the game after playing it a couple of times and after understanding how it worked (Table 3). People who do not work in the area of ontology building initially had difficulties to understand the meaning of "adding a more general term". Most wanted to *tag* Wikipedia pages at first try, instead of adding super-class relationships. This shows that we need to improve the help section.

Table 3. Gaming fun and experiences reported by the participants

Player	Summary of the experiences (quotes)
1	Fun but demanding
2	Super, nice game
3	Fun to play, not boring because demanding
4	Fun to play
5	Fun to play
6	Hard to understand at first, had difficulties to grasp the meaning of super-class
7	Once I understood the game, it was really fun to play, addictive
8	Fun to play
9	Hard to grasp the meaning of "a more general term", fun to play

It helped a lot to guide the players by stressing that they have to be able to apply a natural language phrase like *"A is an example of the type of things B"* for instanceOf relations and *"Each example of type A is always also an example of type B"* for subClassOf relations. We could also observe that with increasing experience, the score users achieved increased. **Output of the Games:** Even though this is early work and the scope of the evaluation was rather small, the preliminary results (Table 4) of the evaluation of OntoGame are promising. The nine individuals paired up in 10 different teams and played a total of 26 rounds. By doing so, they processed 116 Wikipedia pages. The pairs were able to reach agreement on the type of conceptual element (class or instance) of 102 articles (88%). According to user comments, they found the distinction between class and instance straightforward. Additionally, many pages in this experiment that were chosen by the Wikipedia random page functionality described a person, which makes the distinction easy. Of the remaining 102 articles, in phase II more than 65% of the pages could be consensually assigned to a super-class (identified by a label at this stage only). The longer the individuals played (2-3 games), the more agreement they could reach with their partners. We have manually validated every consensual (1) class/instance choice and (2) super-class/instance-of relationship which was the output of OntoGame. We found that the class/instance choice of articles was in 99% of the cases correct. Furthermore, the super-class or instance-of relationships that players agreed on during playing, was appropriate in more than 92% of the cases. Table 4 summarizes the results.

Table 4. Results of the 116 rounds

Number of consensual class/instance choices	103 of 116	88.79%
Number of consensual super-class/instance-of relations	67 of 103	65.05%
Number of correct class/instance choices	102 of 103	99.03%
Number of correct super-class/instance-of relations	62 of 67	92.54%

5 Discussion and Outlook

We have suggested applying the idea of "games with a purpose" by von Ahn to the tasks typically found in the construction of ontologies and in the annotation of resources. In particular, we defined suitable gaming scenarios and deployed a first prototype. From preliminary experiments with a first set of users, we can see that the resulting games are both fun to play and produce reasonable and useful ontological data. In particular, the high quality of the consensual conceptual choices is surprising. We are currently in the process of extending OntoGame in the following directions: (1) integration of lexical resources, which increases the gaming fun and facilitates consensus between users, (2) adding new scenarios from Table 1, namely such for proposing attributes and for mapping between different ontologies retrieved from Swoogle or Watson, (3) integrating the PROTON ontology into the challenge, and (4) general improvements of the usability and user interface. In parallel, we are preparing an extended evaluation of our approach, both in a controlled environment and on a Web scale.

Acknowledgments. The authors would like to thank Michael Waltl and Andreas Klotz. This work has been funded by the Austrian BMVIT/FFG under the FIT-IT project myOntology (grant no. 812515/9284). Martin Hepp has also support from a Young Researcher's Grant (Nachwuchsförderung 2005–2006) from the Leopold-Franzens-Universität Innsbruck, which is thankfully acknowledged.

References

1. Hepp, M., Bachlechner, D., Siorpaes, K.: OntoWiki: Community-driven Ontology Engineering and Ontology Usage based on Wikis. In: International Symposium on Wikis (WikiSym 2005), San Diego, California, USA (2005)
2. Gómez-Pérez, A., Fernández-López, M., Corcho, O.: Ontological Engineering. Springer, Heidelberg (2004)
3. Marlow, C., et al.: Position Paper, Tagging, Taxonomy, Flickr, Article, ToRead. In: World Wide Web Conference (WWW2006), Edinburgh, Scotland (2006)
4. Emigh, W., Herring, S.: Collaborative authoring on the web: A genre analysis of online encyclopedias. In: Hawai International Conference on System Sciences, Hawaii (2005)
5. Kuznetsov, S.: Motivations of Contributors to Wikipedia. ACM SIGCAS Computers and Society 36(2) (2006)
6. Von Ahn, L.: Games with a Purpose. IEEE Computer 29(6), 92–94 (2006)
7. Von Ahn, L., Dabbish, L.: Labeling Images with a Computer Game. In: Conference on Computer/Human Interaction (CHI 2004), ACM, New York (2004)

8. Von Ahn, L., Peekaboom: A Game for Locating Objects in Images, SIGCHI Conference on Human Factors in Computing Systems, ACM Montréal, Québec, Canada (2006)
9. Von Ahn, L., Kedia, M., Blum, M.: Verbosity: a game for collecting common-sense facts. In: Conference on Human Factors in Computing Systems, Montréal, Canada (2006)
10. Von Ahn, L., et al.: Improving Accessibility of the Web with a Computer Game. In: SIGCHI Conference on Human Factors in Computing Systems, ACM, New York (2006)
11. Uschold, M., King, M.: Towards a Methodology for Building Ontologies. In: Proceedings of the Workshop on Basic Ontological Issues in Knowledge Sharing, Montreal, Canada (1995)
12. Dean, M., Schreiber, G.: OWL Reference, W3C Working draft (2003)
13. Hepp, M., Siorpaes, K., Bachlechner, D.: Harvesting Wiki Consensus: Using Wikipedia Entries as Vocabulary for Knowledge Management. IEEE Internet Computing 11(5), 54–65 (2007)
14. Van Damme, C., Hepp, M., Siorpaes, K.: FolksOntology: An Integrated Approach for Turning Folksonomies into Ontologies. In: Bridging the Gap between Semantic Web and Web 2.0 Workshop at the ESWC, 2007, Innsbruck, Austria (2007)
15. Ding, L., et al.: Swoogle: A Search and Metadata Engine for the Semantic Web. In: 13th ACM Conference on Information and Knowledge Management (2004)
16. Swoogle. Swoogle Semantic Web Search Engine, http://swoogle.umbc.edu/
17. Terziev, I., Kiryakov, A., Manov, D.: BULO: Guidance, SEKT Deliverable 1.8.1 (2005)
18. Fellbaum, C.: Wordnet: an electronic lexical database. MIT Press, Cambridge (1998)

Towards Social Network Based Approach for Software Engineering Ontology Sharing and Evolution

Pornpit Wongthongtham, Elizabeth Chang, and Ahmed A. Aseeri

Digital Ecosystems and Business Intelligence Institute
Curtin University of Technology, Perth, Australia
{Pornpit.Wongthongtham,Elizabeth.Chang,
Ahmed.Aseeri}@cbs.curtin.edu.au

Abstract. In this paper, we present a multi-agent social network approach that integrates the software engineering ontology and expert recommendation facilities for communities of software engineers remotely working on related software engineering projects. The software engineering ontology enables an active ecology of agents to convey, consume and act on project information (semi) autonomously, according to explicit software engineering domain knowledge. Recommendation techniques are addressed to make progress – ability to recommend useful project information, solution(s) for project issues that arise as experts.

Keywords: Software Engineering, Ontology, Social Network based Approach.

1 Introduction

Recently, there is an explosion of interest in ontologies as artefacts to represent human knowledge and as a critical component in several applications. One unique area of research is the software engineering ontology. Over the last three years, we have developed the world's first Software Engineering Ontology which is available online at www.seontology.org. The software engineering ontology defines common sharable software engineering knowledge including particular project information [1]. Software engineering ontology typically provides software engineering concepts – what they are, how they are related, and can be related to one another – for representing and communicating over software engineering knowledge and project information through the internet [2]. These concepts facilitate common understanding of software engineering project information to all the distributed members of a development team in a multi-site development environment. This should not be confused with the distributed systems, such as CORBA where the development is centralised but deployment is distributed. The ontology enables effective ways of sharing and reusing the knowledge and the project information for remote software engineers and software developers. Reaching a consensus of understanding is of benefit in a distributed multi-site software development environment. Software engineering knowledge is represented in the software engineering ontology whose instantiations are undergoing evolution. Software engineering ontology instantiations

R. Meersman, Z. Tari, P. Herrero et al. (Eds.): OTM 2007 Ws, Part II, LNCS 4806, pp. 1233–1243, 2007.
© Springer-Verlag Berlin Heidelberg 2007

signify project information which is shared and has evolved to reflect project development, changes in software requirements or in the design process, to incorporate additional functionality to systems or to allow incremental improvement, etc. This evolution of instances provides many new challenges to an ability to design and deliver project information.

In this paper, we present a multi-agent based recommender approach that integrates the software engineering ontology and expert recommendation facilities for communities of software engineers and software developers remotely working on related software engineering projects. The integration of the software agents and the software engineering ontology is an innovative technology that can significantly improve recommender approach for multi-site distributed software development. Particularly, the software engineering ontology enables an active ecology of agents to convey, consume and act on project information (semi) autonomously, according to explicit software engineering domain knowledge. Recommendation techniques are addressed to make progress – ability to recommend useful project information, solution(s) for project issues that arise as experts.

The aims of the project are summarised as follows.

1) The development of a new multi-agent based recommender system architecture that facilitates meaningful communication, discussion, negotiation, and information exchange through collaborative agents.

2) The development of the notion of collaborative agents enabling semantic interoperability.

3) The development of a new project management approach of sharing project artefacts, experts, progress, document and so on.

4) The development of recommendation techniques to manage project issues that arise as experts.

5) The implementation of a new prototype that is a realisation of system architecture and utilises the software engineering ontology and the multi-agent based system.

In the next section, we discuss the related work. In the third section, we highlight agents for remote collaboration. The forth section is dedicated to multi-agent social network approach. In the section five, we describe the current state of the implementation of a prototype of the system. The paper ends by drawing conclusions.

2   Related Work

Some work has been carried out in the areas of expert recommendation including development of tools and systems to support e-business, e-learning and the like. Most of the proposed systems are applied to the recommendation of web pages or documents e.g. GroupLens [3], Adaptive Web Site Agent [4], Remote Assistant for Programmers [5]. The systems recommend documents using different criteria: similarities among users, user preferences for the subject area, similarity between document, frequency of citation and frequency of access, etc. A referral system which is based on the idea of social networks [6] is an agent based system. The agents who preserve the privacy and autonomy of their users build social network learning models with each other in terms of expertise (ability to produce correct domain answers), and sociability (ability to produce accurate referrals), and take advantage of the information derived from such a social network for helping their users to find other users on the basis of their interests [5]. Additionally, there is some work in the development of tools and systems for supporting e-learning e.g. Web Based Teaching (WBT) [7], I-MINDS [8]. These systems enable students to actively participate in a virtual classroom which is centred on a website containing teaching materials and an electronic bulletin board for question answering. The systems are based on different kinds of agents e.g. teacher agents, student agents, remote proxy agents. Agents interact with humans and are responsible for disseminating information, maintaining student profiles, generating individual quizzes and exercises, filtering students' questions and messages to reduce traffic and manage classroom sessions progress.

However, these current researchers do not focus on a multi-agent social network approach to support software engineering ontology sharing and management. There has not been any expert approach expertise in a field of software engineering applied to the recommendation of particular software engineering project information. Tackling the disadvantages associated with remote communication over software engineering project information is, therefore, still a longstanding problem in remote collaborative software development. Awareness of what work is being done according to the plan, what work is being done to co-operate between teams, what issues have been raised, what issues have been clarified and how to get together to discuss and to make a decision on those issues, are challenging. Different teams might not be aware of what tasks are being carried out by others, potentially leading to problems such as two groups overlapping in some work or other work not being performed due to misinterpretation of the task. Wrong tasks may be carried out due to ignorance of who to contact to get the proper details. If everyone working on a certain project is located in the same area, then situational awareness is relatively straightforward but the overheads in communications to get together to discuss the problems, to raise issues, to make decisions and to find answers in a distributed environment can become very high. Consequently, these problems cause developmental delays, as outstanding issues are not resolved and issues cannot be discussed immediately or in time over a distributed team environment.

3   Agents for Remote Collaboration

In this paper, we present a multi-agent social network approach to support software engineering ontology sharing and management for remote software developers during common software projects or activities. The ability to make use of software engineering knowledge described in the software engineering ontology enables agents to have the capability of extraction and conveyance semantic rich project information from online dedicated repositories. This assists remote members to have a clear understanding of project information conveyed by the agents. Automatic reasoning capacity of autonomous agents helps manage project issues that arise. Agents are not only able to convey any involved project information according to explicit software engineering domain knowledge but also recommend solution(s) for any project issues as experts on a constant and autonomous basis.

Consider the following scenario from the not too far distant future: Matthew, who is a programmer working in a team in Australia, realises that the project design does not make sense. Matthew then tells his software agent to raise this issue with the design team and request changes. Matthew's agent goes to online project repositories to find out who to contact. After verifying that changes Matthew has requested are conceptually correct according to software engineering domain knowledge described in the software engineering ontology, the agent then raises the issue with the involved design agents. The design agents inform their clients residing in the US and the UK and start collecting their clients' ideas and pass on to a recommender agent. After all, the recommender agent makes a tentative solution for the project issue, informs everyone and updates project information repositories. Team members residing at different sites cannot have a face-to-face meeting. Team members' agents make a virtual meeting on behalf of their clients. As we see from this scenario, in the near future there will be an active ecology of agents that will collaborate with each other and access information sources to fulfil users' needs. Having efficient use of information, organisations can operate virtually and collaborate across huge distances in Australia and around the world. This creative approach is among the fastest growing sectors of the new economy and needed to exploit the huge potential in the multi-site distributed software development industry.

There are currently two sources of knowledge: 'SoftWare Engineering Body of Knowledge (SWEBOK)' and 'Software Engineering Ontology'. The SWEBOK [9] developed by SEI, Canigei Melon University, USA, is a glossary of terms. It provides a definition of each term. It does not define the concepts and the relationship between concepts. We have the Software Engineering Ontology which is a big extension of the SWEBOK. It defines the concepts, and inter-links between the concepts and the relationship between each other and the organisation of the whole body of knowledge, rather than definition of each discrete term.

Nevertheless, either the Software Engineering Ontology or the SWEBOK will provide a benefit for a team of people working together, particularly if a person wants to find the terms and has a good querying facility to do that. These are still passive structures. The passive nature arises from the fact that the user has to come to the ontology with enough knowledge to know what to look for. We move beyond this point and are looking at creating an active support using ontology as the background. We now create Agents situated in the foreground that interact and mediate between the ontology and human agents. The innovation lies in that we are creating an active support for software engineers, especially for multi-site teams working on one project, so that when a software engineer comes to this platform, he/she would get an advisor or a recommender rather than just a Dictionary or the organisation of Knowledge.

For example, if a software engineer knows the terms, he can go ahead and look for it from software engineering body of knowledge to get clarification of the terms. If he/she wants to know the relationships between the terms, he can use the Software Engineering Ontology. On the other hand, if he/she wants advice on what to do in a certain situation, then we need some active components which are intelligent enough to utilise the ontology to advise the user. It is the creation of these active components that is the subject of this proposal. This approach represents a big shift from the previously available approach to software development in a multi-site environment. This represents a significant break-through for Software Engineering studies.

In order to achieve this, we will have to extend the existing knowledge on interaction between agents and ontologies and between agents and software engineers. However, we will do it purely on the context of supporting the Software Engineering Ontology. Generally, this approach can be used in a generic sense for any ontology. The result of this project should be capable of being used in any other ontologies. These are the classes of Agents that we need: user agents, ontology agents, and recommender agent.

Note, an important thing to realise that we are NOT creating agent-oriented software, which is software developed by utilising the Agent Paradigm as a primary construct of the software, which is legitimate work and quite a lot of work has been done e.g. [10], [11] and so forth. But what we are doing here is that we are using multi-agent paradigm in conjunction with ontology to support software developers and engineers, no matter what development methodology they use, whether they use Object-Oriented, Agent-oriented, or classical Data flow techniques. This is an important step forward.

Our particular approach is the integration of the agent technology with the semantic web technology. Our approach is based on a dynamic network of platforms managing teams of geographically members.

4 Multi-agent Social Network Approach

A key feature of the system is that it is an agent-based system integrated with another key emerging technology - semantic web. It is an innovative agent-based system supporting remote collaboration applications.

4.1 Multi-agent Based Recommender System Architecture

The multi-agent based system architecture is grounded on the notion of a set of software agents: User agents, Ontology agents and a Recommender agent. Fig. 1 represents main components in the social network based system architecture.

User agents are agents that allow the interaction between the member and the different parts of the system and among the members themselves. This agent serves as the communication media for members of the team. The agents are initiators based

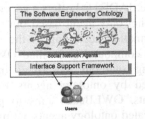

Fig. 1. – Main components in the social network based system architecture

on the member actions to the system and responsible for building the member profile and maintaining it while the member is online. Interactions are performed when the member is active in the system through a web-based interface. The user agents carry out the specific operations accordingly; all the operations have different logic involved. The user agents' life cycle depends on completeness or satisfaction of their task objectives. In other words, the user agents are created for a purpose and once the purpose is achieved, the agents are terminated i.e. when the member is offline.

Ontology agents are responsible for maintaining the software engineering ontology and for serving the right pieces of project information to members' queries. This agent is responsible for updating the software engineering project information into the software engineering ontology and forwarding the update to user agents for updating the right members.

In a software engineering project, the project information over a period of time needs to be modified to reflect project development, changes in the software requirements or in the design process, in order to incorporate additional functionality to systems or to allow incremental improvement. Since changes are inevitable during project development, the software engineering ontology is continuously confronted with many versions of them. Thus, each version is taken care of, by each ontology agent. Such changes are maintained by ontology agents. The ontology agents facilitate updating tasks and ensure reliability and consistency. Note that it does not change the original concepts and relationships in the ontology, rather instantiations of the software engineering ontology change conforming to the ontology change.

Recommender agent is responsible for recommending tentative solutions on project issue. The decision recommended by the agent is based on members in the team agreeing to vote, along with the reputation of each individual member involved in the project.

4.2 Ontological Support for Agent Communication

A major aim of our approach is to enable the integration process from syntactic interoperability to semantic interoperability. The FIPA standard organisation produced a comprehensive set of specifications for interoperable multi-agent systems [12, 13]. It defines the Agent Communication Language (ACL) which provides the syntax for interoperability enabling agents to manage ontologies. The project information in our project is represented as instances of the software engineering ontology defining concepts and their relations. The software engineering ontology is used as an agreement between agents.

The agent platform used i.e. JADE [14] offers a general support for ontologies. It is necessary that the system is not only agent based but also is support for ontologies. Agents convey project information according to the explicit software engineering ontology. We have chosen a compound tool called OWLBeans [15] that allows the use of software engineering ontology described in OWL DL. The software engineering ontology is used by ontology agents for performing their tasks in cooperation with other agents. OWLBeans aims to provide software engineering ontology management. Dedicated ontology agents acting as ontology servers are able to use and manage the complete software engineering ontology described in OWL and provide services to the other agents that need it. These ontology servers use the

Jena toolkit [16, 17] to load, maintain, and reasoning facilities. These ontology servers provide facilities to load maintain and reason the software engineering ontology. They also provide a set of simple actions for querying and manipulating the software engineering ontology. Additionally, proper authorisation mechanisms are taken into account. In particular, the underlying JADE security support has been leveraged to implement a certificate-based access control. Only authenticated and authorised members will be granted access to manage the software engineering ontology.

After some analysis of possible query languages, we have come to the decision of utilising OWL-QL [18], a formal query language targeting the Ontology Web Language (OWL). The reason is simply because the software engineering ontology is represented in OWL and semantic-aware agents use knowledge represented in the OWL software engineering ontology.

4.3 Member Profile Management

An important requirement that has guided the design of our approach has been the support for distributed multi-site team members. The management of experts and project information can provide a great advantage for new members or new teams to catch up with the current project. The team is distributed in multiple sites. Each team can exist and operate isolated sharing of the experts and the project information repositories. The multi-site distributed development nature provides the best conditions for sharing and retrieving project information but also entails some significant techniques in member profile evaluation.

Every member in the teams involved in a given software engineering project has a right to vote for proposed solutions. Everybody's vote is worth points. Below is a list of requirements:

1)   A member can work on a project or multiple projects at the same time
2)   A member can work in a team or in multiple teams
3)   A member can work in different teams in different projects
4)   A project involves multiple teams and multiple members
5)   A member has a reputation value for a particular area or domain in a given project
6)   A member can have a different reputation value for a different area or domain in a different project
7)   The reputation value of the team member continues to increase if the team member votes for the chosen (or correct) solution and vice versa
8)   The reputation value of the team member decreases if the team member did not vote for the chosen (or correct) solution and vice versa.

The vote cast by each team member is mathematically weighted by the factor of which 'members who actually work on a task have the best understanding of that task'. In other words, if a member votes on an issue which arises within the area he/she is working on, presumably this falls within his/her area of expertise, then his/her vote carries more weight than that of a member who does not have expertise in the issue area, or who does not really work on it. It is assumed, for example, that the designers of a project who work on the project design, have expertise in project design, or know more than others do about this aspect of a project. Thus, if the project issue relates to project design, the votes of members in the design team carry more weight than others.

The reputation value of members may change with time. In other words, over a period of time and in a particular area or domain of expertise, the reputation value of a given member may increase or decrease or in some instances may remain the same. We need a means to account for this dynamic changes in the reputation value of a given user over a duration of time, when taking into account the vote cast by that user [19]. In order to the account for the dynamic nature of reputation, we propose to use the Markov Model. Additionally, using the Markov Model, we can determine what could be the most probable future reputation value of a given team member in the category of the issue at a time in which the decision has to be made. Additionally it may the case the past sequence of reputation values of a given user may exhibit a trend (upwards trend or downward trend or a steady trend) or seasonality pattern (periodical seasonality pattern or non-periodical seasonality pattern) or random noise. The Markov Model and its variants, provide a very good means in order to model the trend components or seasonality components or noise in a given reputation series, and predict the future reputation value while accounting for these components.

The adaptation of the Markov Model for modelling reputation is based on a finite state process. There a total of seven trust states, with one trust state denoting ignorance. We make use of the integers in the domain of {0, 6} to represent the trust states, with 0 denoting ignorance. The values from 0 to 6 represent the trust states. The past duration of time over which the reputation of a given user is to be modelled is termed as time space. The time space is divided into non-overlapping, mutually exclusive and non-intersecting finite number of slots of time, each of which is termed as time slot. In order to model the dynamic nature of reputation and in order to predict the future reputation value of a given user, its reputation value of over the sequence of the past time slots is determined, resulting in the reputation series.

Firstly, the Markov Model determines if the reputation series is a non-stationary reputation series. It does so by determining if the reputation series exhibits (a) seasonality components (b) trend components or (c) noise components. Broadly speaking, two scenarios arise here: (a) the given reputation series exhibits just one of the aforesaid components and (b) the reputation series exhibits more than one of the aforesaid mentioned components.

We would develop algorithms to handle both of these scenarios. In this case (when the reputation series exhibits one or more than one of the above components listed above), we would develop an algorithm that would be capable of modelling, the dynamic nature of reputation taking into account the corresponding components present in the reputation series. Additionally, the developed algorithm would be capable of predicting the future reputation value of the given user, taking into account the corresponding components present in the reputation series. Furthermore, we develop specialized algorithms in order to address the scenario where in the reputation series exhibits only one of the above mentioned components. The above process is used when the reputation series is basically a non-stationary reputation series. However in case the reputation series is stationary (which is rarely the case) [19], then the process of modelling the dynamic nature of reputation and the process of predicting the future reputation value of a given user is straightforward, relative to the process of modelling a non-stationary reputation series and subsequently predicting the future reputation value of a non-stationary reputation series. In such a case it is based on a discrete time state process and would be capable of modelling the stationary aspect of the reputation series and subsequently predicting the future reputation values well.

4.4 Reputation Based Recommendation Technique

Whenever issues arise, the recommender agent sends a message to every involved member through user agents requesting for an opinion. Subsequently, the agent gathers and stores the possible solutions. Of all the possible solutions, one solution is recommended for one of the proposed solutions. Each vote is weighted by the expertise of the member casting it in the area of the problem. The reputation value of individual member who votes is weighted.

Assume that a weighting value for member who his/her expertise is not in the area of the issue is 0.2 and a weighting value for member who his/her expertise is in the area of the issue is 0.8. For example, three possible solutions named A, B, and C on the project design issue. Let's assume that for the design area, the reputation point of a member who votes for solution A is 1. Since this member's expertise is in the area of design, which is the area where the issue is raised, (i.e. project design), this member's vote would have a value of 0.8 (1 multiplied by 0.8). If the reputation value of a member who votes for solution B is 2, then this member's vote would have of value 0.4 (2 multiplied by 0.2) because this member's expertise area is requirement (this member is an analyst) while the issue is about project design. Similarly, if the reputation value of a member who votes for solution C is 1, then this member's vote would have a value of 0.2 (1 multiplied by 0.2) because this member's expertise area is construction (this member is a programmer) while the issue is about project design.

For a particular issue, whichever solution has the highest vote value will be recommended. Therefore, from this example, solution A that has the highest vote value is recommended as a final solution. Once a solution has been recommended and finalised, the project data in the software engineering ontology will be updated along the lines of the recommended solution. The system advises all team members of the final decision and records the event. The users' reputation points are also updated for future use.

5 System Implementation

A first prototype of the multi-agent based recommender system is under development using JADE. JADE (Java Agent Development framework) is a software framework that simplifies the implementation of agent applications in compliance with the FIPA specifications for interoperable intelligent multi-agent systems [20, 21].

Given the distributed nature of the JADE based multi-agent systems, teams can be distributed through multiple sites connecting usually through the internet and situated in different parts of the globe. Each system platform can be distributed on different computation nodes and it is connected to a web server where agents reside allowing direct interactions with the members. In the web server, there might be more than one user agent and ontology agent but there is a unique recommender agent. In order to cope with performance due to the number of the members to be managed, each user agent is assigned to each member. For reliability reasons, ontology agents manage versioning of the software engineering ontology. It is necessary to develop an ontology server that maintains, uploads and versions of the ontologies but can work in conjunction with the multi-agent system. This requires an extension of an existing single agent ontology server.

6 Conclusion

In this paper, we have defined the multi-agent based collaboration architecture and analysed the roles of the three types of software agents within the architecture. We have also defined the nature and mechanisms of the interaction between the software agents and ontology within collaboration architecture. The reputation based recommendation technique is determined. A first prototype system is under development and evaluation.

References

1. Wongthongtham, P., et al.: Ontology-based multi-site software development methodology and tools. Journal of Systems Architecture 52(11), 640–653 (2006)
2. Wongthongtham, P.: A methodology for multi-site distributed software development. In: School of Information Systems. 2006, Curtin University of Technology: Perth (2006)
3. Resnick, P., et al.: GroupLens: An open architecture for collaborative filtering of netnews. In: Computer Supported Cooperative Work, Chapel Hill (1994)
4. Pazzani, M.,, Billsus, D.: Adaptive Web Site Agents. Autonomous Agents and Multi-agent Systems 5, 205–218 (2002)
5. Mari, M., Poggi, A., Turci, P.: Ontology-Based Remote Collaboration for the Development of Software System. In: the 2nd Italian Semantic Web Workshop (SWAP 2005), University of Trento, Trento, Italy (2005)
6. McDonald, D.W.: Evaluating expertise recommendations. In: International ACM SIGGROUP Conference on Supporting Group Work, Boulder, CO (2001)
7. Ishikawa, T., Matsuda, H., Takase, H.: Agent Supported Collaborative Learning Using Community Web Software. In: International Conference on Computers in Education, Auckland, New Zealand (2002)
8. Liu, X., et al.: I-MINDS: An Application of Multi-agent System Intelligence to On-line Education. In: IEEE International Conference on Systems, IEEE, Washington DC, USA (2003)
9. Bourque, P.: SWEBOK Guide Call for Reviewers (2003) [cited 29 May 2003]; Available from: http://serl.cs.colorado.edu/ serl/seworld/database/3552.html
10. Henderson-Sellers, B., Giorgini, P.: Agent-Oriented Methodologies 2005, p. 413. Idea Group Publishing (2005)
11. Zhang, Z., Zhang, C.: Agent-Based Hybrid Intelligent Systems: An Agent-Based Framework for Complex Problem Solving. Springer, Germany (2004)
12. FIPA Specifications. 2005 [cited; Available from: http://www.fipa.org
13. Bellifemine, F., Poggi, A., Rimassa, G.: Developing multi-agent systems with a FIPA-compliant agent framework. Software Practice and Experience 31, 103–128 (2001)

14. JADE home page. [cited; Available from: http://jade.tilab.com
15. Bergenti, F., et al.: An Ontology Support for Semantic Aware Agents. In: Seventh International Bi-Conference Workshop on Agent-Oriented Information Systems (AOIS-2005 @AAMAS). 2005. Utrecht, The Natherlands (2005)
16. Carroll, J.J., et al.: Jena: Implementing the Semantic Web Recommendations, Digital Media Systems Laboratory, HP Laboratories Bristol (2004)
17. McCarthy, P.: Introduction to Jena: use RDF models in your Java applications with the Jena Semantic Web Framework, SmartStream Technologies, IBM developerWorks (2004)
18. Fikes, R., Hayes, P., Horrocks, I.: OWL-QL - A Language for Deductive Query Answering on the Semantic Web. in Technical Report KSL-03-14, Knowledge Systems Lab, Stanford University: CA, USA
19. Chang, E., Dillon, T., Hussain, F.K.: Trust and Reputation for Service Oriented Environment: Technologies For Building Business Intelligence And Consumer Confidence. John Wiley and Sons, Chichester (2006)
20. Bellifemine, F., JADE: Java Agent DEvelopment Framework, Telecom Italia Lab: Torino, Italy (2001)
21. Bellifemine, F., Poggi, A., Rimassa, G.: JADE: a FIPA2000 compliant agent development environment. In: The fifth International Conference on Autonomous Agents, Montreal, Quebec, Canada, ACM Press, New York (2001)

An Outlook on Semantic Business Process Mining and Monitoring

A.K. Alves de Medeiros[1], C. Pedrinaci[2], W.M.P. van der Aalst[1],
J. Domingue[2], M. Song[1], A. Rozinat[1], B. Norton[2], and L. Cabral[2]

[1] Eindhoven University of Technology, P.O. Box 513,
5600MB, Eindhoven, The Netherlands
{a.k.medeiros,w.m.p.v.d.aalst,m.s.song,a.rozinat}@tue.nl
[2] Knowledge Media Institute, The Open University, Milton Keynes, UK
{c.pedrinaci,j.b.domingue,b.j.norton,l.s.cabral}@open.ac.uk

Abstract. Semantic Business Process Management (SBPM) has been
proposed as an extension of BPM with Semantic Web and Semantic
Web Services (SWS) technologies in order to increase and enhance the
level of automation that can be achieved within the BPM life-cycle. In
a nutshell, SBPM is based on the extensive and exhaustive conceptualization of the BPM domain so as to support reasoning during business
processes modelling, composition, execution, and analysis, leading to important enhancements throughout the life-cycle of business processes. An
important step of the BPM life-cycle is the analysis of the processes deployed in companies. This analysis provides feedback about how these
processes are actually being executed (like common control-flow paths,
performance measures, detection of bottlenecks, alert to approaching
deadlines, auditing, etc). The use of semantic information can lead to
dramatic enhancements in the state-of-the-art in analysis techniques. In
this paper we present an outlook on the opportunities and challenges on
semantic business process mining and monitoring, thus paving the way
for the implementation of the next generation of BPM analysis tools.

1 Introduction

Nowadays many companies use information systems to support the execution of
their business processes. Examples of such information systems are ERP, CRM
or Workflow Management Systems. These information systems usually generate
events while executing business processes [9] and these events can be recorded
in logs (cf. Figure 1). The competitive world we live in requires companies to
adapt their processes in a faster pace. Therefore, continuous and insightful feedback on how business processes are actually being executed becomes essential.
Additionally, laws like the Sarbanes-Oxley Act force companies to show their
compliance to standards. In short, there is a need for good analysis tools that
can provide feedback information about how business process are actually being
executed based on the observed (or registered) behavior in event logs.

Business Process Management (BPM) systems aim at supporting the whole
life-cycle (design, configuration, execution and analysis) necessary to deploy and

R. Meersman, Z. Tari, P. Herrero et al. (Eds.): OTM 2007 Ws, Part II, LNCS 4806, pp. 1244–1255, 2007.
© Springer-Verlag Berlin Heidelberg 2007

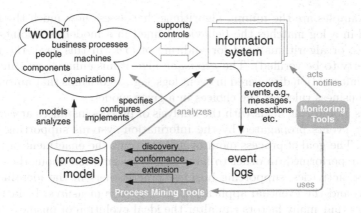

Fig. 1. Overview of process mining and monitoring

maintain business process in organizations. However, current approaches to BPM suffer from a lack of automation that would support a smooth transition between the business world and the IT world [14]. The difficulties for automating the transition between both worlds is due to a lack of machine processable semantics. Therefore, [14] proposes the creation of SBPM systems. Such systems combine Semantic Web and SWS technologies with BPM as a solution for overcoming these difficulties. In a nutshell, SBPM targets accessing the process space (as registered in event logs) of an enterprize at the *knowledge level* so as to support reasoning about business processes, process composition, process execution, etc. The driving force behind SBPM is the use of ontologies [12].

A key aspect of maintaining systems and the processes they support is the capability to analyze them. This analysis can be real-time and may eventually lead to some action or can just be used to inform the involved systems/people. When going SBPM, the main opportunity is that this analysis can be enhanced because it is based on concepts rather than syntax. This semantic perspective is captured by annotating the elements in the systems. So, two challenges arise in this aspect: (i) *how to make use of this semantic data*, and (ii) *how to mine this semantic information* and, consequently, help in the migration of current systems to SBPM environments. In this paper we show how process mining and monitoring techniques successfully utilize semantic data in SBPM systems.

Process mining techniques are especially suitable to analyze event logs. The analysis provided by current process mining techniques [2,4] can be seen as from three types: *discovery*, *conformance* and *extension* (cf. Figure 1). The techniques that focus on *discovery* mine information based on data in an event log only. This means that these techniques do not assume the existence of pre-defined models to describe aspect of processes in the organization. Examples are *control-flow mining* algorithms that extract a process model based on the dependency relations that can be inferred among the tasks in the log. The algorithms for *conformance* checking verify if logs follow *prescribed* behaviors and/or rules. Therefore, besides a log, such algorithms also receive as input a model (e.g., a Petri net or a set of rules) that captures the desired property or behavior to check. Examples are the mining algorithms that assess how much the behavior expressed in a log matches the behavior defined in a model and points out the differences, or algorithms used for auditing of logs (in this case, the model is the property to be verified). The *extension* algorithms enhance existing models based on information discovered in event logs, e.g., algorithms that automatically discover business rules for the choices in a given model.

Process monitoring deals with the analysis of process instances at *runtime* by processing events propagated by the information systems supporting business processes. The goal of process monitoring is to track the enactment of processes as they are performed, in order to have timely information about the evolution of business activities, supporting business practitioners in the identification of deviations and the eventual application of corrective measures. In fact, experience shows that many factors can alter the ideal evolution of business processes (e.g., human intervention, mechanical problems, meteorological adversities, etc) and the quick adoption of special measures can mitigate to an important extent the eventual consequences, thus reducing or even avoiding derived economical losses. The importance of process monitoring in BPM is widely acknowledged and in fact all the main vendors in this sector provide their own solution. Two kinds of monitoring are usually distinguished: (i) *active monitoring* which is concerned with "real time" propagation of relevant data concerning the enactment of business processes, such as the status or the execution time; and (ii) *passive monitoring* which delivers information about process instances upon request.

The ideas presented in this paper are currently being implemented in the context of the European project SUPER [1]. As stated in [1], SUPER "aims at providing a semantic-based and context-aware framework, based on Semantic Web Services technology that acquires, organizes, shares and uses the knowledge embedded in business processes within existing IT systems and software, and within employees' heads, in order to make companies more adaptive". This semantic framework will support the four phases of the BPM life-cycle.

The remainder of this paper provides an outlook about semantic business process mining (Section 2) and monitoring (Section 3), discusses related work in the area of semantic analysis (Section 4), and presents the conclusion and future steps (Section 5).

2   Semantic Business Process Mining

The use of ontologies in SBPM yields two opportunities for process mining techniques. The first opportunity is to make use of the ontological annotations in logs/ models to develop more robust process mining techniques that analyze the logs/ models at the concept level. In this case, it is assumed that event logs and models indeed link to ontologies. The second opportunity is to use process mining techniques to discover or enhance ontologies based on the data in event logs.

Developing Semantic Process Mining Techniques
As explained in Section 1, current process mining techniques focus on the *discovery* of models, the *conformance* between models and logs, and *extension* of

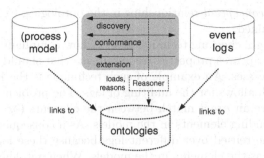

Fig. 2. Semantic process mining: basic elements

models based on information derived from event logs (cf. Figure 1). However, the analysis they support is purely syntactic. In other words, these mining techniques *are unable to reason over the concepts behind the labels in the log*, thus the actual semantics behind these labels remain in the head of the business analyst which has to interpret them. Leveraging process mining to the conceptual layer can enhance state-of-the-art techniques towards more advanced, adaptable and reusable solutions.

The basic elements to build semantic process mining tools are: *ontologies*, *references from elements in logs/models to concepts in ontologies*, and *ontology reasoners* (cf. Figure 2). *Ontologies* [12] define the set of shared concepts necessary for the analysis, and formalize their relationships and properties. The *references* associate meanings to labels (i.e., strings) in event logs and/or models by pointing to concepts defined in ontologies. The *reasoner* supports reasoning over the ontologies in order to derive new knowledge, e.g., subsumption, equivalence, etc. The use of ontologies and reasoners causes an immediate benefit to process mining techniques: the level of abstraction is raised from the syntactical level to the semantical level. The following paragraphs sketch some of the ways in which semantics can aid process mining (some of which have been implemented in ProM [3]).

The *discovery* techniques mine models based on event logs. Control-flow mining techniques are prominent in this perspective. These techniques focus on the discovery of a business model that capture the control-flow structure of the tasks in the log. Currently, these techniques mainly discover a *flat* model showing all the tasks encountered in the log, i.e., a single large model is shown without any hierarchy or structuring. However, if the tasks in these instances would link to concepts in ontologies, subsumption relations over these ontologies could be used to aggregate tasks and, therefore, mine *hierarchical* process models supporting different levels of abstraction. Other discovery techniques focus on organizational mining, which target the discovery of organizational related aspects in event logs. These algorithms are based on the tasks in the logs and the performers of these tasks. The main driving force here is the concept of task similarity. In a nutshell, tasks are considered to be similar based on their names, performers and context (neighboring tasks in the process instances). When these concepts are linked to tasks/performers in logs, more robust similarity criteria can be inferred that make use of the *conceptual* relationships in the ontologies. Consequently, better models can be mined.

The *conformance* checking techniques verify how compliant a model and a log are. This model captures properties/requirements that should be fulfilled by the execution of processes. An example of such technique is the LTL Conformance Checker [2] which allows for the auditing of logs. The problem here is that these techniques require an *exact* match between the elements (or strings) in the log and the corresponding elements in the models. As a consequence, many defined models cannot be reused over different logs because these logs do not contain the same strings as the elements in the models. When ontologies are used, these models can be defined over *concepts* and, as far as the elements in different logs link to the same concepts (or super/sub concepts of these concepts), the conformance can be assessed without requiring any modification of the models or the logs.

The *extension* techniques enhance models based on information mined from event logs. Like the conformance checking techniques, the enhancements are only possible with an exact match between elements in models and logs. Thus, the use of ontologies would bring this match to the concept level and, therefore, models could also be extended based on different logs.

As mentioned before, several of these ideas are currently being implemented as semantic plug-ins in the ProM tool. Actually, the *Semantic LTL Checker* analysis plug-in is already publicly available [1]. This plug-in extends the original LTL Checker [2] by adding the option to provide concepts as input to the parameters of LTL formulae. All the semantic plug-ins developed in ProM are based on the following concrete formats for the basic building blocks (cf. Figure 2): (i) *event logs* are in the SA-MXML file format, which is a semantically annotated version of the MXML format already used by ProM [2]; (ii) *ontologies* are defined in WSML [10]; and (iii) the *WSML 2 Reasoner Framework* [3] is used to perform all the necessary reasoning over the ontologies.

Using Process Mining to Discover/Enhance Ontologies

So far we have focussed on using semantics to enhance process mining techniques. However, there are opportunities in the other direction too because process mining techniques can be used to (i) discover or enhance ontologies and (ii) automatically infer concepts to elements that are not semantically annotated but that belong to partially annotated logs/models. When deploying SBPM systems, a core requirement is that (some of) the elements in the configured models should link to concepts in ontologies because that is how the semantic perspective is embedded in such systems. Therefore, if companies want to go in this direction, they need to add these semantic annotations to their systems. Here,

---

[1] This plug-in can be downloaded together with the nightly build for the ProM tool at http://ga1717.tm.tue.nl/dev/prom/nightly/. It can be started by clicking the menu option "Analysis → Semantic LTL Checker".

[2] The schema for the SA-MXML format is available at http://is.tm.tue.nl/research/processmining/SAMXML.xsd

[3] This framework is publicly available at http://tools.deri.org/

three options are possible. The first one is to *manually* (i) create all the necessary ontologies and (ii) annotate the necessary elements in the SBPM systems. The second option is to use tools to (semi-)*automatically* discover ontologies based on the elements in event logs. Note that, if necessary, these mined ontologies can be manually improved. The third option is a combination of the previous two in which models/logs are partially annotated by a person and mining tools are used to discover the other missing annotations for the remaining elements in logs/models. Discovery and extension process mining techniques can play a role in the last two options.

Basically, three opportunities exist to extract semantics from logs. First, process mining techniques can be created to derive relationships between concepts for activities and performers. This scenario assumes that the subsumption relationships for the concepts in an ontology have not been defined. A task is usually only executed by a group of performers who have certain properties (e.g. organizational units, skills) for a given process, and these properties can be expressed by the concepts linked to these performers. This way subsumption relationships can be discovered from event logs that contain semantic information. Second, if the log is partially annotated then mining techniques can be developed to automatically annotate the tasks and/or performers that do not link to any concepts. Third, if there are no semantic annotations, concepts that describe tasks or performers can be discovered from process logs by applying the existing mining techniques to discover these concepts/ontologies. The mined organizational structures such as roles and teams can be good candidates for concepts. Note that a group of performers executing a same task might belong to the same role and have the same role concept. Performers involved in the same instances might have the same team concept.

3   Semantic Business Process Monitoring

Reaching the level of automation demanded by current businesses requires reasoning over the knowledge gained by applying mining techniques combined with pre-existing contextual domain knowledge about some specific business process. We refer as *Semantic Business Process Monitoring* to the enhancement of Business Process Monitoring with formal semantic descriptions to achieve this. We propose a 5-phases approach, *Observe - Evaluate - Detect - Diagnose - Resolve*, structured around an extensive use of ontologies as the core means for defining formal conceptualizations, and Problem-Solving Methods (PSM) as composable SWS encapsulating the expertise of the monitoring tool [6,20].

Figure 3 depicts our approach to Semantic Business Process Monitoring. The process starts with the *Observe* phase, which is in charge of gathering information populated by the IT infrastructure. The *Evaluate* phase uses this information for computing process metrics such as the execution time, the number of failures, etc. The *Detect* phase follows and uses previously computed metrics and monitoring data in order to detect or predict process deviations and special situations one might want to track. Finally, once a process deviation has been identified, the

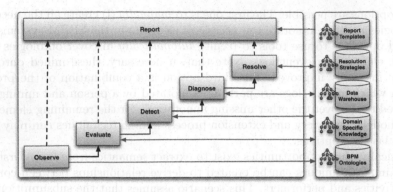

Fig. 3. Phases of Semantic Business Process Monitoring

Diagnose phase is in charge of determining the cause which can eventually be used during the *Resolve* step for defining and applying corrective actions. In parallel, at any time, we have to present information to the user about the overall monitoring process. Each of these monitoring phases, but in particular Detection, Diagnosis, and Resolution, present interesting challenges that have to be addressed and where knowledge-based techniques can help to improve the current state-of-the-art. We shall next identify the main opportunities that arise and depict the approach we envision for Semantic Business Process Monitoring.

Observe. The Observe phase is concerned with obtaining *monitoring* information and lifting it into a semantic form. This phase covers the so-called Extract-Transform-Load step which requires integrating a large amount of disparate information coming from several distributed and heterogeneous systems. Ontologies [12] are therefore a particularly well-suited candidate for supporting this task. An initial version of such an ontology has been defined in [18] based on the MXML format defined within the ProM framework [3]. Once in an ontological form, the monitoring information supports navigation, manipulation, and querying at the knowledge level, which is closer to human understanding and can potentially lead to important improvements in the user interface. In fact, in a recent report by Gartner [19] metadata management is presented as the most important capability that Business Intelligence tools should integrate. Ontologies are therefore a key enabling technology for achieving this. Additionally, semantic monitoring data is amenable to automated reasoning thus enabling the application of knowledge-based technologies as described next. Among the possibilities brought, consistency checking can be applied in this phase for detecting anomalies in the monitoring data itself thus reducing the noise for subsequent analysis and potentially enhancing quality of the analysis results.

Evaluate. This phase is in charge of the timely computation of process metrics, such as the execution time or the number of failures. We can distinguish between generic metrics that can be computed for every process, and domain-specific metrics [8]. To support business practitioners, we envision the definition of domain-specific metrics using a metric ontology, and the capability for users to define SWS that can be invoked by platforms like the IRS-III [6] to perform the metric computation. In a somewhat recursive way, we envisage *formalizing the analysis results themselves*. This provides independence with respect to the engines or algorithms utilized for performing the calculations, and supports a semantically enhanced view over the results. More importantly, an ontological definition of analysis results, enhances the overall body of knowledge for supporting further reasoning. In fact, it is quite usual that taking a decision requires performing and correlating diverse analysis, e.g., by combining the processes that did not perform well, with the resources involved in them, one could identify the bottlenecks. Formalizing the results enables reasoning over the computationally expensive analysis results within runtime monitoring tasks, as well as it allows for automatically combining them in order to perform more complex evaluations. In this sense we envision the use of SWS technologies for supporting the definition of analysis processes as the orchestration of different analysis techniques.

Detect. The Detect phase is in charge of identifying or predicting deviations with respect to the expected behavior of a process. The simplest approach is based on the definition of thresholds with respect to certain metrics. More complex solutions can be applied by approaching detection as a classification problem [8]. Our approach can support the seamless application of knowledge-based algorithms, e.g., classification PSMs [20], the enhancement of existing algorithms with semantic information, or even the runtime adaptation of the detection process. It is known that selecting the appropriate algorithm to apply given the task at hand is particularly important [5,8]. Having an extensive conceptualization of the BPM domain can indeed be particularly beneficial in order to select the presumably most suitable algorithm. This can be achieved by performing dynamic selection of SWS implementing some algorithm on the basis of the characteristics of the domain. For example, knowing the kind of process analyzed, e.g., shipping process, we can identify the typical or more relevant deviations, e.g., deadline exceeded, and select the algorithm accordingly. Additional advantages can be gained if relations between metrics and domain data, as well as mining results are modelled, allowing the system to overcome the lack of information earlier in the execution of the process. Finally, contextual knowledge can also strengthen existing algorithms like data mining approaches to symptoms detection [8] where this knowledge can play an important role supporting the enhancement of the algorithm with semantic feature selection.

Diagnose. Once any deviation has been detected or predicted, we have to *diagnose* the origin of the problem. In the BPM community, diagnosis often depends on the actual interpretation of the data by the user [8,16]. In order to do so the detection phase is often based on some structured approach that can be relatively easily understood by humans, e.g., decision trees. Diagnosis has been a popular topic in Artificial Intelligence, and has led to a quite exhaustive characterization of the task as well as to a wide range of implementations [5,20] which it would be desirable to benefit from. Knowledge-based methods have been applied to diagnosing automated systems (where some behavioral model typically exists), as well as to the diseases (where this kind of model is typically missing). It is therefore safe to assume that we can make use of the wealth of research on diagnosis for Semantic Business Process Monitoring. It is worth noting in this respect that a close integration between monitoring and mining can allow us to reuse mined process models for informing the diagnosis algorithm. This can be of great advantage when no prescribed process model exists or when the prescribed model differs to an important extent from the actual mined model.

Resolve. The final phase is concerned with the design and application of a resolution strategy for addressing some previously diagnosed process deviation. Resolution is by far the most complex task within our approach and in fact little work besides ad-hoc exception handling or *undo and retry* has been done within the BPM community [13,16]. These approaches cannot cope with the wide range of deviations that can arise during the enactment of a process and fully automated handling of any process deviation is simply not realistic due to unforeseen situations affecting user-defined and process-specific conditions [16]. Hence, in a similar vein to [16] we contemplate the application of Case-Based Reasoning for retrieving, adapting, and applying resolution strategies in an attempt to deal with previously diagnosed deviations. Like in the previous phases, the resolution strategies will be defined as orchestrations of SWS, allowing users to specify their own strategies by reusing and combining problem-solving expertise over their domain specific terms. This approach is inline with that proposed by [15] that can in fact serve as a basis for defining general resolution templates. We expect however that the capability for executing PSMs and our extensive conceptualization of the BPM domain will enable the creation of more complex strategies. For instance, Organizational knowledge can support the escalation of tasks [22], Rescheduling based on Configuration Problem-Solving can allow adapting resource allocation, or even Planning and Scheduling using reusable and equivalent process fragments can support the implementation of process escalations by degrading the Quality of Service [22].

4   Related Work

The idea of using semantics to perform process analysis is not new [7,11,14,17,21]. In 2002, Casati et al. [7] introduced the *HPPM intelligent Process Data Warehouse (PDD)*, in which taxonomies are used to add semantics to process execution data and, therefore, support more business-like analysis for the provided reports. The work in [11] is a follow-up of the work in [7]. It presents a complete architecture for the analysis, prediction, monitoring, control and optimization of process executions in Business Process Management Systems (BPMSs). This set of tools suite is called *Business Process Intelligence (BPI)*. The main difference of these two approaches to ours is that (i) taxonomies are used to capture the semantic aspects (in our case, ontologies are used), and (ii) these taxonomies are flat (i.e., no subsumption relations between concepts are supported). Hepp et al. [14] proposes merging Semantic Web, Semantic Web Services , and Business Process Management (BPM) techniques to build Semantic BPM systems. This visionary paper pinpoints the role of ontologies (and reasoners) while performing semantic analysis. However, the authors do not elaborate on the opportunities and challenges for semantic process mining and monitoring. The works by Sell et al. [21] and O'Riain et al. [17] are related to ours because the authors also use ontologies to provide for the semantic analysis of systems. The main difference is the kind of supported analysis, since their work can be seen as the extension of OLAP tools with semantics. The work in [17] shows how to use semantics to enhance the business analysis function of detecting the core business of companies. This analysis is based on the so-called Q10 forms. Our paper is the first one to provide an outlook on semantic process mining and monitoring techniques.

5   Conclusions and Future Work

This paper has presented several directions for the development of semantic process mining and monitoring tools. These tools can be used to analyze SBPM systems. The main opportunity provided by such systems is the *link between the generated events* (necessary for analysis) *and the actual concepts they represent*. This link is achieved by annotating the elements (models, events etc) in SBPM systems with concepts in ontologies. However, this same opportunity also raises two challenges. The first one is *how to make use of this semantic perspective* in process mining and monitoring tools. For the development of semantic process mining tools, we have proposed a framework composed of three building blocks (annotated event logs, ontologies and reasoners) and have discussed different ways in which techniques aiming at the discovery, conformance or extension perspectives could go semantic. For the monitoring tools, we have explained a five-phase approach (observe, evaluate, detect, diagnose and resolve) in which knowledge-based techniques play an essential role. The second challenge is *how to mine the semantic information* and, therefore, help in the migration of current information systems to SBPM environments. Here we have illustrate how process mining techniques could use events relating to tasks and performers to (i) automatically discover or enhance ontologies, and (ii) help in the semantic annotation of the elements in information systems.

As indicated throughout the paper, some of the presented ideas have already been implemented in the context of the SUPER European project. In fact, our future work will proceed in this direction (the development of further ideas in the SBPM environment defined in SUPER).

Acknowledgements. This research is supported by the European project SUPER [1].

References

1. European Project SUPER - Semantics Utilised for Process Management withing and between Enterprises. http://www.ip-super.org/
2. van der Aalst, W.M.P., de Beer, H.T., van Dongen, B.F.: Process Mining and Verification of Properties: An Approach Based on Temporal Logic. In: Meersman, R., Tari, Z. (eds.) On the Move to Meaningful Internet Systems 2005: CoopIS, DOA, and ODBASE. LNCS, vol. 3760, pp. 130–147. Springer, Heidelberg (2005)
3. van der Aalst, W.M.P., van Dongen, B.F., Günther, C.W., Mans, R.S., Alves de Medeiros, A.K., Rozinat, A., Rubin, V., Song, M., Verbeek, H.M.W., Weijters, A.J.M.M.: ProM 4.0: Comprehensive Support for Real Process Analysis. In: Kleijn, J., Yakovlev, A. (eds.) Application and Theory of Petri Nets and Other Models of Concurrency (ICATPN 2007). LNCS, vol. 4546, pp. 484–494. Springer, Heidelberg (2007)
4. van der Aalst, W.M.P., van Dongen, B.F., Herbst, J., Maruster, L., Schimm, G., Weijters, A.J.M.M.: Workflow Mining: A Survey of Issues and Approaches. Data and Knowledge Engineering 47(2), 237–267 (2003)
5. Benjamins, R.: Problem-solving methods for diagnosis and their role in knowledge acquisition. Int. Journal of Expert Systems: Research & Applications 8(2), 93–120 (1995)
6. Cabral, L., Domingue, J., Galizia, S., Gugliotta, A., Tanasescu, V., Pedrinaci, C., Norton, B.: IRS-III: A Broker for Semantic Web Services based Applications. In: Cruz, I., Decker, S., Allemang, D., Preist, C., Schwabe, D., Mika, P., Uschold, M., Aroyo, L. (eds.) ISWC 2006. LNCS, vol. 4273, pp. 201–214. Springer, Heidelberg (2006)
7. Casati, F., Shan, M.-C.: Semantic Analysis of Business Process Executions. In: Jensen, C.S., Jeffery, K.G., Pokorný, J., Šaltenis, S., Bertino, E., Böhm, K., Jarke, M. (eds.) EDBT 2002. LNCS, vol. 2287, pp. 287–296. Springer, Heidelberg (2002)
8. Castellanos, M., Casati, F., Dayal, U., Shan, M.-C.: A Comprehensive and Automated Approach to Intelligent Business Processes Execution Analysis. Distributed and Parallel Databases 16(3), 239–273 (2004)
9. Dumas, M., van der Aalst, W.M.P., ter Hofstede, A.H.M.: Process-Aware Information Systems: Bridging People and Software through Process Technology. Wiley & Sons, Chichester (2005)
10. Fensel, D., Lausen, H., Polleres, A., de Bruijn, J., Stollberg, M., Roman, D., Domingue, J.: Enabling Semantic Web Services: The Web Service Modeling Ontology. Springer, Heidelberg (2007)
11. Grigori, D., Casati, F., Castellanos, M., Dayal, U., Sayal, M., Shan, M.-C.: Business Process Intelligence. Computers in Industry 53(3), 321–343 (2004)
12. Gruber, T.R.: A translation approach to portable ontology specifications. Knowledge Acquisition 5(2), 199–220 (1993)
13. Hagen, C., Alonso, G.: Exception handling in workflow management systems. IEEE Transactions on Software Engineering 26(10), 943–958 (2000)
14. Hepp, M., Leymann, F., Domingue, J., Wahler, A., Fensel, D.: Semantic business process management: A vision towards using semantic web services for business process management. In: ICEBE, pp. 535–540 (2005)
15. Klein, M., Dellarocas, C.: A Knowledge-based Approach to Handling Exceptions inWorkflow Systems. Comput. Supported Coop. Work 9(3-4), 399–412 (2000)
16. Luo, Z., Sheth, A., Kochut, K., Miller, J.: Exception Handling in Workflow Systems. Applied Intelligence 13(2), 125–147 (2000)

17. O'Riain, S., Spyns, P.: Enhancing the Business Analysis Function with Semantics. In: Meersman, R., Tari, Z. (eds.) On the Move to Meaningful Internet Systems 2006: CoopIS, DOA, GADA, and ODBASE. LNCS, vol. 4275, pp. 818–835. Springer, Heidelberg (2006)
18. Pedrinaci, C., Domingue, J.: Towards an Ontology for Process Monitoring and Mining. In: Semantic Business Process and Product Lifecycle Management (SBPM 2007), Innsbruck, Austria (June 2007)
19. Schlegel, K., Hostmann, B., Bitterer, A.: Magic Quadrant for Business Intelligence Platforms, 1Q07. Gartner RAS Core Research Note G00145507, Gartner (January 2007)
20. Schreiber, G., Akkermans, H., Anjewierden, A., de Hoog, R., Shadbolt, N., van de Velde, W., Wielinga, B.: Knowledge Engineering and Management: The CommonKADS Methodology. MIT Press, Cambridge (1999)
21. Sell, D., Cabral, L., Motta, E., Domingue, J., Pacheco, R.: Adding Semantics to Business Intelligence. In: DEXA Workshops, pp. 543–547. IEEE Computer Society Press, Los Alamitos (2005)
22. van der Aalst, W.M.P., Rosemann, M., Dumas, M.: Deadline-Based Escalation in Process-Aware Information Systems. Decision Support Systems 43(2), 492–511 (2007)

A Role and Attribute Based Access Control System Using Semantic Web Technologies*

Lorenzo Cirio[1], Isabel F. Cruz[1], and Roberto Tamassia[2]

[1] University of Illinois at Chicago
{lcirio,ifc}@cs.uic.edu
[2] Brown University
rt@cs.brown.edu

Abstract. We show how Semantic Web technologies can be used to build an access control system. We follow the role-based access control approach (RBAC) and extend it with contextual attributes. Our approach provides for the dynamic association of roles with users. A Description Logic (DL) reasoner is used to classify both users and resources, and verify the consistency of the access control policies. We mitigate the limited expressive power of the DL formalism by refining the output of the DL reasoner with SPARQL queries. Finally, we provide a proof-of-concept implementation of the system written in Java.™

1 Introduction

In this paper, we present an access control system for context-aware environments designed and built using Semantic Web technologies. We adopt the *role based access control* (RBAC) model [12]. In an RBAC system, roles are assigned to users statically using a procedure performed by the security administrators. Although revisions are possible, they are not supposed to be frequent nor to be done at run time [12]. This approach can be restrictive in various situations, including in mobile situations that are common in *context-aware* or *pervasive computing*, where the identity of the users is not known in advance. For example, we may wish to grant a role to a visitor, or to a first time client, without permanently registering the client's data.

Recently, various proposals have appeared in the literature to extend RBAC with concepts like team membership, users' tasks, organizational hierarchy, and contextual information (such as position and time) [16]. Each of these approaches captures an interesting (yet partial) aspect to be directly integrated in the access control system.

To introduce flexibility into the procedure of role assignment, we borrow ideas from attribute-based access control systems (ABAC) [1,19]. In an ABAC system, permissions are associated with a set of rules expressed on measurable parameters and are granted to users who can prove compliance with these rules.

---

* Work supported in part by NSF grants IIS-0326284, IIS-0324846, IIS-0513553, IIS-0713403, and OCI-0724806.

R. Meersman, Z. Tari, P. Herrero et al. (Eds.): OTM 2007 Ws, Part II, LNCS 4806, pp. 1256–1266, 2007.

However, unlike ABAC, we do not directly associate the permissions with the attributes, but instead borrow the concept of role from RBAC, which we use as an intermediate structure between attributes and permissions.

In an RBAC system, the first interaction between the user and the system is an identification procedure, followed by the retrieval of the roles leased to the user from a database. We replace this phase with a handshake procedure in which the roles that can be claimed by the user are determined on the basis of the provided attributes. These roles are then enabled as usual in the activation phase.

Using this approach, we can move from an authentication system based just on identity to one that takes into account attribute values. For example, in the case of our visitor, a credential issued by a trusted third party or GPS reading can supply the value for an attribute that will give access to a particular role.

We summarize the contributions of our work as follows:

Access model. By combining RBAC, which supports static roles, with ABAC, we introduce a new access control model that enables dynamic assignments of roles to users in two different ways. First, privileges associated with resources are dynamically assigned depending on the attribute values of the resources. Second, attribute values associated with users determine the association of users with privileges. This kind of approach is especially suited to context-aware or pervasive computing.

Semantic Web technologies. We develop a framework that supports our security model using Semantic Web technologies and in particular OWL-DL. To this end, we produce a high level OWL-DL ontology that expresses the elements of a Role Based Access Control system, and build a domain specific ontology that captures the features of the application. Inferencing as supported by OWL-DL determines containment among classes, for example, how policies relate to resources or how instances are classified into their correct categories.

Expressiveness and design. We encountered several expressiveness problems and design choices. In particular, one limitation of the actual OWL classifier is the limited or missing support of concrete data types, which prevents stating conditions that involve data type comparison. To address this problem, we use SPARQL queries to express path properties. We use ontology design "best practices" [11] for developing complex structures in OWL.

Implementation. We provide a proof-of-concept implementation of the system, integrating standard elements with our application specific code. Our code is written in Java.™ The RDF data are processed using the Jena™ framework, which also provides a SPARQL query engine. In addition, our system does not depend on a specific DL-reasoner. Therefore, by leveraging existing technologies such as secure network infrastructures, RDF data processors, and DL-reasoners, our implementation can quickly adapt to new standards and new implementations as they become available.

This paper is organized as follows. In Section 2 we describe an OWL-DL ontology that expresses the elements of a Role Based Access Control system to define privileges and associate them with roles and resources in a domain ontology. In Section 3 we describe briefly the OWL design patterns we use to build the ontologies and the software architecture of our implementation. We discuss related work in Section 4 and show how our work advances the state of the art when compared to that work. Finally, we draw conclusions and point to future work in Section 5.

2   RBAC Modeling

In this section, we describe our method for creating an OWL-DL ontology that expresses the modeling abstractions of RBAC. We then show how we attach this ontology to a domain ontology and explain the tasks that are performed by the security administrator and by the DL classifier. We also discuss the limitations in expressiveness of OWL-DL, the use of SPARQL queries, and how we integrate SPARQL with OWL-DL.

In the development of the RBAC ontology, we have followed these principles: (1) the access control ontology should limit itself to expressing the modeling abstractions of RBAC; (2) no hypothesis is made about the domain knowledge, neither semantically nor syntactically—the resulting role ontology shall not depend on external factors like the type of organization or type of procedures nor on the structure of the domain ontology (which could be either monolithic or a layered system composed of different ontologies); and (3) we do not capture any workflow procedure inside the model so as to preserve both simplicity and generality.

In agreement with these hypotheses, we provide a tool that can express RBAC constructs, which are intended to be imported and used by the domain ontology, such as the library ontology of Figure 1. In this figure, we represent three ontologies: the RBAC ontology and two domain ontologies: *general* and *custom*. The former is an "off-the-shelf" (reused) ontology and the latter an ontology specifically developed for the access control system. Higher level abstractions or extensions can be implemented outside of the role ontology, either directly in the domain ontology or in additional, intermediate layers.

In providing an RBAC classification for all the classes of the domain ontologies, there are two actors at play: the security administrator and the DL classifier. We assume that the security administrator explicitly defines: (1) privileges (e.g., *ConsultInLibrary* = (*NonCirculatingItem*, *Read*)); (2) the association of privileges with classes in the domain ontology (e.g., *ConsultInLibrary* with *LibraryCardHolder*). The DL classifier propagates the RBAC elements according to the axioms that were originally stated in the domain ontologies using any valid inference, for instance, inheritance (e.g., every subclass of role *Student* will be classified as *Role* and will enjoy the privileges of *Student*).

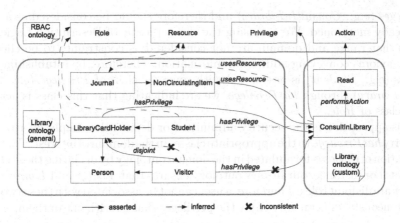

Fig. 1. Relationships between the domain ontologies and the RBAC ontology

The RBAC ontology has the following four classes:

Action. This is a partial, or self-standing, class that represents an action that can be performed by a user on a resource. It is intended to be the super class of the actual actions that can be performed in the system (e.g., *Read*).

Resource. This is a defined class, representing the authorization objects. The DL classifier will place under *Resource* all the classes that match the condition

$$\exists subject\_to.Privilege \sqcap \neg \{Action, Role, Privilege\}$$

that is, every instance that has relationship *subject_to* with an entity of type *Privilege* and that is not an *Action*, a *Role*, or a *Privilege*. Using the classifier, we can identify all the objects that have been treated like a *Resource* in the domain ontology. In our example of Figure 1, this is the case for *Journal* and *NonCirculatingItem*.

Privilege. This is a partial (self standing) class representing a pair (a, r) with $a \in Action$, $r \in Resource$. The one-to-one association is imposed by a cardinality constraint. In our example, privilege *ConsultInLibrary*, represents the pair *(NonCirculatingItem, Read)*. By using inheritance between *Journal* and *NonCirculatingItem*, the DL classifier will also infer the privilege *(Journal, Read)*.

Role. This is a defined class. The DL classifier will place under *Role* all the classes that match the condition $\exists hasPrivilege.Privilege$. We then expect that any class that is declared to have a privilege in the domain ontology to be classified as *Role*, as is the case with *Student* and *LibraryCardHolder*.

The RBAC ontology has the following properties:

hasPrivilege $\subset$ *Role* $\times$ *Privilege*. It is a many-to-many association of roles and privileges. Figure 1 shows how the property is imported into a domain ontology. It is used to indicate that a specific class in the domain (e.g., *Student*) has some privileges (e.g., *ConsultInLibrary*) and therefore should be considered as a *Role*, as already mentioned. We are using the characteristic of Description Logic that states that range and domain are not considered as constraints to be checked, but as axioms conveying additional information. Therefore, by establishing that a class (e.g., *Student*) is associated with some subclass of *Privilege* (e.g., *ConsultInLibrary*) through *hasPrivilege*, we are indicating that this class is actually a subclass of *Role*.

This approach lets the security administrator define roles simply by attaching property *hasPrivilege* to the appropriate classes in the domain of interest. The static separation of duties is formulated in the domain ontology by declaring those classes disjoint. For instance, our library ontology ensures that *Visitor* and *LibraryCard-Holder* are disjoint roles. Therefore, a user cannot be associated with these two roles simultaneously and can only enjoy the privileges that go with one of them.

notTogetherWith $\subset$ *Role* $\times$ *Role*. It is a many-to-many association of roles with roles, used to express the dynamic separation of duties. Therefore if *RoleA* is active, the system will refuse to activate any instance of a subclass of *Role* that is in a *notTogetherWith* relationship with *RoleA*. The property is declared to be symmetric, therefore the DL classifier will compute the symmetric closure of the relation, ensuring that two roles are dynamically separated even when just one of them is declared incompatible with the other.

performsAction : *Privilege* $\rightarrow$ *Action*. It is a total function that associates each privilege with the action it allows to perform. Together with *usesResource*, this construct is used to ensure that each privilege is a pair (a, r) with $a \in$ *Action* and $r \in$ *Resource*.

usesResource : *Privilege* $\rightarrow$ *Resource*. It is a total function associating to each privilege the resource on which it operates. Together with *performsAction*, this construct is used to ensure that each privilege is a pair (a, r) with $a \in$ *Action*, $r \in$ *Resource*. The inverse of *usesResource* is a property named *subject_to*. It is not a function because one resource can be managed by different privileges.

An issue that we must address is that of limitations of expressiveness that arise in OWL-DL. Consider the example where we want to state that a *Candidate* is a *Student* that prepares a *Thesis* and is advised by an *Adviser*:

$$Candidate \equiv Student \sqcap \exists prepare.Thesis \sqcap \exists is\_advised.Adviser$$

Likewise, we can state that an *Adviser* advises a *Candidate* and reviews a *Thesis*:

$$Faculty\_Member \sqcap \exists review.Thesis \sqcap \exists advise.Student$$

These statements only involve classes, therefore there is no way for us to specify that the actual instance of thesis prepared by a specific student is the same instance of thesis that is reviewed by the intended adviser [7]. Figure 2 shows how the model described is interpreted by the DL reasoner. The classifier will infer that the instance of *Adviser* has to be linked to an instance of *Thesis*, but cannot infer that this instance is the one linked to the student who is supervised by the adviser.

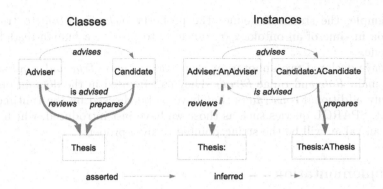

Fig. 2. An ontology showing a path property

A possible solution to this problem is to first apply a DL classification and then a set of Horn clauses [17]. However, our problem of verifying path properties between instances can be reduced to querying the ontology to verify if such a path exists. Queries expressed in SPARQL provide us with the solution to this problem. The language adopts a closed world assumption, following the usual convention for database systems. This is in contrast with the open world assumption adopted by Description Logic that does not imply that there exist particular instances of the classes involved that satisfy a particular constraint.

In particular, we are interested in ensuring that some constraints expressed at design time are met at run time, therefore the ASK construct fits our needs. The query of Figure 3 is the path query that solves the example previously discussed and illustrated in Figure 2. The WHERE section of the query is used to specify the graph pattern that we are looking to match. Alternative paths can be taken into account, using the keyword UNION to indicate them.

In expressing queries, there is information that could be interesting to refer to, such as the session identifier, which is known only at run time. To overcome this limitation, we introduce the following convention: the variable ?session can be used as a reference to the session identifier. At run time, the execution environment will take care of replacing the symbol with its actual value, wherever it appears.

We seamlessly integrate SPARQL queries with our OWL ontology using annotation properties, which carry meta-information about the ontology.

```
ASK WHERE {
    ACandidate
        prepares ?thesis;
        is_advised AnAdviser.
    AnAdviser reviews ?thesis
}
```

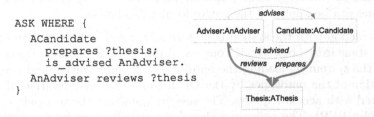

Fig. 3. Path property expressed in SPARQL

For example, the standard defines the property *owl:versionInfo* to trace the evolution in time of an ontology or *rdfs:label* to provide a human-readable tag for a node.

We have declared two annotation properties, *requiresTrue* and *requiresFalse*, with domain *Role* and range *String*. They can be used in the domain ontology to specify additional constraints that have to be checked by the runtime environment. SPARQL queries such as those we have just introduced, which return a Boolean value, will be the string pointed to by a property.

3   Implementation

In this section, we describe briefly our prototype implementation, focusing on OWL design patterns and software architecture. We use three OWL design patterns: value partition, linked list, and recursive composition, which are summarized below.

Value partition. This OWL design pattern addresses the problem of assigning values to attributes of a class and is recommended by the W3C Semantic Web Best Practices and Deployment Working Group [11].

Linked list. Inspired by Drummon *et al.* [6] and leveraging the reasoning capabilities of OWL, we have developed this OWL design pattern to represent a sequence of nodes connected by links.

Recursive composition. Adapting the composite design pattern in software engineering, we have developed this OWL design pattern to model a containment tree without using the subclassing mechanism. The navigation from child to parent is given by property *containedIn*, which is declared to be transitive. To avoid unexpected inferences, we add an axiom that limits the containment to instances of a parent class [14].

Our access control system is based on the following assumptions:

- The background knowledge used to make the access decisions is available as Semantic Web ontologies. Several techniques exist for converting relational databases and XML databases into ontologies (see, e.g., [4,20]).
- The organization has established a secure infrastructure to register, propagate, request and verify user attributes. This functionality is offered by network services that support single sign-on across domains and trusted user directories. See, for example, PERMIS [2] and Shibboleth [13].
- We use the ontology format also for the attributes. Again, conversion techniques can be applied if the native format is different.

The interaction with the system is performed through a Java API and is based on web standards. Logically, it consists of the following phases: (1) configuration of the system by loading the ontologies; (2) initialization and consistency verification of the ontologies by the DL reasoner; (3) instantiation of sessions associated with access requests. The session manager acts as a policy enforcement point(PEP). The policy decision point (PDP) is implemented with code that wraps the DL-reasoner.

The code is organized into classes associated with the elements of the RBAC ontology. Additional functionality is introduced for the purpose of managing the sessions. The library is completed by several helper classes dealing, for example, with cryptographic operations and the management of name spaces. We leverage the Jena™ library freely available from HP.

4   Related Work

Neumann and Strembeck have designed and implemented *xoRBAC*, a network service offering role based access control [10]. They also extend this service to take into account contextual information, which is fed into the system via software sensors whose output can enter into the formulation of the policies [15]. Users are authenticated using digital certificates in standard X.509 format. The RBAC information is stored in RDF-XML format, but the authors never mention the use of reasoners, therefore we assume that RDF is only used as a way to store data. The policies are written in a dedicated language that is later converted into an XOTcl (eXtended Object Tcl) script and executed by the system.

Kagal *et al.* propose *Rein*, a distributed framework based on ontologies to share and compose access control policies [8]. They reuse reference policies [9] and adapt them to their needs. While policies may be expressed in different ontology languages, their prototype uses the rule language N3 and the reasoner CWM. The authors motivate the choice of a rule based language over a DL-based language, with a higher expressive power. They note that this choice presents a couple of difficulties as it prevents classifying the different policies and detecting incompatibilities in their definitions.

Toninelli *et al.* describe a context-aware access control framework for pervasive computing [17]. They present a scenario where participants from different organizations attend a meeting. In their approach, users can safely exchange information and share resources, without knowing their identities beforehand. This method relies on sensors that capture the contextual information. Context is represented with a context ontology, which needs to be imported and is specialized from the ontologies modeling the domain of interest. Description Logic is used to classify the context models and discover their relationships (such as equivalence or generalization). The context models developed at design time need to be associated with the data from the sensors to form enforceable policies. This instantiation procedure depends on rules expressed in logic programs. As compared to our approach, they do not provide the decoupling capability that is achieved by having roles.

Di *et al.* [5] represent RBAC entirely in Description Logic. Thus, their access control model is entirely contained in a DL reasoner. They offer constructs to model the role hierarchy, the static and dynamic separation of duties and some form of cardinality constraints. They do not enrich the system with additional languages and are therefore limited to the constraints of open world reasoning. Our approach contrasts with this one because by adding another language, we are not limited to the expressiveness of a DL-reasoner.

KAoS is a rich component-based framework, for expressing, administrating and enforcing policies [18]. It is especially targeted to distributed computing environments as grid computing and Semantic Web Services. It offers a high level ontology that explicitly provides constructs to model processes and transactions. To offer this expressive power, the system integrates OWL-DL with the SWRL rule language. While our approach leverages existing technologies and can quickly adapt to new standards and new implementations as they become available, KAoS relies on the construction of many of the system components.

Damiani *et al.* propose *GEO-RBAC*, a formal framework for "spatially-aware" RBAC for location-based applications, where roles are activated based on the position of the user [3]. The framework uses spatial entities to model objects, user positions, and geographically bounded roles and further consider hierarchies to model permission, user, and activation inheritance. Properties concerning satisfiability, implications, and evaluation of their proposed classes of constraints are proved. While role-based, GEO-RBAC does not use Semantic Web technologies.

5   Conclusions and Future Work

We have shown how available Semantic Web technologies, namely OWL and Description Logic, can be used to build an access control system. To this end, we have developed a high level OWL-DL ontology that expresses the elements of a role based access control system and we have built a domain-specific ontology that captures the features of a sample scenario. Finally, we have joined these two artifacts to take into account attributes in the definition of the policies and in the access control decision. The use of attributes is twofold: to classify users into access control roles and to classify resources as access control objects.

We have used Description Logic to express inferences, for example, to detect the containment among classes that show how policies relate to resources. The limited expressive power of Description Logic has been mitigated by introducing SPARQL queries that can check additional constraints against an available knowledge base. Thanks to the annotation properties offered by OWL, we have embedded the SPARQL queries into the ontologies, integrating the policies in one entity. Our system prototype uses Java™and can be easily extended and integrated in a networked architecture to be offered as a service.

The forthcoming OWL 1.1 standard will allow us to simplify and add expressive power to our system. In the area of security, the effectiveness of rules to express policies is demonstrated by the many existing access control systems based on this mechanism. However, while rule-based languages are widely debated in the Semantic Web community, their integration with Description Logic is still an open issue. Therefore, it is interesting to investigate how supporting a rule language will change our framework, especially considering that rule languages have complementary strengths and weaknesses with respect to Description Logic. Finally, the task of the ontology developer will be made easier by improvements to the ontology debugger. In particular, it would be interesting to increase the precision in identifying the axioms that cause an inconsistency.

References

1. Al-Kahtani, M.A., Sandhu, R.S.: Induced role hierarchies with attribute-based RBAC. In: 8th ACM Symposium on Access Control Models and Technologies (SACMAT), pp. 142–148. ACM Press, New York (2003)

2. Chadwick, D.W., Otenko, A.: The PERMIS X.509 role based privilege management infrastructure. Future Generation Computer Systems 19(2), 277–289 (2003)

3. Damiani, M.L., Bertino, E., Catania, B., Perlasca, P.: GEO-RBAC: A spatially aware RBAC. ACM Trans. on Information and System Security 10(1), 2 (2007)

4. de Laborda, C.P., Conrad, S.: Bringing relational data into the Semantic Web using SPARQL and Relational.OWL. In: 3rd Int. Workshop on Semantic Web and Databases (SWDB), IEEE, Los Alamitos (2006)

5. Di, W., Jian, L., Yabo, D., Miaoliang, Z.: Using semantic web technologies to specify constraints of RBAC. In: 6th Int. Conf. on Parallel and Distributed Computing Applications and Technologies (PDCAT), pp. 543–545. IEEE, Los Alamitos (2005)

6. Drummond, N., Rector, A., Stevens, R., Moulton, G., Horridge, M., Wang, H.H., Seidenberg, J.: Putting OWL in order: Patterns for sequences in OWL. In: *OWL: Experiences and Directions (OWLED) ISWC Workshop* (2006)

7. Horrocks, I., Kutz, O., Sattler, U.: The even more irresistible SROIQ. In: 10th International Conference on Principles of Knowledge Representation and Reasoning (KR), pp. 57–67 (2006)

8. Kagal, L., Berners-Lee, T., Connolly, D., Weitzner, D.: Self-describing delegation networks for the Web. In: 7th IEEE Int. Workshop on Policies for Distributed Systems and Networks (POLICY), pp. 205–214. IEEE, Los Alamitos (2006)

9. Kagal, L., Berners-Lee, T., Connolly, D., Weitzner, D.J.: Using Semantic Web technologies for policy management on the Web. In: 21st National Conference on Artificial Intelligence (AAAI), AAAI Press (2006)

10. Neumann, G., Strembeck, M.: Design and implementation of a flexible RBAC-service in an object-oriented scripting language. In: 8th ACM Conference on Computer and Communications Security (CCS), pp. 58–67 (2001)

11. Rector, A.: Representing specified values in OWL: "value partitions" and "value sets". Note NOTE-swbp-specified-values-20050517, W3C (May 2005)

12. Sandhu, R.S., Coyne, E.J., Feinstein, H.L., Youman, C.E.: Role-based access control models. Computer 29(2), 38–47 (1996)

13. Scavo, T., Cantor, S.: Shibboleth Architecture, Technical Overview, Working Draft 02. Technical report, Internet2 Consortium (June 2005)

14. Seidenberg, J., Rector, A.L.: Representing transitive propagation in OWL. In: Embley, D.W., Olivé, A., Ram, S. (eds.) ER 2006. LNCS, vol. 4215, pp. 255–266. Springer, Heidelberg (2006)

15. Strembeck, M., Neumann, G.: An integrated approach to engineer and enforce context constraints in RBAC environments. ACM Trans. on Information and System Security 7(3), 392–427 (2004)

16. Tolone, W., Ahn, G.-J., Pai, T., Hong, S.-P.: Access control in collaborative systems. ACM Computing Surveys 37(1), 29–41 (2005)

17. Toninelli, A., Montanari, R., Kagal, L., Lassila, O.: A semantic context-aware access control framework for secure collaborations in pervasive computing environments. In: 5th International Semantic Web Conference, pp. 473–486 (2006)

18. Uszok, A., Bradshaw, J.M., Johnson, M., Jeffers, R., Tate, A., Dalton, J., Aitken, S.: KAoS policy management for semantic web services. IEEE Intelligent Systems 19(4), 32–41 (2004)
19. Wang, L., Wijesekera, D., Jajodia, S.: A logic-based framework for attribute based access control. In: ACM Workshop on Formal Methods in Security Engineering (FMSE), pp. 45–55. ACM Press, New York (2004)
20. Xiao, H., Cruz, I.F.: Integrating and Exchanging XML Data Using Ontologies. In: Spaccapietra, S., Aberer, K., Cudré-Mauroux, P. (eds.) Journal on Data Semantics VI. LNCS, vol. 4090, pp. 67–89. Springer, Heidelberg (2006)

Detecting Semantic Relations Between Nominals Using Support Vector Machines and Linguistic-Based Rules[*]

Isabel Segura-Bedmar, Doaa Samy, Jose L. Martínez-Fernández, and Paloma Martínez

Universidad Carlos III de Madrid, Computer Science Departament,
Avd. Universidad, 30, Leganes, 28911, Madrid, Spain
{isegura,dsamy,jlmferna,pmf}@inf.uc3m.es

Abstract. This paper describes the improvement of an automatic system for detecting semantic relations between nominals by the use of linguistically motivated knowledge combined with machine learning techniques. A previous version of the system using a Support Vector Machine classifier was evaluated in the 4th International Workshop on Semantic Evaluations, SEMEVAL [5]. The performance of the system improved significantly by the application of the linguistic based rules.

Keywords: computational semantics, semantic relations, classification, Support Vector Machines, Sequential Minimal Optimization.

1 Introduction

The enormous growth of Internet in the last decade has produced a growing interest in the automatic processing of written text, especially from the semantic perspective. Ontology development to support Semantic Web has lead to the improvement of techniques capable of managing underlying semantic relations between concepts. Detecting semantic relations is a key issue in automating the ontology learning process. Techniques to automatically locate these concepts and the relations between them have been the object of many studies. Following this research line, [3] identifies four techniques to automatically extract semantic relations from texts: pattern-based techniques [1], association rules [3], conceptual clustering and ontology learning from schemata [2]. In many cases, combined approaches are used to detect concepts and relations among them [6].

Following a linguistic perspective, studies on classification of semantic relations can be grouped in two major classes. The first focuses on "complex nominals", while the second studies the relation between nouns in more complex and longer constructions either at the phrase, clause or sentence level. The majority of approaches lie in the first group.

At the linguistic level, semantic relationships represent a number of challenges. They are highly ambiguous and context dependent, where information on the syntactic or the

---

[*] This work has been partially supported by the Regional Government of Madrid under the Research Network MAVIR (S-0505/TIC-0267).

R. Meersman, Z. Tari, P. Herrero et al. (Eds.): OTM 2007 Ws, Part II, LNCS 4806, pp. 1267–1273, 2007.
© Springer-Verlag Berlin Heidelberg 2007

lexical level are usually not enough to discover the underlying relations. These relations are not syntactically, but semantically governed.

2 The System

The system described in the present paper is a hybrid supervised system integrating different resources: machine learning algorithms (support vector machines) and language resources (tokenizers, POS taggers, WordNet as a lexical-semantic resource and linguistic-based rules).

Our scope only considers nominals where constituents of the relation are formally tagged as nouns. The nominals can occur either on the phrase, clause or sentence level. This fact constitutes the major challenge in this task since most of the previous studies limited their approaches to certain types of nominals, namely noun compounds, nominalizations and nouns on the phrase level.

2.1 The Data

The 4th International Workshops on Semantic Evaluation, SEMEVAL 2007[1], created a benchmark dataset and evaluation task, Classification of Semantic Relations between Nominals, enabling researches to compare their algorithms.

In this task, a nominal is a noun, excluding named entities, or base noun phrase,. A base noun phrase is a noun and its premodifiers (nouns, adjetives, determiners).

The dataset included seven semantic relatons: Cause-Effect, Instrument-User, Product-Producer, Origin-Entity, Purpose-Tool, Part-Whole, Content-Container.

For each relation, 140 training sentences, 70 testing sentences and a precise definition of the relation were provided by the organizers. Each sentence was manually annotated including the following information: the nominal boundaries (e.g., *Put <e1>tea</e1> in a <e2>heat-resistant jug</e2>*), the used query to search the sentence in the web (*e.g.*, "* *in a* *") and the Wordnet sense key for each nominal (*e.g.*, *WordNet(e1) = "tea%1:13:00::", WordNet(e2) = "jug%1:06:00::"*).

2.2 Support Vector Machines

The set of features used for the classification of semantic relations includes information from different levels: word tokens, POS tags, verb lemmas, semantic information from WordNet, etc.

Information regarding word tokens, POS tags and lemmas are automatically extracted using the using infrastructure for developing software components for Language Engineering GATE[2]. Lexical features are related to both the nominals and the contextual information where certain features keep track of the words occurring before, after and between the nominals. Contextual features include the two words before the first nominal, the two words after the second nominal and the word list in-between. Another feature, called *numinter,* indicating the number of words between nominals is also considered.

---

[1] http://nlp.cs.swarthmore.edu/semeval/
[2] http://gate.ac.uk/

Features regarding the POS level focus on the contextual information. The system extracts the POS tags of the words occurring before and after the nominals. In addition, another feature includes the path from the first nominal to the second nominal. This path feature is built by the concatenation of the POS tags of the words occurring between both nominals. In case the list of words occurring in between the two nominals include a verb or a preposition, more morphological features are considered such as the lemma and the voice in case of a verb and the lemma and type in case of a preposition.

At the semantic level, we used features obtained from WordNet and already provided in the training and testing data. This information includes the synset number and the lexical file number for each nominal. For the Part-Whole relation, we use an additional feature indicating the meronymy relation between the two nominals.

The set of WordNet lexical files can help to determine if the nominals satisfy the restrictions for each relation. For example, in the relation Theme-Tool, the theme should be an object, an event, a state of being, an agent, or a substance. Else, it is possible to affirm that the relation is false.

An additional feature, called *WordNet vector* was designed. It is a 13 dimension vector where each coordinate represents one of the 13 nodes in the third level of depth in WordNet. We use a binary representation, i.e. if the synset is ancestor of the nominal it is assigned the value 1, else it is assigned the value 0. Our initial hypothesis considers that this representation for the nominals could perform well on unseen data. Finally, features related to the query provided for each sentence are also taken into consideration.

Support Vector Machines (SVM) model was selected for its good performance in many classification tasks such as Semantic Role Labeling and Document Categorization. The data is represented as an *n*-dimensional vector, and the final goal is finding the maximal separating hyperplane of dimension *n-1* between the two classes (*true* and *false*). There are several algorithms for quickly finding the parameters of the hyperplane. We decided to use Sequential Minimal Optimization (SMO) [4]. SMO breaks the large quadratic programming (QP) optimization problem needed to be resolved in SVM into a series of smallest possible QP problems. These small QP problems are analytically solved, avoiding, in this way, a time-consuming numerical QP optimization as an inner loop. We used Weka [7] an implementation of the SMO.

Both run time and accuracy depend critically on the values given to two parameters: the upper bound on the coefficient's values in the equation for the hyperplane (-C), and the degree of the polynomials in the non-linear mapping (-E) [7]. The best settings for a particular dataset can be found only by experimentation. We did not experiment with the polynomial kernel. The default value is set to 1. However, we made numerous experiments to find the best value for the parameter C (C=1, C=10, C=100, C=1000, C=10000), but the results were not remarkably affected. Probably, this is due to the small size of the training set.

2.3 Rules

Parallel to the work on training the machine learning model, a set of linguistic and knowledge based constraint rules were manually developed. Our main concern, in this respect, is not the coverage of the rules, but the improvement achieved by applying them. Thus, the constraint rules do not pretend to achieve a total coverage of the whole dataset, but to improve the results of the learning algorithm in cases where it is obvious to decide if the relation is true or false according to simple semantic and knowledge constraints.

For each semantic relation, a number of rules were developed based on the definition given by the SEMEVAL-task 4 and the analysis of the training dataset. In this analysis, we considered simple heuristics and linguistic features on the different levels (POS, lexical, syntactic and semantic). We counted on three basic sources of information, considered as basic criteria:

- WordNet Information concerning synsets and lexical file numbers of each nominal in the WordNet structure.

- POS information mainly lemmas, grammatical categories and lemmas, provided from the basic processing module.

- Lexical information regarding the word tokens occurring in the sentences are also provided in the basic processing module.

It is important to highlight the fact that we did not use a syntactic analyzer or a chunker. Thus, the syntactic information, included in the rules, consists basically of simple sequences of POS tags. In this respect, we believe that integrating information concerning chunks or basic syntactic analysis in future versions would be highly valuable as we will discuss later.

Taking into account the above mentioned criteria, we can state that our rules combine semantic information with lexical and morpho-syntactic information. These perspectives are not mutually exclusive. On the contrary, they do overlap in many ways, revealing the complexity of natural language since it adopts a variety of mechanisms to reflect the relations between concepts. These mechanisms are manifested in different linguistic levels varying from the most abstract conceptual to the more precise lexical or morphological (nouns, verbs and prepositions) levels.

Rules developed can be divided into four basic classes: 1) semantic rules dealing with the most abstract level. These rules depend mainly on the lexical file numbers. 2) Lexico-semantic rules where special lemmas and tokens of the contextual information are taken into account. 3) morphosyntactic rules where special patterns consisting of sequences of POS are considered. 4) Complex rules combining two or more of the above mentioned criteria or addressing special grammatical phenomena such as the passive voice. In the following paragraphs, we will give examples of these rules among the different semantic relationships.

For the CAUSE-EFFECT relation, a set of four rules was developed. For example, the rule 1 belong to the first class of simple semantic rules. It states the following:

Rule Cause-Effect 1: IF Cause IS (*TIME* OR *LOCATION* OR *PERSON* OR *ANIMAL* OR *GROUP* OR *COMMUNICATIVE ACT*) THEN R(Cause-Effect)=False
However, rule 3 and rule 4 are complex rules reflecting certain patterns where two sources of information are combined: information about tokens, lemmas and synsets of the nominals in concern and information concerning the words occurring in between.

`Rule Cause-Effect 3:` IF (CAUSE_TOKEN=EFFECT_TOKEN) AND
(TOKEN_BETWEEN_NOMINALS= "after") THEN   R(Cause-Effect)=False
For example, *"frustration after frustration"* this pattern does not imply a cause-effect relation. However, it indicates a frequency.

In many cases, the CAUSE-EFFECT relation might be expressed through the use of the preposition *from* indicating an EFFECT *from* a CAUSE. However, according to the case indicated by rule 4, the use of the preposition is governed and implied by the noun and, thus, it looses its semantic connotation of CAUSE-EFFECT.

Rule Cause-Effect 4: If (CAUSE_TOKEN=="protection" (synset= 03969138 OR synset = 00805831) OR (CAUSE_TOKEN=="abstinence" (synset= 01054457 OR synset= 04827378)) AND (TOKEN_BETWEEN_NOMINALS= "from") THEN R(Cause-Effect)=False

Other types of complex rules deal with grammatical phenomena such as the passive voice. The role of passiveness is crucial in indicating agency features since passiveness implies patience and denies agency. This linguistic fact is represented in a special rule for the INSTRUMENT-AGENCY relation indicating that using the passive voice negates the agency and, thus, the relation is false.

Rule Instrument-Agency 1: If ((NOMINAL_AGENCY + Verb Participle+ "with"+ NOMINAL_INSTRUMENT) OR (NOMINAL_AGENCY + Verb(be) + Verb_Participle+ "with" +NOMINAL_INSTRUMENT)) Then R(Instrument-Agency)= False

For the PRODUCT-PRODUCER relation, other features were considered in developing the set of rules. The analysis of the training dataset revealed that this relationship is frequently represented in English language through the use of transitive verbs indicating actions like: {make, produce, manufacture}, {develop, compose, prepare} or {create, procreate}. A complex rule is developed (combining lexical, semantical and syntactic through a sequence of POS tags) using morpho-syntactic and lexical information together with semantic information regarding the synsets of these verbs and simple semantic constraints indicating that PRODUCT cannot be TIME, STATE or ACTIVITY.

Pattern: <NOMINAL_ PRODUCER> verb <NOMINAL_ PRODUCT>
Rule Product-Producer 2: IF (verb (active voice) = {make| produce| manufacture| fabricate} {cultivate| grow| farm for} {develop| compose| prepare} {create| procreate} {come from} {provide} {emit} AND PRODUCT!=*STATE* AND PRODUCT!=*ACTIVITY* AND PRODUCT!=*TIME* Then R(Product-Producer)=TRUE

The order of executing the rules is essential; it goes from the most specific to the most generic in order to avoid false generalizations. If no lexico-syntactic patterns are available, we apply the general semantic-based rules For example; the above rule in the PRODUCT-PRODUCER relation implies another generic rule indicating the following:

Rule Product-Producer 1: IF PRODUCT IS (STATE OR ACTIVITY OR TIME) THEN R(Product-Producer)=FALSE

As we previously mentioned, the semantics of prepositions are, in many cases, responsible of determining the relation between the nominals. However, this could lead to ambiguous cases where the use of the preposition is governed by the preceding noun. For example, the preposition *from* is frequently used to indicate ORIGIN-ENTITY relation where an ENTITY comes *from* an ORIGIN. Nevertheless, in cases where the noun preceding the preposition demands the use of *from* and the semantic content of the noun doesn't imply an ORIGIN-ENTITY relation, the relation is said to be false. This fact is expressed in a simple lexical rule:

Pattern: VERB <NOMINAL_ ENTITY> + PREP= "FROM" + <NOMINAL_ORIGIN>
Rule Origin-Entity 1: IF verb = {separate| prohibit| protect|
save| avoid| shield| isolate} Then R(Origin-Entity)=FALSE
Following these guidelines, a number of rules were developed according to the
semantics implied in each relation. The coverage of these rules varied from one
relation to another and the improvements obtained over the scores of the machine
learning algorithm also varied as we will discuss in the next section.

3   Experimental Results and Conclusions

Introducing the rules enhanced overall scores in all relations except CONTENT-
CONTAINER as shown in Table1. Best results were obtained in the PRODUCT-
PRODUCER relation where the improvement reached almost 15% raising the overall
accuracy from 60.8% to 75%. This is due to the relatively high coverage of the rules
(48%) and their high precision. In CAUSE-EFFECT, rules improved the results
approximately 4.5% raising the overall accuracy from 56.2% to 60.75%. Again, the
rules proved a high precision. For the INSTRUMENT-AGENCY, improvement was
trivial (only 1.3%) and, thus, we believe that more rules should be implemented.
Results for ORIGIN-ENTITY relation was improved by a percentage of 5% reaching
an overall accuracy of 70.5% in comparison with 65.4% achieved by the machine
learning algorithm alone. For the THEME-TOOL relation, enhancement obtained is
almost 6% improving the total accuracy from 54.2% to 60%. It is also important to
point out that the precision of the rules in this case was optimum 100% over the rule.

Finally in the PART-WHOLE relation, rules covered 36% of the test data. The
improvement obtained reaches 6% enhancing the total accuracy to 76.38% after
70.8% obtained by the machine learning module (for this relation, our system was the
first in the SEMEVAL task#4).

The graph 1 shows the comparison between results obtained by the machine-
learning module and those obtained after combining the rules.

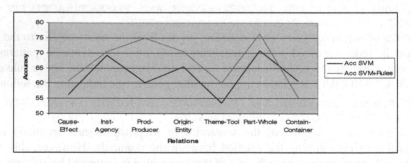

Graph 1. Compared Results between SVM and Hybrid Approach

Results obtained demonstrate that applying a hybrid approach on a supervised
system achieves good results in a semantic classification task between nominals.
Improvement obtained varies from 1.2% till 15% over the total accuracy scores
obtained by the support vector machine algorithm, except for the relation CONTENT-
CONTAINER.

Table 1. Number of Rules, Coverage and Results

Relations	Rules (num)	Coverage Train	Coverage Test	Acc. SVM	Acc. SVM+Rules
Cause-Effect	4	25%	27%	56,2%	60,75%
Inst-Agency	3	8%	8%	69,2%	70,5%
Prod-Producer	7	50%	48%	60,2%	75%
Origin-Entity	15	36%	38%	65,4%	70,5%
Theme-Tool	6	38%	25%	53,5%	60%
Part-Whole	45	62%	36%	70,8%	76,38%
C-Container	15	16%	18%	60.8%	55.40%

We want to introduce further training and enhancement. Improving results across the whole dataset requires wider use of semantic information such as hypernym for each synset or information regarding the entity features. Also, introducing more syntactic information through syntactic chunking could affect significantly the results. In the rule-based module, we consider increasing the set of rules for a broader coverage.

Semantic classification proposed by this system can be integrated in a Question Answering system or ontology building. For QA systems, to answer complex questions, a deeper understanding of the question is necessary. For example, to answer questions such as "list of all causes from asthma" (cause-effect), "list of all country producer of olive oil" (product-producer) or "what are the chemical components of aspirin? (Part-Whole). Detecting these types of semantic relations could help to find the exact answer.

References

1. Kietz, J.U., Maedche, A., Volz, R.: A method for semi-automatic ontology acquisiton from a corporare Intranet. In: Dieng, R., Corby, O. (eds.) EKAW 2000. LNCS (LNAI), vol. 1937, pp. 1–4. Springer, Heidelberg (2000)
2. Lee, D., CHU, W.W.: Constraints-Preserving Transformation from XML Document Type Definition to Relational Schema. In: Laender, A.H.F., Liddle, S.W., Storey, V.C. (eds.) ER 2000. LNCS, vol. 1920, Springer, Heidelberg (2000)
3. Maedche, A., Staab, S.: Measuring similarity between ontologies. In: Gómez-Pérez, A., Benjamins, V.R. (eds.) EKAW 2002. LNCS (LNAI), vol. 2473, Springer, Heidelberg (2002)
4. Platt, J. Sequential Minimal Optimization: A Fast Algorithm for Training Support Vector MachinesMicrosoft Research Technical Report MSR-TR-98-14 (1998)
5. Segura Bedmar, I., Samy, D., Martinez-Fernández, J.L.: UC3M: Classification of Semantic Relations between Nominals using Sequential Minimal Optimization. In: Proc. of SEMEVAL, ACL O7. Prague (2007)
6. Specia, L., Motta, E.: A hybrid approach for extracting semantic relations from texts. In: 2nd Workshop on Ontology Learning and Population (2006)
7. Witten, H., Frank, E.: Data Mining: Practical machine learning tools and techniques. Morgan Kaufmann, San Francisco (2005)

The Effect of Context on Semantic Similarity Measurement

Carsten Keßler[1], Martin Raubal[2], and Krzysztof Janowicz[1]

[1] Institute for Geoinformatics, University of Münster, Germany
{carsten.kessler,janowicz}@uni-muenster.de
[2] Department of Geography, University of California at Santa Barbara, USA
raubal@geog.ucsb.edu

Abstract. Similarity measurement is currently being established as a method to explore content on the Semantic Web. Semantically annotated content requires formal concept specifications. Such concepts are dynamic and their semantics can change depending on the current context. The influence of context on similarity measurement is beyond dispute and reflected in recent similarity theories. However, the systematics of this influence has not been investigated so far. Intuitively, the results of similarity measurements should change depending on the impact of the current context. Particularly, such change should converge to 0 with a decreasing impact of the respective contexts. To hold up to this assertion, a quantification of the impact of context on similarity measurements is required. In this paper, we use a combination of the SIM-DL theory, which measures similarity between concepts represented using description logic, and a context model distinguishing between internal and external context to quantify this impact. The behavior of similarity measurements within an ontology specifying geospatial feature types is observed under varying contexts. The results are discussed with respect to the corresponding impact values.

1 Introduction and Motivation

Information integration and retrieval are central aspects of the Semantic Web. It has been argued that cognitively plausible methods for achieving such integration and retrieval require the employment of semantic similarity measurement [1]. The influence of context on semantic similarity measurement is a well-known phenomenon that has long been observed in psychological experiments [2]. As an example, consider an information retrieval scenario where a user is looking for buildings that are similar to *churches*. For a similarity-based ranking of the results, the best matches depend on the context: within a *religion* context these could be *cathedral, chapel* and *convent*, whereas for a *sightseeing* context they could be *palace, castle* and *museum*.

Although the results of human similarity ratings depend on the context, and this dependency is also reflected in recent similarity theories [3,4,5], the nature of this influence and its actual impact have not been subject to thorough research yet. Previous work looked primarily at how contextual information can

R. Meersman, Z. Tari, P. Herrero et al. (Eds.): OTM 2007 Ws, Part II, LNCS 4806, pp. 1274–1284, 2007.

be employed to adjust the results of a similarity measurement. In this paper, we present observations of similarity measurements under changing contexts using the description logic based SIM-DL theory [5]. The long-term objective of our research is a generic context model for similarity measurement, which represents the important characteristics of context. This envisioned model can then be used for the development of similarity-based applications, particularly for the assessment of the impact of different context parameters. Building such context model requires a better understanding of the interaction between context and similarity measurement.

The focus of this paper lies on observing and quantifying the change in the results of a similarity measurement under different contexts. We introduce and formalize the notion of *impact* for contexts: the more the context changes the involved concepts, the larger its impact. The results of a similarity measurement should directly depend on the impact the context has on the compared concepts. In other words, if the same similarity measurement is performed under a series of contexts with decreased impact, the change in the similarity values will converge to 0. To enable such behavior, an impact measure for contexts is required. In the following, we introduce such an impact measure, based on the SIM-DL theory and a context model consisting of an internal context and a set of external context rules. A scenario from geographic feature type lookup is used to test the model with varying concept pairs and for different contexts.

The remainder of this paper is organized as follows: we first present related work on semantic similarity measurement and context, followed by an introduction of the SIM-DL theory, and the formal specification of context impact. Section 3 describes the use case and its context formalization. In Section 4, the resulting similarity and impact values are evaluated, and the results discussed. Section 5 presents conclusions and gives directions for future research.

2 Similarity Measurement and Context

This section presents previous work on similarity measurement and context, and introduces the SIM-DL theory. The context representation and the according formalization of context impact are demonstrated.

2.1 Semantic Similarity Measurement

The notion of similarity was established in psychology to determine why and how entities are grouped to categories, and why some categories are comparable to each other while others are not [2,6]. The main challenge with respect to semantic similarity measurement is the comparison of meanings. A language must be specified to represent the nature of entities and metrics are needed to determine how (conceptually) close these compared entities are. While entities can be expressed in terms of attributes, the representation of entity types is more complex [7]. Depending on the expressivity of the representation language, types are specified as sets of features, dimensions in a multidimensional space, or formal restrictions specified on sets using various kinds of description logic. Similarity is measured between entity types, which are representations of concepts in human minds, therefore the results depend on what is said (in terms of computational representation) about these types. This again depends on the employed representation language, and therefore most similarity measures cannot be compared. Besides the question of representation, context is another major challenge for similarity assessments. In general, meaningful notions of similarity cannot be determined without defining in respect to what similarity is measured [2,8,9].

Similarity measurements have been investigated as a method for information retrieval in the semantic web over the last years [1,10]. Stroulia and Wang [11] developed a context-free similarity measure for Web services based on Web Service Description Language specifications. Based on Tversky's feature model [12], Rodríguez and Egenhofer [3] built an extended model called Matching Distance Similarity Measure (MDSM) that supports a basic context theory, automatically determined weights, and asymmetry. Raubal and Schwering [13,14] used conceptual spaces [15] to implement models based on distance measures within geometric space. Several measures [5,16,17] were developed to close the gap between ontologies described by various kinds of description logic, and similarity theories that had not been able to handle the expressivity of such languages.

2.2   Similarity Theory: SIM-DL

SIM-DL [5,18,19] is an asymmetric and context aware similarity measurement theory used for information retrieval within an ontology (or several ontologies using the same shared vocabulary [20]). An early implementation of SIM-DL as DIG-compliant [21] server is available at http://sim-dl.sf.net. The latest version supports comparison between concepts specified using the expressive description logic $\mathcal{ALCHQ}$ [19].

In SIM-DL, similarity between concepts in canonical form [5,22] is measured by comparing their definitions for overlap, where a high level of overlap indicates high similarity and vice versa. In description logic (complex) concepts are specified based on primitive concepts and roles using language constructors such as intersection, union, and existential quantification. Hence, similarity is defined as a polymorphous, binary, and real-valued function $X \times Y \rightarrow R[0,1]$ providing implementations for all language constructs offered by the used description logic. The overall similarity between concepts is the normalized (and weighted) sum of the single similarities calculated for all parts (i.e., superconcepts) of the concept definitions. A similarity value of 1 indicates that the compared concepts cannot be differentiated, whereas 0 implies total dissimilarity. As most feature and geometric approaches, SIM-DL is a asymmetric measure, i.e. the similarity $sim(C_s, C_t)$ is not necessarily equal to $sim(C_t, C_s)$. The comparison of two concepts depends therefore not only on their descriptors but also on the direction in which both are compared. Further details on SIM-DL and the involved similarity functions are given in [5,19].

2.3 Previous Work on Context

Context has been investigated from the perspectives of different research areas such as ubiquitous computing, interoperability, automatic metadata generation and web search. Accordingly, any definition of context largely depends on the field of application[1]. Concerning research on context for similarity measurement [3,4,5], existing context definitions are often tailored to specific similarity theories, and the context is mostly used to select the domain of application, i.e. a set of concepts that is taken into account for the similarity measurement. Moreover, context is used to assign weights to (parts of) the different concepts or instances within the domain of application. To clarify what we refer to as a similarity measurement's context, we use the following definition from [9]:

Definition 1. *A similarity measurement's context is any information that helps to specify the similarity of two entities more precisely concerning the current situation. This information must be represented in the same way as the knowledge base under consideration, and it must be capturable at maintainable cost.*

While capturability and cost are not relevant for this research, it is important that context only refers to such information that has an impact on the similarity measurement. Furthermore, it must be represented in the same way as the given knowledge base. For reasons of simplicity, we assume here that the context only refers to parts of the knowledge base at hand, as described in the following section.

2.4 Context and Impact Specification

The domain of application alone is not always sufficient to reflect the context of a similarity measurement; for example, a domain of application, such as transportation, cannot be used to specify that, by law, it is not allowed for trucks to drive on the German *Autobahn* on Sundays—hence, an external rule is required that removes the superconcept *NavigableByTrucks* from the *Autobahn* concept on Sundays. Accordingly, we define a context K as a combination of *internal* and *external* context (eq. 1): The internal context c_{int} is a concept which specifies the domain of application, i.e. all concepts subsumed by c_{int}, such as *NavigableByTrucks*. The external context is a set of rules R that allows for the modification of the concepts selected via c_{int}. Every rule consists of a condition that specifies the circumstances under which the rule is activated[2], a number of modifying concepts c_m, and the affected concepts c_a to which these modifications apply (eq. 2). Every modification either adds (+) a superconcept to the affected concepts by intersection, or removes it (−). In the special case where a negated superconcept is present, and the same

---

[1] See [23] for an overview of different research areas investigating context, and the according definitions.

[2] A mechanism for the automatic selection of the appropriate rules R based on the conditions is required for an implementation, but out of scope for this paper.

(non-negated) superconcept is added via +, the negation is overridden. Note that these changes are only temporary and revoked after the similarity measurement.

$$K = \langle c_{int}, \{R_1, ..., R_n\} \rangle \tag{1}$$

$$R : condition \longrightarrow \langle \{\pm c_{m_1}, ..., \pm c_{m_n}\}, \{c_{a_1}, ..., c_{a_n}\} \rangle \tag{2}$$

To allow statements about the impact of a context on a similarity measurement, we introduce a formal measure that quantifies this impact (eq. 3): A context's impact on a similarity measurement is defined as the overall change the corresponding context rules cause to the search- and target concepts c_s, c_t. For every modifying concept c_m, the absolute change is quantified in terms of how many of the superconcepts are changed (added or removed) in c_s or c_t. Beyond this, the kind of change is reflected: if the application of c_m makes c_s and c_t more similar, the absolute value is counted positive; if it makes them less similar, the value is counted negative. Whether a single c_m is positive or negative is determined according to Table 1, which lists all possible combinations for adding or removing superconcepts, depending on whether they are already part of c_s or c_t (or both). The combined measure (eq. 3) takes into account that a rule which makes two concepts more similar, and one which makes them less similar, may compensate for each other if both appear in the same context. The outcoming impact measures range from 0, where the concepts under consideration are not changed, to 1, where all of the original superconcepts are removed.

$$Imp(K, c_s, c_t) = \sum \frac{\pm c_m}{|\{c_a | c_a \sqsupseteq c_s \sqcup c_t\}|} \tag{3}$$

As mentioned in the introduction, the results of a similarity measurement should intuitively depend on the impact the context has on the compared concepts: when the context's impact increases, the change in the measurement's result should also increase. In other words, if the same similarity measurement is performed under a

Table 1. Possible combinations of adding or removing superconcepts from search concepts, and target concepts respectively. The contents of the table show the development of similarity under the preconditions given in the header; an increase in similarity is marked +, decrease is marked −, no change by 0.

	$c_m \sqsupseteq c_s, c_m \sqsupseteq c_t$	$c_m \sqsupseteq c_s, c_m \not\sqsupseteq c_t$	$c_m \not\sqsupseteq c_s, c_m \sqsupseteq c_t$	$c_m \not\sqsupseteq c_s, c_m \not\sqsupseteq c_t$
$+c_s$	0	0	+	−
$+c_t$	0	+	0	0
$-c_s$	−	+	0	0
$-c_t$	−	0	0	0
$+c_s, +c_t$	0	+	+	+
$+c_s, -c_t$	−	0	−	−
$-c_s, +c_t$	−	+	0	0
$-c_s, -c_t$	−	+	0	0

series of contexts with decreased impact, the change in the similarity values should converge to 0. The specification of this behavior (eq. 4) requires a standard context K_{std}, consisting of totality (i.e. the whole ontology, $\top$) as the internal context and without external context (i.e. without any rules): $\langle \top, \{\emptyset\} \rangle$.

$$\lim_{Imp(R_K) \to 0} sim_K(a, b) = sim_{K_{std}}(a, b) \qquad (4)$$

3   Use Case: Geographic Feature Type Lookup

In this section, we present an application scenario involving a feature type ontology. Different geographic feature types, as represented in the Web Ontology Language (OWL), are compared under different contexts.

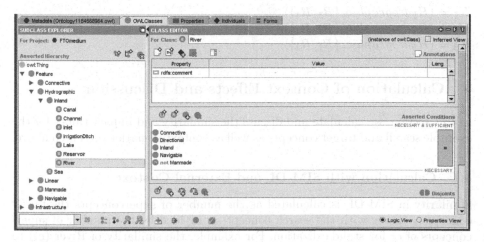

Fig. 1. Snapshot of the feature type ontology (see http://sim-dl.sf.net/downloads/)

3.1   Feature Type Ontology

Gazetteers are place name directories that make use of type lookup functionality to determine a geographic feature's type, such as *Road*, *Country* or *City*. The feature types are often classified in semi-formal feature type thesauri. However, similarity measurement among different feature types requires a formal description of the types. In the following, we refer to a feature type ontology for hydrographic features (figure 1) that was created based on the Alexandria Digital Library Feature Type Thesaurus[3]. Once complete, such an ontology allows for similarity-based queries to a semantic geo-webservice [19]. Since any complex similarity tasks build upon the comparison of concept pairs, we use SIM-DL and the context definition presented in section 2.4 to measure similarities between *River* and *Lake*, *Lake* and *Reservoir*, *Canal* and *River*, and between *Reservoir* and *Canal*.

---

[3] http://www.alexandria.ucsb.edu/gazetteer/FeatureTypes/ver070302/index.htm.

3.2   Contexts Within the Scenario

The internal context c_{int} is set fixed to *Hydrographic* for the use case. We introduce a set of context rules that can be combined to different contexts[4]; for example, R_1 states that seas and lakes are not navigable during a storm, whereas R_4 states that it is allowed to navigate in reservoirs in the case of an emergency:

R_1: Storm $\longrightarrow \langle\{-Navigable\}, \{Sea, Lake\}\rangle$
R_2: Flooding $\longrightarrow \langle\{-Linear\}, \{River, Channel, IrrigationDitch\}\rangle$
R_3: Night-time $\longrightarrow \langle\{-Navigable\}, \{Canal\}\rangle$
R_4: Emergency $\longrightarrow \langle\{+Navigable\}, \{Reservoir\}\rangle$

Based on R_1-R_4, we define the following scenarios referring to different situations such as emergency at night (K_1), storm at night (K_2) and stormtide (K_3):

$K_1 = \langle Hydrographic, \{R_3, R_4\}\rangle$
$K_2 = \langle Hydrographic, \{R_1, R_3\}\rangle$
$K_3 = \langle Hydrographic, \{R_1, R_2\}\rangle$

4   Calculation of Context Effects and Discussion

In this section, we calculate and discuss the similarity and impact values for the example search and target concepts as well as context scenarios presented above.

4.1   Calculation with SIM-DL and External Context

Similarity in SIM-DL is calculated as the number of superconcepts the target concept c_t shares with the search concept c_s, divided by the number of superconcepts of c_s for standardization. For example, the similarity of River (c_s) to Lake (c_t) is $\frac{3}{6}$ (0.5, see first line of results in table 2), as River is represented by 6 superconcepts, of which Lake shares 3. This calculation is valid under K_{std}; if the context changes, the similarity may also change. For example, K_2 removes the superconcept Navigable from Lake, so that there are only two common superconcepts left and similarity changes to $\frac{2}{6}$ (0.33). On the other hand, K_1 does not affect the similarity of River to Lake because these concepts are not part of R_3 and R_4.

All similarity results in table 2 are also annotated with the impact values as calculated according to eq. 3. For example, the impact of K_2 on the similarity of River to Lake is $-\frac{1}{6}$ (−0.17), since one of the 6 superconcepts of River ⊔ Lake is removed. The impact is negative, because a superconcept is removed from c_t which is a superconcept of both c_s and c_t, resulting in a negative impact on the overall similarity (see first column, fourth line in table 1). Table 2 gives an overview of all similarity and impact values for the concept pairs under consideration.

---

[4] For reasons of readability, we only use combinations of two context rules in this paper; however, an arbitrary number of rules can be combined in principle.

Table 2. Similarity results for the three different external contexts. The internal context was set fixed to Hydrographic.

Search concept C_s	Target concept C_t	K_1	K_2	K_3	K_{std}
River	Lake	0.5	0.33	0.4	0.5
Impact		0	-0.17	-0.03	0
Lake	River	0.6	0.5	0.5	0.6
Impact		0	-0.17	-0.17	0
Lake	Reservoir	0.8	0.75	0.75	0.6
Impact		+0.17	+0.17	+0.17	0
Reservoir	Lake	0.8	0.6	0.6	0.6
Impact		+0.17	0	0	0
Canal	River	0.75	0.75	0.6	0.8
Impact		-0.14	-0.14	-0.14	0
River	Canal	0.5	0.5	0.6	0.66
Impact		-0.14	-0.14	-0.17	0
Canal	Reservoir	0.38	0.38	0.3	0.3
Impact		0	+0.17	0	0
Reservoir	Canal	0.5	0.3	0.3	0.3
Impact		+0.17	0	0	0

4.2   Discussion of Results

The similarity values produced by SIM-DL using the combined context model introduced in section 2.4 generally appear plausible. For example, $sim(Lake, Reservoir)$ increases when the Reservoir becomes Navigable (K_1), or if the Lake is not Navigable (K_1, K_2). Note that K_2 and K_3 do not affect the similarity in the inverse direction, since the according rule R_1 only makes the Lake not navigable, i.e. the rule removes a superconcept from c_s that did also not match c_t before the rule was applied, therefore there is no change. It must be pointed out that leaps in the similarity values under different contexts, such as in the comparison of Reservoir and Lake, are for the most part due to the small number of concepts contained in the ontology: since every concept is only described by a small number of superconcepts, every modification to this set of superconcepts causes comparably large changes in the similarity values. The observations made in this paper need to be verified for more complex ontologies in the future.

Concerning the calculation of the impact values, the behavior as described in eq. 4 cannot be observed. While the tendency for the impact is generally correct, the actual impact value does not correlate with the corresponding change in similarity. For example, K_1 has an impact of +0.17 on $sim(Reservoir, Canal)$, as shown in the last line of table 2. The according similarity value increases by 0.2 with respect to the same comparison under K_{std}. At the same time, $sim(River, Lake)$ decreases by 0.1 under K_3, where the corresponding impact value is as low as -0.03. These results point to the fact that the impact model introduced in section eq. 3 may be too loosely coupled to the similarity theory.

5 Conclusions and Future Work

This paper has introduced a method, which quantifies the impact of context on the results of semantic similarity measurements. SIM-DL produces plausible results, even for ontologies that use only a small number of primitives for concept description. The separation between internal and external context allows for the addition of conditional rules that cannot be expressed within the ontology. The notion of context impact presented in this paper, however, does not fully correspond to the expected behaviour. Although the general tendency of the impact values corresponds to the trend of the change in similarity values, the quantification of this change is not yet reflected in the impact value. To solve this issue, further research on different strategies for impact specification is required. For example, knowledge about individuals (via assertions) [16] could be used to make more precise statements about the impact of a given context.

Beyond this, the limitation of the current model to the methods for adding and removing concepts needs to be enhanced by more sophisticated ways of assigning weights to the superconcepts. This would allow for rules such as "Glaze $\longrightarrow \langle \{Navigable : 0.3\}, \{Road\} \rangle$". This weighting would also allow for resolution of contradicting rules in the same context—a case that cannot be handled so far. To allow such rules, methods for the determination of the weights and the corresponding application rules are needed. Moreover, future work should focus on how the developed strategies can be generalized and transferred to other concept representations and similarity theories. The behavior observed for concept pairs in this paper also needs to be compared to similarity rankings and tested for cognitive plausibility in a human subjects test.

Acknowledgements

This research is partly funded by the German Research Foundation (DFG) under the project title "Semantic Similarity Measurement for Role-Governed Geospatial Categories", see http://sim-dl.sf.net.

References

1. Gärdenfors, P.: How to make the semantic web more semantic. In: Varzi, A.C., Vieu, L. (eds.) Formal Ontology in Information Systems, Proceedings of the Third International Conference (FOIS 2004), pp. 17–34. IOS Press, Amsterdam (2004)
2. Medin, D., Goldstone, R., Gentner, D.: Respects for similarity. Psychological Review 100(2), 254–278 (1993)
3. Rodríguez, A., Egenhofer, M.: Comparing geospatial entity classes: an asymmetric and context-dependent similarity measure. International Journal of Geographical Information Science 18(3), 229–256 (2004)

4. Albertoni, R., Martino, M.D.: Semantic similarity of ontology instances tailored on the application context. In: Meersman, R., Tari, Z. (eds.) On the Move to Meaningful Internet Systems 2006: CoopIS, DOA, GADA, and ODBASE. LNCS, vol. 4275, pp. 1020–1038. Springer, Heidelberg (2006)
5. Janowicz, K.: SIM-DL: Towards a semantic similarity measurement theory for the description logic $\mathcal{ALCNR}$ in geographic information retrieval. In: Meersman, R., Tari, Z., Herrero, P. (eds.) On the Move to Meaningful Internet Systems 2006: OTM 2006 Workshops. LNCS, vol. 4278, pp. 1681–1692. Springer, Heidelberg (2006)
6. Goldstone, R., Son, J.: Similarity. In: Holyoak, K., Morrison, R. (eds.) Cambridge Handbook of Thinking and Reasoning, pp. 13–36. Cambridge University Press, Cambridge (2005)
7. Donnelly, M., Bittner, T.: Spatial relations between classes of individuals. In: Cohn, A., Mark, D. (eds.) COSIT 2005. LNCS, vol. 3693, pp. 182–199. Springer, Heidelberg (2005)
8. Goodman, N.: Seven strictures on similarity. In: Goodman, N. (ed.) Problems and projects, pp. 437–447. Bobbs-Merrill, New York (1972)
9. Keßler, C.: Similarity measurement in context. In: Kokinov, B., Richardson, D., Roth-Berghofer, T., Vieu, L. (eds.) Sixth International and Interdisciplinary Conference on Modeling and Using Context. LNCS (LNAI), vol. 4635, pp. 277–290. Springer, Heidelberg (2007)
10. Rissland, E.: Ai and similarity. IEEE Intelligent Systems 21(3), 39–49 (2006)
11. Stroulia, E., Wang, Y.: Structural and semantic matching for assessing web-service similarity. International Journal of Cooperative Information Systems 14, 407–437
12. Tversky, A.: Features of similarity. Psychological Review 84(4), 327–352 (1977)
13. Raubal, M.: Formalizing conceptual spaces. In: Varzi, A., Vieu, L. (eds.) Formal Ontology in Information Systems, Proceedings of the Third International Conference (FOIS 2004). Frontiers in Artificial Intelligence and Applications, vol. 114, pp. 153–164. IOS Press, Amsterdam, NL (2004)
14. Schwering, A., Raubal, M.: Spatial relations for semantic similarity measurement. In: Akoka, J., Liddle, S.W., Song, I.-Y., Bertolotto, M., Comyn-Wattiau, I., van den Heuvel, W.-J., Kolp, M., Trujillo, J., Kop, C., Mayr, H.C. (eds.) Perspectives in Conceptual Modeling. LNCS, vol. 3770, pp. 259–269. Springer, Heidelberg (2005)
15. Gärdenfors, P.: Conceptual Spaces - The Geometry of Thought. MIT Press, Cambridge (2000)
16. d'Amato, C., Fanizzi, N., Esposito, F.: A dissimilarity measure for $\mathcal{ALC}$ concept descriptions. In: Proceedings of the 2006 ACM Symposium on Applied Computing (SAC), Dijon, France, pp. 1695–1699 (2006)
17. Borgida, A., Walsh, T., Hirsh, H.: Towards measuring similarity in description logics. In: Proceedings of the 2005 International Workshop on Description Logics (DL2005), EUR, Edinburgh, Scotland, UK, vol. 147 (2005)
18. Janowicz, K.: Similarity-based retrieval for geospatial semantic web services specified using the web service modeling language (wsml-core). In: Scharl, A., Tochtermann, K. (eds.) The Geospatial Web - How Geo-Browsers, Social Software and the Web 2.0 are Shaping the Network Society, pp. 235–246. Springer, London (2007)
19. Janowicz, K., Keß ler, C., Schwarz, M., Wilkes, M., Panov Espeter, M., Bäumer, B.: Algorithm, implementation and application of the sim-dl similarity server. In: Fonseca, F., Rodríguez-Tatsets, A. (eds.) Second International Conference on GeoSpatial Semantics (GeoS 2007), Mexico City, Mexico, Springer (forthcoming, 2007)

20. Lutz, M., Klien, E.: Ontology-based retrieval of geographic information. International Journal of Geographical Information Science 20(3), 233–260 (2006)
21. Bechhofer, S.: The dig description logic interface: Dig/1.1. In: DL2003 Workshop, Rome (2003)
22. Horrocks, I.: Implementation and optimization techniques. In: Baader, F. (ed.) The description logic handbook: theory, implementation, and applications, pp. 306–346. Cambridge University Press, New York (2003)
23. Bazire, M., Brézillon, P.: Understanding context before using it. In: Dey, A.K., Kokinov, B., Leake, D.B., Turner, R. (eds.) CONTEXT 2005. LNCS (LNAI), vol. 3554, pp. 29–40. Springer, Heidelberg (2005)

Using Ontologies to Map Concept Relations in a Data Integration System

Paolo Ceravolo, Ernesto Damiani, Alex Gusmini,
and Marcello Leida

Università degli studi di Milano,
Dipartimento di Tecnologie dell'Informazione
via Bramante, 65
26013 Crema (CR), Italy
{ceravolo,damiani,leida}@dti.unimi.it,
agusmini@crema.unimi.it
http://ra.crema.unimi.it/kiwi

Abstract. In this paper we propose a Data Integration System based on an ontology as Global Representation. The paper briefly introduces the motivations and the benefits that the use of an ontology brings to a Data Integration System, which is also formally defined. The paper, in particular, focuses on the limitation of the proposed system to handle the different relations that can exist between concepts of an ontology. We divide the relations that can be defined in an ontology in two different sets: *Mapped* relations, which we consider atomic relations and *Derived* relations, which can be generated by combining mapped relations using SWRL rules. Some examples of the different kind of relations are reported to clarify the concepts and a new definition of Mapping in the Data Integration System is proposed, in order to define the atomic relations. The paper ends with considerations about the problems that still need to be considered.

Keywords: Data Integration, Ontology, SWRL, Rule Engine, Mapping Generation, Ontology Relations, Semantic.

1 Introduction

Nowadays, a huge amount of data is available and is accessible from different sources (data bases, web sites, e-mails, documents, ...) and is stored in different formats, in different locations and for different purposes. This situation leads to redundancy of information and splitting of related information over several data sources.

A system that analyses, collect and aggregate this data is therefore crucial for distributed applications that rely on the data to perform their tasks.

A Data Integration System (DIS) aims at solving the problem of accessing different data sources through a common representation G. The correspondences between elements of the data sources and the elements of the common representation G are modeled by means of a mapping.

R. Meersman, Z. Tari, P. Herrero et al. (Eds.): OTM 2007 Ws, Part II, LNCS 4806, pp. 1285–1293, 2007.

Several works in literature describe DIS from different point of views: declarative and procedural approach ([5]), *Local As View* and *Global As View* approach ([6], [1], [4], [8]) that refers to the definition of the common representation G.

The system that we consider in this paper is based on *Global as View* approach, because the G is given through a combination of elements of L by associating concepts of G with the correspondent set of elements of L.

Formally we can define a DIS as the following triple:

$$DIS =< G, L, M >$$

Where G is the global representation, L the set of local representations (the data sources) composed by n single representations $s_1, s_2, ..., s_n$ and M is the mapping between L and G. The mapping M can be derived by a complex process. We define M as another triple dependent on a set of matching relations M_t:

$$M(M_t) =< CM, IC, I_lC >$$

Where M_t is the matching among the elements $\{e_{s_a}^1, e_{s_a}^2 ... e_{s_m}^n\}$ belonging to the local representations $s_a, ..., s_n$ in L and a set of elements $\{e_g^1, e_g^2, ... e_g^n\}$ belonging to the global representation G. M_t can be defined as a set of relations:

$$e_{s_k}^i \cong^\delta e_g^j$$

or

$$e_{s_k}^i \cong^\delta e_{s_h}^j$$

Where $\cong$ is a binary relation from the following set of the relations: *equality, string inclusion,* and *concept inclusion* $(=, <, \subset)$. We consider this set of relations that allows us to model the matching between elements that are not only equivalent but also partially related. For example let's consider a table in a data base that is composed by the columns Name, Surname, Street, City and Country and the attributes Fullname and Address of a concept in the global representation G. The matching relations between these elements are the following:

Name < Fullname

Surname < Fullname

We associate the elements with the relation of string inclusion $<$ because the istances of Name and Surname are part of the instances of Fullname.

Street ⊂ Address

City ⊂ Address

Country ⊂ Address

We associate the elements with the relation of concept inclusion $\subset$ because Street, City and Country are part of the conceptualization of the element

Address, but not necessarily the instance of **Address** is given by aggregating the source elements.

The index δ, associated to $\cong$, can be a binary ($[0,1]$) or a fuzzy value ($[0..1]$) depending on the method chosen to implement the matching operator and it represents the strength of the relation.

CM (Concept Mapping) is a set of mappings between elements of the local representation L and elements of the global representation G. CM is represented by sets of relations as follows:

$$e_g^k \rightarrow \phi(T)$$

where e_g^k is an element of the global representation G, ϕ is an arbitrary formula that combines the argument T that is a series of elements of the local representation L defined as:

$$T = \{e_{s_a}^1, e_{s_a}^2, \phi(e_{s_c}^5, e_{s_b}^1), \phi(e_{s_b}^2), ..., e_{s_m}^n\}$$

IC (Integrity Constraints) is a mapping between elements of L: we consider relations between objects of the same source schema s_a in the local representation L, such as the typical *primary-key→foreign-key*, but also correspondences between elements of different source schemas s_i, s_j of L that are semantically related. It is defined like CM but is not associated to any element of the Global Schema G:

$$\phi(T) = \omega(S)$$

Where S is another series defined like T. The elements of IC are used to ensure that the instances of the concepts are assembled correctly and the instances retrieved are univocal.

$I_l C$ (Instance-level Constraints) is a set of constraints that are applied to the queries that retrieve the instances of the elements of the global representation. Each set of constraints is associated to a concept c_i of the global schema G:

$$I_l C = \{c_i \rightarrow C_i, c_j \rightarrow C_j, ...\}$$

Where C is a series of constraints on elements of the source schema L, defined for example as:

$$C_i = \{e_{s_a}^1 = e_{s_a}^2, e_{s_b}^1 = STRING, ..., e_{s_m}^n > TODAY\_DATE\}$$

Instance-level Constraints are necessary to retrieve instances of concepts that can not be defined only by means of combination of elements of the local schema L.

Let's consider for example a table **Countries** in the source schema L defined as in Figure 1 and a concept **EUCountry** in the global representation G. The mapping expressed by means of CM and IC does not provides enough information for retrieving the right instances of the concept **EUCountry**, therefore we need additional indications provided by the set of $I_l C$, that in the case of the example are defined as:

$$CM = \{NAME_g^{EUCountry} \rightarrow NAME_{s_a}^{Countries}\}$$

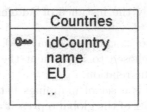

Fig. 1. A table in the source schema that represents the countries with a column of Boolean values (EU), which indicates if a country is part of the European Union

$$I_l C = \{EUCountry \rightarrow \{EU_{s_a}^{Countries} = TRUE\}\}$$

After this brief introduction to our DIS the paper proceeds as follows: Section 2 introduces the problems that the use of an ontology as Global Schema introduces, in Section 3, by means of some examples, we introduce a new element in the definition of mapping in order to solve the problem of representing relations between concepts in the DIS. In Section 4, conclusions and open problems are reported.

2 Use of Ontology as Global Schema

The use of an ontology[1] as a Global Schema G has been already considered in literature [4], it brings several benefits [2] to a DIS: representing the information using a First Order Logic formalism (OWL [7] and SWRL [9]) provides the possibility to perform reasoning on the data for discovering intrinsic information and for checking the consistency and the integrity of the global representation, for discovering more semantic information about the underlying data.

Due to its logical grounding, the expressive power of an ontology is higher than the local representation L: with an ontology we can represent information derived from logical reasoning on the data, additional knowledge that is not directly accessible from the local representation L but that can be derived from it (by mean of queries, aggregating elements, formulas...).

An ontology can be viewed as a *Directed Acyclic Graph* (DAG) of concepts that are connected each other by means of *relations*. A relations is defined from one or more *Domain* concepts to one or more *Range* concepts. Using on ontology as Global representation in a DIS, introduces the problem of how the relations between concepts are considered n the system.

Referring to our DIS, we can divide relations in two main categories:

1. *Atomic relations*: these are relations that can not be derived from other relations but from an analysis of the Local representation L. Consequently they need to be represented by means of a mapping to the local schema.

---

[1] The term ontology has several meanings ranging from philosophy to computer science. In this paper we use the term referring to the Semantic Web standard for Knowledge Representation, such as for instance W3C's OWL specification.

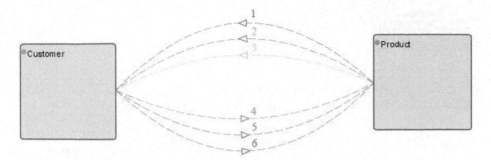

Fig. 2. two concepts `Customer` and `Product` and some examples of different relations that can occur between them: 1. `acquiredBy`, 2. `hadProblemsWith` and 3. `canBeProposedTo` and their opposite relations: 5. `acquiredProduct`, 6. `requestAssitance`, and 7. `probablyIntrestedIn`. Note that all of this are $1 : n$ relations.

Referring to Figure 2: let's consider the relation 5 (`acquiredProduct`) having as Domain the concept `Customer` and as Range the concept `Product`, when an instance x of the concept `Customer` is selected, the instances $y_1, y_2, ..., y_n$ of the concept `Product` that have been acquired by the costumer x are showed. In order to know which instances of `Product` are associated with x we have to define the relation `acquiredProduct` as a mapping.

2. *Composed relations*: these kind of relations are derived relations. They are defined by combining basic relations into new relations or modeling new relations by constraining atomic relations. The instances of the concepts involved in these relations can be derived applying reasoning techniques on the ontology, or defined using rules and process them with a rule engine. Referring to figure 3: the concept `Manager` is a sub class of the concept `Employee` with an additional relation `works Indirectly On` to the concept `Project`. The two concepts `Employee` and `Manager` are connected by the relations `manages` that goes form `Manager` to `Employee` and `managed by` vice-versa. The instances that are involved in the relation `works Indirectly On` can be classified modeling the relation as a SWRL rule, where the relation is defined as "if a manager manages an employee that is working on a project, then the manager is working indirectly on the same project" ($worksDirectlyOn(?employee, ?project) \wedge manages(?manager, ?employee) \rightarrow worksIndirectlyOn(?manager, ?project)$). The use of a rule engine is necessary to process the rule and reclassify the instances.

Composed Relations are not strongly related to the DIS because they are processed after the instances are retrieved from the data sources, but the *Atomic Relations* need to provide the necessary information to the DIS on how the instances of the concepts are related in the local representation L.

In the next Section, we extend the definition of mapping in order to consider also the *Atomic Relations*.

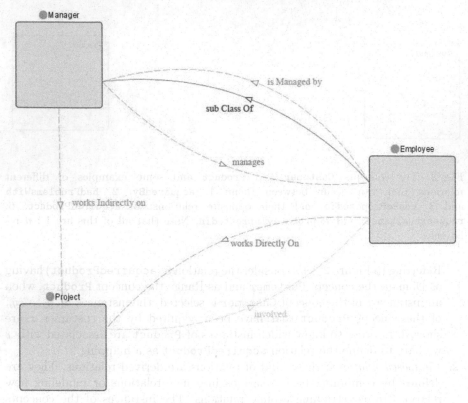

Fig. 3. caption

3 Relations Mapping

The DIS defined in the previous section can retrieve instances of ontology's concepts from the source schemas L but it is not expressive enough if we want to associate instances of concepts considering the relations between them.

To obtain instances of concepts associated by relations, we need to add to the definition of mapping, that we provide in Section 1, a new set of mappings RC (Relation Constraints) which represents how a specific relation between concepts is mapped in the data source. The mapping is modified as follow, adding the new element RC:

$$M(M_t) = < CM, IC, RC, I_lC >$$

where RC is a set of equivalences defined like IC associated to a relation in the global representation G:

$$r_g \rightarrow \{\phi_1(T_1) = \omega_1(S_1), \phi_2(T_2) = \omega_2(T), ..., \phi_n(T_n) = \omega_n(T_n)\}$$

As clarifying example let's consider the situation in Figure 4: the instances of the concept **Manager** that are related to the concept **Employee** by the relation **manages** can be retrieved modeling the relation with the constraint:

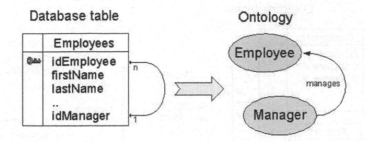

Fig. 4. A simple table in a data source that provides the information for the relation manages

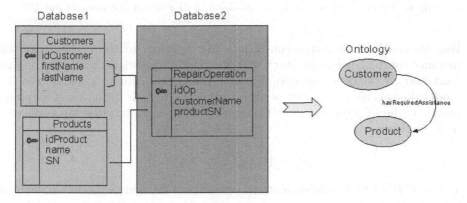

Fig. 5. Semantic relations between tables in two different data sources that provides the information for the relation **hasRequiredAssistance**

$$RC = \{manages_g \rightarrow \{idManager_{s_1}^{Employees} = idEmployee_{s_1}^{Employees}\}\}$$

Let's consider the example in Figure 5 the relation **hasRequiredAssistance** is modeled by semantic relations between the two data sources involved:

$$RC = \{hasRequiredAssistance_g \rightarrow \{$$

$$concatenate(lastName_{s_1}^{Customer}, firstName_{s_1}^{Customer}) = customerName_{s_2}^{RepairOperation},$$

$$productSN_{s_2}^{RepairOperation} = SN_{s_1}^{Product}\}\}$$

Now let's consider the situation in Figure 6: the definition that we provided of our DIS is not expressive enough for modeling this relation. Intuitively we model the relation with the following constraints:

$$RC = \{idManager_{s_1}^{Employees} = idEmployee_{s_1}^{Employees}\}$$

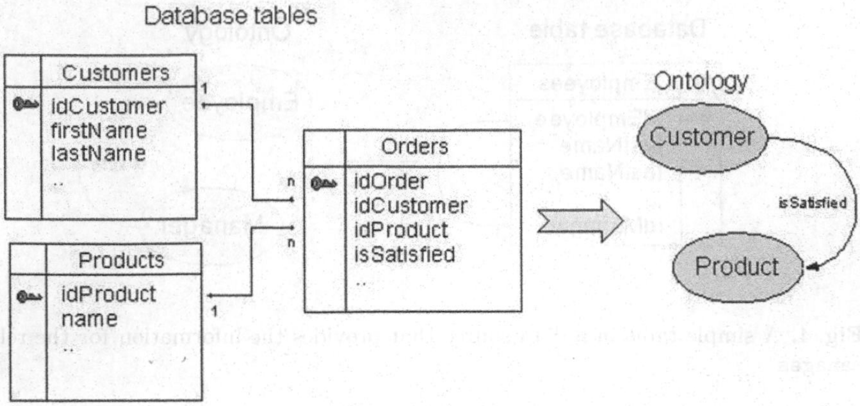

Database tables

Fig. 6. A relation that can not be modeled by the actual definition of our DIS

But we need to add a new constraints that specifies to the system that the instances of the concept Product that the relation associates with the concept Customer are the ones that have the value TRUE in the column isSatisfied. To model this specific relation we need to further modify our definition of mapping adding to the $I_l C$ (Instance-level Constraints) the constraints that are associated to a relation r_i:

$$I_l C = \{c_i \rightarrow C_i, r_i \rightarrow C_j, ...\}$$

Where C is a series of constraints on elements of the source schema L, defined for example as:

$$C_i = \{e^1_{s_a} = e^2_{s_a}, e^1_{s_b} = STRING, ..., e^n_{s_m} > TODAY\_DATE\}$$

Considering this simple addiction to the DIS now is possible to model the relation in Figure 6 with the constrain:

$$I_l C = \{isSatisfied_g \rightarrow \{isSatisfied^{ProductOrder}_{s_1} = TRUE\}\}$$

The system described so far is now expressive enough for modeling the mapping of elements in the local representation L to concepts and relations of an ontology used as global representation G. We already developed a system for the automatic generation of mapping from the local source L to concepts of the ontology. Now we are working on a method to automatically discover mapping from the local source L to relations between concepts but, at the moment, the system still need a domain expert intervention in order to associate the discovered mappings to the correct relation of the ontology.

We plan to discuss these aspects in a future paper.

4   Conclusions

The work done on this paper is not meant to be exhaustive, it intends to highlight the problem of extending our definition of mapping in data integration system considering also the relations that intercour between concepts.

Anyway there are some problems that remains to be solved: the first is how to associate a relation to its correspondent element in M_r. Another problem that need to be considered is to decide which relations should modeled in DL (Open World Assumption) and which ones should be modeled in SWRL (Close World Assumption).

Acknowledgements

This work was partly funded by the Italian Ministry of Research under FIRB contract n. RBNE05FKZ2_004, TEKNE.

References

1. Braun, P., Lotzbeyer, H., Schatz, B., Slotosch, O.: Consistent integration of formal methods, 48–62 (2000)
2. Cui, Z., Damiani, E., Leida, M.: Benefits of Ontologies in Real Time Data Access. In: proceedings IEEE-DEST Conference on Digital Ecosystems and Technologies (February 2007)
3. Duschka, O.M., Genesereth, M.R., Levy, A.Y.: Recursive query plans for data integration. Journal of Logic Programming 43(1), 49–73 (2000)
4. Hakimpour, F., Geppert, A.: Global schema generation using formal ontologies (2002)
5. Hammer, J., Garcia-Molina, H., Widom, J., Labio, W.: The Stanford Data Warehouse Project. IEEE Data Eng. Bulletin, Specail Issue on Materialized Views (1995)
6. Lenzerini, M.: Data Integration: A Theoretical Perspective. In: PODS pp. 233–246 (2002)
7. Web Ontology Language defintion, http://www.w3.org/TR/owl-features/
8. Parent, C., Spaccapietra, S.: Issues and approaches of database integration. Commun. ACM 41(5es), 166–178 (1998)
9. A Semantic Web Rule Language Combining OWL and RuleML, http://www.w3.org/Submission/SWRL/

Management of Inconsistent Data

Sylvia Encheva[1] and Sharil Tumin[2]

[1] Stord/Haugesund University College, Bjørnsonsg. 45, 5528 Haugesund, Norway
sbe@hsh.no
[2] University of Bergen, IT-Dept., P. O. Box 7800, 5020 Bergen, Norway
edpst@it.uib.no

Abstract. Automated tests evaluating students mastering of a skill usually involve only one question related to that skill. To attain a higher level of certainty in the evaluation process we propose a test with four different questions related to application of a single skill. Such a test is intended to facilitate the self-assessment process and can be suggested to students after a skill has been obtained. Lattice theory and higher-order logic are further applied for presenting a structure that can serve as a building block of an intelligent tutoring system.

Keywords: five-valued logic, automated tests.

1 Introduction

A lot of research has been done on evaluating students knowledge during the last few decades. Considerable part of it is aiming at developing automated tests providing immediate feedback. A common drawback of such tests is that they employ first-order logic in the process of decision making. While first-order logic appears to be sufficient for most everyday reasoning, it is certainly unable to provide meaningful conclusions in presence of inconsistent and/or incomplete input [5], [8]. This problem can be resolved by applying many-valued logic. Another interesting issue regarding assessment of knowledge is related to the number of questions one should ask while establishing students' level of mastering of a particular skill. We propose short tests that can be incorporated in an intelligent tutoring system, where each test contains four questions requiring a particular skill application. Such a test can be suggested to students after a skill has been obtained.

Introducing more than one question helps to attain a higher level of certainty in the process of decision making. At the same time the number of possible answer combinations increases rapidly. In practice it appears to be quite difficult and timeconsuming to find a proper response to each answer combination. In order to overcome this obstacle we propose grouping of all possible answer combinations into meaningful sets first. Application of many-valued logic, for drawing conclusions and providing recommendations, is then suggested.

The rest of the paper is organized as follows. Related work, basic terms and concepts are presented in Section 2. The management model is described in

R. Meersman, Z. Tari, P. Herrero et al. (Eds.): OTM 2007 Ws, Part II, LNCS 4806, pp. 1294–1302, 2007.
© Springer-Verlag Berlin Heidelberg 2007

Section 3. The paper ends with a description of the system in Section 4 and a conclusion in Section 5.

2 Background

Let P be a non-empty ordered set. If $sup\{x, y\}$ and $inf\{x, y\}$ exist for all $x, y \in P$, then P is called a *lattice*, [2]. In a lattice illustrating partial ordering of knowledge values, the logical conjunction is identified with the meet operation and the logical disjunction with the join operation.

A three-valued logic, known as Kleene's logic is developed in [9] and has three truth values, truth, unknown and false, where unknown indicates a state of partial vagueness.

The semantic characterization of a four-valued logic for expressing practical deductive processes is presented in [1]. The Belnap's logic has four truth values 'True, False, Both, None'.

Nested line diagrams are used for visualizing large concept lattices, emphasizing sub-structures and regularities, and combining conceptual scales, [14]. A nested line diagram consists of an outer line diagram, which contains in each node inner diagrams.

The five-valued logic introduced in [3] is based on the following truth values:

uu - unknown or undefined,
kk - possibly known but consistent,
ff - false,
tt - true,
ww - inconsistent.

A truth table for the ontological operation $\wedge$ is presented in Table 1.

Table 1. Truth table for the ontological operation $\wedge$

$\wedge$	uu	kk	ff	tt	ww
uu	uu	uu	ff	uu	uu
kk	uu	kk	ff	kk	ww
ff	ff	ff	ff	ff	ff
tt	uu	kk	ff	tt	ww
ww	uu	ww	ff	ii	ww

A level-based instruction model is proposed in [11]. A model for student knowledge diagnosis through adaptive testing is presented in [6]. An approach for integrating intelligent agents, user models, and automatic content categorization in a virtual environment is presented in [12].

The Questionmark system [10] applies multiple response questions where a set of options is presented following a question stem. The final outcome is in a binary form, i.e. correct or incorrect because the system is based on Boolean logic [4], [13].

3   Answer Combinations

For establishing the current level of mastering a particular skill we propose a multiple choice test with four questions. Every answer can be correct c, incorrect i or partially correct p. The case where no answer is provided is denoted by n.
Thus we obtain the following 35 answer combinations:

cccc - four correct answers
cccp - three correct answers and one partially correct answer
cccn - three correct answers and one unanswered question
ccci - three correct answers and one incorrect answer
ccpp - two correct answers and two partially correct answers
ccpn - two correct answers, one partially correct answer and
      one unanswered question
ccpi - two correct answers, one partially correct answer and
      one incorrect answer
ccni - two correct answers, one unanswered question and
      one incorrect answer
ccnn - two correct answers and two unanswered questions
ccii - two correct answers and two incorrect answers
cppp - one correct answer and three partially correct answers
cppn - one correct answer, two partially correct answers, and
      one unanswered question
cppi - one correct answer, two partially correct answers, and
      one incorrect answer
cpni - one correct answer, one partially correct answer,
      one unanswered question, and one incorrect answer
cnnn - one correct answer and three unanswered questions
cnnp - one correct answer, two unanswered questions, and
      one partially correct answer
cnni - one correct answer, two unanswered questions and d
      one incorrect answer
cnii - one correct answer, one unanswered question and d
      two incorrect answers
ciip - one correct answer, two incorrect answers, and
      one partially correct answer
ciii - one correct answer and three incorrect answers
pppp - four partially answers
pppn - three partially answers and one unanswered question
pppi - three partially answers and one incorrect answer
ppnn - two partially answers and two unanswered questions
ppni - two partially answers, one unanswered question and
      one incorrect answer
ppii - two partially answers and two incorrect answers
pnnn - one partially answer and three unanswered questions
pnni - one partially answer, two unanswered questions and
      one incorrect answer

pnii   - one partially answer, one unanswered question and
          two incorrect answers
piii   - one partially answer and three incorrect answers
nnnn - four unanswered questions
nnni - three unanswered questions and one incorrect answer
nnii - two unanswered questions and two incorrect answers
niii - one unanswered question and three incorrect answers
iiii   - four incorrect answers

3.1   Lattice Representation

Inner lattices - all answer combinations are grouped in five lattices with respect
to the five truth values, [3].

Any two nodes, in Fig. 1, Fig. 2, Fig. 3, Fig. 4, connected by an edge differ
in one answer only. Going upwards from one level to another in the lattices in
Fig. 1, Fig. 2, Fig. 3, Fig. 4 increases the level of certainty.

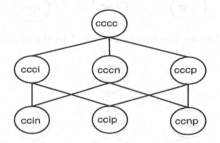

Fig. 1. Answer combinationsrelated to the truth value *tt*

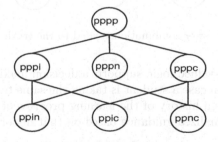

Fig. 2. Answer combinationsrelated to the truth value *kk*

Any two nodes, in Fig. 5, connected by an edge differ in exactly two answers.
Going upwards from one level to another in the lattices in Fig. 5 increases the
level of knowledge.

The nested lattice - the previously described 35 answer combinations are
arranged in a nested lattice, Fig. 6. This is possible since the ontological operation ∧ is both commutative and associative, [3].

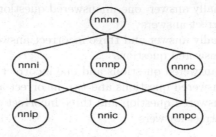

Fig. 3. Answer combinationsrelated to the truth value uu

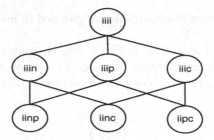

Fig. 4. Answer combinationsrelated to the truth value ff

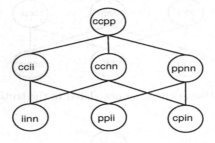

Fig. 5. Answer combinationsrelated to the truth value ww

Application of five-valued logic supports a decision making process related to drawing conclusions in case a student is taking the same test several times. This way the system keeps a history of the learning progress of every student and is able to provide personalized guidance based on that history.

4   System Implementation

The assessment model described in Section 3 provides important rules to be used in a Web-based intelligent tutoring system supporting self-study. After each skill obtained a student is presented with a set of four questions where each question is followed by a number of alternative answers (options).

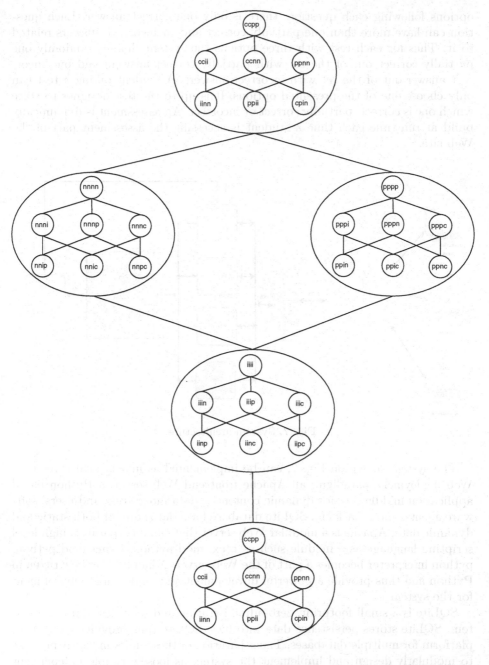

Fig. 6. All answer combinations

The four tests' questions of a particular assessment are randomly taken from a pool of questions related to a particular skill. The bigger number of questions in a pool, the better suited the system is for multiple assessments. Within the options following each question, there is only one correct answer. Each question can have more than one partially correct and/or incorrect answers related to it. Thus for each test with three options the system chooses randomly one partially correct out of the set with partially correct answers and one incorrect answer out of the set with incorrect answers. A student taking a test can only choose one of the presented options. It is up to the test designer to state which one is correct, partially correct or incorrect. An assessment is dynamically build at run time each time a student is accessing the assessment part of the Web site.

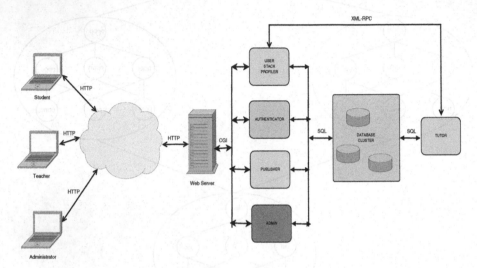

Fig. 7. System architecture

The system shown on Fig. 7, can be implemented as in a typical three-tiers Web deployment paradigm: an Apache front end Web server, a Python based application middleware for dynamic content, a data integration and users' software agents, and a back end SQLite database for a data store of both static and dynamic data. Apache is a modular Web server that can incorporate a high level scripting language as a module such as f. ex. mod_python. Using mod_python, python interpreter becomes a part of the Web server. A handler can be written in Python and thus provide a powerful development and deployment environment for the system.

SQLite is a small footprint, zero-administration and serverless database system. SQLite stores persistence data into files. SQLite thus provides a database platform for multiple databases. The separation of these units makes it possible to modularly design and implement the system as loosely couple independent sub-systems. Communication framework based on XML-RPC is used to connect the Web application middle-ware and the intelligent assessment/diagnostic system together.

Each user has her own database to store session and status data. An authenticated user receives a unique session key that is used to identify the user in the system for that particular session. This session key is saved in the user's Web browser cookie. This session key identifies which database is connected to this user session. The session key cookie and dynamic data stored in user's database are used together to keep user's states of interaction with the system within otherwise a stateless HTTP protocol.

The system is composed of sub-systems, all of which can be written in Python. The dynamic document publisher sub-system compiles a page to be presented to the user. How the page will look is defined in a template object in relation to the user response, current state variables and activities history. Documents and template objects are read from a common documents database. Tests questions data is read from a common tests database.

The authenticator sub-system authenticates a user during login and creates initial session contact in the system if the user provides correct credentials. This sub-system also provides user authorization during an active user's session and is also responsible for session cleanup at user's log off.

The stack profiler sub-system keeps track of user activities history in a stack like data structure in the user database. Each event, like for example response/result of a test or a change of learning flow after following a hint given by the system, is stored in this database. This sub-system provides the percepts to the intelligent tutorial sub-system. The stack profiler receives student's test responses and calculates the final result of the test. The stack profiler provides a user with immediate feed-back. Analysis of the user's profile is given to the tutor sub-system. This sub-system runs in a different processes but operates on the same user database.

The tutor sub-system provides a student with intelligent diagnostics and the optimal study flow for that particular student based on her current profile. The hints and diagnostics are written into the user database. The stack profiler which is accessible by the user's browser, reads the most recent messages from the tutor sub-system whenever hints or diagnostics are requested by the user. Using a rule-based method the tutor sub-system provides the student with the best study advice to facilitate students' progress in mastering new skills. Since the system is built on different sub-systems the independent sub-system can be maintained and improved separately. These sub-systems are operationally bounded with each other by the information stored in the multiple databases.

5 Conclusion

This paper is devoted to assessing students mastering of skills applaying automated testing procedure. Responses to four questions addressing one particular skill are used in the decision making process. This contributes to attaining higher level of certainty in the provided feedback. Employing many-valued logic for drawing a conclusion after a test is taken second time allows real use of student's previous tests results.

References

1. Belnap, N.J.: A useful four.valued logic. In: Dunn, J.M., Epstain, G. (eds.) Modern uses of multiple-valued logic, pp. 8–37. D. Reidel Publishing Co. Dordrecht (1977)
2. Davey, B.A., Priestley, H.A.: Introduction to lattices and order. Cambridge University Press, Cambridge (2005)
3. Ferreira, U.: A Five-valued Logic and a System. Journal of Computer Science and Technology 4(3), 134–140 (2004)
4. Goodstein, R.L.: Boolean Algebra. Dover Publications (2007)
5. Gradel, E., Otto, M., Rosen, E.: Undecidability results on two-variable logics. Archive of Mathematical Logic 38, 313–354 (1999)
6. Guzmàn, E., Conejo, R.: A model for student knowledge diagnosis through adaptive testing. In: Lester, J.C., Vicari, R.M., Paraguaçu, F. (eds.) ITS 2004. LNCS, vol. 3220, pp. 12–21. Springer, Heidelberg (2004)
7. Huffman, D., Goldberg, F., Michlin, M.: Using computers to create constructivist environments: impact on pedagogy and achievement. Journal of Computers in mathematics and science teaching 22(2), 151–168 (2003)
8. Immerman, N., Rabinovich, A., Reps, T., Sagiv, M., Yorsh, G.: The boundery between decidability and undecidability of transitive closure logics. In: Marcinkowski, J., Tarlecki, A. (eds.) CSL 2004. LNCS, vol. 3210, Springer, Heidelberg (2004)
9. Kleene, S.: Introduction to Metamathematics. D. Van Nostrand Co., Inc., New York (1952)
10. http://www.leeds.ac.uk/perception/v4_mrq.html
11. Park, C., Kim, M.: Development of a Level-Based Instruction Model in Web-Based Education. In: Luo, Y. (ed.) CDVE 2004. LNCS, vol. 3190, pp. 215–221. Springer, Heidelberg (2004)
12. Santos, C.T., Osòrio, F.S.: Integrating intelligent agents, user models, and automatic content categorization in virtual environment. In: Lester, J.C., Vicari, R.M., Paraguaçu, F. (eds.) ITS 2004. LNCS, vol. 3220, pp. 128–139. Springer, Heidelberg (2004)
13. Whitesitt, J.E.: Boolean Algebra and Its Applications, Dover Publications (1995)
14. Wille, R.: Concept lattices and conceptual knowledge systems. Computers Math. Applic. 23(6-9), 493–515 (1992)

Approaches to Inconsistency Handling in Description-Logic Based Ontologies

David Bell[1], Guilin Qi[2], and Weiru Liu[1]

[1] School of Electronics, Electrical Engineering and Computer Science
Queen's University Belfast
Belfast, BT7 1NN, UK
{w.liu,da.bell}@qub.ac.uk
[2] Institute AIFB
Universität Karlsruhe
D-76128 Karlsruhe, Germany
gqi@aifb.uni-karlsruhe.de

Abstract. The problem of inconsistency handling in ontologies has recently been attracting a lot of attention. When inconsistency occurs in an ontology, there are mainly two ways to deal with it - we either resolve it or reason with the inconsistent ontology.

In this paper, we give a survey of the existing approaches for handling inconsistency in ontologies and analyze their usability. We focus on Description Logics. We give clear examples to illustrate how to use these approaches to deal with practical problems.

1   Introduction

Knowledge representation for the Semantic Web (SW) requires analysis of the universe of discourse for concepts, definitions, objects, roles, properties,etc, and then selecting a computer-usable version of the results. The sharing of heterogeneous information requires agreed carefully-specified terms to describe the dispersed resources. Ontologies play a core role for the success of the Semantic Web as they provide shared vocabularies for different domains, such as Medicine and Bioinformatics. More generic high-level ontologies are also required for general-purpose SW use (see below). There are many representation languages for ontologies, such as Description Logics (DLs) [1], which have clear semantics and formal properties. The quality of ontologies is important for SW technology. However, in practice, it is often difficult to construct an ontology which is error-free. Inconsistency can occur due to several reasons, such as modeling errors, migration or merging ontologies, and ontology evolution. For example, if ontologies such as such as DOLCE, SUMO and CYC are used in a single document, hundreds of mis-alignments of concepts can be detected, and contradictory statements ('unsatisfiable concepts') might be introduced in particular applications. This will cause logical inconsistency. For example, an unusual individual which does not satisfy some of initial assumptions might be encountered in a running application (see Example 2 below). Do we simply flag that individual as an exception, or do we remove some of the clauses that encode our assumptions?

R. Meersman, Z. Tari, P. Herrero et al. (Eds.): OTM 2007 Ws, Part II, LNCS 4806, pp. 1303–1311, 2007.

Current DL reasoners, such as RACER [5] and FaCT [7], can detect logical inconsistency. However, they only provide lists of unsatisfiable classes. The process of *resolving* inconsistency is left to the user or ontology engineers. The need to improve DL reasoners to reason with inconsistency is becoming urgent to make them more applicable.

There are mainly two ways to deal with inconsistent ontologies [8]. One way is to simply avoid the inconsistency and to apply a non-standard reasoning method to obtain meaningful answers. A general framework for reasoning with inconsistent ontologies based on *concept relevance* was proposed in [8]. The idea is to select from an inconsistent ontology some consistent sub-theories based on a *selection function*, which is defined on the syntactic or semantic relevance. Then standard reasoning on the selected sub-theories is applied to find *meaningful* answers. In [18,25,14,13], four-valued logics have been applied to reason with inconsistent ontologies.

The second way to deal with logical contradictions is to resolve logical modeling errors whenever a logical problem is encountered. For example, several methods have been proposed to debug erroneous terminologies and have them repaired when inconsistencies are detected [23,22,17,4].

In this paper, we give a survey of the existing approaches for handling inconsistency in ontologies and analyze their usability. We focus on Description Logics. We give clear examples to illustrate how to use these approaches to deal with practical problems.

The paper is organized as follows. Section 2 provides some basic notions of terminology debugging. We then give an overview of approaches for handling inconsistency in Section 3. Finally, we conclude and discuss the paper in Section 4.

2 Preliminaries

2.1 Description Logics

We now give a brief introduction of Description Logics (DLs) and refer the reader to the DL handbook [1] for more details.

A DL-based ontology (or ontology) $O = (\mathcal{T}, \mathcal{A})$ consists of a set $\mathcal{T}$ of concept axioms (TBox) and role axioms, and a set $\mathcal{A}$ of assertional axioms (ABox). Concept axioms have the form $C \sqsubseteq D$ where C and D are (possibly complex) concept descriptions, and role axioms are expressions of the form $R \sqsubseteq S$, where R and S are (possibly complex) role descriptions. We call both concept axioms and role axioms as terminology axioms. The ABox contains *concept assertions* of the form $C(a)$ where C is a concept and a is an individual name, and *role assertions* of the form $R(a, b)$, where R is a role and a and b are individual names.

The semantics of DLs is defined via a model-theoretic semantics, which explicates the relationship between the language syntax and the model of a domain: An interpretation $\mathcal{I} = (\triangle^{\mathcal{I}}, \cdot^{\mathcal{I}})$ consists of a non-empty domain set $\triangle^{\mathcal{I}}$ and an interpretation function $\cdot^{\mathcal{I}}$, which maps from individuals, concepts and roles to elements of the domain, subsets of the domain and binary relations on the domain, respectively.

Given an interpretation $\mathcal{I}$, we say that $\mathcal{I}$ satisfies a concept axiom $C \sqsubseteq D$ (resp., a role inclusion axiom $R \sqsubseteq S$) if $C^{\mathcal{I}} \subseteq D^{\mathcal{I}}$ (resp., $R^{\mathcal{I}} \subseteq S^{\mathcal{I}}$). Furthermore, $\mathcal{I}$ satisfies a concept assertion $C(a)$ (resp., a role assertion $R(a, b)$) if $a^{\mathcal{I}} \in C^{\mathcal{I}}$ (resp., $(a^{\mathcal{I}}, b^{\mathcal{I}}) \in R^{\mathcal{I}}$).

An interpretation $\mathcal{I}$ is called a *model* of an ontology ontology, iff it satisfies each axiom in the ontology.

2.2 Incoherence in DL-Based Ontologies

We introduce the notion of incoherence in DL-based ontologies defined in [3].

Definition 1 (Unsatisfiable Concept). *A concept name C in an ontology O, is unsatisfiable iff, for each interpretation $\mathcal{I}$ of O, $C^{\mathcal{I}} = \emptyset$.*

That is, a concept name is unsatisfiable in an ontology iff it is interpreted as an empty set by all models of O.

Definition 2 (Incoherent Ontology). *An ontology O is incoherent iff there exists an unsatisfiable concept name in O.*

For example, an ontology $O = \{A \sqsubseteq B, A \sqsubseteq \neg B\}$ is incoherent because A is unsatisfiable in O. As pointed out in [3], incoherence does not provide the classical sense of the inconsistency because there might exist a model for an incoherent ontology. We first introduce the definition of an inconsistent ontology.

Definition 3. *[Inconsistent Ontology] An ontology O is inconsistent iff it has no model.*

However, incoherence and inconsistency are related with each other. According to the discussion in [3], incoherence is a potential cause of inconsistency. That is, suppose C is an unsatisfiable concept in O, by adding an instance a to C will result in an inconsistent ontology. For example, the ontology $O = \{A \sqsubseteq B, A \sqsubseteq \neg B\}$ is incoherent but consistent (any interpretation which interprets A as an empty set and B as an nonempty set is a model of O). However, $O' = \{A(a), A \sqsubseteq B, A \sqsubseteq \neg B\}$ is both incoherent and inconsistent.

In most current work on debugging ontologies, the incoherence problem is discussed at the terminology level. That is, ABoxes are usually considered as irrelevant for incoherence. Therefore, when we talk about an axiom in an ontology, we mean only the terminology axiom.

In the following, we introduce some definitions which are useful to explain logical incoherence.

Definition 4. *[23] Let A be a concept name which is unsatisfiable in a TBox $\mathcal{T}$. A set $\mathcal{T}' \subseteq \mathcal{T}$ is a minimal unsatisfiability-preserving sub-TBox (MUPS) of $\mathcal{T}$ if A is unsatisfiable in $\mathcal{T}'$, and A is satisfiable in every sub-TBox $\mathcal{T}'' \subset \mathcal{T}'$.*

A MUPS of $\mathcal{T}$ and A is the minimal sub-TBox of $\mathcal{T}$ in which A is unsatisfiable. For example, given TBox $\mathcal{T} = \{C \sqsubseteq A, A \sqsubseteq B, A \sqsubseteq \neg B\}$. C is an unsatisfiable concept and it has one MUPS, i.e., $\mathcal{T}$. Based on MUPS, we can classify unsatisfiable concepts into derived unsatisfiable concepts and root unsatisfiable concepts as follows:

Definition 5. *[23] Let $\mathcal{T}$ be an incoherent TBox. A TBox $\mathcal{T}' \subseteq \mathcal{T}$ is a minimal incoherence-preserving sub-TBox (MIPS) of $\mathcal{T}$ if $\mathcal{T}'$ is incoherent, and every sub-TBox $\mathcal{T}'' \subset \mathcal{T}'$ is coherent.*

A MIPS of $\mathcal{T}$ is the minimal sub-TBox of $\mathcal{T}$ which is incoherent. Let us consider the example used to illustrate Definition 4, there is only one MIPS of $\mathcal{T}$: $\{A \sqsubseteq B, A \sqsubseteq \neg B\}$. We say a terminology axiom is *in conflict* in $\mathcal{T}$ if there exists a MIPS of $\mathcal{T}$ containing it.

3   Overview of Approaches for Inconsistency Handling

3.1   Resolving Inconsistency

Debug and Repair DL-Based Ontologies. The first approach for debugging of terminologies is originally proposed in [23]. Their debugging approach is restricted to *unfoldable ALC TBoxes*, i.e., the left-hand sides of the concept axioms (the defined concepts) are atomic and if the right-hand sides (the definitions) contain no direct or indirect reference to the defined concept. Suppose T is an incoherent unfoldable TBox and A is an unsatisfiable in it. To calculate a MUPS of T w.r.t A, we can construct a tableaux from a branch B initially containing only *labelled formula* $(a : A)^\emptyset$ (for a new individual name a) by applying the tableaux rules as long as possible.

Terminological diagnosis, as defined in [22], is an instance Reiter's diagnosis from first principles. Therefore, we can use Reiter's algorithms to calculate terminological diagnoses. An important notion in diagnosis is called a *conflict set*, which is an incoherent subset of a TBox. Given a TBox T, a subset T' of T is a diagnosis for an incoherent T if T' is a minimal set such that $T \setminus T'$ is not a conflict set for T.

A drawback of the debugging approach in [23] is that it is restricted to unfoldable ALC TBoxes. Furthermore, it is based on the tableaux algorithms for DLs. Therefore, it is dependent on the tableaux reasoner. In [17,9], two orthogonal debugging approaches are proposed to detect the clash/sets of support axioms responsible for unsatisfiable classes, and to identify root/derived unsatisfiable classes. The first one is a glass box approach which is based on description logic tableaux reasoner-Pellet. This approach is closely related to the top-down approach to explanation in [24]. However, the approach proposed in [17] is not limited to DL ALC and is designed for OWL DL. The second one is a black box approach [9] which is better suitable to identify dependencies in a large number of unsatisfiable classes. The approach is reasoner-independent, in the sense that the DL reasoner is solely used as an oracle to determine concept satisfiability with respect to a TBox. It consists of two main steps. In the first step, it computes a *single* MUPS of the concept and then it utilizes the Hitting Set algorithm to retrieve the remaining ones. This approach is closely related to the bottom up approach to explanation. Based on the debugging approach, in [10], the authors give a tool to repair unsatisfiable concepts in OWL ontologies. The basic idea is to rank erroneous axioms and then to generate a plan to resolve the errors in a given set of unsatisfiable concepts by taking into account the axiom ranks. In [19], a score ordering on terminology axioms is defined to help the user to select axioms to delete.

Example 1. Suppose that we have an ontology $DICE^1$ which contains the following terminologies:

$ax_1 = Brain \sqsubseteq CentralNervousSystem \sqcap BodyPart \sqcap \forall region.HeadAndNeck$
$ax_2 = CentralNervousSystem \sqsubseteq NervousSystem$
$ax_3 = Disjoint(NervousSystem, BodyPart)$.

ax_1 says that Brain is part of Central Nervous System and part of Body and it is in the region of Head and Neck. ax_2 says that Central Nervous System is part of Nervous System and ax_3 tells us that Nervous System and Body Part are disjoint. The ontology

---

[1] The ontology $DICE$ is under development at the Academic Medical Center in Amsterdam.

is incoherent because $Brain$ is claimed to be a part of Nervous System and Body (according to ax_1 and ax_2), whilst Nervous System and Body Part are claimed to be disjoint (according to ax_3). To resolve the incoherence, we can delete any of the axioms ax_i $(i = 1, 2, 3)$.

Maximal Satisfiable Terminologies. Another way to resolve incoherence in an ontology is to find some subsets of the TBox which are consistent and are maximal w.r.t. set-inclusion. In [16], a tableau-like procedure was proposed to these subsets and some optimization techniques were given to improve the runtime behavior of the procedure. A drawback of the approach in [16] is that it removes an axiom when it is involved in conflict, even if only part of the axiom is responsible for the conflict. So a fine-grained approach was proposed in [12] to generalize the approach in [16]. This approach can not only pinpoint the axioms which are involved in conflict, but also trace which parts of the axioms are responsible for the conflict. Based on the algorithm, a tool was developed for debugging and repairing an incoherent ontology.

In Example 1, there are three maximal satisfiable subsets: $\{ax_1, ax_2\}$, $\{ax_2, ax_3\}$ and $\{ax_1, ax_3\}$. However, ax_1 can be split into three parts:
$Brain \sqsubseteq CentralNervousSystem$
$Brain \sqsubseteq BodyPart$ and
$Brain \sqsubseteq \forall region.HeadAndNeck$.
Only $Brain \sqsubseteq CentralNervousSystem$ and $Brain \sqsubseteq BodyPart$ are involved in the conflict. Therefore, by applying the approach in [12], we can repair the ontology to give the following one: $\{Brain \sqsubseteq \forall region.HeadAndNeck, ax_2, ax_3\}$.

Consistent Ontology Evolution. The work in [6] describes a process to support the consistent evolution of OWL DL based ontologies, which ensures that the consistency of an ontology is preserved when changes are applied to the ontology. The process consists of two main phases: (1) *Inconsistency Detection*, which is responsible for checking the consistency of an ontology with the respect to the ontology consistency definition and identifying parts in the ontology that do not meet consistency conditions; (2) *Change Generation*, which is responsible for ensuring the consistency of the ontology by generating additional changes that resolve detected inconsistencies. The authors define methods for detecting and resolving inconsistencies in an OWL ontology after the application of a change. As for some changes there may be several different consistent states of the ontology, *resolution strategies* allow the user to control the evolution.

The methods for detecting inconsistencies rely on the idea of a selection function are to identify the *relevant* axioms that contribute to the inconsistency. In the most simple case, syntactic relevance – considering how the axioms of the ontology are structurally connected with the change – is used. Based on the selection function, algorithms to find *minimal inconsistent subontologies* and *maximal consistent subontologies* are presented.

The approach only supports repairs by removing complete axioms from the ontology, a weakening based on a finer granularity is mentioned as an extension, but no algorithms are proposed. The approach does not make a distinction between ABox and TBox axioms, as such both ABox and TBox inconsistencies are trivially supported. Further, the approach does not provide any explicit support for dealing with networked ontologies.

Let us consider an example given in [6].

Example 2. Given a University ontology which contains the following terminology axioms and assertional axioms:

$Researcher \sqsubseteq Person$ (researchers are persons)

$PhDStudent \sqsubseteq Student$ (PhD students are students)

$Student \sqsubseteq \neg Researcher$ (students are not researchers)

$Article \sqsubseteq Publication$ (articles are publications)

$Researcher(Johanna)$ (Johanna is a researcher).

Suppose we now receive a new assertion which says that Johanna is a PhD student, i.e., PhDStudent(Johanna). This assertion is in conflict with the existing ontology because students cannot be researchers whilst Johanna is claimed to be both a PhD student thus a student and a researcher. According to the algorithm in [6], we can delete $Student \sqsubseteq \neg Researcher$ to restore consistency.

Knowledge Base Revision in Description Logics. In [20], the revised AGM postulates for belief revision in [11] were generalized and two revision operators which satisfy the generalized postulates were given. One operator is the weakening-based revision operator which is defined by weakening of statements in a DL knowledge base. The idea of weakening a terminology axiom is similar to weakening an uncertain rule in [2]. That is, when a term is involved in conflict, instead of dropping it completely, we remove those individuals which cause the conflict. The weakening-based revision operator may result in counterintuitive results in some cases, so another operator was proposed to refine it. It was shown that both operators capture some notions of minimal change.

Let us go back to Example 2. To resolve inconsistency, we can now weaken the terminology axiom $Student \sqcap \neg\{Johanna\} \sqsubseteq \neg Researcher$, where $\{Johanna\}$ is a *nominal*. That is, all students except Johanna are not researchers. In this way, we can add PhDStudent(Johanna) to the ontology consistently.

The weakening-based approach is more fine-grained than the approach in [6]. However, it is computationally harder because we need to find the individuals which are responsible for the conflict.

Knowledge Integration for Description Logics. In [15], an algorithm, called *refined conjunctive maxi-adjustment* (RCMA for short) was proposed to weaken conflicting information in a *stratified* DL knowledge base and obtain some consistent DL knowledge bases. To weaken a terminological axiom, they introduced a DL expression, called *cardinality restrictions* on concepts. However, to weaken an assertional axiom, they simply delete it. In [21], the authors first define two revision operators in description logics, one is called a weakening-based revision operator and the other is its refinement. The revision operators are defined by introducing a DL constructor called *nominals*. The idea is that when a terminology axiom or a value restriction is in conflict, they simply add explicit exceptions to weaken it and assume that the number of exceptions is minimal. Based on the revision operators, they then propose an algorithm to handle inconsistency in a *stratified* description logic knowledge base. It was shown that when the weakening-based revision operator is chosen, the resulting knowledge base of their algorithm is semantically equivalent to that of the RCMA algorithm. However, their syntactical forms are different.

3.2   Reasoning with Inconsistent Ontologies

Coherence-Based Approaches. This kind of approach is based on the idea of removing contradictory information before applying classical reasoning algorithms-especially for question-answering. This can be realized e.g. by starting with an empty (thus consistent) ontology and incrementally selecting and adding axioms to that ontology, which do not result in inconsistency. A general framework for reasoning with inconsistent ontologies based on *concept relevance* was proposed in [8]. The idea is to select from an inconsistent ontology some consistent sub-theories based on a *selection function*, which is defined on the syntactic or semantic relevance. Then standard reasoning on the selected sub-theories is applied to find *meaningful* answers.

Example 3. Given our updated University ontology which contains the following terminology axioms and assertional axioms:

$Researcher \sqsubseteq Person$ (researchers are persons)
$PhDStudent \sqsubseteq Student$ (PhD students are students)
$Student \sqsubseteq \neg Researcher$ (students are not researchers)
$Article \sqsubseteq Publication$ (articles are publications)
$Researcher(Johanna)$ (Johanna is a researcher)
$PhDStudent(Johanna)$ (Johanna is a PhD student).

As we have seen, this ontology is inconsistent. Suppose we want to query if Johanna is a person, i.e. $Person(Johanna)$. We first select the axioms which are directly relevant to the query $Person(Johanna)$, i.e., axioms where the concept $Person$ and/or individual $Johanna$ appear(s). They are $Researcher \sqsubseteq Person$, $Researcher(Johanna)$ and $PhDStudent(Johanna)$. From these axioms, we apply a DL reasoner and we can infer that $Person(Johanna)$. So the answer is "yes" for the query.

Paraconsistent Reasoning on Inconsistent Ontologies. The second approach does not modify the knowledge base but changes the semantics under which it is reasoned with, employing a so-called *paraconsistent semantics*. Unlike the semantics of classical two-valued semantics, the semantics employed in this case uses four truth values, namely for *true* (t), *false* (f), *undetermined* (u) and *overdetermined* (o). The fourth truth value, *overdetermined*, stands for contradictory information. That is, if an assertion $C(a)$ gets assigned the truth value *overdetermined*, then this assertion is considered to be *true* and *false* at the same time.

Let us consider Example 3 again. By applying the four-valued semantics in [13], we can infer that $Student(Johanna)$, i.e. Johanna is a student. At the same time, we can infer that both $Researcher(Johanna)$ (Johanna is a researcher) and $\neg Researcher$ $(Johanna)$ (Johanna is not a researcher). Therefore, by applying the paraconsistent semantics, we may infer contradictory conclusions.

4   Discussion and Conclusion

In this paper we give a brief survey of existing work on inconsistency handling in DL-based ontologies. We divide the existing approaches into two families: the approaches that resolve inconsistency and the approaches that reason with inconsistent ontologies.

When resolving inconsistency, we differentiate logical inconsistency from incoherence. The former is in the sense of inconsistency in classical logic and the latter is more DL-specific. The debugging and repair approaches are usually applied to resolve incoherence. When resolving inconsistency, we can either delete an axiom in the ontology or weaken it. The weakening-based approaches are usually more fine-grained than the deletion-based ones. However, the computational complexity of the former approaches are usually greater.

When reasoning with inconsistent ontologies, we either select some consistent subontologies and apply a DL-reasoner to answer the query or change the semantics of the ontology languages. The first kind of approaches do not have good semantical explanation and are usually syntax-dependent. However, they are proven to be very efficient for reasoning with some real life ontologies. The second kind of approaches have semantical definitions. However, we may draw contradictory conclusions using paraconsistent semantics.

Acknowledgements. The second author was supported by the European Commission NeOn (IST-2006-027595, http://www.neon-project.org/). We also want to thank Yue Ma for her valuable comments on the draft paper.

References

1. Baader, F., Calvanese, D., McGuinness, D.L., Nardi, D., Patel-Schneider, P.F. (eds.): The Description Logic Handbook: Theory, Implementation, and Applications. Cambridge University Press, New York (2003)
2. Benferhat, S., Kaci, S., Le Berre, D., Williams, M.-A.: Weakening conflicting information for iterated revision and knowledge integration. Artif. Intell. 153(1-2), 339–371 (2004)
3. Flouris, G., Huang, Z., Pan, J.Z., Plexousakis, D., Wache, H.: Inconsistencies, negations and changes in ontologies. In: Proc. of AAAI 2006 (2006)
4. Friedrich, G., Shchekotykhin, K.M.: A general diagnosis method for ontologies. In: Gil, Y., Motta, E., Benjamins, V.R., Musen, M.A. (eds.) ISWC 2005. LNCS, vol. 3729, pp. 232–246. Springer, Heidelberg (2005)
5. Haarslev, V., Möller, R.: RACER system description. In: Goré, R.P., Leitsch, A., Nipkow, T. (eds.) IJCAR 2001. LNCS (LNAI), vol. 2083, pp. 701–706. Springer, Heidelberg (2001)
6. Haase, P., Stojanovic, L.: Consistent evolution of owl ontologies. In: Gómez-Pérez, A., Euzenat, J. (eds.) ESWC 2005. LNCS, vol. 3532, pp. 182–197. Springer, Heidelberg (2005)
7. Horrocks, I.: The fact system. In: de Swart, H. (ed.) TABLEAUX 1998. LNCS (LNAI), vol. 1397, pp. 307–312. Springer, Heidelberg (1998)
8. Huang, Z., van Harmelen, F., ten Teije, A.: Reasoning with inconsistent ontologics. In: Proc. of 19th International Joint Conference on Artificial Intelligence(IJCAI 2005), pp. 254–259. Morgan Kaufmann, San Francisco (2005)
9. Kalyanpur, A., Parsia, B., Grau, B.C., Sirin, E.: Justifications for entailments in expressive description logics. Technical report, University of Maryland Institute for Advanced Computer Studies (UMIACS) (2006)
10. Kalyanpur, A., Parsia, B., Sirin, E., Grau, B.C.: Repairing unsatisfiable concepts in owl ontologies. In: Sure, Y., Domingue, J. (eds.) ESWC 2006. LNCS, vol. 4011, pp. 170–184. Springer, Heidelberg (2006)
11. Katsuno, H., Mendelzon, A.O.: Propositional knowledge base revision and minimal change. Artif. Intell. 52(3), 263–294 (1992)

12. Lam, J., Pan, J.Z., Seeman, D., Vasconcelos, W.: A fine-grained approach to resolving unsatisfiable ontologies. In: Proc. of WI 2006 (2006)
13. Ma, Y., Hitzler, P., Lin, Z.: Algorithms for Paraconsistent Reasoning with OWL. In: Proceedings of ESWC2007 (to appear, 2007)
14. Ma, Y., Lin, Z., Lin, Z.: Inferring with inconsistent OWL DL ontology: A multi-valued logic approach. In: Grust, T., Höpfner, H., Illarramendi, A., Jablonski, S., Mesiti, M., Müller, S., Patranjan, P.-L., Sattler, K.-U., Spiliopoulou, M., Wijsen, J. (eds.) EDBT 2006. LNCS, vol. 4254, pp. 535–553. Springer, Heidelberg (2006)
15. Meyer, T., Lee, K., Booth, R.: Knowledge integration for description logics. In: Proc. of 20th National Conference on Artificial Intelligence (AAAI 2005), pp. 645–650. AAAI Press (2005)
16. Meyer, T., Lee, K., Booth, R., Pan, J.Z.: Finding maximally satisfiable terminologies for the description logic alc. In: Proc. of AAAI 2006 (2006)
17. Parsia, B., Sirin, E., Kalyanpur, A.: Debugging OWL ontologies. In: Proc. of WWW 2005, pp. 633–640 (2005)
18. Patel-Schneider, P.F.: A four-valued semantics for terminological logics. Artificial Intelligence 38, 319–351 (1989)
19. Qi, G., Hunter, A.: Measuring incoherence in description logic-based ontologies. In: Proc. of ISWC 2006. Springer Verlag (to appear, 2007)
20. Qi, G., Liu, W., Bell, D.A.: Knowledge base revision in description logics. In: Fisher, M., van der Hoek, W., Konev, B., Lisitsa, A. (eds.) JELIA 2006. LNCS (LNAI), vol. 4160, pp. 386–398. Springer, Heidelberg (2006)
21. Qi, G., Liu, W., Bell, D.A.: A revision-based algorithm for handling inconsistency in description logics. In: Proc. of NMR 2006 (2006)
22. Schlobach, S.: Diagnosing terminologies. In: Proc. of AAAI 2005, pp. 670–675 (2005)
23. Schlobach, S., Cornet, R.: Non-standard reasoning services for the debugging of description logic terminologies. In: IJCAI 2003, pp. 355–362 (2003)
24. Schlobach, S., Huang, Z., Cornet, R.: Inconsistent ontology diagnosis: Evaluation. Technical report, Department of Artificial Intelligence, Vrije University Amsterdam; SEKT Deliverable D3.6.2 (2006)
25. Straccia, U.: A sequent calculus for reasoning in four-valued description logics. In: Galmiche, D. (ed.) TABLEAUX 1997. LNCS, vol. 1227, pp. 343–357. Springer, Heidelberg (1997)

Fuzzy-DL Reasoning over Unknown Fuzzy Degrees

Stasinos Konstantopoulos and Georgios Apostolikas

Institute of Informatics and Telecommunications
NCSR 'Demokritos'
Ag. Paraskevi 153 10, Athens, Greece
{konstant,apostolikas}@iit.demokritos.gr

Abstract. In this paper we describe a fuzzy Description Logic reasoner which implements resolution in order to provide reasoning services for expressive fuzzy DLs. The main innovation of this implementation is the ability to reason over assertions with abstract (unspecified) fuzzy degrees. The answer to queries is, consequently, an algebraic expression involving the (unknown) fuzzy degrees and the degree of the query. We describe the implementation and discuss a use case in the domain of semantic meta-extraction where conventional DL reasoning is not applicable.

1 Introduction

One of the most active areas of research around the construction of the Semantic Web is the bridging of the *semantic gap*, the difference between the automatically extracted, concrete features of web objects and the abstract, semantic features necessary for semantic browsing and querying.

In the domain of multimedia analysis, conceptual modelling technologies are used to bridge this gap, by defining abstract concepts like 'interview' in terms of concrete features, e.g., the recognition of two human figures and speech in a video. Such definitions are made in the context of *ontologies*, representation technologies that capture conceptual knowledge about a domain.

There are two sources of error in this scenario: the video analysis tools used to extract the concrete features (e.g., the appearance of human figures or of a microphone in the video) and the logic rules used to infer abstract features from concrete ones. Since neither of these two levels performs perfectly, erroneous features are going to be assigned at some point; negative feedback from the users of the system is invaluable for improving the system, but it is neither reasonable nor reliable to expect that users accurately identify the source of the error, so that feedback is directed to the party responsible.

In our example, requiring that users giving feedback know about the system-internal definition of the concept 'interview' and are able to tell if the definition is not applicable (e.g., a video showing a conversation between a shopkeeper and a customer) or the recognition was faulty (the video was, in fact, a documentary where a single person describes a statue) is only going to result in sparser and less reliable feedback due to the increased complexity of user input required.

R. Meersman, Z. Tari, P. Herrero et al. (Eds.): OTM 2007 Ws, Part II, LNCS 4806, pp. 1312–1318, 2007.

The problem we are tackling in the work described here is exactly this: given that a user has flagged an abstract feature as wrong, decide whether it is more appropriate to direct this feedback to the video analysis tools or to the ontological definitions. The problem is further convoluted by the fact that it is often desired that such definitions are not absolute, but vague. In our example, consider an ontological rule that states that a video of two people talking *looks like* an interview to a degree of 80%.

This paper is organized as follows: we first provide an overview of vague ontological reasoning, and then proceed to describe a methodology for utilizing negative user feedback in the context of semantic meta-data extraction. Subsequently, we concentrate on our proposed reasoning system, which supports the requirements of this methodology. Finally we draw conclusion and outline future research directions.

2   Reasoning over Vague Knowledge

Ontologies are representation technologies that capture conceptual knowledge about a domain by defining a hierarchy of *concepts*, where more general concepts subsume more specific ones. Concepts are sets of *instances*, or individual objects of the domain. Instances have *properties*, which relate them either to other instances of the domain or to concrete values (e.g. numbers or strings). Properties of the former kind are called *relations* and of the latter *data properties*.

One of the most prominent formalisms for representing ontological knowledge is OWL [1]. OWL is closely coupled to *Description Logics* (DLs) [2], a fragment of first-order predicate logic. DLs give up expressivity in favour of lower computational complexity, but care is taken that their expressivity is sufficient to reason over OWL ontologies [3].

In order to be able to capture 'vague' knowledge—note the 'looks like' in our example above—multi-valued logics replace the binary yes-no valuation of logical formulae with a numerical one, denoting the *degree* to which the formula holds. *Multi-valued DLs* have been successfully used in multimedia information extraction [4] and the ability to model vague concepts has been explicitly stated as a desideratum by the Semantic Web community [5].

Logical formalisms like Description Logics are typically interpreted with *set-theoretic semantics* which define logical connectives and operators in terms of set theory. We shall not here re-iterate these formal foundations, but refer to handbooks of Description Logics [2, Chapter 2]. Informally, unary predicates (concepts) are interpreted as sets of individuals, binary predicates (relations) as sets of pairs of individuals, and the logical connectives as set operations; i.e., concept disjunction is interpreted as set union, concept conjunction as set intersection, and so on.

Binary logics base their interpretations on *crisp* set theory, where an individual's membership in a set gets a binary (true-false) valuation. Multi-valued logics, on the other hand, base their interpretations on vague set-theoretic semantics, where an individual's membership in a set gets associated with one of many (instead of two) possible values, typically real values between 0 and 1.

Fuzzy set theory [6] is such a multi-valued set theory, where the valuation denotes the *degree* to which an individual is a member of the set, or, in other words, the degree to which an individual is a typical member of a set. Fuzzy interpretations are based on algebraic *norms* that provide multi-valued semantics for the logical connectives; the norm that applies to conjunction is called *triangular norm* or *t-norm* and the one for implication is the *implication norm* or *i-norm*. In the work described here we use Łukasziewicz semantics, where the t-norm of the expression $X \wedge Y$ is given by $\max(0, X + Y - 1)$ and the i-norm of $X \rightarrow Y$ by $\max(1, 1 - X + Y)$

The most common approach to implementing multi-valued reasoners is to combine proof algorithms, like *resolution* or *tableaux*, with numerical methods [7,8]. It should noted, however, that all existing reasoning algorithms and implementations require that the degrees of all assertions in the knowledge base are numerical constants, a restriction which renders them unsuitable for our back-propagation methodology described in Section 3 below.

3   Error Back-Propagation

As mentioned in the introduction, we are addressing the issue of analysing erroneous results by a blame assignment system, in order to provide corrective feedback to the level that would be more likely to have introduced the error.

We do this by providing a simple cost measure for the desired changes in the degrees of the concrete features, so that the new system correctly tags the video instance. Given the current system parameters and a new instance of erroneous feature assignment, we need to identify the first-level feature fuzzy degrees that (a) would have yielded 'acceptable' output; and (b) are as close as possible to the degrees calculated by the first-level (video analysis) tools, with respect to Euclidean distance.

User feedback does not give any information for the intermediate level of the system, neither does it include any specific fuzzy degree. Instead it is a binary correct/incorrect opinion, or, at most, a qualitative estimate like 'clearly incorrect', 'almost incorrect', etc. Either way, the system has prior thresholds for translating quantitative membership (fuzzy degrees) to such qualitative descriptions of membership to a concept. In this context, we define as 'acceptable' in point (a) above, a value that satisfies such thresholds for the user's qualitative estimation.

We refer to this scheme as *back-propagation*, since it propagates the error observed at the results of the second level (logic rules) back to the intermediate results of the first level (video analysis). We shall here only briefly outline this method, as it is discussed in detail elsewhere [9]. The goal is to find the concrete feature degree values (first level output) that result in abstract feature degrees (second level output) that satisfy points (a) and (b) above. The method selects a first-level feature and makes its degree an unknown variable, then uses the reasoner to calculate the algebraic relation between this degree and the degree of the abstract feature that was found to be erroneous; this relation has the form of inequalities constraining the possible value assignments. Given the thresholds corresponding to the qualitative description provided by the user, these inequalities are used to calculate thresholds for the concrete feature degrees.

This procedure is repeated for all concrete features, but it should be noted that not all features contribute overall constraints since some proofs might not involve unknown degree values. The goal is now to find the vector of first-level feature degrees that yields an acceptable output and at the same time has the smallest Euclidean distance from the original first-level features. This is achieved by employing an iterative method. An initial acceptable first-level feature vector is found using a coarse search in the feature vector space. Afterwards, the algorithm iterates through the elements in the vector searching for alternate vectors with higher proximity to the original first-level feature vector. The procedure terminates when no more optimization is possible.

4   Reasoning over Unknown Fuzzy Degrees

In order to overcome the limitation of existing fuzzy DL reasoners that all assertions in the knowledge base (KB) are numerical constants, we have designed a novel reasoning methodology which can reason over KBs where some of the degrees as left as variables.

YADLR is a prototype implementation of this methodology, written in Prolog. Its architecture specifies three modules, for all of which multiple implementations are possible. The central module is the *inference engine*, which implements a deduction method like *resolution* or *tableaux*. The inference engine relies on an *algebraic norm* module, which provides semantics for the logical operators. Finally, the *clause representation* module acts as a front-end which translates assertions and queries into YADLR's internal representation, and utilises the inference engine in order to calculate the answer to the query posed.

In its current state of development[1] YADLR implements an SLG Resolution inference engine, the Łukasziewicz set of norms, and a Prolog-term front-end. This last module accepts logical statements in the form of nested Prolog terms, optionally coupled with a fuzzy degree. If the fuzzy degree is omitted it is assumed that its value is unknown and should treated as a variable when calculating derivative fuzzy degrees.

The front end provides reasoning services by converting calls to the service to equivalent logic queries. The three services provided are checking if a given instance is a member of a given concept, retrieving all members of a concept, and calculating all concepts an instance is a member of. All services admit two calling modes, one where the fuzzy degree of the answer to the query is returned, and one which accepts as input a minimum degree and checks whether the query can be answered at a smaller or equal degree of vagueness.

---

[1] See http://sourceforge.net/projects/yadlr

4.1   Fuzzy SLG Resolution

Many common resolution calculi and algorithms are based on the *resolution rule* [10]. The resolution rule specifies the conditions under which a clause R can be deduced from two clauses C_1 and C_2. Resolution is a sound and complete deduction rule for first-order logic, which is to say that it identifies *all* and *only* the cases where two clauses semantically support a third clause.

Selection Linear resolution for General Logic Programming (SLG Resolution, [11]) is a deduction algorithm based on the resolution rule. Given a *query* or *goal* G to prove, SLG resolution makes a left-to-right pass on the literals l_i of G, identifying which (if any) clause in the Knowledge Base (KB) can resolve against l_i. Effectively, each l_i is replaced by the premises of a clause that has l_i as its conclusion. The repeated application of this process takes G through a series of transformations G', G'', etc. until a clause is reached that is either obviously true, i.e., a known logical tautology or a (set of) ground fact(s)—or obviously false, i.e., a logical contradiction.

Literals that, at a given step, can be neither proven nor disproven are placed in a *delay list*. If further down the proof a delayed literal is proven, it is removed from the list; if contradicted the proof fails. A successful proof with an empty delay list means that the formula is satisfied by the background. If, on the other hand, the delay list cannot be closed, a conditional answer is returned which means that the formula is *satisfiable* (subject to the items remaining in the delay list) but not necessarily *satisfied* by this particular KB.

Crisp SLG Resolution is the inference apparatus behind deductive database systems like DATALOG and disjunctive DATALOG. In the domain of Description Logics, the KAON2 system[2] reasons by reducing DL programs to their disjunctive DATALOG equivalents and then using resolution-based reasoning services originally designed for disjunctive DATALOG [12].

In YADLR a fuzzy variation of the SLG algorithm is implemented, where each transformation checks whether the fuzzy degree of the result is above a threshold, and only admits transformation steps that pass this threshold; a successful proof is one that proves that the degree at which the goal is supported by the knowledge base, is above a user-specified threshold. A further refinement of this algorithm, discussed in the following section, also handles unknown fuzzy degrees (in the knowledge base as well as in the goal) and proves algebraic relations between these unknown values.

4.2   Handling Unknown Fuzzy Degrees

When fuzzy values are left as variables in either the knowledge base or the goal G, YADLR effectively restricts the range of fuzzy values that the original clause G admits. More specifically, when ground facts are used in some transformation, the degree of the derivative of G cannot assume certain values if the transformation is to be valid, as specified by the set of norms in use.

---

[2] See also http://kaon2.semanticweb.org/

If all fuzzy degrees are known in advance, this restriction can be immediately checked and the transformation can, accordingly, be accepted of rejected, as shown in the first approximation of the YADLR algorithm in the preceding section. If unknown fuzzy degrees are involved, then each application of the t-norm builds up an ever more restrictive set of inequalities that must be satisfied. At each application of the basic resolution step, the inference module uses the i-norm found in the algebraic norms module, and the latter returns this new set of restrictions.

As seen above, the inference engine defers to the norms module the calculation of the degree of each derivation of G. The i-norm implementation should be able to handle KB assertions where the degree is not specified, but left as a free variable.

In fact, such degrees are not completely unspecified, but possibly restricted by previous iterations, in the same way that the degree of the goal gets restricted. As the proof proceeds, the degree of each node gets calculated using the algebraic norms; whenever assertions with unbound fuzzy degrees are encountered, the admissible values for these variables get restricted within the range that would yield the required valuation for the overall expression, which builds up to a system of linear constraints that is solved using the $clp(\mathcal{Q}, \mathcal{R})$ constraint linear programming library [13].

At the end of this process, and if there are any open branches, one collects at the leaves of the open branches a system of inequations. This system specifies the admissible values for the unbound degrees, so that the original formula at the root of the tree is satisfiable.

It should be noted that some transformations might be invalid even when unknown values are involved, as we might, for example, end up requiring that $x < 0.3$ and $x > 0.5$ simultaneously, but in general a lot more (conditional) solutions will be admitted than in a situation where all values are known.

The answer to the logical query is a disjunction of sets of inequalities, involving the unknown fuzzy degrees in the KB and the query.

5   Conclusions and Future Work

In this work we have proposed a novel method for reasoning over fuzzy Description Logics and demonstrated how it can be useful in improving the accuracy of meta-data extraction from multimedia content. More particularly, we have discussed how to reason over knowledge bases that include assertions of an unknown fuzzy degree, and how this is useful to error back-propagation in a fuzzy DL system.

At its current state of development, the system implements the general SLG resolution algorithm with the Łukasziewicz norms for multi-valued semantics. Planned future development includes designing and implementing a resolution methodology that is optimized for reasoning over Description Logic knowledge bases, as well as implementing more of the various algebraic norms proposed.

We are also planning to look for further use cases for our methodology besides error back-propagation for correcting meta-data extraction systems. Such use cases might include decision support systems where not all the parameters of the problem are known, but discovering the relation between the unknown parameters can provide important information.

Acknowledgements

The work described here was partially supported by DELTIO,[3] a project funded by the Greek General Secretariat of Research & Technology. DELTIO focuses on evolving ontologies to analyse multimedia content, with an application to news broadcasts.

References

1. Smith, M.K., Welty, C., McGuinness, D.L.: OWL web ontology language. Technical report, World Wide Web Consortium (February 2004) W3C Candidate Recommendation, http://www.w3.org/TR/owl-guide/
2. Baader, F., Calvanese, D., McGuinness, D., Nardi, D., Patel-Schneider, P.: The Description Logic Handbook: Theory, Implementation and Applications. Cambridge University Press, Cambridge (2003)
3. Horrocks, I., Patel-Schneider, P.F., van Harmelen, F.: From $\mathcal{SHIQ}$ and RDF to OWL: The making of a web ontology language. Journal of Web Semantics 1(1), 7–26 (2003)
4. Meghini, C., Sebastiani, F., Straccia, U.: A model of multimedia information retrieval. Journal of the ACM 48(5), 909–970 (2001)
5. Kifer, M.: Requirements for an expressive rule language on the semantic web. In: Proc. of W3C Workshop on Rule Languages for Interoperability (2005)
6. Zadeh, L.A.: Fuzzy sets. Information and Control 8(3), 338–353 (1965)
7. Stoilos, G., Stamou, G., Tzouvaras, V., Pan, J.Z., Horrocks, I.: The fuzzy description logic f-SHIN. In: Proc. of the International Workshop on Uncertainty Reasoning For the Semantic Web (2005)
8. Straccia, U.: Reasoning within fuzzy Description Logics. Journal of Artificial Intelligence Research 14, 137–166 (2001)
9. Apostolikas, G., Konstantopoulos, S.: Error back-propagation in multi-valued logic systems. In: Proc. International Conf. on Computational Intelligence and Multimedia Applications (ICCIMA 2007), Sivakasi, India, December 13–15 (2007)
10. Robinson, J.A.: A machine-oriented logic based on the resolution principle. Journal of the ACM 12(1), 23–41 (1965)
11. Chen, W., Warren, D.S.: Tabled evaluation with delaying for general logic programs. Journal of the ACM 43, 20–74 (1996)
12. Hustadt, U., Motik, B., Sattler, U.: Reasoning for description logics around SHIQ in a resolution framework. Technical Report 3-8-04/04, Forschungszentrum Informatik, Karlsruhe (2004)
13. Holzbaur, C.: OFAI clp(Q,R) manual, edition 1.3.3. Technical Report TR-95-09, Austrian Research Institute for Artificial Intelligence, Vienna (1995)

---

[3] See also http://www.atc.gr/deltio/

Application of Tree Mining to Matching of Knowledge Structures of Decision Tree Type

Fedja Hadzic and Tharam S. Dillon

Digital Ecosystems and Business Intelligence Institute,
Curtin University of Technology, Perth, Australia
fedja.hadzic@postgrad.curtin.edu.au,
tharam.dillon@cbs.curtin.edu.au

Abstract. Matching of knowledge structures is generally important for scientific knowledge management, e-commerce, enterprise application integration, etc. With the desire of knowledge sharing and reuse in these fields, matching commonly occurs among different organizations on the knowledge describing the same domain. In this paper we propose a knowledge matching method which makes use of our previously developed tree mining algorithms for extracting frequent subtrees from a tree structured database. Example decision trees obtained from real world domains are used for experimentation purposes whereby some important issues that arise when extracting shared knowledge through tree mining are discussed. The potential of applying tree mining algorithms for automatic discovery of common knowledge structures is demonstrated.

1 Introduction

Knowledge discovery task in general can be hard and time consuming, and hence sharing the already developed knowledge representations is desirable. This particularly occurs when a number of organizations coming from the same knowledge domain would like to have a knowledge basis on which they can integrate their organization specific knowledge. Having a shared knowledge model would save time and costs associated with acquiring general knowledge about the domain at hand.

Matching of knowledge structures has been of interest for a long time and many useful applications can be found in e-commerce, enterprise application integration and the general management of scientific knowledge. The emergence of semi-structured data sources (such as XML) which are commonly used for describing domain knowledge has called for the development of methods for matching of tree structured documents. The TreeDiff software package [1] takes two documents as input, represents them as ordered labeled trees and finds sequence of edit operations to transform one document tree into another. Finding common structures among semi-structured documents is useful for document clustering methods, since the structure is usually ignored by traditional clustering methods [2]. An algorithm which finds largest approximate common substructures between ordered labeled trees has been presented in [3]. A tree inclusion algorithm presented in [4] checks for the inclusion of a query tree that can be of an embedded subtree type. Another related area where

R. Meersman, Z. Tari, P. Herrero et al. (Eds.): OTM 2007 Ws, Part II, LNCS 4806, pp. 1319–1328, 2007.

matching of tree structured knowledge models will be useful is in a classification ensemble which is a multi-classifier system where each classifier is developed for the same domain problem [5, 6]. Hence, if the classifiers are of decision tree type the shared knowledge structure would indicate the general knowledge of the domain. The new classifier is then expected to have better generalization capability and hence be more accurate in classifying future unseen data objects.

The method presented in this paper is based on the use of our previously developed tree mining algorithms in order to automatically extract common knowledge structures. While in this paper we focus on matching of decision tree, the method is generally applicable to matching of tree structured knowledge representations. By using the tree mining approach to knowledge matching many of the structural differences among the knowledge representations can be efficiently detected and the largest common structure is automatically extracted. The implications of mining different subtree types are discussed and for knowledge matching task the most suitable subtree type within the current tree mining framework is indicated. Decision trees obtained from real world datasets are used for experimental purposes. We have previously applied our algorithms on large and complex tree structures and experimentally demonstrated their scalability [7, 8]. In this paper we consider smaller decision trees in order to illustrate the underlying concept in a more comprehensible manner.

2   Frequent Subtree Mining

This section starts by providing the tree mining related definitions necessary for the purpose of describing the present work. We then proceed with an overview of our contributions in the area of frequent subtree mining. Please note that due to space and scope limitations we do not explain the details of our algorithms and the various issues involved when developing tree mining algorithms. For a more detailed overview of the area including related works and algorithm comparisons one may refer to our previous works [7, 8, 9], where such information has been provided

A tree can be denoted as $T(v_0, V, L, E)$, where (1) $v_0 \in V$ is the root vertex; (2) V is the set of vertices or nodes; (3) L is the set of labels of vertices, for any vertex $v \in V$, $L(v)$ is the label of v; and (4) $E = \{(x,y)| x,y \in V \}$ is the set of edges in the tree. A *root* is the topmost node in the tree. The *Parent* of node v is defined as the predecessor of node v. A node v can only have one parent while it can have one or more *children*. A node without any child is a *leaf* node; otherwise, it is an *internal* node. If for each internal node, all the children are ordered, then the tree is an *ordered tree*. The number of children of a node is commonly termed as *fan-out/degree* of the node. A path from vertex v_i to v_j, is defined as the finite sequence of edges that connects v_i to v_j. The length of a path p is the number of edges in p. If p is an *ancestor* of q, then there exists a path from p to q.

The problem of frequent subtree mining can be generally stated as: given a tree database T_{db} and minimum support threshold (σ), find all subtrees that occur at least σ times in T_{db}. Within this framework the two most commonly mined types of subtrees are induced and embedded. An induced subtree preserves the parent-child relationships of each node in the original tree. In addition to this, an embedded subtree allows a parent in the subtree to be an ancestor in the original tree and hence ancestor-descendant relationships are preserved over several levels. Formal definitions follow.

Induced subtree. A tree $T'(r', V', L', E')$ is an *induced subtree* of a tree T (r, V, L, E) *iff* (1) $V' \subseteq V$, (2) $E' \subseteq E$, (3) $L' \subseteq L$ and $L'(v)=L(v)$, (4) $\forall v' \in V'$, $\forall v \in V$, v' is not the root node, and v' has a parent in T', then parent(v')=parent(v),

Embedded subtree. A tree $T'(r', V', L', E')$ is an embedded subtree of a tree $T(r, V, L, E)$ iff: (1) $V' \subseteq V$; (2) *if* $(v_1, v_2) \in E'$ *then* parent$(v_2) = v_1$ *in* T' *and* v_1 *is ancestor of* v_2 *in* T; if $v' \in V'$, $v \in V$ and v' is not the root node, set *ancestor*$(v') \subseteq$ set *ancestor*(v) and set *ancestor*$(v') \neq \emptyset$; $L' \subseteq L$ and $L'(v)=L(v)$.

Level of embedding. If $T'(r', V', L', E')$ is an embedded subtree of T, and there is a path between two nodes p and q, the *level of embedding* (δ) is defined as the length of the path between p and q, where $p \in V'$ and $q \in V'$, and p and q form an ancestor-descendant relationship. A *maximum level of embedding* (Φ) is the limit on the level of embedding between any p and q. In other words, given a tree database T_{db} and Φ, then any embedded subtree to be generated will have the maximum length of a path between any two ancestor-descendant nodes equal to Φ.

In addition to the previous definitions, the subtrees can be further distinguished based upon the ordering of sibling nodes. An ordered subtree preserves the left-to-right ordering among the sibling nodes in the original tree while in an unordered subtree this ordering is not preserved. In other words, for an unordered subtree the order of the siblings (and the subtrees rooted at sibling nodes) can be exchanged and the resulting subtree would be considered the same. Examples of different subtree types are given in Fig. 1 below.

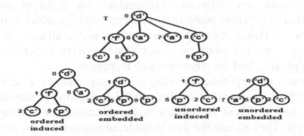

Fig. 1. Example of a tree T and its different subtree types

To determine the frequency of a subtree, most commonly used support definitions are transaction based and occurrence match support [7] and the choice is application dependant. Transaction based support (TS) is used when only the existence of items within a transaction is considered important, whereas occurrence match (OC) support takes the repetition of items in a transaction into account and counts the subtree occurrences in the database as a whole. Recently, we have proposed the hybrid support definition [10]. Using hybrid support threshold of xly, a subtree is considered frequent iff it occurs in x transactions and it occurs at least y times in each of those x transactions. Hence, it keeps the extra information about the intra-transactional occurrences of a subtree.

Our work in the area of frequent subtree mining is characterized by adopting a Tree Model Guided (TMG) [11, 7] candidate generation. This non-redundant systematic enumeration model ensures only valid candidates are generated which conform to the actual tree structure of the data. Embedding List [11] representation of the tree structure has allowed for an efficient implementation of the TMG approach which resulted in efficient algorithm MB3-Miner [11] for mining of ordered embedded subtrees, which was accompanied with the TMG mathematical model for estimating the worst case complexity of enumerating all embedded subtrees. The large complexity of mining embedded subtrees motivated our Level of Embedding [7] constraint so that one can decrease the level of embedding constraint gradually down to 1, from which all the obtained subtrees are induced. Razor algorithm [12] was a further extension developed for mining embedded subtrees where the distance of nodes relative to the root of the subtree needs to be considered. Using the general TMG framework the UNI3 [9] and U3 [8] algorithms mine unordered induced and embedded subtrees, respectively. From the application perspective, in [13] we have indicated the potential of the tree mining algorithms in providing interesting biological information when applied to tree structured biological data.

3 Overview of the Approach

In this section we provide a brief overview of our approach to knowledge matching. It is motivated by the real world scenario of different organizations from the same domain obtaining their knowledge models separately and then in future wanting to match their models in order to obtain a shared understanding of the domain. In real world, it is common that different organizations use different names for same concepts but this is a different problem of concept matching which is out of the scope of the current work. Hence, the current assumption is that all concept (attributes and constraints) names are the same or if not some concept matching algorithm has already been applied to find the corresponding mappings.

Fig. 2 illustrates the experimental setup in this paper used for demonstrating how knowledge matching can be achieved through the application of tree mining. Suppose that there are four different organizations in possession of database (DB) for a particular domain. They all use the data considered relevant for the domain to build a knowledge model (KM) by using some existing data mining tools. Each KM is most

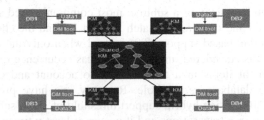

Fig. 2. Illustration of the knowledge matching approach

likely to differ in the way the knowledge is structured and the amount of concept granularity. However since they are describing the same domain there are usually parts of knowledge equivalent among different KMs. This is where the tree mining algorithms will prove useful since they are capable of efficiently extracting substructures from large tree databases. More specifically, the shared knowledge model will be obtained by extracting the largest common sub-structure among the available knowledge models.

The obtained knowledge model could be less specific than the knowledge model from a particular data source, but it is therefore valid for all organizations. Furthermore, each of the different organizations could have their own specific part of knowledge which is only valid from their perspective, and which can be added to the shared knowledge model so that every aspect for that organization is covered. Hence, the shared knowledge model can be used as the basis for structuring the knowledge for that particular domain and different communities of users can extend this model when required for their own organization specific purposes.

In this section we have described a real world scenario which we aim to mimic in this paper and details are provided in the sections that follow. However, the core of our proposed method is the application of tree mining for extracting the common knowledge structures from knowledge models provided by different organizations.

4   Experimental Setup

The publicly available datasets from the UCI machine learning repository [14] were used. In real world scenarios, the data collected by different organizations can differ in various aspects. The instance collection may differ and certain attributes may be found important by one organization and irrelevant by another. To account for these possible differences in our experiments the subset of data used as input to the data mining tools was different in some cases, and at times we used a feature selection criterion [15] to include only a subset of domain attributes for learning. The end result is that certain models will be more general or specific than others. The described methods of adding variability among knowledge representations were performed only if the classification accuracy remained sufficiently high. In some cases the application of these methods did not have much effect on the obtained decision tree and hence in our domains the number of decision trees considered may differ.

Once the knowledge models are obtained they are all represented as a separate subtree in the tree database. In this regard each of the knowledge models can be viewed as a separate transaction within the tree database. This kind of representation is suitable for the application of tree mining algorithms for the analysis of independent knowledge structures. Since in this work we are not considering the problem of concept matching we are assuming that the concepts describing the same aspect (attribute or constraint) of a domain have the same label. Next, we describe each of the domains considered and show the different knowledge models obtained.

Wisconsin Breast Cancer Database. This dataset consists of 10 attributes describing the clinical cases. The task is to predict whether a clinical instance is classified as

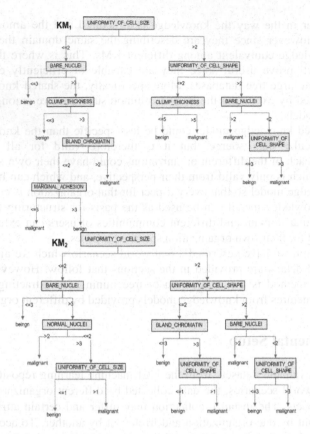

Fig. 3. Knowledge models for breast-cancer domain (class values: benign, malignant)

benign or malignant. For this domain we could only obtain two knowledge models that are different while still being sufficiently accurate. These are displayed in Fig. 3.

Voting Records. This data set describes the 1984 United States Congressional Voting Records Database [14]. It consists of 16 Boolean attributes indicating a certain policy and the classification task is to predict whether the instance describes a democrat or a republican. For this domain we obtained three different and accurate knowledge models, which are displayed in Fig.4.

Please note that for the ease of representation the edges in the displayed knowledge models are labeled, rather than being considered as separate nodes themselves. However, within the current tree mining framework the edge labels are usually not considered. Each of the labeled edges from the picture above becomes a node itself with its parent node being the node from which the edge is emanating, whereas its child node is the node to which the edge is connecting.

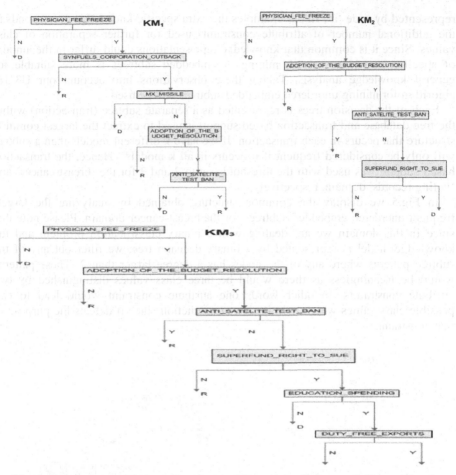

Fig. 4. Knowledge models for voting domain (class values: D-democrat, R- republican)

5 Results and Discussion

When comparing decision trees the order of sibling nodes is not considered important, and hence unordered subtree mining is more suitable. One further choice to make is whether we are interested in mining induced or embedded subtrees. The difference occurs in the fact that if embedded subtrees are mined we are allowing structures to be considered similar even if they occur at different levels in the tree. In an embedded subtree the relationship is not limited to parent-child and hence by allowing ancestor-descendant relationships enables the extraction of more sub-structures where the levels of embeddings between the nodes are not limited to one and can be different. For example if in knowledge representation 'A' the level of embedding between the nodes 'a' and 'b' representing some domain concepts is much larger than the level of embedding between the nodes representing the same concepts in knowledge representation 'B', then 'A' stores more specific knowledge about the concept represented by node 'a'. In our examples the extra specific knowledge corresponds to the additional number of attribute constraints used for further separation of class values. Since it is common that knowledge representations could differ in the amount of specific knowledge stored, mining of embedded subtrees is more suitable for general knowledge analysis. Taking these observations into account our U3 [8] algorithm for mining unordered embedded subtrees will be used.

Each of the decision trees is represented as a separate subtree (transaction) within the tree database and transaction based support is used to extract the largest common structure that occurs in each transaction. If we have k different models then a subtree will only be considered frequent if it occurs in all k models. Hence, the transaction based support was used with the thresholds set to 2 and 3 for the 'breast-cancer' and 'voting records' domain, respectively.

In Fig.5 we display the common structure obtained by analyzing the largest frequent unordered embedded subtrees for the breast-cancer domain. Please note that since in this domain we are dealing with a binary classification problem and the knowledge model is represented by a binary decision tree, we filter out any of the subtree patterns where any of the nodes has a degree larger than 2. These patterns would be meaningless as there would be three class values distinguished by two attribute constraints. In other words one attribute constraint would lead to two possible class values which is in itself a contradiction since it defeats the purpose of that constraint.

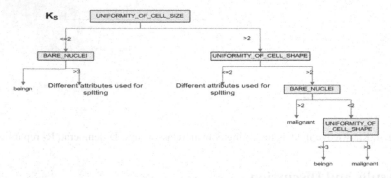

Fig. 5. Shared knowledge structure (K$_S$) of the 'breast-cancer' domain

When analyzing the filtered results there were four largest common subtrees. At the points where both KMs from Fig.3 match the two variations occur with nodes 'benign' and 'malignant' (i.e. class values). This indicates that at these points in the knowledge structures different attributes are used for further class separation (see Fig. 5). In fact if induced subtrees are mined then there would be only one largest common subtree since no further match would be made past the point where concept nodes are different. This however may not always be desirable as will be shown for the 'voting records' domain.

The largest frequent unordered embedded subtree for the voting records domain is displayed in Fig. 6. In this case there was only one largest common subtree, K$_S$. By comparing K$_S$ with the knowledge models from Fig.4 we can see that it was necessary to mine embedded subtrees in order to detect the common knowledge structure.

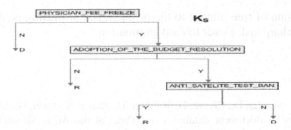

Fig. 6. Shared knowledge structure (K_S) of the 'voting records' domain

Looking back at the knowledge models from Fig. 4 it could be said that the KM_2 most closely matches the shared structure Ks from Fig.6. KM_2 only differs to the Ks in the sense that it has one extra attribute after the last node which splits the class values further into 'D' and 'R', rather than generalizing it to 'R'. The knowledge model KM_1 stores two additional attributes after the first attribute node while KM_3 stores additional three attributes at the last node. We can see that KM_2 is more general than KM_3, but KM_2 is only more general than KM_1 between the first and second attribute while it is more specific at the last node.

The examples used indicate the differences between mining induced and embedded subtrees. In the example from Fig. 3 largest induced subtree (Fig. 5) was more suitable whereas for the second example in Fig.4 largest embedded subtree (Fig. 6) allowed for the similarity in structure to be detected besides the differences in specificity. As a guide, one could start with mining of embedded subtrees and if most of the large frequent patterns are meaningless in the sense described earlier, one could only concentrate on induced subtrees. However, contrasting to our examples, the level of embeddings at which the knowledge structures can differ could in reality be very large. To make the large change from mining embedded to induced subtrees runs the risk of missing many other common structures where the difference in the level of embedding is not so great. In these cases the maximum level of embedding (Φ) [7] could be useful so that Φ is progressively decreased until some differences are resolved. In fact it probably will not provide only one matching KM as for the induced case, but larger common structures would be detected. Which method to adapt is again dependent on the type of knowledge that is being matched as for some applications induced subtrees may be sufficient and the level of embedding can be ignored while for others it is important as it indicates the extra specific knowledge stored in a different knowledge model. Either way even if the user is not a domain expert different options can be tried with respect to Φ and this should reveal some more detail about the similarities and differences among the knowledge structures.

6   Conclusions and Future Work

In this paper we have described a way in which the tree mining algorithms can be effectively used for obtaining the shared knowledge from separate knowledge models of decision tree type. Implications of mining different subtree types were discussed with the unordered embedded subtree type being the most appropriate for the task.This is our preliminary work in the area and as such the aim was to demonstrate that the application of tree mining to the problem is a promising method. It takes the knowledge matching task closer toward automation.

References

1. Wang, J.T.L., Shasha, D., Chang, G., Relihan, L., Zhang, K., Patel, G.: Structural matching and discovery in document databases. In: Proc. of the ACM SIGMOD Int'l Conf. on Management of Data, pp. 560–563 (1997)
2. Wang, K., Liu, H.: Discovering Structural Association of Semistructured Data. In: IEEE Transactions on Knowledge and Data Engineering (1999)
3. Wang, J.T.L., Shapiro, B.A., Shasha, D., Zhang, K., Currey, K.M.: An algorithm for finding the largest approximately common substructures of two trees. IEEE Transactions on Pattern Analysis and Machine Intelligence 20(8), 889–895 (1998)
4. Chen, Y., Shi, Y., Chen, Y.: Tree inclusion algorithm, signatures and evaluation of path-oriented queries. In: ACM symposium on Applied computing, Dijon, France, pp. 1020–1025 (2006)
5. Zhang, Y., Street, W.N., Burer, S.: Sharing Classifiers among Ensembles from Related Problem Domains. In: Proc. of the 5th IEEE Int'l Conf. on Data Mining (2005)
6. Prodromidis, A., Chan, P., Stolfo, S.: Metalearning in distributed data mining systems: Issues and approaches. In: KDD'07 (1997)
7. Tan, H., Dillon, T.S., Hadzic, F., Feng, L., Chang, E.: IMB3-Miner: Mining Induced/Embedded subtrees by constraining the level of embedding. In: Ng, W.-K., Kitsuregawa, M., Li, J., Chang, K. (eds.) PAKDD 2006. LNCS (LNAI), vol. 3918, Springer, Heidelberg (2006)
8. Hadzic, F., Tan, H., Dillon, T.S., Chang, E.: U3 – mining unordered embedded subtrees using TMG candidate generation. In: The 1st ACM Int'l Conf. on Web Search and Data Mining, San Francisco Bay Area, California, USA (submitted, 2008)
9. Hadzic, F., Tan, H., Dillon, T.S.: UNI3 – Efficient Algorithm for Mining Unordered Induced Subtrees Using TMG Candidate Generation. In: IEEE Symposium on Computational Intelligence and Data Mining (CIDM 2007), Honolulu, Hawaii (2007)
10. Hadzic, F., Tan, H., Dillon, T.S., Chang, E.: Implications of frequent subtree mining using hybrid support definition. Data Mining & Information Engineering, June 18-20, The New Forest, UK (2007)
11. Tan, H., Dillon, T.S., Hadzic, F., Chang, E., Feng, L.: MB3-Miner: mining eMBedded sub-TREEs using Tree Model Guided candidate generation. In: Proc. of the 1st Int'l Workshop on Mining Complex Data, in conj. with ICDM 2005, Houston, Texas, USA (2005)
12. Tan, H., Dillon, T.S., Hadzic, F., Chang, E.: Razor: mining distance constrained embedded subtrees. In: IEEE ICDM 2006 Workshop on Ontology Mining and Knowledge Discovery from Semistructured documents, December 28-22, Hong Kong (2006)
13. Hadzic, F., Dillon, T.S., Sidhu, A., Chang, E., Tan, H.: Mining Substructures in Protein Data. In: IEEE ICDM 2006 Workshop on Data Mining in Bioinformatics (DMB 2006), December 18-22, Hong Kong (2006)
14. Blake, C., Keogh, E., Merz, C.J.: UCI Repository of Machine Learning Databases. Irvine, CA: University of California, Department of Information and Computer Science (1998), http://www.ics.uci.edu/ mlearn/MLRepository.html
15. Hadzic, F., Dillon, T.S.: Using the Symmetrical Tau (τ) Criterion for Feature Selection in Decision Tree and Neural Network Learning. In: Proc. of the 2nd FSDM Workshop, in conj. with 2006 SIAM Int'l Conf. on Data Mining, Bethesda (2006)

A New Expanding Tree Ontology Matching Method

Feiyu Lin and Kurt Sandkuhl

School of Engineering, Jönköping University, Gjuterigatan 5,
551 11 Jönköping, Sweden
{feiyu.lin,kurt.sandkuhl}@jth.hj.se

Abstract. Currently, many ontologies are available even through they are created by different organizations. Although these ontologies are developed for various application purpores and areas, they often contain overlapping information. In order to achieve interoperability between ontologies, we need to find a way to integrate various ontologies. For example, ontology engineers would like to create new ontologies based on existing ontologies, adapt or extend existing ontologies for the same or different domains. In this context, it is important to find the corresponding entities in different ontologies. The main ideas we contribute to the research field are introducing the expanding ontology tree to determine the semantic similarity of ontologies.

1 Introduction

When people or machines must communicate among themselves, they need a shared understanding of the concepts. When for example the term "match" is mentioned, does it mean "game" or "equal"? An ontology can be used to express the meaning of concepts using hierarchical and other relationships, instances and axioms in order to exactly describe the concept. For instance, if we describe the "match" concept in the ontology with a subclass "football match", with a "has playing team" relationship (connecting to the concept "team"), with a "has referee" relationship, and with a "play time" relationship, people or machines can unambiguously understand that the concept "match" here is referring to a game.

Currently, many ontologies are available even though they are created by different organizations. Although these ontologies are developed for various application purposes and areas, they often contain overlapping information. In order to achieve interoperability between ontologies, we need find a way to integrate various ontologies. For example, ontology engineers would like to create new ontologies based on existing ontologies, adapt or extend existing ontology for the same or different domains. In this context, it is important to find the corresponding entities in different ontologies. In this paper, we introduce the expanding ontology tree to determine the semantic similarity of ontologies.

The rest of this paper is organized as follows. In section 2, we give some definitions that are used in the paper. In section 3, we introduce different methods that can be used to calculate distance between concepts and we discuss related work. In

R. Meersman, Z. Tari, P. Herrero et al. (Eds.): OTM 2007 Ws, Part II, LNCS 4806, pp. 1329–1337, 2007.
© Springer-Verlag Berlin Heidelberg 2007

section 4, we describe our similarity method based on the expanding ontology tree. In the final section we discuss future work.

2 Terminology

2.1 Ontology

There are many different definitions of the term "ontology" in computer science related research fields. We use Gruber's proposal as our ontology definition:

"*An Ontology is a formal, explicit specification of shared conceptualization.*" [7].

Furthermore, we use the W3C recommendation ontology language OWL (Web Ontology language) [19] to represent ontology. An OWL ontology consists of Individuals, Properties, Classes and Axioms.

2.2 Merging / Integration / Alignment / Mapping / Matching /

In our earlier work [12], the terms mapping, matching and alignment are defined as follows (based on [16] or [10]):

Ontology merging: combine two ontologies from the same subject area into a new ontology.

Ontology integration: combine two ontologies from the different subject areas into a new ontology.

Ontology alignment: identify correspondences between the source ontologies.

Ontology mapping: find equal parts in different source ontologies.

Ontology matching: find similar parts in the source ontologies or finding translation rules between ontologies.

In the context of OWL ontologies, alignment, mapping and matching concerns classes and properties in the different OWL ontologies.

The paper [23] separates different types output of ontology matching (i.e. each correspondence is in the range 0 to 1, relations between entities of a system can be more expressive result, like equivalence, subsumption, incompatibility). Here we focus on correspondences in the range 0 to 1.

3 Similarity Computation and Related Work

This chapter will summarize different methods for distance calculation between concepts (3.1) and discuss selected systems for ontology matching (3.2) according our earlier work [12].

3.1 Distance Calculation

Different strategies can be used to calculated the distance between concepts in ontologies to determine correspondence, e.g. string similarity, synonyms, structure similarity and based on instances.

For string similarity, [2] has good survey of the different methods to calculate string distance from edit-distance like functions (e.g. Levenstein distance, Monger-Elkan distance, Jar-Winkler distance) to token-based distance functions (e.g. Jaccard similarity, TFIDF or cosine similarity, Jense-Shannon distance).

Treating synonyms (with the help of dictionary or thesaurus) can help to solve the problem of using different terms in the ontologies for the same concept. For example, an ontology might use "diagram", the other could use "graph" referring to the same concern. The famous WordNet dictionary [25] can support improving the similarity measure.

Measures of structure similarity are usually based on the is-a or part-of hierarchy of the ontology in the graph. For example, if two classes' super classes and sub classes are same, we may say these two classes are the same. The Similarity flooding [15] matching algorithm uses graphs to find corresponding nodes in the graphs based on a fix-point computation. Two nodes are similar if their neighbors are also similar.

Similarity based on instances are for examples used in GLUE [3] and FCA-Merge [24]. GLUE uses multiple machine learners and is exploiting information in concept instances and taxonomic structure of ontologies. GLUE uses a probabilistic model to combine results of different learners. FCA-Merge is a method for comparing ontologies that have a set of shared instances or a shared set of documents annotated with concepts from source ontologies. Based on this information, FCA-Merge uses mathematical techniques from Formal Concept Analysis to produce a lattice of concepts which relates concepts from the source ontologies.

3.2 Ontology Matching Systems

There are some ontology matching systems available using some or all of the above methods.

One example is PROMPT [17], which is a semi-automatic tool and a plug-in for the open-source ontology editor PROTÉGÉ [21]. It determines string similarity and analyzes the structure of ontology. It provides guidance for the user for merging ontologies. It suggests the possible mapping and determines the conflicts in the ontology and proposes solutions for theses conflicts.

Another example is Chimaera [1] that is a tool for the Ontolingua editor. It supports merging multiple ontologies and diagnosing individual or multiple ontologies. If string matches are found, the merge is done automatically, otherwise the user is prompted for further action.

Next, we look at FOAM [6] which is a tool to fully or semi-automatically align two or more OWL ontologies. It is based on heuristics (similarity) of the individual entities (concepts, relations, and instances). These entities are compared using string similarity and SimSet for set comparisons.

Another tool, OLA [4] takes into account all the possible characteristics of ontologies (i.e., terminological, structural and extensional). String similarity is used to calculate similarity between labels. Structural constraints are considered during the matching.

ASCO [11] uses as much available information in the ontology as possible (e.g. concepts, relations, structure). It applies string similarity. TF/IDF is used for calculating a similarity value between descriptions of the concepts or relations. WordNet is integrated to find synonyms. Structure matching is used for modifying or asserting the similarity of two concepts or relations.

Finally, S-MATCH [5] takes the structures as input and gives semantic relationships like equivalence, more general, less general, mismatch and overlapping between the nodes of the graphs as output.

4   Semantic Similarity Based on Expanding Ontology Tree

Despite the variety of distance methods used, most of the above methods do not consider the weight of a concept based on the subset of the ontology. We use an expanding tree to represent a subset of the ontology that emphasizes the concept weight.

The following section 4.1 introduces a simple example with two ontologies. Section 4.2 introduces the rules for creating polygons and applies them on the example ontologies. In section 4.3, the example ontologies string similarities are calculated. Section 4.4 presents the result of the similarity.

4.1   Example Ontologies

Two onotlogies, Biblio.owl and BibTex.owl which are developed in the Maponto [14] project are chosen as our examples. Biblio ontology is a bibliographic ontology based on FRBI [20]. It has 12 classes, 37 object properties and 64 data type properties in Biblio ontology. BibTex ontology is a bibliographic ontology based on the bibTeX record. It has 43 classes, 22 object properties and 24 data type properties. Both ontologies have no instances.

For example, both ontologies have the class Agent (see Figure 1 and Figure 2). Figure 1 and Figure 2 are generated by the OntoViz [18] tool in the Protégé. The subclasses are connected is-a in the figures (e.g. Person and Corproate_Body). The domain and range of the relationships (object properties) are illustrated through arrows.

The Agent class similarity will be determined by three levels, based on expanding trees. Section 4.2 will describe how to draw expanding trees.

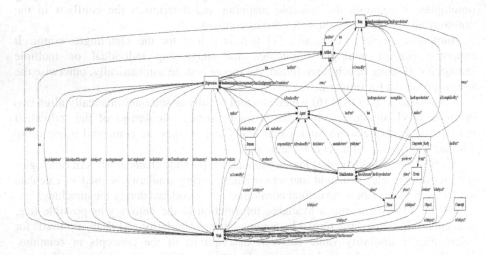

Fig. 1. Agent in Biblio Ontology

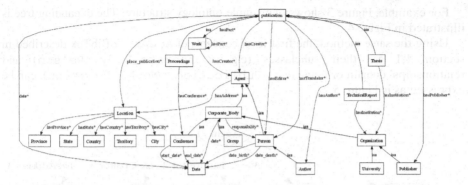

Fig. 2. Agent in BibTex Ontology

4.1.1   Expanding Ontology Tree with Three Levels

We try to go through every path that relate compared classes in the expanding trees. The different levels are given different weights depending on the depth of the compared classes. The first level concepts, which get the weight as 3 are the class' subclasses and each relationship where it is domain or range. The second level concepts which get weight 2, are depending on the first level concepts' subclasses and their relationship's ranges. Similarity we can get the third level concepts, with weight 1, based on the second level concepts.

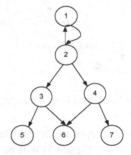

Fig. 3. Simple ontology structure

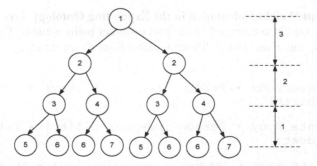

Fig. 4. Expanding tree example

For example, Figure 3 shows one simple ontology structure. The expanding tree is illustrated in Figure 4.

Using the same method, the first level concepts of `Agent` in BibTex described in section 4.1 are their subclasses (e.g. `Corporate_Body`, `Person`) and relationships' domain or range (e.g. `Publication`, `Work`, `Location`), can be expressed as in Figure 5.

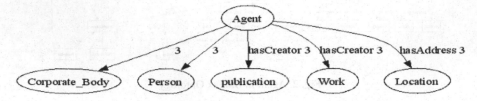

Fig. 5. The first level concepts of Agent in BibTex

The second level for `Corporate_Body` in Figure 5 includes subclasses (e.g. `Confernce`, `Group`, `Organizatin`) and relationships' domains (e.g. `Person`, `Date`). Using the same rules we can get the third level as shown in Figure 6.

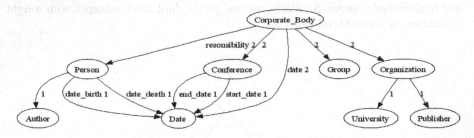

Fig. 6. The second and third level concepts of Corporate_Body

In the same way, the other concepts in the first level (e.g. `Person`, `publication`, `Work`, `Location`, see Figure 5) will be represented in the ontology tree with different paths and weights.

4.1.2 Concept Weight Calculation in the Expanding Ontology Tree
The concepts weight is summed up to from different paths weights. For example, in the Figure 6, there are three different paths from `Corporate_Body` through `Person`:

- `Corporate_Body` • `Person (responsibility)` • `Date` `(date_birth)`

- `Corporate_Body` • `Person (responsibility)` • `Date` `(date_death)`

- `Corporate_Body` • `Person (responsibility)` • `Author (is-a)`

The weight for `Corporate_Body` related to the `Person` part is:

$$3 * 3 + 2 * 3 + 1 * 3 = 18 \tag{1}$$

The weight for the whole `Corporate_Body` is 52. Similarly, we can get different concepts weights in the expanding tree.

4.2 String Similarity Calculation

SecondString [22] is an open-source package of string matching methods based on the Java language. Based on the string similarity methods comparison result in [8], The Slim TFIDF string-distance method is the most appropriate method for our approach. Slim is the same-letter index mixture distance and TFIDF is a token-based distance metric where two strings are considered as multisets of tokens. Slim TFIDF is a hybrid-distance function. First the compared strings are broken into substrings. Then Slim operates over the broken substrings.

4.2.1 Example Ontologies String Similarity Result

First, the classes' string-distances illustrated in Figure 1 and 2 will be compared. After calculating the Slim TFIDF string-distance using the SecondString tool with threshold 0.95, we get the results shown in Table 1 for the example ontologies presented in section 4.1. The scores below 0.95 will not be used.

Table 1. Classes' String-distance in Agents

Similarity	Biblio	BibTex
1.0	Person	Person
1.0	Corporate_Body	Corporate_Body
1.0	Work	Work

Table 2 shows the relationships' string-distances in Figure 1 and 2.

Table 2. Relationships' String-distance in Agents

Similarity	Biblio	BibTex
0.95	hasImitation	hasInsititution
0.96	publisher	hasPublisher
0.97	hasTranslation	hasTranslator
1.0	responsibility	responsibility

4.3 Calculation of Ontology Similarity

We treat ontology matching as asymmetric. For example, a small ontology may perfectly match some parts of large ontology, the similarity between the small ontology and large ontology is 1.0 then, but not vice versa.

The similarity between two concepts is computed as:

$$sim\,(x, y) = \frac{\sum w_{matched-concepts}}{\sum w_{x_i}}$$ (2)

For example, there are three classes matched between `Agent` of Biblio and Bib-Tex (see Section 4.3). Based on the rules in section 4.2.2, the weight of the matched classes in Biblio is 1200, the weight for the whole `Agent` is 3984. Similarly, the results for BibTex are 73 and 209 respectively. Since the ranges or domains for the matched relationships don't match each other, the relationships are not counted in the examples. So the similarity of Agent between Biblio and BibTex is:

$$sim\,(Agent_{Biblio}, Agent_{BibTex}) = \frac{1200}{3984} = 0.3$$ (3)

The similarity of Agent between BibTex and Biblio is:

$$sim\,(Agent_{BibTex}, Agent_{Biblio}) = \frac{73}{209} = 0.35$$ (4)

5 Summary and Future Work

We expand an ontology tree and set weights in the tree to calculate ontology concept similarity. According to our method, more important concepts can get a higher value. But there are still some problems that need to be solved in future work:

- How to combine our approach with other ontology matching methods, like synonyms, instance matching, etc.
- Use of an alternative method (e.g. [13]) to compare relationships instead of comparing relationship ranges and domains directly.
- Consider more reasonable weight through theoretical analysis and experiments, instead of setting 1, 2 and 3 directly.

The approach presented is still under development, i.e. the paper presents work in progress. It will be evaluated in our MediaILOG project and ontology matching benchmark [9]. We will adapt and extend our methods according to the experiment results.

Acknowledgment. Part of this work was financed by the Hamrin Foundation (Hamrin Stiftelsen), project Media Information Logistics.

References

1. Chimera: http://www.ksl.stanford.edu/software/chimaera/
2. Cohen, W.W., Ravikumar, P., Fienberg, S.E.: A Comparison of String Distance Metrics for Name-Matching Tasks IJCAI-2003 (2003)
3. Doan, A., Madhavan, J., Dhamankar, R., Domingos, P., Halevy, A.: Learning to match ontologies on the Semantic Web The VLDB Journal. 12(4), 303–319

4. Euzenat, J.R., Valtchev, P.: Similarity-based ontology alignment in OWL-lite. In: Proc. 15th ECAI, Valencia (ES (2004)
5. Fausto, G., Pavel, S., Mikalai, Y.: S-Match: an algorithm and an implementation of semantic matching (2004)
6. Foam, http://www.aifb.uni-karlsruhe.de/WBS/meh/foam/
7. Gruber, T.R.: A translation approach to portable ontology specifications. Knowledge Acquisition 5(2), 199–220 (1993)
8. Güemes, A.H.: A Prototype System for Automatic Ontology Matching Using Polygons. Computer and Electrical Engineering. Jönköping University (2006)
9. Initiative, O.A.E., http://www.ontologymatching.org/evaluation.html
10. Keet, C.M.: Aspects of Ontology Integration Unpublished (2004), available at: http://www.meteck.org/AspectsOntologyIntegration.pdf
11. Le, B.T., Dieng-Kuntz, R., Gandon, F.: Ontology Matching: A Machine Learning Approach for building a corporate semantic web in a multi-communities organization. In: ICEIS 2004, Porto, Portugal (2004)
12. Lin, F., Sandkuhl, K.: Polygon-based Similarity Aggregation for Ontology Matching. In: International Workshop on Semantic and Grid Computing (SGC 2007) in ISPA07, Niagara Falls, Canada (2007)
13. Maedche, A., Staab, S.: Measuring Similarity between Ontologies. In: Gómez-Pérez, A., Benjamins, V.R. (eds.) EKAW 2002. LNCS (LNAI), vol. 2473, Springer, Heidelberg (2002)
14. Maponto: http://www.cs.toronto.edu/semanticweb/maponto/
15. Melnik, S., Garcia-Molina, H., Rahm, E.: Similarity Flooding: A Versatile Graph Matching Algorithm. In: Proc. 18th International Conference on Data Engineering (ICDE), San Jose, CA US (2002)
16. Noy, N.F., Musen, M.A.: Evaluating Ontology-Mapping Tools: Requirements and Experience. In: Proc. OntoWeb-SIG3 Workshop at the 13th International Conference on Knowledge Engineering and Knowledge Management, Siguenza, Spain (2002)
17. Noy, N.F., Musen, M.A.: The PROMPT Suite: Interactive Tools for Ontology Merging and Mapping. In: SMI (ed.). Stanford University, CA, USA (2003)
18. OntoViz, http://protege.cim3.net/cgi-bin/wiki.pl?OntoViz
19. Owl, http://www.w3.org/TR/owl-features/
20. Plassard, M.-F.: Functional Requirements for Bibliographic Records. (1998)
21. Protégé, http://protege.stanford.edu/
22. SecondString, http://secondstring.sourceforge.net/
23. Shvaiko, P., Euzenat, J.E.O.: A Survey of Schema-based Matching Approaches. Data Semantics (2005)
24. Stumme, G., Maedche, A.: FCA-Merge: Bottom-up merging of ontologies. In: 7th Intl. Conf. on Artificial Intelligence (IJCAI 2001), Seattle, WA (2001)
25. WordNet, http://wordnet.princeton.edu

Deductive Web Services:
An Ontology-Driven Approach for Service Interoperability in Life Science

Nadia Yacoubi Ayadi[1,2], Zoé Lacroix[1], María-Esther Vidal[3],
and Edna Ruckhaus[3]

[1] Scientific Data Management Lab
Arizona State University
Tempe AZ 85287-5706, USA
{nadia.yacoubi,zoe.lacroix}@asu.edu
[2] RIADI Lab, National School of Computer Science
Campus universitaire, 2010 La Manouba, Tunisia
[3] Dept. of Computer Science and Information Technology
Universidad Simón Bolívar, Venezuela
{mvidal,ruckhaus}@ldc.usb.ve

Abstract. We present an ontology-driven approach to service composition with deductive databases. We formalize Web services as *Deductive Web services* (DWS) where the *Extensional Service Base* (ESB) stores all input and output signatures of services. The *Intensional Service Base* (ISB) corresponds to a set of deductive rules that express the semantics of the ESB captured by a domain ontology. We provide a framework for the composition of Web services represented as deductive Web services. In particular, we show that an approach that combines *schema mapping* and *deductive databases* can provide the missing link in service composition to implement scientific workflows.

1 Introduction

In recent years, the emergence of Web services has stimulated organizations from different domains to provide access to their core applications on the Web [2]. In the context of life sciences Web services are made available to implement complex tasks through service composition within *scientific workflows* [4]. Scientists need to address three issues when they wish to design and implement scientific workflows. The first problem is related to the increasing difficulties of identifying and accessing existing services and understanding how the results should be interpreted and contribute to scientific discovery. Secondly, the lack of interoperability among existing services make their composition error-prone and effort and time-consuming [13]. Finally, the selection of resources to achieve scientific tasks may lead to multiple implementation attempts often failing because of poor performance [5]. It is critical to develop technology to assist scientists in the selection and composition of services.

R. Meersman, Z. Tari, P. Herrero et al. (Eds.): OTM 2007 Ws, Part II, LNCS 4806, pp. 1338–1347, 2007.

Resource discovery approaches [6,3,15] guide scientists in the selection of resources while providing various meta-data that ease their composition. Scientific workflow systems such as Taverna [11], Kepler [10], and SIBIOS [7] enable the construction and execution of workflows over distributed Web services. Taverna proposes a list of *"shims"*, i.e., services that resolve basic syntactical mismatches in order to reconciliate closely related inputs and outputs. Such a solution is not scalable as a new shim needs to be manually created for each pair of services that need to interoperate. [10] describes a scalable framework that uses mappings to one or more ontologies for reconciling two services. Similarly, our approach exploits solutions developed by the database community to assist scientists in the development of scalable and sound scientific workflow implementations [14]. A similar approach, which describes a generic and scalable architecture for service composition within a wider data transformation framework based on XML data format, is presented in [15].

In this paper, we focus on the reasoning side offered by our approach. We propose an ontology-based framework for the automatic composition of Web services, and an approach to generate composite services from high-level declarative descriptions. Our main contribution in this paper is to propose a precise definition of the problem of Web services composition by introducing a novel formalism to represent both Web services and domain ontologies. We formalize Web services as a deductive database called *Deductive Web services* (DWS). The *Extensional Service Base* (ESB) comprises meta-level predicates which represent Web service descriptions in a declarative way. The *Intensional Service Base* (ISB) is composed by a set of predicates defined by deductive rules that express the properties or semantics of the ESB. Finally, we show how our formalism enables the automatic composition of services.

The remaining of the paper is organized as follows. Section 2 introduces a motivating example related to protein sequence analysis. Section 3 defines the proposed formalism. Section 4 formalizes the problem of schema mapping and service composition by providing a model-theoretic semantics of our formalism; an example illustrates how our approach leads to the automatic generation of mappings. We conclude in Section 5.

2   Motivating Example

We consider a simple workflow whose objective is to identify a protein family for a nucleotide sequence. The workflow consists of two steps: the retrieval of the nucleotide sequences of interest followed by the alignment against a protein database (see Figure 1). The implementation of the workflow (i.e., selection of resources to achieve those two tasks) raises syntactic and semantic conflicts that need to be addressed. Suppose that the scientist chooses an EMBL Web service[1] to retrieve a nucleotide sequence given a keyword or a sequence accession number. Nucleotide sequences retrieved from EMBL have a format specific to EMBL.

---

[1] http://www.ebi.ac.uk/Tools/webservices/

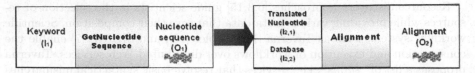

Fig. 1. Example of Scientific Workflow

The scientist may select BLASTx, a variant service of the Basic Local Alignment Search Tool (BLAST)[2] to implement the alignment task. BLASTx requires, as input, sequences in FASTA format. An adapter is needed to perform the syntactic transformation from one format (EMBL format) to another (FASTA format). Semantic issues are raised by conflicts between the semantic types of parameters. For instance, EMBL provides as output a `nucleotide sequence` while BLASTx declares as input parameter a `translated nucleotide`. The concept of `translated nucleotide` defines the translation of a DNA or a RNA sequence into an Amino Acid sequence. Note that the translation is done automatically by BLASTx. However, it is necessary to make the semantics of the concept `translated nucleotide` explicit to avoid the failure of the execution of the protocol. In this paper, we present a framework to infer semantically compatible services by the means of available domain ontologies. Reasoning capabilities offered by the formalism of DWS are proposed as the main contribution of the paper.

3   Deductive Web Services

In this section, we present an unambiguous model of Web services with well-defined semantics. We consider Web services as computational entities semantically described and enhanced by reasoning capabilities. We model Web services as Deductive Web Services (DWS) and we assume that Web services are defined in terms of types whose properties are expressed in different ontologies. The knowledge represented in an ontology is also modeled as a deductive base called Deductive Ontology Base (DOB), as proposed in [12]. First, a DOB is composed of an Extensional Ontology Base (EOB) and an Intensional Ontology Base (IOB). The knowledge explicitly described in the ontology that defines the types of arguments of the services is represented in the EOB. On the other hand, implicit knowledge and properties that can be inferred are represented as intensional predicates in the IOB. Meta-level predicates are provided in both bases to describe explicit and implicit knowledge. For example, the description of classes, properties, and the hierarchies they participate in, are modeled in the EOB. Predicates as `isClass` and `subClassOf` are used to express explicitly that something is a class and that the relationship subclass relates two classes. The transitive closure of a class in a subclass hierarchy is represented by a rule that defines the intensional predicate `areClasses`.

---

[2] http://www.ncbi.nlm.nih.gov/BLAST/

Table 1. Example of DOB ontology

EOB predicate
isOntology(O)
isClass(Sequence)
subClassOf(NucleotideSequence, Sequence)
subClassOf(TranslatedNucleotide, AAsequence)
subClassOf(AAsequence, Sequence)
subClassOf(DNAsequence, NucleotideSequence)
subClassOf(RNAsequence, NucleotideSequence)
isOProperty(IsTranscribedIn,DNAsequence, RNAsequence
isOProperty(has-a, Protein, AASequence)
isOProperty(IsTranslatedIn, RNAsequence, TranslatedNucleotide)
isOProperty(has-a-protein, TranslatedNucleotide, Protein)
maxCardinality(TranslatedNucleotide, has-a-protein, 1)
minCardinality(TranslatedNucleotide, has-a-protein, 0)

A set of predicates in the EOB and IOB corresponds to a canonical representation of the knowledge that can be expressed in a particular ontological language. In [12], we modeled a subset of the Web Ontology Language OWL Lite [8] as a DOB. Note that some of the predicates refer to domain concepts (e.g., isClass, areClasses), and some to individuals (e.g., isIndividual, areStatements).

Example 1. *Table 1 illustrates a portion of an OWL Lite ontology representing a classification of different types of sequences and their inter-relationships.*

Web services are mapped to a domain ontology and are defined as an Extensional Service Base (ESB) and an Intensional Service Base (ISB).

Definition 1 (Deductive Web services). *A Deductive Web service = ⟨ESB, ISB⟩ is a deductive database composed of a set built-in Extensional Service Base (ESB) and of an Intensional Service Base (ISB). ESB is composed of ground predicates representing the description of the service, and ISB built-in predicates that express the semantics of ESB built-in predicates.*

ESB predicates are *ground*, i.e., variables are not allowed as arguments. An ESB models knowledge explicitly represented in a service ontology. Thus, an ESB is composed of a set of meta-level predicates that specify the input and output types of a service, the pre-conditions and effects of each service, and the subsumption relationship between services. A set of ISB built-in predicates represents the semantics of the ESB predicates. Additionally, ISB predicates may express equivalence of input and output types, or whether two services can be composed or mapped. Table 2 illustrates some built-in ESB and ISB predicates. ISB predicates are defined by a set of Datalog rules that establish the meaning of the predicates. Note that Table 2 contains a canonical representation of a subset of OWL-S.

Table 2. Some built-in ESB and ISB Predicates

ESB predicate	Description
isService(S)	S is a Web service name
subService(S1, S2)	S1 is a sub-service of S2
impOntology(S,O)	The Web service S imports the ontology O
serviceCategory(S,C)	S belongs to the category C
parameterName(P)	P is a parameter name
hasInputParameter(S,I)	I is an input parameter name of the Service S
hasOutputParameter(S,O)	O is an output parameter name of the Service S
isParameterClass(P,C)	P is a parameter belonging to the ontology class C
hasInputType(S,C)	C is an input parameter type of the service S
hasOutputType(S,C)	C is an output parameter type of the service S
isPProperty(Pr,D,R)	Pr is a parameter property with domain D and range R
isStatement(P,Pr,J)	P has the property Pr with the value J
isDatatypeProperty(P,D)	The parameter P has as datatype D
ISB predicate	
areParameterClasses(P,C2) :- isParameterClass(P,C1),	
areSubClasses(C1,C2).	
areParameterClasses(P2,C2) :- isOProperty(Pr,C1,C2),	
areParameterClasses(P1,C1),	
isPProperty(Pr,P1,P2).	
inputParameterService(S,P) :- areParameterClasses(P,C),	
areInputType(S,C),isService(S).	
outputParameterService(S,P) :- areParameterClasses(P,C),	
areOutputType(S,C), isService(S).	
areInputType(S,C1) :- hasInputType(S,C2),	
isService(S), areSubClasses(C1,C2).	
AreOutputType(S,C1) :- hasOutputType(S,C2),	
isService(S), areSubClasses(C1,C2).	

Any request for a service or any question related to services is expressed as a *query*. For example, a query can ask whether two service arguments are equivalent, or if they can be composed.

Definition 2 (Query). *A DWS query is defined as a rule q:* $Q(\overline{X}) \leftarrow \exists \overline{Y} B$ $(\overline{X}, \overline{Y})$, *where B is the query's goal.*

Example 2. *BLAST finds regions of local similarity between gene and protein sequences. It compares gene or protein sequences to sequence databases and calculates the significance of alignment matches. Table 3 represents BLAST as a DWS. The submission of the sequence* seq123 *to BLAST could be expressed in our formalism by the following query.*

```
Q    ←isParameter(seq123, RNASequence) ∧
     isStatement (seq123, SequenceFormat, ">AB183428.1
     14-JUL-2007gaaaagat...") ∧
     isService(BLAST).
```

However, because the type of `seq123` *is declared as* `RNASequence` *in* `Q` *the query would not return any results. Indeed, the description of BLAST as a DWS (see Table 3) requires a parameter of type* **Sequence***. Domain knowledge captured in an ontology such as expressed in the DOB in Table 1 allows the correspondence of the two types through the subclass hierarchy. Given the following ISB predicate:*

```
areInputType(S,C1)    :- hasInputType(S,C2), isService(S),
                         areSubclasses(C1,C2).
```

and the following IOB predicate:

```
areSubclasses(C1,C2)  :- subClassOf(C1,C2).
areSubclasses(C1,C2)  :- subClassOf(C1,C3),
                         areSubclasses(C3,C2)
```

We can infer that all subclasses of **Sequence** *map the input type to BLAST. Thus, the classes* `NucleotideSequence`, `DNAsequence`, `RNAsequence`, `AAsequence`, *and* `TranslatedNucleotide` *are subclasses of* **Sequence** *and are suitable input types for BLAST. This simple example illustrates that the use of ontologies as deductive databases augments significantly the input and output types of a service.*

Table 3. BLAST Web service represented as a Deductive Web Service

ESB predicate
`isService(BLAST)`
`hasInputType(BLAST, Sequence)`
`hasOutputType(BLAST, BLASTreport)`
`subService(BLASTx, BLAST)`
`hasInputType(BLASTx, TranslatedNucleotide)`
`isPProperty(Has-Format, seq123, FASTA)`
`isParameterClass(Seq123, RNASequence)`
`isStatement(seq123, SequenceFormat,>AB183428.1 14-JUL-2007gaaaagat..)`
`isPProperty(IstranslatedIn, seq123, seq'123)`

The benefits of using ontologies to describe Web services properties and behavior are twofold: ontologies allow the representation of inherent semantics behind services that are not explicitly defined in data format descriptions [9]; and *Open World* reasoning on Web services and ontologies may augment the space of services composition and matching choices. In our framework, we suppose that to define the semantics of a DWS, we may import a domain ontology O using the predicate `impOntology(S,O)`. We can infer in our deductive service base when two services are considered equivalent or can be composed to create a new composite service. We introduce the intensional predicates equivServices(S1,S2) and compositeService(S), which are defined as follows: The predicate areSimilar(I1,I2) is an EOB predicate that indicates that the types I1 and I2 are similar (same name or similar subclasses and super-classes). On the other hand, the predicate map(O1,I2) expresses that the set of parameters in O1 can be mapped to parameters of I2.

```
equivServices(S1,S2) :-  compositeService(S1), areInputType(S1,I1),
                         areOutputType(S1,O1),
                         compositeService(S2), areInputType(S2,I2),
                         areOutType(S2,O2),
                         areSimilar(I1,I2), areSimilar(O1,O2).
compositeService(S)   :-  isService(S).
compositeService(S)   :-  compositeService(S1), areOutputType(S1,O1),
                         compositeService(S2), areInputType(S,I2),
                         map(O1,I2).
```

4   Mapping-Based Service Composition

In this section, we show how we can exploit reasoning capabilities offered by the DWS formalism to generate composite services through the efficient (semi-) automatic inference of mappings. The use of the deductive database paradigm offers a persistent storage of knowledge and scalable queries on the semantic Web. To diminish the impact of large sets of explicit and implicit ontology facts in the performance of reasoning and querying tasks, we use query optimization and evaluation techniques previously defined in [12]. Experimental studies have shown that these techniques are able to identify query plans whose cost is a small portion of the cost of the worst plan. Therefore, the application of these techniques to the approach described in this paper, provides a solution to the problem of efficiently composing services and identifying schema mappings. Given two deductive Web services WS_1 and WS_2, a query q projects the output parameters of Web service WS_1. The query q is formulated in terms of concepts belonging to WS_1 that can be used to fire WS_2, thus, we need to map the output parameters of WS_1 to the input structure of WS_2. A valuation consists in mapping the input structure of WS_2 to any of all possible instantiations obtained through evaluating WS_1. The set of all possible valid instantiations consists of the *space of mappings*. In this sense, the Web service composition is realized through the space of mappings which should be organized in an optimal way in order to choose the most *suitable mapping*, if it exists. The semantics of DWS is specified with Definitions 3 and 4.

Definition 3 (Interpretation). *An Interpretation* $I = (\Delta^I, \mathcal{P}^I, .^I)$ *is composed of:*

- *A non-empty interpretation domain* Δ^I *corresponding to the union of the sets of valid URIs of ontologies, classes, services, parameters, object and datatype properties, and individuals. These sets are pairwise disjoint.*
- *A set of interpretations* $\mathcal{P}^I$, *of the extensional predicates EP (ESB or EOB) and intensional predicates IP (ISB or IOB).*
- *An interpretation function* $.^I$ *which maps each n-ary built-in predicate* $p^I \in \mathcal{P}^I$ *to an n-ary relation* $\prod_{i=1}^{n} \Delta^I$.

For an EP predicate $p(t_1, ..., t_n)$, and an interpretation I, $I \models p(t_1, ..., t_n)$, i.e., I satisfies p, iff $(t_1, ..., t_n) \in p^I$.

For an IP predicate $R{:}H(\overline{X}) \leftarrow \exists \overline{Y} B(\overline{X}, \overline{Y})$, and an interpretation I, $I \models R$, i.e., I satisfies R, iff whenever I satisfies the body B, it also satisfies the head H. $I \models B$ if B is empty or I satisfies every DWS or DOB atom in B. $I \models H$ if H is non-empty and I satisfies the atom in H.

Definition 4 (Model). *Given a Deductive Base $\mathcal{D}$ and an interpretation I, $I \models \mathcal{D}$ iff I models every EP and IP predicate in $\mathcal{D}$.*

We are able to directly perform reasoning tasks on Web services parameters, i.e., inputs and outputs, using ISB predicates, e.g., `areInputType`. Moreover, we have meta-level predicates that express the semantics of Web services concepts, e.g., `isPProperty`. Additionally, we propose predicates to infer relations between Web services, e.g., *Equivalence* or *Composite* ISB predicates.

In the following, we provide a model theoretic semantics for equivalence and composition of services.

Definition 5 (Equivalent Services). *Given two deductive web services WS1 and WS2 represented by the ESB predicates IsService(WS1) and IsService(WS2), respectively. Let T_{WS1} and T_{WS2} be sets of related intensional predicates of WS1 and WS2, respectively, where a predicate P belongs to T_{WS1} (resp. T_{WS2}) if IsService(WS1)(resp. IsService(WS2)) appears in the body of the rule that defines P. WS1 and WS2 are equivalent iff all interpretation I that satisfies T_{WS1} also satisfies T_{WS2} and conversely.*

Definition 6 (Composite Services). *Given two deductive web services WS1 and WS2, such that, the output of WS1 is Out1 and the input of WS2 is In2, WS1 can be composed with WS2 iff every valid valuation [1] of the variables in Out1 and In2 that satisfies WS1 also satisfies WS2.*

Example 3. *Consider the motivating example described in Section 1. EMBL and BLASTx Web services are considered for composition, whereby EMBL is the data provider and BLASTx is the data consumer. In Table 3, BLASTx is modeled as the sub-service of BLAST that takes a specified kind of sequences declared as translated nucleotide. Assume that we submit to BLASTx a nucleotide sequence retrieved from EMBL. The corresponding query is formulated as follows and should be evaluated by the DWS (Table 3) and the corresponding DOB (Table 1):*

```
q(I2)    :-compositeService(EMBL), hasOutputParamater(EMBL, seq123),
            compositeService(BLASTx), hasInputParameter(I2).
```

Because all possible valid instantiations of the output parameters of the service EMBL need to be considered during the sideways-passing of the bindings, a large space of possible mappings may be generated. Note that some ISB predicates are used to express the inheritance of properties and relationships between ontological concepts and Web service parameters. For instance, the following ISB predicate express the fact that if a relationship R exists between two concepts $C1$ and $C2$, then every parameters $P1$ and $P2$ that are respectively instances of $C1$ and $C2$ are related by the relationship R.

```
areParameterClasses(P2,C2)- isOProperty(Pr,C1,C2),
                            isPProperty(Pr,P1,P2),
                            areParameterClasses(P1,C1).
```
In that case, we can infer the following:

```
areParameterClasses(seq'123, TranslatedNucleotide):-
   isOProperty(IsTranslatedIn, RNAsequence, TranslatedNucleotide),
   isPProperty(IsTranslatedIn, seq123, seq'123),
   areParameterClasses(seq123, RNAsequence).
```

Among all the facts inferred the most relevant mapping is the one where the predicate representing the input structure of BLASTx, i.e., InputParameterService
(Blastx,I2), *is the head of the mapping rule:*

```
InputParameterService(BLASTx, seq'123):-
            areParameterClasses(seq'123,TranslatedNucleotide),
            AreInputType(BLASTx,TranslatedNucleotide),
            isService(BLASTx).
```

Finally, to reduce the complexity of the evaluation of queries, the space of mappings should be organized in an optimal way to choose the most suitable ones, if such criteria exist to order them.

5 Conclusions and Future Work

We present a deductive approach to represent domain ontologies and Web services, as well as the conditions that have to be satisfied by two services in order to be composed or mapped. We introduce Deductive Web Services a new formalism to express Web services and to achieve service composition exploiting a domain ontology. The approach presented in this paper is exploited in ProtocolDB [16], a system that stores scientific protocols and assists scientists in the expression of executable workflows. We are currently conducting experimental studies to evaluate the approach and analyze the behavior of different strategies for a variety of *scientific workflows*. In the future, we aim at exploiting this approach to reason about data provenance.

Acknowledgments. This research was partially supported by the National Science Foundation[3] (grants IIS 0223042, IIS 0431174, IIS 0551444, and IIS 0612273). The research done by the Universidad Simón Bolívar team, has been partially supported by the DID-USB. The authors would like to thank Christophe Legendre, Maliha Aziz, and Ismail Khalil Ibrahim for their valuable input.

[3] Any opinion, finding, and conclusion or recommendation expressed in this material are those of the authors and do not necessarily reflect the views of the National Science Foundation.

References

1. Abiteboul, S.: Querying semi-structured data. In: Afrati, F.N., Kolaitis, P.G. (eds.) ICDT 1997. LNCS, vol. 1186, pp. 1–18. Springer, Heidelberg (1996)
2. Alonso, G., Casati, F., Kuno, H., Machiraju, V.: Web Services: Concepts, Architecture and Applications. Springer, Heidelberg (2004)
3. Cohen-Boulakia, S., Davidson, S., Froidevaux, C., Lacroix, Z., Vidal, M.-E.: Path-Based Systems to Guide Scientists in the Maze of Biological Resources. Journal of Bioinformatics and Computational Biology 4(5), 1069–1095 (2006)
4. Conery, J.S., Catchen, J., Lynch, M.: Rule-based workflow management for bioinformatics. VLDB Journal 14(3), 318–329 (2005)
5. Lacroix, Z., Legendre, C.: Analysis of a Scientific Protocol: Selecting Suitable Resources. In: IEEE International Workshop on Service Oriented Technologies for Biological Databases and Tools, Salt Lake City, UT, July 9-13, pp. 130–137 (2007)
6. Lord, P.W., Alper, P., Wroe, C., Goble, C.A.: Feta: A light-weight architecture for user oriented semantic service discovery. In: Gómez-Pérez, A., Euzenat, J. (eds.) ESWC 2005. LNCS, vol. 3532, pp. 17–31. Springer, Heidelberg (2005)
7. Mahoui, M., Ben-Miled, Z., Srinivasan, S., Dippold, M., Yang, B., Li, N.: SIBIOS Ontology: A Robust Package for the Integration and Pipelining of Bioinformatics Services. In: Leser, U., Naumann, F., Eckman, B. (eds.) DILS 2006. LNCS (LNBI), vol. 4075, pp. 104–113. Springer, Heidelberg (2006)
8. McGuinness, D.L., van Harmelen, F.: Owl web ontology language overview (2006), available at http://www.w3.org/TR/2004/REC-owl-features-20040210/
9. McIlraith, S.A., Son, T.C., Zeng, H.: Semantic web services. IEEE Intelligent Systems 16(2), 46–53 (2001)
10. McPhillips, T.M., Bowers, S., Ludäscher, B.: Collection-oriented scientific workflows for integrating and analyzing biological data. In: Leser, U., Naumann, F., Eckman, B. (eds.) DILS 2006. LNCS (LNBI), vol. 4075, pp. 248–263. Springer, Heidelberg (2006)
11. Oinn, T., Addis, M., Ferris, J., Marvin, D., Senger, M., Greenwood, M., Carver, T., Glover, K., Pocock, M.R., Wipat, A., Li, P.: Taverna: a tool for the composition and enactment of bioinformatics workflows. Bioinformatics 20(17), 3045–3054 (2004)
12. Ruckhaus, E., Vidal, M.-E., Ruiz, E.: Query evaluation and optimization in the semantic web. In: Workshop on Applications of Logic Programming in the Semantic Web and Semantic Web Services, pp. 17–32 (2006)
13. Stein, L.: Creating a bioinformatics nation. Nature 417, 119–120 (2002)
14. Yacoubi, N., Lacroix, Z.: Resolving Scientific Service Interoperability With Schema Mapping. In: IEEE 7th International Symposium on Bionformatics and Bioengineering October 14-17, Boston, MA (2007)
15. Zamboulis, L., Martin, N., Poulovassilis, A.: Bioinformatics Service Reconciliation By Heterogeneous Schema Transformation. In: Data Integration in the Life Sciences. LNCS, vol. 4544, pp. 89–104 (2007)
16. Kinsy, M., Lacroix, Z., Legendre, C., Wlodarczyk, P., Yacoubi, N.: ProtocolDB: Storing Scientific Protocols with a Domain Ontology. In: Workshop on Approaches and Architectures for Web Data Integration and Mining in Life Sciences in conjunction with Web Information Systems Engineering. LNCS, vol. 4832 (2007)

Protein Ontology Instance Store

Amandeep S. Sidhu, Tharam S. Dillon, and Elizabeth Chang

Digital Ecosystems and Business Intelligence Institute,
Curtin University of Technology Perth, Australia
{Amandeep.Sidhu,Tharam.Dillon,
Elizabeth.Chang}@cbs.curtin.edu.au

Abstract. Biomedical Knowledge of Proteomics Domain is represented in the Protein Ontology, whose instantiations, which are undergoing evolution, need a good management and maintenance system. Protein Ontology instantiations signify data and information about proteins that is shared and has evolved to reflect development in Protein Ontology Project and Proteomics Domain itself. Protein Ontology needs to be populated with data and information from data and information sources proteomics domain. In this chapter we explore the implementation methodology for Data and Knowledge Management in Proteomics Domain by mapping various protein data and knowledge sources to the concepts defined in the protein ontology and developing a Protein Instance Store.

1 Introduction

For proteomics domain users frequently use web sites of protein data sources, but often fail to retrieve the correct information due to the heterogeneous and complex structure of the data formats. Protein Ontology provides a vocabulary for representing knowledge about proteomics domain and describes specific data sources therein (Sidhu et al., 2005a). The role of protein ontology is that of making explicit specified conceptualizations that can be shared, reused, and integrated in the analysis and design stages of information and knowledge systems for bioinformatics.

Protein Data and Knowledge captured in Protein Ontology Concepts and Instantiations represents abstraction of data sources and expertise in proteomics domain. Abstraction is divided into generic and derived concepts of protein ontology. Protein Ontology instantiations are derived as a result of populating protein data and information and are referred to as instances of protein ontology classes. Instantiations are also known as instance knowledge of the protein ontology. The instantiations of PO represent knowledge about respective proteins. Concrete data instances about various proteins from underlying diverse protein data and knowledge sources are stored as PO instantiations in PO Instance Store.

The Protein Ontology Instance Store is created for entering existing protein data using the PO format. PO uses data sources include new proteome information resources like PDB, SCOP, and RESID as well as classical sources of information where information is maintained in a knowledge base of scientific text files like OMIM and

R. Meersman, Z. Tari, P. Herrero et al. (Eds.): OTM 2007 Ws, Part II, LNCS 4806, pp. 1348–1354, 2007.
© Springer-Verlag Berlin Heidelberg 2007

from various published scientific literature in various journals. PO Instance Store is represented using OWL. PO Instance Store at the moment contains data instances of following protein families: (1) Prion Proteins, (2) B.Subtilis, and (3) Chlorides. More protein data instances will be added as PO is more developed. All the Protein Ontology Instances are available for download (http://proteinontology.info/proteins. htm) in OWL format that can be read by any popular editor like Protégé (http://protege.stanford.edu/).

In reality protein data sources are updated over a period of time to reflect developpment in proteomics. Since changes are inevitable during proteomics experimentations, the protein ontology instance store is continuously confronted with the evolution problem. If such changes are not properly traced or maintained, this would impede the use of the protein ontology. Therefore a semi-automatic process becomes increasingly necessary to facilitate updating tasks and to ensure reliability. PO Instance Store evolution problem can be handled partly by using the Difference Operator of PO Algebra (Sidhu et al., 2006). It will suggest instances that are not entered properly or if there is change in underlying protein data and knowledge sources from which PO Instance Store is populated. Note that this is not a protein ontology evolution process because it does not change the original concepts and relationships in the ontology; rather instantiations of the ontology change or that conform to the ontology change.

2 PO Instantiations Transformation

In this section, we particularly report on how protein data are transformed or mapped into concepts formed in the protein ontology as instance knowledge. Protein Ontology Web Retrieval System (PO-WEB) manages connection between Protein Ontology Conceptual Framework and the Protein Ontology Instance Store. PO-WEB is built on top of Jena (Carroll et al., 2004) , which we would like to gratefully acknowledge. Jena developed by the Hewlett-Packard Company is a Java framework having capacity of manipulating ontologies. The version of Jena used is Jena 2.1. PO-WEB provides acquisition, navigation, and, querying of the Protein Ontology Instance Store. The process of acquiring data and knowledge from proteomics domain is described in this stage, which applies algorithms and methods analyzing protein data files and proteomics domain texts. The terminology used by domain experts is defined in protein ontology. In this study, to collect a glossary of concepts (classes) for the proteomics domain, first, an analysis was performed on 4 major protein data sources: PDB (Wesbrook et al., 2002), SCOP (Murzin et al., 1995), SWISS-PROT (Bairoch and Apweiler, 1997) and PIR (Barker et al., 1998). A web interface is used to parse the data from various protein data sources like PDB and unify them in the PO format. Protein data is parsed according OWL schema specifications of Protein Ontology (Sidhu et al., 2007, Sidhu et al., 2005b).

Protein Ontology concept structures are formulated so that they can easily be navigated. The knowledge is provided in hierarchical form so upper level concepts or lower level concepts or adjacent concepts can easily be navigated. Technically for this function, PO-WEB focuses on the Protein Ontology Schema in OWL, and the set of statements that comprises the abstraction and instantiations. To navigate the PO

Instance Store, PO-WEB reads Protein Ontology Schema in OWL and then accesses the individual instances of the elements. PO Conceptual Hierarchy with their brief description is at: http://proteinontology.info/hierarchy.htm

3 Case Study

In this section we will discuss case study of protein family that we populated as instance of Protein Ontology drawing information from multiple protein data sources. Bacillus Subtilis is a bacterium commonly found in soil. B. Subtilis has the ability to form a tough, protective endospore, allowing the organism to tolerate extreme environmental conditions. B. Subtilis has proven highly amenable to genetic manipulation, and has therefore become widely adopted as a model organism for laboratory studies, especially of sporulation, which is a simplified example of cellular differentiation. It is also heavily flagellated, which gives B. Subtilis the ability to move quite quickly. With the assistance of our colleagues in the Science Faculty we have populated PO Instance Store with all 13 major protein complexes that belong to B. Subtilis family. All the PO Instance files are available for download at the PO website (http://proteinontology.info/proteins.htm).

We will integrate information about a B. Subtilis septum formation MAF protein complex with D - (UTP) from Protein Data Bank and UniProt in this example. Basic Information about B. Subtilis Protein, which is described in terms of description of entry, molecules, source cell and references, is fetched from Protein Data Bank (PDB ID: 1EXC). This information is represented in PDB format as follows:

```
TITLE     CRYSTAL STRUCTURE OF B. SUBTILIS MAF PROTEIN COMPLEXED WITH
TITLE     2 D-(UTP)
COMPND    MOL_ID: 1;
COMPND    2 MOLECULE: PROTEIN MAF;
COMPND    3 CHAIN: A, B;
COMPND    4 ENGINEERED: YES
SOURCE    MOL_ID: 1;
SOURCE    2 ORGANISM_SCIENTIFIC: BACILLUS SUBTILIS;
SOURCE    3 ORGANISM_COMMON: BACTERIA;
SOURCE    4 EXPRESSION_SYSTEM_COMMON: BACTERIA;
SOURCE    5 EXPRESSION_SYSTEM_VECTOR_TYPE: PLASMID;
SOURCE    6 EXPRESSION_SYSTEM_PLASMID: PQE30
KEYWDS    B.SUBTILIS MAF PROTEIN COMPLEXED WITH DUTP, STRUCTURAL
KEYWDS    2 GENOMICS, PSI, PROTEIN STRUCTURE INITIATIVE, MIDWEST CENTER
KEYWDS    3 FOR STRUCTURAL GENOMICS, MCSG
EXPDTA    X-RAY DIFFRACTION
AUTHOR    G.MINASOV,M.TEPLOVA,G.C.STEWART,E.V.KOONIN,W.F.ANDERSON,
AUTHOR    2 M.EGLI,MIDWEST CENTER FOR STRUCTURAL GENOMICS (MCSG)
REVDAT    2   18-JAN-05 1EXC    1        JRNL    AUTHOR KEYWDS REMARK
REVDAT    1   14-JUN-00 1EXC    0
JRNL      AUTH   G.MINASOV,M.TEPLOVA,G.C.STEWART,E.V.KOONIN,
JRNL      AUTH 2 W.F.ANDERSON,M.EGLI
JRNL      TITL   FUNCTIONAL IMPLICATIONS FROM CRYSTAL STRUCTURES OF
JRNL      TITL 2 THE CONSERVED BACILLUS SUBTILIS PROTEIN MAF WITH
JRNL      TITL 3 AND WITHOUT DUTP.
JRNL      REF    PROC.NATL.ACAD.SCI.USA        V.  97  6328 2000
JRNL      REFN   ASTM PNASA6  US ISSN 0027-8424
```

Fig. 1. Protein Entry Information for B. Subtilis in PDB.

Title, Keywords, Experimental Method (EXPDTA), and Authors from PDB are described in Description Concept of Protein Ontology in a Web Ontology Language representation as follows:

```
<Description rdf:ID="DescriptionInstance11101">
  <Keywords rdf:datatype="http://www.w3.org/2001/XMLSchema#string"
  >B.SUBTILIS MAF PROTEIN COMPLEXED WITH DUTP, STRUCTURALGENOMICS, PSI, PROTEIN STRUCTURE INITIATIVE, MIDWEST CENTERFOR STRUCTURAL GENOMICS, MCSG</Keywords>
  <Experiment rdf:datatype="http://www.w3.org/2001/XMLSchema#string"
  >X-RAY DIFFRACTION</Experiment>
  <Title rdf:datatype="http://www.w3.org/2001/XMLSchema#string"
  >CRYSTAL STRUCTURE OF B. SUBTILIS MAF PROTEIN COMPLEXED WITH D-(UTP)</Title>
  <Authors rdf:datatype="http://www.w3.org/2001/XMLSchema#string"
  >G.MINASOV,M.TEPLOVA,G.C.STEWART,E.V.KOONIN,W.F.ANDERSON,M.EGLI,MIDWEST CENTER FOR STRUCTURAL GENOMICS (MCSG)</Authors>
</Description>
```

Fig. 2. Instance of Description Concept for B. Subtilis in PO.

Information about molecules present in the protein complex from PDB are described in Molecule Concept of Protein Ontology represented using Web Ontology Language as follows:

```
<Molecule rdf:ID="MoleculeInstance25">
  <Engineered rdf:datatype="http://www.w3.org/2001/XMLSchema#string">: YES </Engineered>
  <MoleculeID rdf:datatype="http://www.w3.org/2001/XMLSchema#string">1 </MoleculeID>
  <MoleculeName rdf:datatype="http://www.w3.org/2001/XMLSchema#string">PROTEIN MAF </MoleculeName>
</Molecule>
```

Fig. 3. Instance of Molecule Concept for B. Subtilis in PO.

Organism and Cellular Source where protein resides from PDB is described using Source Cell Concept of Protein Ontology, and represented in Web ontology Language as follows:

```
<SourceCell rdf:ID="SourceCellInstance29">
  <ExpressionSystemVector rdf:datatype="http://www.w3.org/2001/XMLSchema#string"
  >PLASMID</ExpressionSystemVector>
  <OrganismCommon rdf:datatype="http://www.w3.org/2001/XMLSchema#string"
  >BACTERIA</OrganismCommon>
  <ExpressionSystem rdf:datatype="http://www.w3.org/2001/XMLSchema#string"
  >BACTERIA</ExpressionSystem>
  <OrganismScientific rdf:datatype="http://www.w3.org/2001/XMLSchema#string"
  >BACILLUS SUBTILIS</OrganismScientific>
  <Plasmid rdf:datatype="http://www.w3.org/2001/XMLSchema#string"
  >PQE30</Plasmid>
</SourceCell>
```

Fig. 4. Instance of Source Cell Concept for B. Subtilis in PO.

Literature References of Protein Complex from PDB are described using Reference Concept of Protein Ontology, and represented in Web ontology Language as follows:

```
<Reference rdf:ID="ReferenceInstance28">
  <CitationReferenceNumbers rdf:datatype="http://www.w3.org/2001/XMLSchema#string"
  >ASTM PNASA6  US ISSN 0027-8424</CitationReferenceNumbers>
  <CitationTitle rdf:datatype="http://www.w3.org/2001/XMLSchema#string"
  >FUNCTIONAL IMPLICATIONS FROM CRYSTAL STRUCTURES OF THE CONSERVED BACILLUS SUBTILIS PROTEIN MAF WITH AND WITHOUT DUTP.</CitationTitle>
  <CitationAuthors rdf:datatype="http://www.w3.org/2001/XMLSchema#string"
  >G.MINASOV,M.TEPLOVA,G.C.STEWART,E.V.KOONIN,W.F.ANDERSON,M.EGLI</CitationAuthors>
  <CitationReference rdf:datatype="http://www.w3.org/2001/XMLSchema#string"
  >PROC.NATL.ACAD.SCI.USA      V.  97  6328 2000</CitationReference>
</Reference>
```

Fig. 5. Instance of Reference Concept for B. Subtilis in PO.

Atomic Coordinates for Standard Residues in a Protein Structure are represented by Atom Records in PDB as follows:

```
ATOM    582  O   CYS A  74       0.035   1.381  47.441  1.00 20.49           O
ATOM    583  CB  CYS A  74       1.976   0.517  45.106  1.00 21.49           C
ATOM    584  SG  CYS A  74       2.365  -0.588  43.705  1.00 22.73           S
```

Fig. 6. Atom Records for B. Subtilis in PDB.

The Atom Record (Atom ID: 583) is described in Protein Ontology as Atom-Instance4484 and is represented in Web Ontology Language as follows:

```
<Atoms rdf:ID="AtomInstance4484">
  <Z rdf:datatype="http://www.w3.org/2001/XMLSchema#float">45.106</Z>
  <Atom rdf:datatype="http://www.w3.org/2001/XMLSchema#string"
  > CB </Atom>
  <X rdf:datatype="http://www.w3.org/2001/XMLSchema#float">0.517</X>
  <Occupancy rdf:datatype="http://www.w3.org/2001/XMLSchema#float"
  >1.0</Occupancy>
  <ATOMResSeqNum rdf:datatype="http://www.w3.org/2001/XMLSchema#int"
  >74</ATOMResSeqNum>
  <TempratureFactor rdf:datatype="http://www.w3.org/2001/XMLSchema#float"
  >21.49</TempratureFactor>
  <AtomID rdf:datatype="http://www.w3.org/2001/XMLSchema#int"
  >583</AtomID>
  <Y rdf:datatype="http://www.w3.org/2001/XMLSchema#float">1.976</Y>
  <Element rdf:datatype="http://www.w3.org/2001/XMLSchema#string"
  >C</Element>
</Atoms>
```

Fig. 7. Instance of Atoms Concept for B. Subtilis in PO.

Although Protein Data Bank (PDB) is very comprehensive it does not provide information about Protein Sequences in traditional FASTA format, which is still widely used, and provides no information about basic functionality of the protein complex. We try to gather this information from UniProt database and integrate it with information gathered from PDB in Protein Ontology. Protein Sequence in FASTA format is represented as follows in UniProt database:

```
>Q02169|MAF_BACSU Septum formation protein Maf - Bacillus subtilis.
MTKPLILASQSPRRKELLDLLQLPYSIIVSEVEEKLNRNFSPEENVQWLAKQKAKAVADL
HPHAIVIGADTMVCLDGECLGKPQDQEEAASMLRRLSGRSHSVITAVSIQAENHSETFYD
KTEVAFWSLSEEEIWTYIETKEPMDKAGAYGIQGRGALFVKKIDGDYYSVMGLPISKTMR
ALRHFDIRA
```

Fig. 8. Protein Sequence for B. Subtilis in UniProt.

Protein Sequence information from UniProt is integrated with Protein Structure information from PDB using ATOMSequence Concept in Protein Ontology ATOMSequence is constructed using generic concepts of Chains, Residues, and Atoms. The reasoning is already there in the underlying Protein Data, as each Chain in a Protein represents a sequence of Residues, and each Residue is defined by a number of three-dimensional atoms in the Protein Structure. ATOMSequence is represented in Web Ontology Language as follows:

```
<ATOMSequence rdf:ID="ATOMSequence_1">
    <_Structure_Chain rdf:resource="#Chains_1"/>
    <_Strucure_ATOM rdf:resource="#AtomInstance4484"/>
    <_Structure_Residue rdf:resource="#Residues_1"/>
</ATOMSequence>
```

Fig. 9. Instance of ATOMSequence Concept for B. Subtilis in PO.

UniProt database describes the functions of B. Subtilis Protein in terms of FUNCTION, SUBUNIT and SUBCELLULAR LOCATION as follows:

FUNCTION: Involved in septum formation.
SUBUNIT: Homodimer.
SUBCELLULAR LOCATION: Cytoplasm *(Potential)*.
SIMILARITY: Belongs to the maf family.

Fig. 10. Protein Functions for B. Subtilis in UniProt.

Information about the functionality of B. Subtilis Protein is described in Protein Ontology as a Physiological Function:

```
<PhysiologicalFunctions rdf:ID="PhysiologicalFunctions_1">
    <CellularFunction rdf:datatype="http://www.w3.org/2001/XMLSchema#string"
    >Involved in septum formation in Cytoplasm</CellularFunction>
</PhysiologicalFunctions>
```

Fig. 11. Instance of PhysiologicalFunctions Concept for B. Subtilis in PO.

4 Conclusion

Protein Ontology Instance Store is a repository developed based on our earlier work of Protein Ontology Conceptual Framework and Protein Ontology Algebra. Protein Ontology Instance Store is populated using process of instantiations transformation from existing Protein Data Sources to Protein Ontology Instance Store to provide data integration. Although at the moment PO Instance Store consists of few protein families, we are currently in process of transforming data about all the proteins available to PO Instance Store. More recent updates of the same will be available on the Protein Ontology Website (http://www.proteinontology.info/).

References

Bairoch, A., Apweiler, R.: The SWISS-PROT protein sequence data bank and its supplement TrEMBL. Nucleic Acids Research 25, 31–36 (1997)

Barker, W.C., Garavelli, J.S., Haft, D.H., Hunt, L.T., Marzec, C.R.: The PIR-International Protein Sequence Database. Nucleic Acids Research 26, 27–32 (1998)

Carroll, J.J., Dickinson, I., Dollin, C., Reynolds, D., Seaborne, A.: Jena: implementing the semantic web recommendations. In: Proceedings of the 13th international World Wide Web conference on Alternate track papers & posters (2004)

Murzin, A.G., Brenner, S.E., Hubbard, T.: SCOP: A Structural Classification of Proteins Database for the Investigation of Sequences and Structures. Journal of Molecular Biology 247, 536–540 (1995)

Sidhu, A.S., Dillon, T.S., Chang, E.: Ontological Foundation for Protein Data Models. In: Meersman, R., Tari, Z. (eds.) On the Move to Meaningful Internet Systems 2005: CoopIS, DOA, and ODBASE. LNCS, vol. 3761, Springer, Heidelberg (2005a)

Sidhu, A.S., Dillon, T.S., Chang, E.: Towards Semantic Interoperability of Protein Data Sources. In: Meersman, R., Tari, Z. (eds.) On the Move to Meaningful Internet Systems 2006: CoopIS, DOA, GADA, and ODBASE. LNCS, vol. 4276, Springer, Heidelberg (2006)

Sidhu, A.S., Dillon, T.S., Chang, E.: Protein Ontology. In: Chen, J., Sidhu, A.S. (eds.) Biological Database Modeling, Artech House, New York (in press, 2007)

Sidhu, A.S., Dillon, T.S., Chang, E., Sidhu, B.S.: Protein Ontology Development using OWL. In: Grau, B.C., Horrocks, I., Parsia, B. (eds.) st Workshop on OWL: Experiences and Directions (OWLED 2005). CEUR Workshop Proceedings (CEUR-WS.org), Galway, Ireland

Wesbrook, J., Feng, Z., Jain, S., Bhat, T.N., Thanki, N., Ravichandran, V., Gilliland, G.L., Bluhm, W.F., Weissig, H., Greer, D.S., Bourne, P.E.: The Protein Data Bank: unifying the archive. Nucleic Acids Research 30, 245–248 (2002)

Westbrook, J., Fitzgerald, P.M.D.: The PDB format, mmCIF formats and other data formats. In: Bourne, P.E., Weissig, H. (eds.) Structural Bioinformatics. Westbrook, John Wiley & Sons, Inc., Hoboken, NJ.

Three Fold System (3FS) for Mental Health Domain

Maja Hadzic and Roberta Ann Cowan

Curtin University of Technology,
Digital Ecosystems and Business Intelligence Institute (DEBII)
GPO Box U1987 Perth,
Western Australia 6845, Australia
{m.hadzic,r.cowan}@curtin.edu.au

Abstract. Along with an increase in the number of mentally ill people, research into all aspects of mental health has increased in recent years. In all disciplines information is the key to success but major problems adversely affect the efficiency and effectiveness that available mental health information is used. These relate to the lack of existing infrastructure to support effective access and information retrieval, and lack of tools to analyze the available information and derive useful knowledge from it. In this paper we explain how the ontology, multi-agent system and data mining technologies can be implemented within the mental health domain to effectively address these issues. The synergy of these frontier technologies may result in an intelligent information infrastructure that provides effective and efficient use of all available information.

Keywords: mental health research, health information systems, ontology-based multi-agent systems, data mining, intelligent information retrieval.

1 Introduction

The World Health Organisation predicted that depression would be the world's leading cause of disability by 2020 [1]. The exact causes of many mental illnesses are unclear. Mental illness is a causal factor in many chronic conditions such as diabetes, hypertension, HIV/AIDS resulting in higher cost to the health system [2]. The recognition that mental illness is costly and many cases may not become chronic if treated early has lead to an increase in research in the last 20 years.

Information technologies must be effectively implemented within health domain. In their paper, Horvitz-Lennon *et al.* [2] state that we need to fully embrace information technology and its potential for improving service efficiency and develop a better information infrastructure for patient care.

In the health portfolios of government the need for information access is understood. The US Department of Health and Human Services (www.hhs.gov) has the major goals of (1) constant access to health information for the health care consumer, and (2) improved tracking of chronic disease management for public health [3]. This has led to the development of a portal www.health.gov for agency information such as Healthfinder.gov (www.healthfinder.gov) and MedlinePlus® (www.nlm.nih.gov/medlineplus/) for information to the public, and ONCHIT (Office of the National

R. Meersman, Z. Tari, P. Herrero et al. (Eds.): OTM 2007 Ws, Part II, LNCS 4806, pp. 1355–1364, 2007.
© Springer-Verlag Berlin Heidelberg 2007

Coordinator for Health Information Technology) to institute heath information transfer. The Australian Federal Government developed a public gateway to health information (www.healthinsite.gov.au) and Health Connect (www.health.gov.au/internet/hconnect/publishing.nsf/Content/home) for health information transfer. The UK Department of Health offers health information to citizens via NHS Direct (www.nhsdirect.nhs.uk/) for information to the public and NHS Connecting for Health (www.connectingforhealth.nhs.uk) for heath information transfer.

Specific and targeted searches are very difficult with current search engines. For example, a search for "genetic causes of bipolar disorder" using Google provides 960,000 hits consisting of a large assortment of well meaning general information sites with few interspersed evidence-based resources. The information provided by the government sites is not necessarily returned on the first page of a 'google' search. A similar search on Medline Plus retrieves all information about bipolar disorder plus information on other mental illnesses. The main problem of the current search engines is that they match specific strings of letters within the text rather than searching by meaning. This is where the ontologies come into their own as they enable searching by meaning of the information rather then its appearance in the text.

Moreover, we believe that the synergy of ontology with multi-agent and data mining systems can result in an intelligent information infrastructure which will meet and efficiently address the needs of the mental health community. Such a 'three-fold system (3FS)' will be an innovative breakthrough within the community and would fit within the goals and criteria set out by Coye and Kell [4]. This innovation is presented as follows. An overview of some of the related works is given in Section 2. The issues related to the volume and management of the existing information and the issues specific to the mental health domain are discussed in Section 3. In Section 4, we discuss the three major technologies that can bring a breakthrough in this complicated situation. We provide a conceptual framework for a possible implementation of such system in Section 5. The benefits of the resulting system are discussed in Section 6. This paper is concluded in Section 7.

2 State of Play

The importance of ontologies has been recognised within the medical community and work has begun on developing and sharing medical ontologies [5]. The Unified Medical Language System (UMLS) [6] project develops and distributes multi-purpose, electronic 'knowledge sources' and associated lexical programs. LinKBase [5, 7] contains over 5 million medical concepts, relationships and terms. The UMLS and LinKBase are to large ontologies for the mental health projects, although a subset of them could be used to design a mental health ontology. Ontology reuse techniques, such as ontology alignment and merging, can be used for this purpose.

Use of agent-based systems enables us to model, design and build complex medical information systems. Multi-agent systems are being used more and more in the medical domain. Agent Cities [8], BioAgent [9], Holonic Medical Diagnostic System HMDS [10] are all agent-based systems. Some agents are designed for a specific hospital and their databases (e.g., Agent Cities). The information available to these systems is institution specific and the agents help with the management of that information within a closed system. The agents cannot gather new knowledge regarding the disease in question from outside. Other agents are designed for the searching the internet, such as BioAgent and HMDS, which search for biochemical and medical diagnostic information respectively. No agents have been designed specifically for mental health domain as yet.

Ontologies were initially designed for agents to perform intelligent actions. Only in the Service Agent Layer (i.e., Ontology Service) of the BioAgent [9], does an ontology exist. In this case, the ontology is only used to provide semantics on data that is locally stored, not to interrogate the wider information domain of the internet. An ontology-based multi-agent system for human disease study and control has been proposed by Hadzic and Chang [11], and is in early implementation stage.

Within the biomedical and health field, data mining techniques have been predominately used for tasks such as text mining, gene expression analysis, drug design, genomics and proteomics [12]. Text mining uses techniques from data mining [13]. The data analysis necessary for microarrays has necessitated data mining [15]. Recently, use of data mining methods has been proposed for the purpose of mapping and identification of complex disease loci [16]. However, the proposed methods are yet to be implemented. Use of data mining techniques within the biomedical field is encouraging and is useful as a point of reference. Some of the existing data mining techniques, with minor modifications, could be used for mental health information seeking.

3   An Alarming Combination: Petabytes of Medical Data and Complexity of Mental Illnesses

New modern techniques in medicine, science and information technology are providing huge, rapidly accumulating amounts of information. To extract and analyze the data poses a much bigger challenge for researchers than to generate the data. Experienced scientists, clinicians, mental health professionals and doctors are overwhelmed with this situation. As the size of the existing corpus of knowledge is so large, the possibility of searching it effectively is very low. In most cases some important information is being neglected.

Information regarding mental illness is dispersed over various resources and it is difficult to link this information, to share it and find specific information when needed. The information covers different mental illnesses with a huge range of results regarding different disease types, symptoms, treatments, causal factors (genetic and environmental) as well as candidate genes that could be responsible for the onset of these diseases. We need to take a systematic approach to making use of enormous amount of available information that has no value unless analysed and linked with other available information from the same mental health domain.

Three major problems are slowing advancement in mental health research: (1) issues related to the lack of an existing infrastructure to support effective *access* to the relevant information (2) lack of an existing infrastructure to support effective information *retrieval* of distributed information with inconsistent structures; and (3) lack of tools to *analyse* this information and derive useful knowledge from it. Currently, there is no lack of information but its sheer volume and its structure is a factor hindering research.

Medical information alone totals several petabyte [17]. In 2007, a researcher setting a genetic human trial for "bipolar disorder" would have needed to sift through a many results, from a variety of sources (for example, Entrez Gene www.ncbi.nlm.nih.gov/entrez/) to have found that the human gene loci 2p13-16 are potential positive sites for this disorder (the information originating from the research reported by Liu *et al.* [18]).

As research continues, new papers or journals are frequently published and added to databases and more and more of this published information is available via the internet. Problematically, no collaborative framework currently exists to help inform mental health researchers and practitioners of the latest research and where and when it will become available.

Portions of the information or data on the internet may be related to each other, portions of the information may overlap, and portions of the information may be semi-complementary between one another. No knowledge based middleware is available to help identify these issues.

The complexity of mental illnesses adds further complications to research and makes control of the illness even more difficult. Mental illnesses do not follow Mendelian patterns but are caused by a number of genes usually interacting with various environmental factors [19]. There are many different types of severe mental illness [2], for example depression, bipolar disorder, schizophrenia. Genetic research has identified candidate loci on human chromosomes 4, 7, 8, 9, 10, 13, 14 and 17 [18]. There is some evidence that environmental factors such as stress, life-cycle events, social environment, economic conditions, climate etc. are important [2, 20]. Usually in any medical research, one research team concentrates on one aspect, be it genetic or environmental, for one type of disorder. But for mental illnesses, in order for research to be inclusive, we need to look at many causal factors simultaneously. We need a tool that combines and examines all the causal factors together and derives relationships between the illness-causing factors and specific illness type.

To overcome the currently complex and complicated situation, an intelligent and efficient information system needs to be designed that does not require researchers to sift through the same or similar results from different databases. This expert system needs to be able to link information from the heterogeneous and disparate databases, and then take the leap forward of providing newly mined data patterns.

4   Strength of the Three-Fold System (3FS)

The combination of multi-agent systems, ontology and data mining, the 'three-fold system (3FS)', has the power to take the data contained within the health information technology systems and release unbounded, unconfined knowledge.

Agents are intelligent software objects capable of autonomous actions and are sociable, reactive and proactive in an information environment [21]. Agents can answer queries, retrieve information from servers, make decisions and communicate with systems, other agents or with users. A multi-agent system provides a distributed collaborative platform and as such determines the system dynamics.

Ontologies were brought into the computer and information society to be used by various agents and for different applications addressing the problems of various knowledge domains. Ontology provides a shared common understanding of a domain and means to facilitate knowledge reuse by different applications, software systems and human resources [22, 23]. Ontologies are highly expressive knowledge models and as such increase expressiveness and intelligence of a system.

Ontologies represent the domain knowledge and can be used to support some important processes within a multi-agent system such as: problem decomposition and task sharing among different agents, result sharing and analysis, information retrieval, selection and integration etc [11]. Ontology and multi-agent technologies can be used together to enable efficient and effective access, extraction and manipulation of the information from various information resources.

An ontology-based multi-agent system would enable us to retrieve a list of all possible causes of mental illness (genetic and environmental). We could use such a system to retrieve the information about "genetic causes of bipolar disorder". The latter result would be a small but very precise subset of the former. This result would be much more informative compared to the results of current search engines which would retrieve ad hoc information about bipolar disorder. Additional application of data mining techniques on the results retrieved through an ontology-based multi-agent system would take us a step further by identifying precise combinations and patterns of illness-causing factors associated with each illness type.

Data mining is a set of processes that is based on automated searching for actionable knowledge buried within a huge body of data to extract information and find hidden patterns and behaviours for the purpose of making predictive models for decision making and new discoveries. Data mining helps find patterns and knowledge that are embedded in the data and requires exploration and analysis of large quantities of target data for the purpose of better understanding and deriving knowledge regarding the problem at hand. Data mining draws work from areas including database technology, machine learning, statistics, pattern recognition, information retrieval, neural networks, knowledge based systems, artificial intelligence, high-performance computing and data visualization [24].

The primary role of data mining technology in the improving the efficiency of the information retrieval process has been recognised by Chen [25]. Some advantages of data mining include:

- efficient processing of large and complex data (scalability)
- automatically analysing, detecting errors and inconsistencies, classifying, and summarizing the data with no human intervention (automation)
- extracting novel and useful patterns which leads to new knowledge and discoveries (knowledge extractions)
- combining the advantages of various disciplines (multi-disciplinary nature)
- reducing the costs and time associated with the data analysis as a result of its automation (cost and time efficiency).

An efficient, intelligent and dynamic information retrieval system can be designed to unify the advantages of the multi-agent system, ontology and data mining technologies. The synergy of the different but complementary techniques gives the resulting system the following important characteristics:

1. dynamic behaviour, through implementation of a multi-agent system
2. intelligent behaviour, through use of ontology
3. reasoning abilities of constructive nature, through implementation of data-mining technologies.

These characteristics enable the 3FS to use the available information effectively and efficiently and to derive new high quality information.

5 An Implementation of the Three-Fold System (3FS)

We propose the 3FS to include Generic Mental Illness Ontology (GMIO), GMIO-based multi-agent system and Data mining agents.

Generic Mental Illness Ontology (GMIO) can be developed to contain general mental health information. Four sub-ontologies can be designed as a part of the GMIO to represent knowledge about illness sub-groups (e.g., clinical depression, postnatal depression), illness causes (such as environmental and genetic), phenotypes (which describe illness symptoms) and possible treatments. The ontology and sub-ontologies will serve as template to generate Specific Mental Illness Ontologies (SHIO), the information specific to an illness in question (e.g., bipolar, depression, schizophrenia).

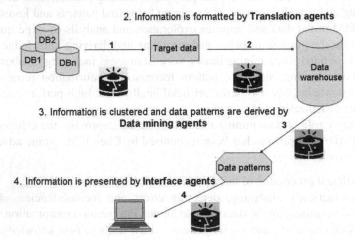

1. Information is retrieved by **Information agents**

2. Information is formatted by **Translation agents**

3. Information is clustered and data patterns are derived by **Data mining agents**

4. Information is presented by **Interface agents**

Fig. 1. GMIO-based Multi-agent system

The GMIO needs to be effectively utilised within a multi-agent system. We need to define a multi-agent system architecture that will be based on GMIO and will enable the agents to collaborate effectively. A possible solution that includes different agent types (Information, Translation, Data mining and Interface agents) is represented in Figure 1. *Information agents* (Arrow 1) have the task of recognising and retrieving relevant information from various information resources. Each Information agent may have a set of assigned information resources that it needs to visit in order to gather requested information. GMIO can be used to annotate the information resources. This will enable the Information agents to understand available information and retrieve requested information effectively. The retrieved information can be represented as instances of GMIO defining concepts and their relations. Information agents hand over the retrieved information to *Translation agents* (Arrow 2). If the information retrieved by Information agents is raw and in various formats, it needs to be put into the same format. This is the task of Translation agents. When all the retrieved information is put into the same format it is locally stored in a data warehouse. *Data mining agents* (Arrow 3) compare and analyse the formatted information simultaneously in order to derive new knowledge and patterns from the information. The results are presented in a meaningful format to the user. *Interface agents* (Arrow 4) can be designed to assist with this.

Two different types of data mining agents can be designed to have different functions within the system. *Clustering agent* (1) may have the task of locating relatively homogenous subgroups in the given set of data. For example, a Clustering agent can cluster all data related to a specific disease type within a cluster. *Pattern discovery agent* (2) may be designed to discover data patterns typical for illness in question. For example, the Pattern discovery agent may provide information about precise combination of genetic and environmental illness causing factors associated with a specific illness type. This results in hidden patterns and new discoveries, results which have a potential to bring a breakthrough in the research, control and prevention of mental illnesses. Currently, no tools exist to provide this or a similar facility in mental health domain.

Protégé [27] developed by Stanford University can be utilized for the modelling of the ontologies. Java Agent DEvelopment framework (JADE) can be used to develop the multi-agent based system. JADE is a software framework that simplifies the implementation of agent applications in compliance with the FIPA (Foundation of Intelligent Physical Agents) specifications for interoperable intelligent multi-agent systems and it offers a general support for ontologies. Data mining application can also be implemented in Java. The ontology can be stored as a computer readable description of knowledge and the agents can use this knowledge for intelligent actions.

6 Benefits for the Mental Health Domain

Mental illness is becoming one of the major problems of our society [28]. The number of mentally ill people is increasing globally each year. This introduces major costs in economic and human terms, to the individual communities and the nation in general, both in rural and urban areas.

Access to and efficient use of, available information through an intelligent information infrastructure will greatly assist mental health research. This research will assist in prevention of mental illness and assist in delivery of effective and efficient mental health services. Such a system will dramatically increase the value of the huge body of information which is currently heterogeneously stored and spatially dispersed. The vision is for this system to provide information for all levels of inquiry; for example, from the high level researcher wanting to know effects of environmental variables on the loci of chromosome 2, through the general practitioner needing to know the dosage of a drug combination for a clinical depressive alcoholic through to the carer of a schizophrenic child needing respite care. Thus the bridges between the mental health and general health sectors, that Horvitz-Lennon et al [2] aspire to will be constructed.

The resulting system will:

- support physicians in early diagnosis and treatment of mental illnesses as well as in provision of high quality health services through greatly increasing the effectiveness and accuracy of such services;
- support patients and their caregivers in dealing with, managing and treating illness;
- provide the general public accurate, reliable and up-to-date information that promotes understanding of mental illness. This will support management and prevention of mental illness.
- provide medical researchers support in advancing their research in identifying the disease causing factors and effective patient treatments as well as in reducing the possibility of redundant research (saving research time, effort and resources). Modern science and technology together with the associated medical research have the opportunity to ameliorate mental illnesses that are afflicting a substantial sector of the population.
- support the development of new technologies and facilitating development of technologies for maintaining good health;
- reduce the cost of the mental health budget and enhance the spending power of money spent on health and medical research by providing better information access [29];

The system described above goes some way to delivering what Patel et al [26] say is 'necessary to transform the quality of mental health care'. The system delivers a quality improvement in resources applicable to a diverse set of mental illnesses, improves the infrastructure for evidence-based interventions and brings in innovation for quality improvement in mental health care.

The 3FS principle could just as easily be applied to other medical and health domains as well as to the knowledge domains from other disciplines. The developed system can be used internally in a specific organization such as a pharmaceutical company or consortium such as Universities Australia, or externally in a government-supported public information service.

7 Conclusion

In this paper we have discussed numerous problems associated with use of the available information within mental health domain. We described the 3FS which unifies multi-agent systems, ontology and data mining technologies to effectively address information access, retrieval and analysis.

The implementation of such synergetic system will result in positive transformation of world-wide mental health research and management to a more effective and efficient regime. This system is highly significant and innovative, and represents a new generation of information-seeking tool. This is not another search engine. The totality of this innovation in information seeking comes from combining multi-agents, ontology and data mining. The beauty of this innovative tool is that the complexity is hidden from the client. Users will simply see targeted answers to their questions. The agents, ontology and data-mining working in concert remove complexity to where it should be, not on the human client's computer screen but within the tool.

References

1. Lopez, A.D., Murray, C.C.J.L.: The global burden of disease, 1990-2020. Nature Medicine 4, 1241–1243 (1998)
2. Horvitz-Lennon, M., Kilbourne, A.M., Pincus, H.A.: From silos to bridges: meeting the general health care needs of adults with severe mental illnesses. Health Affairs 25(3), 659–669 (2006)
3. Anderson, G.F., Frogner, B.K., Johns, R.A., Reinhardt, U.E.: Health care spending and use of information technology in OECD countries. Health Affairs 25(3), 819–831 (2006)
4. Coye, M.J., Kell, J.: How hospitals confront new technology. Health Affairs 25(1), 163–173 (2006)
5. Ceusters, W., Martens, P., Dhaen, C., Terzic, B.: LinkFactory: an advanced formal ontology management System. In: Proceedings of Interactive Tools for Knowledge Capture (KCAP 2001) (2001)
6. Bodenreider, O.: The Unified Medical Language System (UMLS): integrating biomedical terminology. Nucleic Acids Research 32(1), 267–270 (2004)
7. Montyne, F.: The importance of formal ontologies: A case study in occupational health. In: Proceedings of the Open Enterprise Solutions: Systems, Experiences, and Organizations (OES-SEO 2001), Luiss Publications (2001)
8. Moreno, A., Isern, D.: A first step towards providing health-care agent-based services to mobile users. In: Proceedings of the First International Joint Conference on Autonomous Agents and Multiagent Systems (AAMAS 2002), pp. 589–590 (2002)
9. Merelli, E., Culmone, R., Mariani, L.: BioAgent: A mobile agent system for bioscientists. In: Proceedings of The Network Tools and Applications in Biology Workshop Agents in Bioinformatics (NETTAB 02) (2002)
10. Ulieru, M.: Internet-enabled soft computing holarchies for e-Health applications. New Directions in Enhancing the Power of the Internet, pp. 131–166. Springer, Heidelberg (2003)
11. Hadzic, M., Chang, E.: Ontology-based Multi-agent systems support human disease study and control. In: Czap, H., Unland, R., Branki, C., Tianfield, H. (eds.) Frontiers in Artificial Intelligence and Applications (special issues on Self-organization and Autonomic Informatics), vol. 135, pp. 129–141. IOS Press, Amsterdam (2005)
12. Zaki, M., Wang, J., Toivonen, J.T.L., Toivonen, H.T.T.: Data mining in bioinformatics: report on BIOKDD 2003. SIGKDD Explorations 5(2), 198–200 (2003)
13. Hearst, M.A.: Untangling text data mining. In: Proceedings of the 37th Annual Meeting of the Association for Computational Linguistics, pp. 3-10 (1999)
14. Shatkay, H., Feldman, R.: Mining the biomedical literature in the genomic era: An Overview. Journal of Computational Biology 10(6), 821–855 (2003)
15. Piatetsky-Shapiro, G., Tamayo, P.: Microarray data mining: Facing the challenges. SIGKDD Explorations 5(2), 1–6 (2003)
16. Onkamo, P., Toivonen, H.: A survey of data mining methods for linkage disequilibrium mapping. Human Genomics 2(5), 336–340 (2006)

17. Goble, C.: The grid needs you. Enlist now. In: Proceedings of the International Conference on Cooperative Object Oriented Information Systems, pp. 589–600 (2003)

18. Liu, J., Juo, S.H., Dewan, A., Grunn, A., Tong, X., Brito, M., Park, N., Loth, J.E., Kanyas, K., Lerer, B., Endicott, J., Penchaszadeh, G., Knowles, J.A., Ott, J., Gilliam, T.C., Baron, M.: Evidence for a putative bipolar disorder locus on 2p13-16 and other potential loci on 4q31, 7q34, 8q13, 9q31, 10q21-24, 13q32, 14q21 and 17q11-12. Molecular Psychiatry 8(3), 333–342 (2003)

19. Smith, D.G., Ebrahim, S., Lewis, S., Hansell, A.L., Palmer, L.J., Burton, P.R.: Genetic epidemiology and public health: hope, hype, and future prospects. The Lancet 366(9495), 1484–1498 (2005)

20. Craddock, N., Jones, I.: Molecular genetics of bipolar disorder. The British Journal of Psychiatry 178(41), 128–133 (2001)

21. Wooldridge, M.: An Introduction to Multiagent Systems. John Wiley and Sons, Chichester (2002)

22. Gómez-Pérez, A.: Towards a framework to verify knowledge sharing technology. Expert Systems with Applications 11(4), 519–529 (1996)

23. Gómez-Pérez, A.: Knowledge sharing and reuse. The Handbook on Applied Expert Systems, 1–36 (1998)

24. Sestito, S., Dillon, T.S.: Automated Knowledge Acquisition. Prentice Hall, Australia, Sydney (1994)

25. Chen, W.L.: Chemoinformatics: Past, Present, and Future. Journal of Chemical Information and Modelling 46, 2230–2255 (2006)

26. Patel, K.K., Butler, B., Wells, K.B.: What Is Necessary To Transform The Quality Of Mental Health Care. Health Affairs 25(3), 681–693 (2006)

27. Protégé: an Ontology and knowledge-based editor, Stanford Medical Informatics, Stanford University, School of Medicine (2006), http://protege.stanford.edu/

28. Australian Government, Department of Health and Ageing Factbook 2006 (accessed February 20, 2007) http://www.health.gov.au/internet/wcms/publishing.nsf/Content/Factbook2006-1

29. Garber, A.M.: PERSPECTIVE: To Use Technology Better. Health Affairs, W51–W53 (2006)

Author Index

Lecture Notes in Computer Science

Sublibrary 3: Information Systems and Application, incl. Internet/Web and HCI

For information about Vols. 1– 4412
please contact your bookseller or Springer